AN ILLUSTRATED GUIDE TO VETERINARY MEDICAL TERMINOLOGY

THIRD EDITION

Janet Amundson Romich, DVM, MS

DELMAR
CENGAGE Learning™

Australia • Brazil • Japan • Korea • Mexico • Singapore • Spain • United Kingdom • United States

An Illustrated Guide to Veterinary Medical Terminology, Third Edition
Janet Amundson Romich, DVM, MS

Vice President, Career and Professional
Editorial: Dave Garza

Director of Learning Solutions: Matthew Kane

Acquisitions Editor: Benjamin Penner

Managing Editor: Marah Bellegarde

Senior Product Manager: Darcy M. Scelsi

Editorial Assistant: Scott Royael

Vice President, Career and Professional
Marketing: Jennifer McAvey

Marketing Director: Debbie Yarnell

Marketing Coordinator: Jonathan Sheehan

Production Director: Carolyn Miller

Production Manager: Andrew Crouth

Senior Content Project Manager:
Elizabeth C. Hough

Senior Art Director: David Arsenault

Technology Project Manager: Benjamin Knapp

For product information and technology assistance, contact us at
Professional & Career Group Customer Support, 1-800-648-7450

For permission to use material from this text or product,
submit all requests online at **cengage.com/permissions**

Further permissions questions can be e-mailed to
permissionrequest@cengage.com

Library of Congress Control Number: 2008938145

ISBN-13: 978-1-4354-2012-0

ISBN-10: 1-4354-2012-8

Delmar
5 Maxwell Drive
Clifton Park, NY 12065-2919
USA

Cengage Learning products are represented in Canada
by Nelson Education, Ltd.

For your lifelong learning solutions, visit **delmar.cengage.com**

Visit our corporate website at **cengage.com**

Notice to the Reader

Publisher does not warrant or guarantee any of the products described herein or perform any independent analysis in connection with any of the product information contained herein. Publisher does not assume, and expressly disclaims, any obligation to obtain and include information other than that provided to it by the manufacturer. The reader is expressly warned to consider and adopt all safety precautions that might be indicated by the activities described herein and to avoid all potential hazards. By following the instructions contained herein, the reader willingly assumes all risks in connection with such instructions. The publisher makes no representations or warranties of any kind, including but not limited to, the warranties of fitness for particular purpose or merchantability, nor are any such representations implied with respect to the material set forth herein, and the publisher takes no responsibility with respect to such material. The publisher shall not be liable for any special, consequential, or exemplary damages resulting, in whole or part, from the readers' use of, or reliance upon, this material.

Printed in the United States of America
5 6 7 12 11

CONTENTS

PREFACE

TO THE STUDENT

Medical terminology may seem like a foreign language to you. Many of the terms are unfamiliar, seem strange, or do not make sense. However, to communicate in the medical world, you need a thorough understanding of the language.

Most medical terms are based on word parts that already may be familiar to you. You may have heard words such as *appendicitis, gastritis,* and *tonsillectomy* or used them in the past. You may not realize how many medical terms you already know. Building on this foundation, new word parts will make learning medical terminology more logical.

This text and the accompanying materials simplify the process of learning medical terminology. Review the introductory sections so that you are familiar with the organizational scheme of the textbook and StudyWARE™. Once you become comfortable with the materials, you will find yourself learning medical terms faster than you ever imagined possible.

Chapter Organization

The chapters in *An Illustrated Guide to Veterinary Medical Terminology, Third Edition,* are organized in the following fashion:

- Introduction to medical terms
- Anatomical foundations
- Body systems
- Species-specific chapters

Chapter 1 provides the basics of how medical terms are formed, analyzed, and defined. Chapter 2 provides terms used in everyday dialogue regarding positioning of animals and relationships between body parts. Chapters 3 and 4 discuss anatomical landmarks both internally (musculoskeletal system) and externally (common terms for landmarks on an animal's body). Chapter 5 consists of terms used in the animal industry to describe males and females of selected species and terms for their young and for groups of their species. Chapters 6 through 15 are organized by body systems. These chapters describe the anatomy of the body system; include clinical terms used in reference to it; and conclude with diagnostic tests, pathology, and procedures for the body system. Chapters 16 and 17 relate tests, procedures, and treatments used in the care of animals in the medical field. Chapters 18 through 23 are species-specific chapters that you can study independently to enhance your knowledge of a particular species or that your instructor

may incorporate into other chapters to assess your progress. Appendix A consists of tables of abbreviations, and Appendix B contains plural forms of medical terms. Appendix C lists prefixes, combining forms, and suffixes.

How to Use StudyWARE™ to Accompany *An Illustrated Guide to Veterinary Medical Terminology, Third Edition*

The StudyWARE™ software helps you learn terms and concepts in *An Illustrated Guide to Veterinary Medical Terminology, Third Edition.* As you study each chapter in the text, make sure you explore the activities in the corresponding chapter of the software. Use StudyWARE™ as your own private tutor to help you learn the material in *An Illustrated Guide to Veterinary Medical Terminology, Third Edition.*

Getting started is easy. Install the software by inserting the CD-ROM into your computer's CD-ROM drive and following the on-screen instructions. When you open the software, enter your first and last name so the software can store your quiz results. Then choose a chapter from the menu to take a quiz or to explore one of the activities.

Menus

You can access the menus from wherever you are in the program. The menus include Quizzes and other Activities.

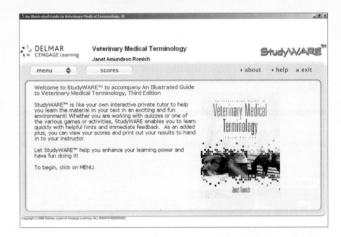

QUIZZES

Quizzes are made up of multiple-choice questions. You can take the quizzes in practice mode and in quiz mode. Use practice mode to improve your mastery of the material. You have multiple tries to get the answers correct. Instant feedback tells you whether you're right or wrong and helps you learn quickly by explaining why an answer was correct or incorrect. Use quiz mode when you are ready to test yourself and keep a record of your scores. In quiz mode, you have one try to get the answers right, but you can take each quiz as many times as you want.

SCORES

You can view your previous score for each quiz and print your results to hand in to your instructor.

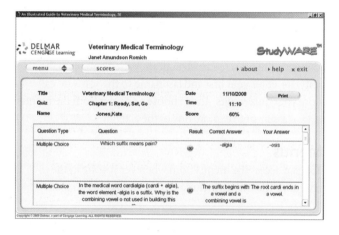

ACTIVITIES

Activities include image labeling, hangman, crossword puzzles, and flash cards. Have fun while increasing your knowledge!

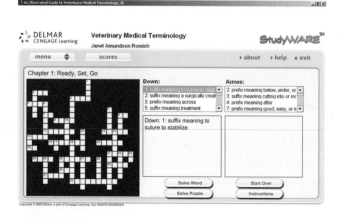

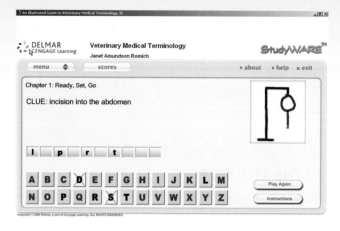

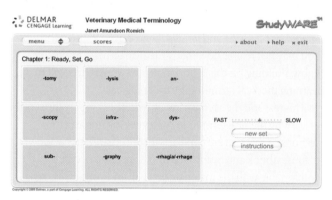

AUDIO LIBRARY

The StudyWARE™ Audio Library is a reference that includes audio pronunciations and definitions. Use the audio library to practice pronunciation and to review definitions for medical terms. You can browse terms by chapter or search by key word. Listen to pronunciations of the terms you select or listen to an entire list of terms.

Note: Instructors should expect students to master the terms in each section before they progress to the next section since the word parts will not be repeated in subsequent chapters. For example, the prefix *hypo-* may first appear in the gastrointestinal chapter but not be repeated in the endocrine chapter. However, words containing the prefix *hypo-* will be found in the endocrine chapter.

TO THE INSTRUCTOR

An *eResource to accompany An Illustrated Guide to Veterinary Medical Terminology, Third Edition,* is available to help you with course instruction. The *eResource* contains an instructor's guide that provides answer keys for all exercises in the text, teaching tips, and activities to enhance your teaching of medical terminology. A test bank contains 1,000 questions in the following formats: multiple choice, short answer, and matching. An image library containing the images from the text can be used to create PowerPoint® slides, transparencies, or handouts for students. PowerPoint® presentations can be used to deliver lectures or to provide as handouts to students.

ACKNOWLEDGMENTS

Special thanks to the following people who helped review this text and answered many questions regarding medical terminology throughout its development. Without their expertise, the text would not have been as complete.

Kevin R. Berry, CVT
Gaska Dairy Health Services, Columbus, WI

Kay Bradley, BS, CVT
Madison Area Technical College, Madison, WI

Kenneth Brooks, DVM, Diplomate ABVP
Lodi Veterinary Hospital, SC, Lodi, WI

Eric Burrough, DVM
Kirkwood Community College, Cedar Rapids, IA

Stephen J. Carleton, DVM
Quinnipiac University, Connecticut

Anne E. Chauvet, DVM, Diplomate ACVIM—Neurology
University of Wisconsin Veterinary Medical Teaching Hospital, Madison, WI

Jane Clark, DVM
Madison Area Technical College, Madison, WI

Michael T. Collins, DVM, PhD
University of Wisconsin School of Veterinary Medicine, Madison, WI

Thomas Curro, DVM, MS
Henry Doorly Zoo, Omaha, Nebraska

Deb Donohoe, LATG
Medical College of Wisconsin, Milwaukee,WI

Wendy Eubanks, CVT
Delafield Small Animal Hospital, Delafield, WI

Ron Fabrizius, DVM, Diplomate ACT
Poynette Veterinary Service, Inc., Poynette, WI

Kelly Gilligan, DVM
Four Paws Veterinary Clinic, LLC, Prairie du Sac, WI

John H. Greve, DVM, PhD
Iowa State University, Ames, IA

Gerald Hackett, DVM
California Polytechnic University, Pomona, CA

Brain J. Heim, DVM
Cedar Valley College, Lancaster, Texas

Mark Jackson, DVM, PhD, Diplomate ACVIM, MRCVS
Glasgow University, Scotland

Linda Kratochwill, DVM
Crowe-Goebel Veterinary Clinic, Scanlon, MN

Amy Lang, RTR
University of Wisconsin Veterinary Medical Teaching Hospital, Madison, WI

Laura L. Lien, CVT, BS
Moraine Park Technical College, Fond du Lac, WI

Carole Maltby, DVM
Maple Woods Community College, Kansas City, MO

A. Edward Marshall, DVM, PhD
Auburn University, Auburn, AL

Sheila McGuirk, DVM, PhD, Diplomate ACVIM
University of Wisconsin Veterinary Medical Teaching Hospital, Madison, WI

James T. Meronek, DVM, MPH
USDA-APHIS Veterinary Services Madison, WI

David Morales, DVM
Oklahoma State University, Oklahoma City, OK

Karl Peter, DVM
Foothill College, Los Altos Hills, CA

Kathrine Polzin, BA, CVT
University of Wisconsin Veterinary Medical Teaching Hospital, Madison, WI

Stuart Porter, VMD
Blue Ridge Community College, Weyers Cave, VA

Teri Raffel, CVT
Madison Area Technical College, Madison, WI

Linda Sullivan, DVM
University of Wisconsin Veterinary Medical Teaching Hospital, Madison, WI

Laurie Thomas, BA, MA
Clinicians Publishing Group/Partners in Medical Communications, Clifton, NJ

Beth Uldal Thompson,VMD
Veterinary Technician/Veterinary Learning Systems, Trenton, NJ

I also would like to express my gratitude to Beth Thompson, VMD, and Laurie Thomas, BA, MA, of Veterinary Learning Systems for their determination in advancing my writing skills through the publication of journal articles for *Veterinary Technician Journal.* Without their guidance I would not have honed my writing skills. I also would like to thank the many veterinary technician and laboratory animal technician students at Madison Area Technical College for their support and continued critique of the veterinary terminology course. A special thank-you goes to the 1998 veterinary technician and laboratory animal technician students at Madison Area Technical College, who learned terminology through my rough draft of the original text. Finally, I would like to thank the excellent staff at Cengage Delmar Learning and my family for their continued support.

ABOUT THE AUTHOR

Dr. Janet Romich received her Bachelor of Science degree in Animal Science from the University of Wisconsin–River Falls and her Doctor of Veterinary Medicine and Master of Science degree from the University of Wisconsin–Madison. Currently, Dr. Romich teaches at Madison Area Technical College in Madison, Wisconsin, where she has taught and continues to teach a variety of science-based courses. Dr. Romich was honored with the Distinguished Teacher Award in 2004 for use of technology in the classroom, advisory and professional activities, publication list, and fund-raising efforts. She received the Wisconsin Veterinary Technician Association's Veterinarian of the Year Award in 2007 for her contributions in educating veterinary technician students and promoting the use of veterinary technicians in the workplace. She is a member of the Biosafety Committee for a biopharmaceutical company, an IACUC member for a hospital research facility, and an advisory board member for a distance learning veterinary technician program. Dr. Romich authored the textbooks *Fundamentals of Pharmacology for Veterinary Technicians* and *Understanding Zoonotic Diseases*, as well as served as a coauthor on *Delmar's Veterinary Technician Dictionary*. Dr. Romich remains active in veterinary practice through her relief practice, where she works in both small- and mixed-animal practices.

HOW TO USE THIS TEXT

An Illustrated Guide to Veterinary Medical Terminology, Third Edition, helps you learn and retain medical terminology using a logical approach to medical word parts and associations. Following are the keys to learning from this text.

Illustrations

Complete with detailed labeling, the text's line drawings clarify key concepts and contain important information of their own. In addition to line drawings, photos are included to enhance the visual perception of medical terms and improve retention of medical terms and use of these terms in the real world. Review each illustration and photo carefully for easy and effective learning.

Charts and Tables

Charts and tables condense material in a visually appealing and organized fashion to ensure rapid learning. Some tables include terms organized by opposites or body systems to facilitate relating the information to various situations.

New Terms

New terms appear in bold type, followed by the pronunciation and definition.

Pronunciation System

The pronunciation system is an easy approach to learning the sounds of medical terms. This system is not laden with linguistic marks and variables, ensuring that students do not get bogged down in understanding the key. Once students become familiar with the key, it is very easy for them to progress in speaking the medical language.

Pronunciation Key

PRONUNCIATION GUIDE

- Pronunciation guides for common words are omitted.
- Any vowel that has a dash above it represents the long sound, as in ā *hay*, ē *we*, ī *ice*, ō *toe*, and ū *unicorn*.
- Any vowel followed by an "h" represents the short sound, as in ah *apple*, eh *egg*, ih *igloo*, oh *pot*, and uh *cut*.
- Unique letter combinations are as follows: oo *boot*, ər *higher*, oy *boy*, aw *caught*, and ow *ouch*.

OTHER PRONUNCIATION GUIDELINES

Word parts are represented in the text as prefixes, combining forms, and suffixes. The notation for a prefix is a word part followed by a hyphen. The notation for a combining form (word root and its vowel to ease pronunciation) is the root followed by a slash and its vowel, as in *nephr/o*. The notation for a suffix is a hyphen followed by the word part. The terms *prefix, combining form*, and *suffix* do not appear in the definitions.

Learning Objectives

The beginning of each chapter lists learning objectives so that students know what is expected of them as they read the text and complete the exercises.

Review Exercises

Exercises at the end of each chapter help you interact with and review the chapter's content. The exercises include several formats: multiple choice, matching, case studies, word building, diagram labeling, and crossword puzzles. The answers to the exercises are found in the Instructor's Manual.

CHAPTER 1

[READY, SET, GO]

Objectives

Upon completion of this chapter, the reader should be able to

- Identify and recognize the parts of a medical term
- Define commonly used prefixes, combining forms, and suffixes presented in this chapter
- Analyze and understand basic medical terms
- Recognize the importance of spelling medical terms correctly
- Practice pronunciation of medical terms
- Recognize the importance of medical dictionary use
- Practice medical dictionary use

INTRODUCTION TO MEDICAL TERMINOLOGY

Medical terms are used every day in medical offices, newspapers, television, and conversational settings. Most people are familiar with many medical terms; however, other medical terms seem complicated and foreign. Learning and understanding how medical terminology developed can help in mastering these terms.

Current medical vocabulary is based on terms of Greek and Latin origin, **eponyms** (words formed from a person's name), and modern language terms. The majority of medical terms are derived from word parts based on Greek and Latin words. Increasing familiarity with these Greek and Latin terms as well as the ability to identify word parts aids in learning common medical terms and recognizing unfamiliar medical terms by word analysis. Medical terminology may seem daunting at first because of the length of medical words and the seemingly complex spelling rules, but once the basic rules of breaking down a word into its constituents are mastered, the words become easier to read and understand.

ANATOMY OF A MEDICAL TERM

Many medical terms are composed of word part combinations. Recognizing these word parts and their meanings simplifies learning medical terminology. These word parts are as follows:

- **prefix:** word part found at the beginning of a word. Usually indicates number, location, time, or status.

- **root:** word part that gives the essential meaning of the word.
- **combining vowel:** single vowel, usually an *o*, that is added to the end of a root to make the word easier to pronounce.
- **combining form:** combination of the root and combining vowel.
- **suffix:** word part found at the end of a word. Usually indicates procedure, condition, disease, or disorder.

Table 1–1   Contrasting Prefixes

Without a prefix, the root **traumatic** means pertaining to injury.	**A-** (ah or ā) means without or no. **Atraumatic** means without injury.
Without a prefix, the root **uria** means urination.	**An-** (ahn) means without or no. **Anuria** means absence of urine.
Ab- (ahb) means away from. **Abduction** means to take away from the midline.	**Ad-** (ahd) means toward. **Adduction** means move toward the midline.
Without a prefix, the root **emetic** means pertaining to vomiting.	**Anti-** (ahn-tī or ahn-tih) means against. **Antiemetics** work against or prevent vomiting.
Dys- (dihs) means difficult, painful, or bad. **Dysphagia** means difficulty eating or swallowing.	**Eu-** (yoo) means good, easy, or normal. **Euthyroid** means having a normally functioning thyroid gland.
Endo- (ehn-dō) means within or inside. **Endocrine** means to secrete internally.	**Ex-** (ehcks) or **exo-** (ehcks-ō) means without, out of, outside, or away from. **Exocrine** means to secrete externally (via a duct).
Endo- means within or inside. **Endoparasite** is an organism that lives within the body of the host.	**Ecto-** (ehck-tō) means outside. **Ectoparasite** is an organism that lives on the outer surface of the host.
Hyper- (hī-pər) means elevated, higher, or more than normal. **Hyperglycemia** means elevated amounts of blood glucose.	**Hypo-** (hī-pō) means depressed, lower, or less than normal. **Hypoglycemia** means depressed amounts of blood glucose.
Inter- (ihn-tər) means between. **Intercostal** means between the ribs.	**Intra-** (ihn-trah) means within. **Intramuscular** means within the muscle.
Poly- (pohl-ē) means many or excessive. **Polyuria** means excessive amount or frequency of urination.	**Oligo-** (ohl-ih-gō) means scant or little. **Oliguria** means scant amount or frequency of urination.
Pre- (prē) means before. **Preanesthetic** means pertaining to before anesthesia.	**Post-** (pōst) means after. **Postanesthetic** means pertaining to after anesthesia.
Sub- (suhb) means below, under, or less. **Sublingual** means under the tongue.	**Super-** (soo-pər) and **supra-** (soo-prah) mean above, beyond, or excessive. **Supernumerary** means more than the regular number. **Suprascapular** means above the shoulder blade.

Understanding the meaning of the word parts allows the dissection of medical terms in a logical way. By breaking down unfamiliar terms into recognizable word parts, the veterinary professional can greatly increase his or her medical vocabulary.

PREFIXES

Prefixes are added to the beginning of a word or root to modify its meaning. For example, the term *operative* can be modified using various prefixes.

- The prefix **pre-** means before. **Preoperative** means before or preceding an operation.
- The prefix **peri-** (pehr-ē) means around. **Perioperative** means pertaining to the period around an operation or the period before, during, and after an operation.
- The prefix **post-** means after. **Postoperative** means after an operation.

Table 1-2 Directional Prefixes and Their Meanings

Prefix	Pronunciation	Definition
epi-	(eh-pē)	upper
extra-	(ehcks-trah)	outside
hyper-	(hī-pər)	above, increased, or more than normal
hypo-	(hī-pō)	below, under, or decreased
infra-	(ihn-frah)	below or beneath
inter-	(ihn-tər)	between
intra-	(ihn-trah)	within
meta-	(meht-ah)	beyond
per-	(pər)	throughout
sub-	(suhb)	below, under, or decreased
super-	(soo-pər)	above, increased, or more than normal
supra-	(soo-prah)	above, increased, or more than normal
trans-	(trahnz)	across
ultra-	(uhl-trah)	above, increased, or more than normal

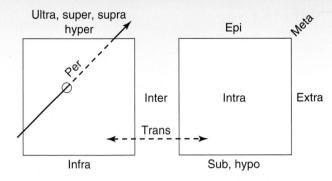

Figure 1–1 Directional prefixes

Many prefixes have another prefix whose meaning is opposite of its own. Initially, when learning prefixes, it is helpful to learn them in these pairs or in similar groups (Table 1–1, Table 1–2, and Figure 1–1).

COMBINING VOWELS

A combining vowel sometimes is used to make the medical term easier to pronounce. The combining vowel is used when the suffix begins with a consonant, as in the suffix **-scope.** An **arthroscope** is an instrument used to visually examine the joint. Because the suffix *-scope* begins with a consonant, the combining vowel *o* is used. *O* is the most commonly used combining vowel; however, *i* and *e* may be used as well. A combining vowel is not used when the suffix begins with a vowel, as in the suffix **-itis. Gastritis** is inflammation of the stomach. Because the suffix *-itis* begins with a vowel, the combining vowel *o* is not used.

A combining vowel is always used when two or more root words are joined. For example, when **gastr/o** (stomach) is joined with **enter/o** (small intestine), the combining vowel is used with **gastr/o,** as in the term **gastroenteritis.** A combining vowel is not used between a prefix and the root word.

COMBINING FORMS

The combining form is a word root plus a combining vowel. Combining forms usually describe a part of the body. New words are created when combining forms are added to prefixes, other combining forms, and suffixes. For example, the term *panleukopenia* is composed of the following word parts:

- **pan-** (pahn), a prefix meaning all
- **leuk/o** (loo-kō), a combining form meaning white
- **-penia** (pē-nē-ah), a suffix meaning deficiency or reduction in number

Panleukopenia is a deficiency of all types of white blood cells.

SUFFIXES

Suffixes are attached to the end of a word part to modify its meaning. For example, the combining form *gastr/o* means stomach and can be modified using various suffixes.

- The suffix **-tomy** means cutting into or incision. **Gastrotomy** is incision into the stomach.
- The suffix **-stomy** means a surgically created opening. **Gastrostomy** is a surgically created opening between the stomach and the body surface.
- The suffix **-ectomy** means surgical removal or excision. **Gastrectomy** is surgical removal of the stomach.

Many suffixes can be grouped together by meaning or by the category they modify. Initially, when learning suffixes, it is easiest if the learner groups them by meaning or category.

"Pertaining To" Suffixes

- **-ac** (ahck), as in **cardiac** (pertaining to the heart).
- **-al** (ahl), as in **renal** (pertaining to the kidney).
- **-an** (ahn), as in **ovarian** (pertaining to the ovary).
- **-ar** (ahr), as in **lumbar** (pertaining to the loin, lower back).
- **-ary** (ahr-ē), as in **alimentary** (pertaining to the gastrointestinal tract).
- **-eal** (ē-ahl), as in **laryngeal** (pertaining to the larynx).
- **-ic** (ihck), as in **enteric** (pertaining to the intestines).
- **-ine** (ihn), as in **uterine** (pertaining to the uterus).
- **-ous** (uhs), as in **cutaneous** (pertaining to the skin).
- **-tic** (tihck), as in **nephrotic** (pertaining to the kidneys).

Surgical Suffixes

- **-ectomy** (ehck-tō-mē) = surgical removal, as in **mastectomy,** surgical removal of the breast or mammary glands.
- **-pexy** (pehck-sē) = suture to stabilize, as in **gastropexy,** surgically stabilizing the stomach to the abdominal wall.
- **-plasty** (plahs-tē) = surgical repair, as in **rhinoplasty,** surgical repair of the nose.
- **-stomy** (stō-mē) = surgically created opening, as in **colostomy,** a surgically created opening between the colon and body surface.
- **-tomy** (tō-mē) = cutting into, as in **laparotomy,** an incision into the abdomen.

Procedural Suffixes

- **-centesis** (sehn-tē-sihs) = surgical puncture to remove fluid or gas (for diagnosis or for treatment to remove excess fluid or gas), as in **cystocentesis,** a surgical puncture of the urinary bladder with a needle to remove fluid (urine).

- **-gram** (grahm) = record of, as in **electrocardiogram,** the electrocardiographic hard copy record.
- **-graph** (grahf) = instrument that records (or used as a record), as in **electrocardiograph,** the machine that records the electrical activity of the heart.
- **-graphy** (grahf-ē) = procedure that records, as in **electrocardiography,** the procedure used to record the electrical activity of the heart.
- **-lysis** (lī-sihs) = separation or breakdown, as in **urinalysis,** separation of the urine into its constituents.
- **-scope** (skōp) = instrument to visually examine, as in **endoscope,** an instrument used to visually examine inside the body.
- **-scopy** (skōp-ē) = procedure to visually examine, as in **endoscopy,** the procedure of visually examining inside the body.
- **-therapy** (thehr-ah-pē) = treatment, as in **chemotherapy,** treatment with chemical substances or drugs.

Double *R* Suffixes

- **-rrhagia** or **-rrhage** (rā-jē-ah or rihdj) = bursting forth, as in **hemorrhage,** bursting forth of blood from the vessels.
- **-rrhaphy** (rahf-ē) = to suture, as in **enterorrhaphy,** suturing of the intestines.
- **-rrhea** (rē-ah) = flow, discharge, as in **diarrhea,** complete discharge of the bowels.
- **-rrhexis** (rehck-sihs) = rupture, as in **myorrhexis,** rupture of the muscle.

What is the difference between human and veterinary medical terminology?

Most times the medical terms used in human medical settings are identical to the ones used in veterinary medical settings. The greater number of species in veterinary medicine and the addition of terms used in animal production greatly expand the vocabulary of veterinary professionals. Species-specific anatomical differences also influence the terms used in a specific area. Do you know where the calf muscle is located on a person? Where is the calf muscle in a calf?

ces do not emphasize syllables. Consult with
re pronouncing a word.

L PRONUNCIATION
INES

short or long (Table 1–3). Consonants are gen-
ced as in other English words.

PELLING COUNT?

lling when using medical terminology. Chang-
o letters can change the meaning of a word.
a liver mass, whereas **hematoma** is a mass
of blood. The **urethra** takes urine from the
r to the outside of the body, whereas **ureters**
om the kidney and transport it to the urinary

ms may be pronounced the same but have dif-
s, so spelling is important. For example, ileum
pronounced the same. However, **ileum** is the
he small intestine (*e* = enter/o or *e* = eating),
is part of the pelvic bone (pelvic has an **i**
edical terms have the same spelling as terms
body parts. For example, the combining form
sents the spinal cord and bone marrow. (It
the term meaning white substance.) Other
ferent spellings depending on how the term
atically. For example, when used as a noun,
my stuff secreted from mucous membranes)
rently than when it is used as an adjective (as
brane).

p a medical term in the dictionary, spelling plays
ole. However, the term may not be spelled the
The following guidelines can be used to find a
tionary:

s like *f*, it may begin with *f* or *ph*.
s like *j*, it may begin with *g* or *j*.
s like *k*, it may begin with *c*, *ch*, *k*, or *qu*.
s like *s*, it may begin with *c*, *ps*, or *s*.
s like *z*, it may begin with *x* or *z*.

Table 1–3   Pronunciation Guide

Vowel	Sound	Example
"a" at the end of a word	ah	idea
"ae" followed by r or s	ah	aerobic
"i" at the end of a word	ī	bronchi
"oe"	eh	oestrogen (old English form)
"oi"	oy	sarcoid
"eu"	ū	euthanasia
"ei"	ī	Einstein
"ai"	ay	air
"au"	aw	auditory

Exceptions to Consonant Pronunciations

Consonant	Sound	Example
"c" before e, i, and y	s	cecum
"c" before a, o, and u	k	cancer
"g" before e, i, and y	j	genetic
"g" before a, o, and u	g	gall
"ps" at beginning of word	s	psychology
"pn" at beginning of word	n	pneumonia
"c" at end of word	k	anemic
"cc" followed by i or y	first c = k, second c = s	accident
"ch" at beginning of word	k	chemistry
"cn" in middle of word	both c (pronounce k) and n (pronounce ehn)	gastrocnemius
"mn" in middle of word	both m and n	amnesia
"pt" at beginning of word	t	pterodactyl
"pt" in middle of word	both p and t	optical
"rh"	r	rhinoceros
"x" at beginning of word	z	xylophone xenograph

Conditional and Structural Suffixes

- **-algia** and **-dynia** (ahl-jē-ah and dihn-ē-ah) = pain, as in **arthralgia** and **arthrodynia,** or joint pain.
- **-itis** (ī-tihs) = inflammation, as in **hepatitis,** inflammation of the liver.
- **-malacia** (mah-lā-shē-ah) = abnormal softening, as in **osteomalacia,** abnormal softening of bone.
- **-megaly** (mehg-ah-lē) = enlargement, as in **cardiomegaly,** enlargement of the heart.
- **-osis** (ō-sihs) = abnormal condition, as in **cardiosis,** an abnormal condition of the heart.
- **-pathy** (pahth-ē) = disease, as in **enteropathy,** a disease of the small intestine.
- **-sclerosis** (skleh-rō-sihs) = abnormal hardening, as in **arteriosclerosis,** abnormal hardening of the arteries.
- **-um** (uhm) = structure, as in **pericardium,** the structure surrounding the heart.

Suffixes may change a word's part of speech. Different suffixes may change the word from a noun (naming people, places, or things) to an adjective (descriptor) (Figure 1–2). Examples of this include the following terms:

- **Cyan***osis* is a noun meaning condition of blue discoloration, whereas **cyano***tic* is an adjective meaning pertaining to blue discoloration.
- **Anem***ia* is a noun meaning a blood condition of deficient red blood cells and/or hemoglobin, whereas **anem***ic* is an adjective meaning pertaining to a blood condition of deficient red blood cells and/or hemoglobin.
- **Muc***us* is a noun meaning a slimelike substance that is composed of glandular secretion, salts, cells, and leukocytes, whereas **muc***ous* is an adjective meaning pertaining to mucus.
- **Ili***um* is a noun meaning a part of the hip, whereas **ili***ac* is an adjective meaning pertaining to the hip.
- **Condyl***e* is a noun meaning a rounded projection on a bone, whereas **condyl***ar* is an adjective meaning pertaining to a rounded projection on a bone.

Noun	Suffix	Adjective	Suffix
cyanosis	**-osis**	cyanotic	**-tic**
anemia	**-ia**	anemic	**-ic**
mucus	**-us**	mucous	**-ous**
ilium	**-um**	iliac	**-ac**
condyle	**-e**	condylar	**-ar**
carpus	**-us**	carpal	**-al**

Figure 1–2 Suffix variation depending on usage

- **Carp***us* is a noun meani
and ulna and metacar
adjective meaning per
radius and ulna and me

ANALYZING ME

Medical terminology can b
the following objectives are a
examined for the first time:

- *Dissect:* Analyze the word
basic components.
- *Begin at the end:* After div
define the suffix first,
root. If there are two
them from left to righ
- *Anatomical order:* Wher
words usually are buil
occur in the body. Fo
proper term for inflam
intestine. Because foo
small intestine, the me
the medical term for *sm*
in a medical term ma
flow through organs.
diagnostic procedure
passed retrograde, or i
cal order. In these case
which the equipment

Using these guidelines, anal
divide the term into its basi
ectomy. Defining from bac
gical removal, one combini
the other combining form
the term **ovariohysterect**
ovaries and uterus. This term
ovaries and uterus are found

WHAT DID YO

Proper pronunciation of m
tice. Listening to how medi
using medical dictionaries
pared audio are the best w
are individual variations
personal preference. Mec
they present pronunciatio
mark the syllable receivin
mary accent (ʹ) and the
emphasis with a seconda
face and capitalize the sy

REVIEW EXERCISES

Multiple Choice

Choose the correct answer.

1. The prefix _____ means away from.

   a. ad-
   b. ab-
   c. ex-
   d. endo-

2. The suffix _____ means an instrument to visually examine.

   a. -ectomy
   b. -scope
   c. -scopy
   d. -graphy

3. The prefix _____ means elevated, while the prefix _____ means depressed.

   a. pre-, post-
   b. endo-, exo-
   c. hyper-, hypo-
   d. inter-, intra-

4. The suffix _____ means pertaining to.

   a. -al
   b. -ary or -ar
   c. -ic
   d. all of the above

5. The suffix _____ means incision.

   a. -ex
   b. -tomy
   c. -ectomy
   d. -graphy

6. The suffix _____ means abnormal condition.

   a. -osis
   b. -rrhea
   c. -rrhagia
   d. -uria

7. The suffix _____ means separation or breaking into parts.

   a. -gram
   b. -pexy
   c. -um
   d. -lysis

8. The prefix _____ means below.

   a. supra-
   b. super-
   c. inter-
   d. sub-

9. The prefix(es) _____ mean(s) many.

   a. olig
   b. a-, an-
   c. poly-
   d. eu-

10. The prefix(es) _____ mean(s) without or no.

    a. a-, an-
    b. olig-
    c. dys-
    d. hyper-

11. The suffix -algia means

    a. pain
    b. excessive
    c. liver
    d. abnormal condition

12. The prefix pre- means

    a. after
    b. around
    c. before
    d. during

13. Which suffix may be part of the term meaning a procedure to visually examine?

    a. -lysis
    b. -scopy
    c. -rrhexis
    d. -scope

14. Which type of word part is always placed at the end of a term?

    a. combining form
    b. prefix
    c. suffix
    d. root

15. Which type of word part is always placed at the beginning of a term?

    a. combining form
    b. prefix
    c. suffix
    d. root

16. Which word part gives the essential meaning of a term?

    a. combining form
    b. prefix
    c. suffix
    d. root

17. Which word association is incorrect?

    a. *inter-* means between
    b. *sub-* means below, under, or less
    c. *an-* means without or no
    d. *ad-* means away from

18. Which suffix means to rupture?

    a. -rrhage
    b. -rrhaphy
    c. -rrhea
    d. -rrhexis

19. Which prefix means around?

    a. hyper-
    b. hypo-
    c. peri-
    d. supra-

20. In the term *panleukopenia,* the *o* between *leuk* and *penia* is called a

    a. combining form
    b. suffix
    c. combining vowel
    d. root

Matching

Match the word parts in Column I with the definition in Column II.

Column I	Column II
1. _____ -itis	a. incision or cutting into
2. _____ -gram	b. before
3. _____ post-	c. surgical puncture to remove fluid or gas
4. _____ -tomy	d. difficult, painful, or bad
5. _____ pre-	e. enlargement
6. _____ -centesis	f. excision or surgical removal
7. _____ -therapy	g. liver
8. _____ dys-	h. kidney
9. _____ peri-	i. inflammation
10. _____ ren/o	j. record
11. _____ hepat/o	k. after
12. _____ -megaly	l. treatment
13. _____ -ectomy	m. around

Fill in the Blanks

Write the medical terms that represent the following definitions.

1. Pertaining to the stomach _____
2. Inflammation of the liver _____
3. Abnormal softening of bone _____
4. Joint pain _____
5. Procedure to visually examine inside the body _____
6. Heart enlargement _____
7. Pertaining to the kidney _____
8. Bursting forth of blood from vessels _____
9. Suturing of stomach to body wall _____
10. Treatment with chemicals or drugs _____

Spell Check

Cross out any misspelled words in the following sentences and replace them with the proper spelling.

1. Thick mucous was evident in the cat with upper respiratory disease. _____
2. Urine was collected via cistocentesis so that the urinanalysis could be performed to determine whether the dog had a urinary tract infection. _____
3. The horse's diarhea was caused by intestinal parasites. _____
4. The cutaneus lesion was not cancerous. _____
5. A local anestetic was used so that the surgery could be performed on the cow. _____

Word Part Identification

Underline the word root(s) in the following terms.

1. hepat/itis
2. gastr/o/intestin/al
3. cardi/o/logy
4. intra/ven/ous
5. nephr/osis

Underline the suffix in the following terms.

6. hepat/itis
7. gastr/o/intestin/al
8. cardi/o/logy
9. intra/ven/ous
10. nephr/osis

Underline the prefix in the following terms.

11. hyper/secretion
12. peri/card/itis
13. endo/cardi/um
14. poly/uria
15. ur/o/lith

CROSSWORD PUZZLES

Prefix Puzzle

Supply the correct prefix in the appropriate space for the definition listed.

Across

1 outside
2 without
3 opposite of good
6 less than normal
8 opposite of towards
9 after
10 outside

Down

1 inside
4 excessive
5 between
7 many
8 against

Suffix Puzzle

Supply the correct suffix in the appropriate space for the definition listed.

Across

1 bursting forth
3 procedure that records
5 abnormal softening
8 cutting into
11 surgical removal
13 surgically create new opening
15 abnormal hardening

Down

1 discharge
2 record of
4 disease
6 structure
7 surgical puncture
 to remove fluid
9 suture to stabilize
10 inflammation
12 breakdown
14 abnormal condition

Medical Terms Puzzle

Supply the correct medical term in the appropriate space for the definition listed.

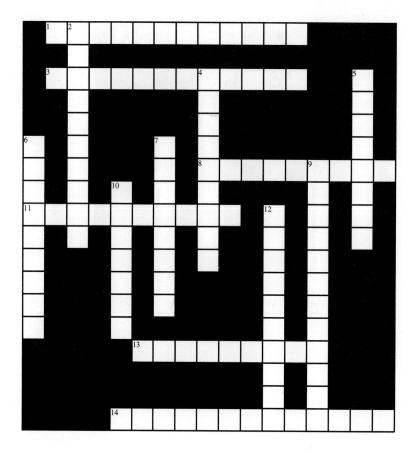

Across

1 enlargement of the heart
3 before an operation
8 pertaining to the skin
11 separation of urine into its components
13 inflammation of the liver
14 increased amount of blood glucose

Down

2 without injury
4 away from midline
5 infrequent urination
6 towards midline
7 frequent urination
9 organism that lives on the outer surface of the host
10 pertaining to the heart
12 surgical removal of the mammary glands

Medical Terms Puzzle

Supply the correct medical term in the appropriate space for the definition listed.

Across

4 disease of intestines
9 structure surrounding the heart
10 instrument to visually examine inside the body
12 excess of regular number

Down

1 between the ribs
2 bursting forth of blood from vessels
3 treatment with drugs
5 difficulty eating or swallowing
6 within muscle
7 pertaining to the gastrointestinal tract
8 abnormal softening of bone
11 pertaining to the loin

WORD SEARCH

Find the following medical terms or word parts in the puzzle below. (Make sure you understand what the terms mean as you find them.)

```
E T I S A R A P O D N E A E S
R E N A L E S O C U L G Y Y G
E N T E R I C P E Y N C S G A
Y M O T C E O R N S O I Y O I
O P Y H Y A E O I P I T X L M
A T C L I P C N R E T A E O E
I E Y Y Y G O U E N A M P N C
Y O U H P A L N T I N U O I Y
H Y O U R S O C U A I A R M L
Y M O T E T S I X E R R T R G
Y L E E F R T A I I U T S E R
M E T U I I O T N K F G A T E
O U T G X T M I M H K F G F P
T K U U C I Y O O I T I U I Y
S O E O E S T N A I H P C S H
```

terminology	uterine
pronunciation	gastritis
penia	colostomy
leuko	gastropexy
hyper	renal
hypo	enteric
endoparasite	prefix
hyperglycemia	suffix
glucose	ectomy
urination	tomy
traumatic	stomy

CASE STUDY

Fill in the blanks to complete the case history.

A 5-yr-old male neutered cat is presented to a veterinary clinic with _____ (painful urination) and _____ (scant urine production). Upon examination the abdomen is palpated and _____ (enlarged urinary bladder) is noted. After completing the examination, the veterinarian suspects an obstruction of the _____ (tube that carries urine from the urinary bladder to outside the body). Blood is taken for analysis, and the cat is admitted to the clinic. The cat is anesthetized, and a urinary catheter is passed. Urine is collected for _____ (breakdown of urine into its components). In addition to the obstruction, the cat is treated for _____ (inflammation of the urinary bladder).

In this case study, the meanings of some unfamiliar medical terms (underlined) cannot be understood by breaking up the term into its basic components. Using a print or online dictionary, define the following medical terms.

   1. palpated   _____

   2. obstruction   _____

   3. catheter   _____

CHAPTER 2

WHERE, WHY, AND WHAT?

Objectives

Upon completion of this chapter, the reader should be able to

- Identify and recognize body planes, positional terms, directional terms, and body cavities
- Identify terms used to describe the structure of cells, tissues, and glands
- Define terms related to body cavities and structure
- Recognize, define, spell, and pronounce medical terms related to pathology and procedures
- Identify body systems by their components, functions, and combining forms
- Identify prefixes that assign numeric value

IN POSITION

Positional terms are important for accurately and concisely describing body locations and relationships of one body structure to another. The terms *forward* and *backward, up* and *down, in* and *out,* and *side to side* are not clear enough descriptions by themselves to have universal understanding in the medical community. Therefore, very specific terms were developed so that there would be no confusion as to the meaning being conveyed. Listed in Table 2–1 and illustrated in Figures 2–1, 2–2, 2–3, and 2–4 are directional terms used in veterinary settings.

Table 2-1   Terms Used to Describe Direction and Surface

Ventral (vehn-trahl) refers to the belly or underside of a body or body part. (*Ventr/o* in Latin means belly.) (*Venture* means to undertake.) (A ventral fin is on the belly.)	**Dorsal** (dōr-sahl) refers to the back. (*Dors/o* in Latin means back.) (*Endorse* means sign on the back.) (A dorsal fin is on the back.) Also refers to the cranial surface of the manus (front paw) and pes (rear paw).
Cranial (krā-nē-ahl) means toward the head. (*Crani/o* in Latin means skull.)	**Caudal** (kaw-dahl) means toward the tail. (*Cauda* in Latin means tail.)
Anterior (ahn-tēr-ē-ər) means front of the body. (*Anteri/o* in Latin means before.) Used more in description of organs or body parts because front and rear are confusing terms in quadrupeds. A quadruped's belly is oriented downward, not forward as in humans.	**Posterior** (pohs-tēr-ē-ər) means rear of the body. (*Posteri/o* in Latin means behind.)
Rostral (rohs-trahl) means nose end of the head. (*Rostrum* in Latin means beak.) **Cephalic** (seh-fahl-ihck) means pertaining to the head. (*Kephale* in Greek means head.)	**Caudal** (kaw-dahl) means toward the tail. (*Cauda* in Latin means tail.)
Medial (mē-dē-ahl) means toward the midline. (*Medi/o* in Latin means middle.)	**Lateral** (laht-ər-ahl) means away from the midline. (*Later/o* in Latin means side.)
Superior (soo-pēr-ē-ər) means uppermost, above, or toward the head. Used more commonly in bipeds. (*Super* in Latin means above.)	**Inferior** (ihn-fēr-ē-ər) means lowermost, below, or toward the tail. Used more commonly in bipeds. (*Inferi* in Latin means lower.)
Proximal (prohck-sih-mahl) means nearest the midline or nearest to the beginning of a structure. (*Proxim/o* in Latin means next.)	**Distal** (dihs-tahl) means farthest from the midline or farthest from the beginning of a structure. (*Dist/o* in Latin means distant.)
Superficial (soop-ər-fihsh-ahl) means near the surface; also called external. (*Super* in Latin means above.)	**Deep** (dēp) means away from the surface; also called internal. (*Deep* means beneath the surface.)
Palmar (pahl-mahr) means the caudal surface of the manus (front paw) including the carpus (from the antebrachial joint distally). (*Palmar* in Latin means hollow of the hand.)	**Plantar** (plahn-tahr) means the caudal surface of the pes (rear paw) including the tarsus (from the tibiotarsal joint distally). (*Plantar* in Latin means sole of the foot.)

THE PLANE TRUTH

Planes are imaginary lines that are used descriptively to divide the body into sections.

- **Midsagittal** (mihd-sahdj-ih-tahl) **plane** is the plane that divides the body into *equal* right and left halves. It also is called the **median** (mē-dē-ahn) plane and the midline (Figure 2–3).

- **Sagittal** (sahdj-ih-tahl) **plane** is the plane that divides the body into unequal right and left parts (Figure 2–4).
- **Dorsal** (dōr-sahl) **plane** is the plane that divides the body into dorsal (back) and ventral (belly) parts (Figure 2–4). It also is called the **frontal** (frohn-tahl) plane or **coronal** (kō-roh-nahl) plane. In humans, the frontal plane is a vertical plane because people stand erect.

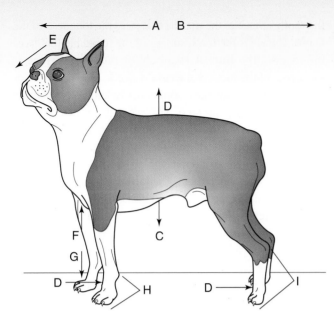

Figure 2–1 Directional and surface terms. The arrows on this Boston terrier represent the following directional terms: A = cranial, B = caudal, C = ventral, D = dorsal, E = rostral, F = proximal, G = distal, H = palmer, I = plantar.

Figure 2–2 Medial versus lateral. The lines on these cats represent the directional terms *medial* and *lateral*. (Photo by Isabelle Francais.)

- **Transverse** (trahnz-vərs) **plane** is the plane that divides the body into cranial and caudal parts (Figure 2–4). It also is called the **horizontal plane** or **cross-sectional plane.** The transverse plane also may be used to describe a perpendicular transection to the long axis of an appendage.

STUDYING

The suffix **-logy** means the study of. Specific terms are used to describe specific branches of study. The study of body structure is called **anatomy** (ah-naht-ō-mē). **Physiology** (fihz-ē-ohl-ō-jē) is the study of body function(s). **Pathology** (pahth-ohl-ō-jē)

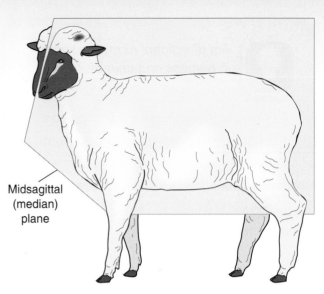

Figure 2–3 Planes of the body. The midsagittal, or median, plane divides the body into *equal* left and right portions.

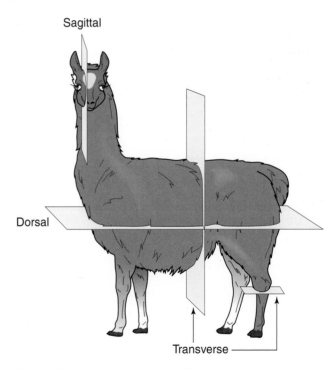

Figure 2–4 Planes of the body. The sagittal plane divides the body into *unequal* right and left parts, the dorsal plane divides the body into back and belly parts, and the transverse plane divides the body into cranial and caudal parts. The transverse plane also describes a perpendicular transection to the long axis of an appendage.

is the study of the nature, causes, and development of abnormal conditions. Combining physiology and pathology results in the term **pathophysiology** (pahth-ō-fihz-ē-ohl-ō-jē), which is the study of changes in function caused by disease. The study of disease causes is **etiology** (ē-tē-ohl-ō-jē).

Are directional terms the same in humans and animals?

Original human anatomy drawings posed a man with his palms forward. Human positional terminology is still based on that pose. If this original drawing had had the palms turned away, human and veterinary anatomical terminology would have been identical.

Directional term confusion

The terms *anterior, posterior, superior,* and *inferior* can be confusing when used with quadrupeds. In quadrupeds, *ventral* is a better term for *anterior* and *dorsal* is a better term than *posterior*.

What does *anterior* mean in a quadruped (cranial or dorsal)?

What does *superior* mean in a quadruped (cranial, dorsal, superficial, or proximal)?

YOU HAVE SAID A MOUTHFUL

Describing positions in the mouth has become increasingly important with the rise of veterinary dentistry. The dental **arcade** (ahr-kād) is the term used to describe how teeth are arranged in the mouth. *Arcade* means a series of arches, which is how the teeth are arranged in the oral cavity. Surfaces of the teeth are named for the area in which they contact (Figure 2–5). The **lingual** (lihng-gwahl) **surface** is the aspect of the tooth that faces the tongue. Remember that linguistics is the study of language, and the tongue is used to make sounds. Some people use lingual surface to describe the tooth surface that faces the tongue on both the maxilla (upper jaw) and mandible (lower jaw). More correctly, the **palatal** (pahl-ah-tahl) **surface** is

the tooth surface of the maxilla that faces the tongue, and the lingual surface is the tooth surface of the mandible that faces the tongue. The **buccal** (buhk-ahl or būk-ahl) **surface** is the aspect of the tooth that faces the cheek. *Bucca* is Latin for cheek. The buccal surface is sometimes called the **vestibular** (vehs-tih-buh-lahr) **surface.** *Vestibule* in Latin means space or cavity at an entrance. The **occlusal** (ō-klū-zahl) **surfaces** are the aspects of the teeth that meet when you chew. Think of the teeth occluding, or stopping, things from passing between them when you clench them. The **labial** (lā-bē-ahl) **surface** is the tooth surface facing the lips. *Labia* is the medical term for lips. **Contact** (kohn-tahckt) **surfaces** are the aspects of the tooth that touch other teeth. Contact surfaces are divided into **mesial** (mē-zē-ahl) and **distal** (dihs-tahl). The mesial contact surface is the one closest to the midline of the dental arcade or arch. The distal contact surface is the one furthest from the midline of the dental arcade (think distance). Each tooth has both contact surfaces, even the last molar, which touches only one tooth surface.

THE HOLE TRUTH

A body **cavity** (kahv-ih-tē) is a hole or hollow space in the body that contains and protects internal organs. The **cranial** (krā-nē-ahl) cavity is the hollow space that contains the brain in the skull. The **spinal** (spī-nahl) cavity is the hollow space that contains the spinal cord within the spinal column. The **thoracic** (thō-rahs-ihck) **cavity,** or **chest cavity,** is the hollow space that contains the heart and lungs within the ribs between the neck and diaphragm. The **abdominal** (ahb-dohm-ih-nahl) **cavity** is the hollow space that contains the major organs of digestion located between the diaphragm and pelvic cavity. The abdominal cavity is commonly called the **peritoneal** (pehr-ih-tohn-ē-ahl) **cavity,** but that is not quite accurate. The peritoneal cavity is the hollow space within the abdominal cavity between the parietal peritoneum and the visceral peritoneum. The **pelvic** (pehl-vihck) cavity is the hollow space that contains the reproductive and some excretory systems (urinary bladder and rectum) organs bounded by the pelvic bones.

Cavities are just one way to segregate the body. Regional terms are also used to describe areas of the body. The **abdomen** (ahb-dō-mehn) is the portion of the body between the thorax and the pelvis containing the abdominal cavity. The **thorax** (thaw-rahcks) is the chest region located between the neck and diaphragm. The **groin** (groyn) is the lower region of the abdomen adjacent to the thigh; it also is known as the **inguinal** (ihng-gwih-nahl) **area.**

Membranes (mehm-brānz) are thin layers of tissue that cover a surface, line a cavity, or divide a space or an organ. The **peritoneum** (pehr-ih-tō-nē-uhm) is the membrane lining the walls of the abdominal and pelvic cavities and it

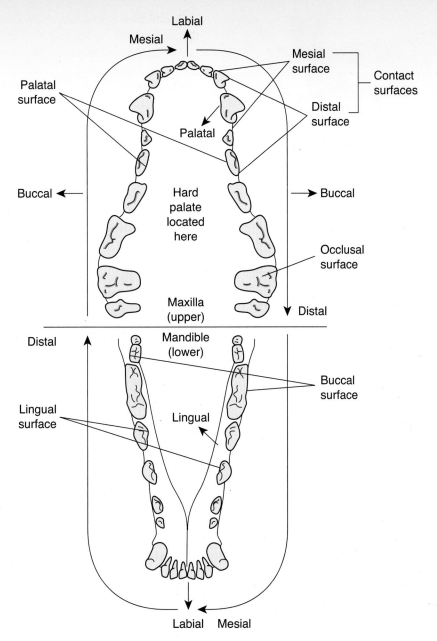

Figure 2–5 Teeth surfaces of the mandible and maxilla. Teeth surfaces are identified by the area they are near.

covers some organs in this area. The peritoneum may be further divided in reference to its location. The **parietal** (pah-rī-eh-tahl) **peritoneum** is the outer layer of the peritoneum that lines the abdominal and pelvic cavities, and the **visceral** (vihs-ər-ahl) **peritoneum** is the inner layer of the peritoneum that surrounds the abdominal organs. Inflammation of the peritoneum is called **peritonitis** (pehr-ih-tō-nī-tihs).

Other terms associated with the abdomen and peritoneum include the umbilicus, mesentery, and retroperitoneal. The

umbilicus (uhm-bihl-ih-kuhs) is the pit in the abdominal wall marking the point where the umbilical cord entered the fetus (Figure 2–6). In veterinary terminology, the umbilicus is also called the **navel** (nā-vuhl). The **mesentery** (mehs-ehn-tehr-ē or mehz-ehn-tehr-ē) is the layer of the peritoneum that suspends parts of the intestine in the abdominal cavity. **Retroperitoneal** (reh-trō-pehr-ih-tō-nē-ahl) means superficial to the peritoneum.

Other membranes of the body are described with the specific body region in which they are found.

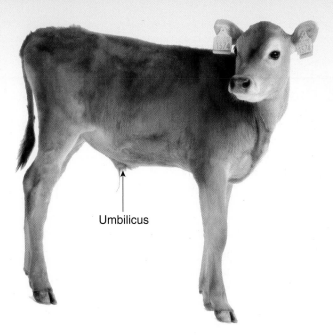

Umbilicus

Figure 2–6 The umbilicus marks the point where the umbilical cord entered the fetus. (Courtesy of iStock photo.)

LYING AROUND

Lay, lie, laid, and *lying* are confusing words in English. However, the only medical term for lying down is **recumbent** (rē-kuhm-behnt). Recumbent is then modified depending on which side is facing down (Figure 2–7).

- **Dorsal recumbency** (dōr-sahl rē-kuhm-behn-sē) is lying on the back.
- **Ventral recumbency** (vehn-trahl rē-kuhm-behn-sē) is lying on the belly = **sternal** (stər-nahl) **recumbency.**
- **Left lateral recumbency** (laht-ər-ahl rē-kuhm-behn-sē) is lying on the left side.
- **Right lateral recumbency** is lying on the right side.

Two less commonly used terms derived from human medical terminology refer to lying down. **Prone** (prōn) means lying in ventral or sternal recumbency; **supine** (soo-pīn) means lying in dorsal recumbency.

To clarify the recumbency terms, remember the following:

- lay = to put, place, or prepare
- laid = past tense of *lay*
- laying = present tense of *lay*
- lie = to recline or be situated
- lain = past tense of *lie*
- lying = present tense of *lie*

MOVING RIGHT ALONG

Medical terms used to describe movement may involve changing prefixes or suffixes to change direction. The terms *adduction* and *abduction* look very similar yet have opposite meanings (Figure 2–8).

Adduction (ahd-duhck-shuhn) means movement toward the midline (think addition to something), and **abduction** (ahb-duhck-shuhn) means movement away from the midline (think child abduction).

Flexion (flehck-shuhn) means closure of a joint angle, or reduction of the angle between two bones. Contracting the biceps involves flexing the elbow. **Extension** (ehcks-tehn-shuhn) means straightening of a joint or an increase in the angle between two bones (Figure 2–9). You extend your hand for a handshake. **Hyperflexion** (hī-pər-flehcks-shuhn) and **hyperextension** (hī-pər-ehcks-tehn-shuhn) occur when a joint is flexed or extended too far. Hyperflexion is the palmar or plantar movement of the joint angles. Hyperextension is the dorsal movement of the joints beyond the reference angle.

Supination and *pronation* are two less commonly used terms in veterinary settings. **Supination** (soo-pih-nā-shuhn) is the act of rotating the limb or body part so that the palmar surface is turned upward, and **pronation** (prō-nā-shuhn) is the act of rotating the limb or body part so that the palmar surface is turned downward. Think of supination as the movement involved with eating soup while cupping the hand. **Rotation** (rō-tā-shuhn) is another term of movement that means circular movement around an axis.

SETTING OUR CYTES AHEAD

Cells are the structural units of the body (Figure 2–10). The combining form for cell is **cyt/o** (sī-tō). Cells are specialized and grouped together to form tissues and organs. **Cytology** (sī-tohl-ō-jē) is the study of cells. The suffix *-logy* means the study of. Cytology involves studying cell origin, structure, function, and pathology.

The cell membrane, cytoplasm, and nucleus are collectively called the **protoplasm** (prō-tō-plahzm). The suffix *-plasm* (plahzm) means formative material of cells, and the combining form *prot/o* means first. The **cell membrane** (also called the plasma membrane) is the structure lining the cell that protects the cell's contents and regulates what goes in and out of the cell. **Cytoplasm** (sī-tō-plahzm) is the gelatinous material located in the cell membrane that is not part of the nucleus. The **nucleus** (nū-klē-uhs) is the structure in a cell that contains nucleoplasm, chromosomes, and the surrounding membrane. **Nucleoplasm** (nū-klē-ōplahzm) is the material in the nucleus, and **chromosomes** (krō-mō-sōmz) are the structures in the nucleus composed of DNA, which transmits genetic information.

IT'S IN THE GENES

Genetic is a term used to denote something that pertains to genes or heredity. A **genetic** (jehn-eh-tihck) **disorder** is any inherited disease or condition caused by defective genes . This term is different from **congenital** (kohn-jehn-ih-tahl), which denotes something that is present at birth. A genetic defect may

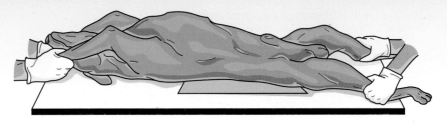

(a) Dorsal recumbency

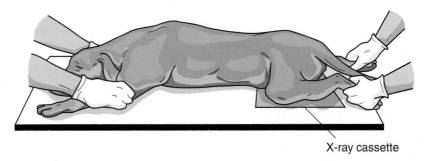

X-ray cassette

(b) Ventral recumbency/sternal recumbency

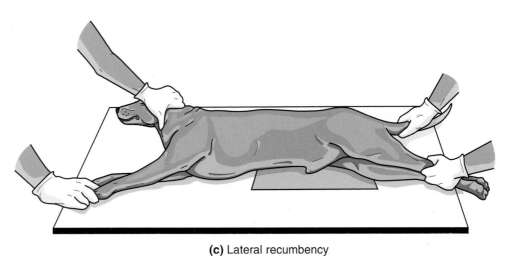

(c) Lateral recumbency

Figure 2–7   Recumbency positions. The position in which an animal lies is important in veterinary medicine, especially in radiographing an animal. (a) This dog is in *dorsal recumbency*. (b) This dog is in *ventral*, or *sternal*, *recumbency*. (c) This dog is in *right lateral recumbency*.

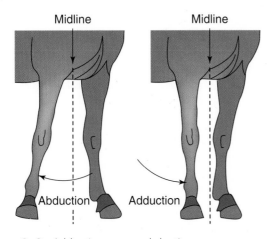

Midline   Midline

Abduction   Adduction

Figure 2–8   *Adduction* versus *abduction*.

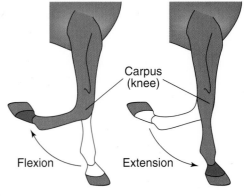

Carpus
(knee)

Flexion   Extension

Figure 2–9   *Flexion* and *extension* of the carpus (knee) of a horse.

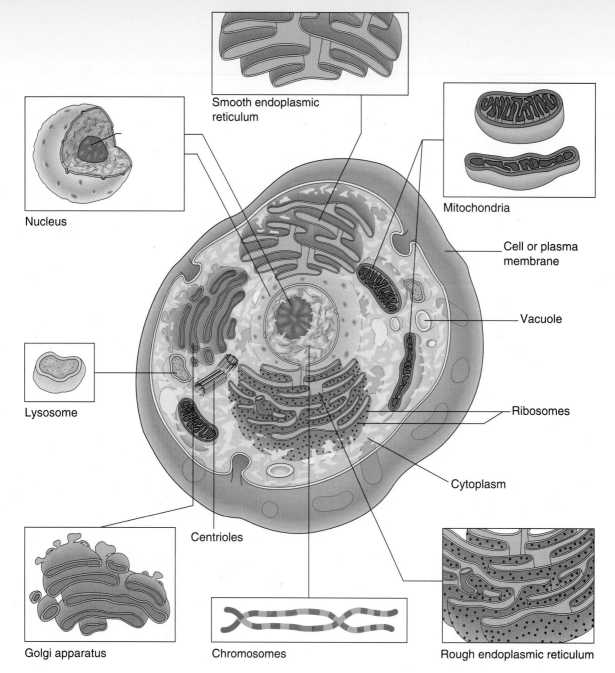

Figure 2–10 Parts of the cell. Parts of the cell include the cell or plasma membrane (serves as the cell's boundary and is semipermeable to allow some things in and out of the cell); nucleus (controls cellular activity and contains genetic material of the cell); nucleolus (produces RNA that forms ribosomes); cytoplasm (semifluid medium containing organelles); chromosomes (structures in the nucleus composed of DNA, which transmits genetic information); centrioles (rod-shaped organelles that maintain cell shape and move chromosomes during cell replication); mitochondria (energy producers of the cell); Golgi apparatus (chemical processor of the cell); endoplasmic reticulum (collection of folded membranes that may contain ribosomes, known as rough endoplasmic reticulum, which synthesize protein, or may be void of ribosomes, known as smooth endoplasmic reticulum, which synthesizes lipids, and some carbohydrates); ribosomes (site of protein synthesis); vacuoles (small membrane-bound organelles containing water, food, or metabolic waste); and lysosomes (digestive system of the cell).

be congenital, but a congenital defect implies only that something faulty is present at birth. An **anomaly** (ah-nohm-ah-lē) is a deviation from what is regarded as normal. *Anomaly* may be used instead of *defect*.

GROUPING THINGS TOGETHER

A group of specialized cells that is similar in structure and function is a **tissue** (tihsh-yoo). The study of the structure, composition, and function of tissue is **histology** (hihs-tohl-ō-jē). **Hist/o** is the combining form for tissue.

There are four types of tissue: epithelial, connective, muscle, and nervous. **Epithelial tissue** (ehp-ih-thē-lē-ahl tihsh-yoo) or **epithelium** (ehp-ih-thē-lē-uhm) covers internal and external body surfaces and is made up of tightly packed cells in a variety of arrangements (Figure 2–11). *Epi-* is a prefix that means above, *thel/o* is a combining form that means nipple but is now used to denote any thin membrane, and *-um* is a suffix that means structure. Epithelial tissue is further divided into mesothelium and endothelium. **Endothelium** (ehn-dō-thē-lē-uhm) is the cellular covering that forms the lining of the internal organs, including the blood vessels. *Endo-* is a prefix

meaning within. **Mesothelium** (mēs-ō-thē-lē-uhm) is the cellular covering that forms the lining of serous membranes such as the peritoneum. The prefix *meso-* means middle.

Connective tissue is another tissue type. Connective tissue adds support and structure to the body by holding the organs in place and binding body parts together (Figure 2–12). Bone, cartilage, dense connective tissue (found in tendons and ligaments), loose connective tissue, and blood are all types of connective tissue. **Adipose** (ahd-ih-pohs) tissue, another form of connective tissue, is also known as fat. **Adip/o** is the combining form for fat.

Muscle tissue is another tissue type that contains cell material with the specialized ability to contract and relax. Three muscle types exist in animals: skeletal, smooth, and cardiac (Figure 2–13). These muscle types are covered in Chapter 3.

Nervous tissue is the last tissue type (Figure 2–14). Nervous tissue contains cells with the specialized ability to react to stimuli and conduct electrical impulses. The nervous system is covered in greater depth in Chapter 13.

Tissue can form normally or abnormally. The suffix **-plasia** (plā-zē-ah) is used to describe formation, development, and growth of tissue and cell *numbers*. The suffix **-trophy** (trō-fē)

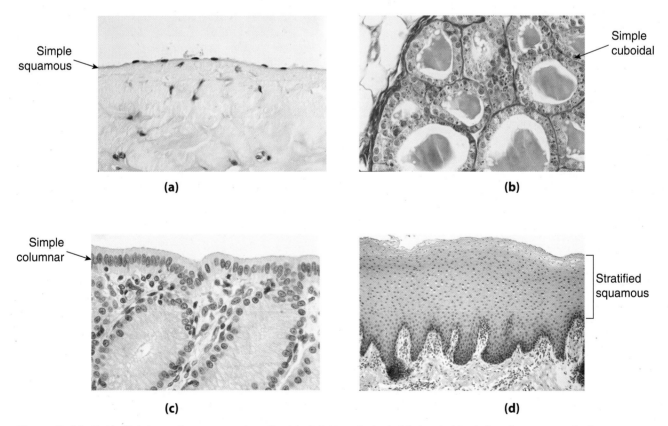

(a)

(b)

(c)

(d)

Figure 2–11 Epithelial tissue. Some examples of epithelial tissue include (a) simple (single layer) squamous (cells are flattened) epithelial tissue, (b) simple cuboidal (cells are cube shaped) epithelial tissue, (c) simple columnar (cells are column shaped) epithelial tissue, and (d) stratified (multilayered) squamous epithelial tissue. (Photomicrographs courtesy of William J. Bacha, PhD, and Linda M. Bacha, MS, VMD.)

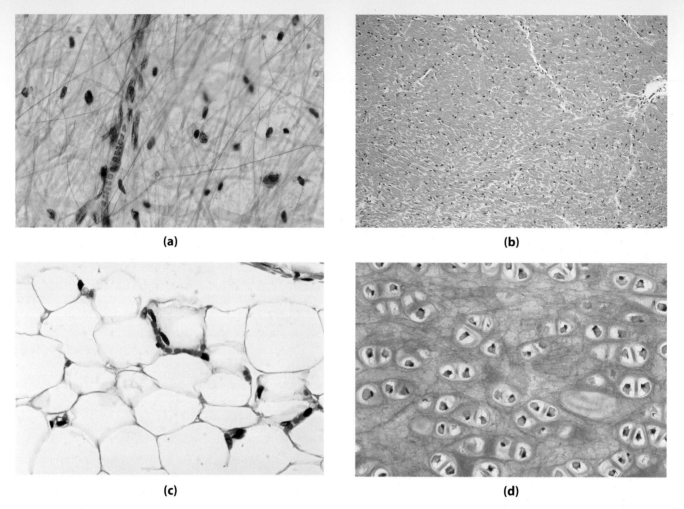

Figure 2–12 Connective tissue. Some examples of connective tissue include (a) loose connective tissue (typically found attached to abdominal organs), (b) dense connective tissue (found in tissues such as ligaments), (c) adipose tissue (lipid or fat tissue), and (d) cartilage (articular cartilage is found on the ends of bones). (Photomicrographs courtesy of William J. Bacha, PhD, and Linda M. Bacha, MS, VMD.)

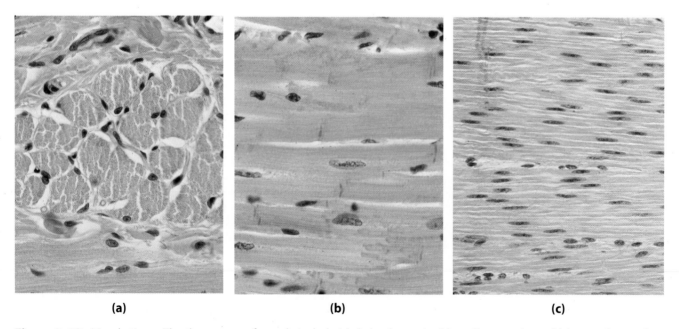

Figure 2–13 Muscle tissue. The three types of muscle include (a) skeletal muscle, (b) cardiac muscle, and (c) smooth muscle. (Photomicrographs courtesy of William J. Bacha, PhD, and Linda M. Bacha, MS, VMD.)

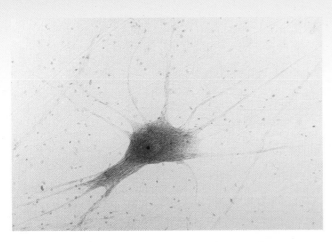

Figure 2–14 Nervous tissue. Photomicrograph of a neuron from the spinal cord of a bovine. (Photomicrographs courtesy of William J. Bacha, PhD, and Linda M. Bacha, MS, VMD.)

means formation, development, and increase in the *size* of tissue and cells. The use of different prefixes describes problems with tissue formation.

- **Anaplasia** (ahn-ah-plā-zē-ah) is a change in the structure of cells and their orientation to each other.
- **Aplasia** (ā-plā-zē-ah) is lack of development of an organ or a tissue or a cell.
- **Dysplasia** (dihs-plā-zē-ah) is abnormal growth or development of an organ or a tissue or a cell.
- **Hyperplasia** (hī-pər-plā-zē-ah) is an abnormal increase in the number of normal cells in normal arrangement in an organ or a tissue or a cell.
- **Hypoplasia** (hī-pō-plā-zē-ah) is incomplete or less than normal development of an organ or a tissue or a cell.
- **Neoplasia** (nē-ō-plā-zē-ah) is any abnormal new growth of tissue in which multiplication of cells is uncontrolled, more rapid than normal, and progressive. Neoplasms usually form a distinct mass of tissue called a **tumor** (too-mər). Tumors may be **benign** (beh-nīn), meaning not recurring, or **malignant** (mah-lihg-nahnt), meaning tending to spread and be life threatening. The suffix **-oma** (ō-mah) means tumor or neoplasm.
- **Atrophy** (ah-tō-fē) is decrease in size or complete wasting of an organ or tissue or cell.

- **Dystrophy** (dihs-trō-fē) is defective growth in the size of an organ or tissue or cell.
- **Hypertrophy** (hī-pər-tō-fē) is increase in the size of an organ or tissue or cell.

The prefix *a-* means without, *hypo-* means less than normal, *hyper-* means more than normal, *dys-* means bad, *ana-* means without, and *neo-* means new.

Glands (glahndz) are groups of specialized cells that secrete material used elsewhere in the body. *Aden/o* is the combining form for gland. Glands are divided into two categories: exocrine and endocrine (Figure 2–15). **Exocrine** (ehck-soh-krihn) **glands** are groups of cells that secrete their chemical substances into ducts that lead out of the body or to another organ. Examples of exocrine glands are sweat glands, sebaceous glands, and the portion of the pancreas that secretes digestive chemicals. **Endocrine** (ehn-dō-krihn) **glands** are groups of cells that secrete their chemical substances directly into the bloodstream, which transports them throughout the body. Endocrine glands are ductless. Examples of endocrine glands are the thyroid gland, the pituitary gland, and the portion of the pancreas that secretes insulin.

An **organ** (ohr-gahn) is a part of the body that performs a special function or functions. Each organ has its own combining form or forms, as listed in Table 2–2. The combining forms have either Latin or Greek origins. If a body part has two combining forms that are used to describe it, how do you know which form to use? In general, the Latin term is used to describe or modify something, as in *renal disease* and *renal tubule*. The Greek term generally is used to describe a pathological finding, as in *nephritis* and *nephropathy*.

1, 2, 3, Go

Medical terms can be further modified by the use of prefixes to assign number value (Table 2–3), numerical order, or proportions. The following prefixes are also used in everyday English, so some of them may be familiar. For example, unicorns are animals with one horn (*uni* = one, *corn* = horn). It would make sense then that a **bicornuate uterus** (*bi* = two, *corn* = horn) is a uterus with two horns. Knowing that *lateral* means pertaining to the side, it would make sense that **unilateral** (yoo-nih-lah-tər-ahl) means pertaining to one side. **Bilateral** (bī-lah-tər-ahl) means pertaining to two sides.

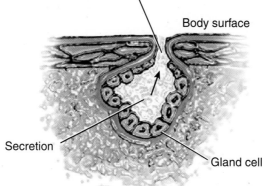

Duct takes secretions out of
body or to another organ

Body surface

Secretion

Gland cell

(a) Exocrine gland (has duct)

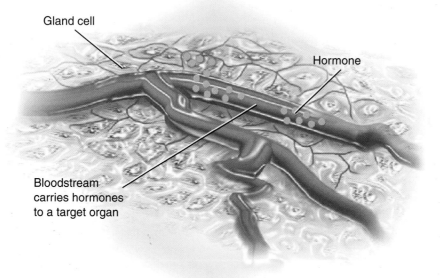

Gland cell

Hormone

Bloodstream
carries hormones
to a target organ

(b) Endocrine gland (ductless)

Figure 2–15 Types of glands. Exocrine glands secrete their chemical substances into ducts that lead out of the body or to another organ. Endocrine glands secrete their chemical substances (hormones) directly into the bloodstream.

Table 2–2   Combining Forms for Organs

Body System	Combining Form	Major Functions
Skeletal system	bones = **oste/o** (ohs-tē-ō), **oss/e** (ohs-ē), or **oss/i** (ohs-ih) joints = **arthr/o** (ahr-thrō) cartilage = **chondr/o** (kohn-drō)	Support and shape, protection, hematopoiesis, mineral storage
Muscular system	muscles = **my/o** (mī-ō) fascia = **fasc/i** (fahs-ē) or **fasci/o** (fahs-ē-ō) tendons = **ten/o** (tehn-ō), **tend/o** (tehn-dō), or **tendin/o** (tehn-dih-nō)	Locomotion, movement of body fluids, body heat generation
Cardiovascular system	heart = **cardi/o** (kahr-dē-ō) arteries = **arteri/o** (ahr-tē-rē-ō) veins = **ven/o** (vēn-ō) or **phleb/o** (fleh-bō) blood = **hem/o** (hē-mō) or **hemat/o** (hē-maht-ō)	Delivers oxygen and nutrients to tissue, transports cellular waste from body, performs immune function and endocrine function
Lymphatic and immune systems	lymph vessels, fluid, and nodes = **lymph/o** (lihm-fō) tonsils = **tonsill/o** (tohn-sih-lō) spleen = **splen/o** (spleh-nō) thymus = **thym/o** (thī-mō)	Provide nutrients to and remove waste from tissues, protect the body from harmful substances
Respiratory system	nose or nares = **nas/o** (nā-zō) or **rhin/o** (rī-nō) pharynx = **pharyng/o** (fahr-ihn-gō) trachea = **trache/o** (trā-kē-ō) larynx = **laryng/o** (lahr-ihng-gō) lungs = **pneum/o** (nū-mō) or **pneumon/o** (nū-mohn-ō)	Brings oxygen into the body for transportation to the cells, removes carbon dioxide and some water waste from the body
Digestive system	mouth = **or/o** (ōr-ō) or **stomat/o** (stō-maht-ō) esophagus = **esophag/o** (eh-sohf-ah-gō) stomach = **gastr/o** (gahs-trō) small intestine = **enter/o** (ehn-tər-ō) large intestine = **col/o** (kō-lō) or **colon/o** (kō-lohn-ō) liver = **hepat/o** (hehp-ah-tō) pancreas = **pancreat/o** (pahn-krē-ah-tō)	Digestion of ingested food, absorption of digested food, elimination of solid waste
Urinary system	kidneys = **ren/o** (rē-nō) or **nephr/o** (nehf-rō) ureters = **ureter/o** (yoo-rē-tər-ō) urinary bladder = **cyst/o** (sihs-tō) urethra = **urethr/o** (yoo-rē-thrō)	Filters blood to remove waste, maintains electrolyte balance, regulates fluid balance
Nervous system and special senses	nerves = **neur/o** (nū-rō) or **neur/i** (nū-rē) brain = **encephal/o** (ehn-sehf-ah-lō) spinal cord = **myel/o** (mī-eh-lō) eyes = **ophthalm/o** (ohf-thahl-mō), **ocul/o** (ohck-yoo-lō), **opt/o** (ohp-tō), or **opt/i** (ohp-tē) sight = **optic/o** (ohp-tih-kō) ears = **ot/o** (ō-tō), **aur/i** (awr-ih), or **aur/o** (awr-ō) **audit/o** (aw-dih-tō), or **aud/i** (aw-dē) external ear = sound = **acoust/o** (ah-koo-stō) or **acous/o** (ah-koo-sō)	Coordinating mechanism, reception of stimuli, transmission of messages

continued

Table 2–2   Combining Forms for Organs *(continued)*

Body System	Combining Form	Major Functions
Integumentary system	skin = **dermat/o** (dər-mah-tō), **derm/o** (dər-mō), or **cutane/o** (kyoo-tā-nē-ō)	Protection of body, temperature, and water regulation
Endocrine system	adrenals = **adren/o** (ahd-reh-nō) gonads = **gonad/o** (gō-nahd-ō) pineal = **pineal/o** (pī-nē-ahl-ō) pituitary = **pituit/o** (pih-too-ih-tō) thyroid = **thyroid/o** (thī-royd-ō) or **thyr/o** (thī-rō)	Integrates body functions, homeostasis, growth
Reproductive system	testes = **orch/o** (ōr-kō), **orchi/o** (ōr-kē-ō), **orchid/o** (ōr-kihd-ō), or **testicul/o** (tehst-tihck-yoo-lō) ovaries = **ovari/o** (ō-vā-rē-ō) or **oophor/o** (ō-ohf-ehr-ō) uterus = **hyster/o** (hihs-tehr-ō), **metr/o** (mē-trō), **metr/i** (mē-trē), **metri/o** (mē-trē-ō), or **uter/o** (yoo-tər-ō)	Production of new life

Table 2–3   Prefixes Assigning Number Value

Number Value	Latin Prefix	Greek Prefix	Examples
1	uni-	mono-	unicorn, unilateral monochromatic, monocyte
2	duo-, bi-	dyo-	duet, bilateral, dyad
3	tri-	tri-	trio, triceratops, triathlon
4	quadri- or quadro-	tetr- or tetra-	quadruplet, tetralogy, tetroxide
5	quinqu-, quint-	pent- or penta-	quintet, pentagon
6	sex-	hex- or hexa-	sexennial, hexose, hexagon
7	sept- or septi-	hept- or hepta-	septuple, heptarchy
8	octo-	oct-, octa-, or octo-	octave, octopus
9	novem- or nonus-	ennea-	nonuple, ennead
10	deca- or decem-	dek- or deka-	decade, dekanem

REVIEW EXERCISES

Multiple Choice

Choose the correct answer.

1. Lateral means

    a. near the beginning

    b. near the front

    c. toward the side

    d. toward midline

2. The sagittal plane divides the body into

    a. cranial and caudal portions

    b. left and right portions

    c. equal left and right halves

    d. dorsal and ventral portions

3. The paw is _____ to the shoulder.

    a. caudal

    b. cranial

    c. proximal

    d. distal

4. The transverse plane divides the body into

    a. cranial and caudal portions

    b. left and right portions

    c. equal left and right halves

    d. dorsal and ventral portions

5. The lining of the abdominal cavity and some of its organs is called the

    a. mesentery

    b. peritoneum

    c. thoracum

    d. membrane

6. The study of structure, composition, and function of tissues is called

    a. cytology

    b. histology

    c. pathology

    d. organology

7. The _____ plane divides the body into dorsal and ventral portions.

    a. sagittal

    b. midsagittal

    c. dorsal

    d. transverse

8. The medical term for lying down is

    a. lateral

    b. sternal

    c. recumbent

    d. surface

9. The medical term for increase in size of an organ, tissue, or cell is

    a. atrophy

    b. hypertrophy

    c. dystrophy

    d. hyperplasia

10. The medical term for the caudal surface of the rear paw, hoof, or foot is

    a. ventral

    b. dorsal

    c. palmar

    d. plantar

11. The term for toward the midline is

    a. medial

    b. lateral

    c. proximal

    d. distal

12. The term for nearest the midline or the beginning of a structure is

    a. medial

    b. lateral

    c. proximal

    d. distal

13. The term for away from the midline is
    a. medial
    b. lateral
    c. proximal
    d. distal

14. The term for farthest from the midline or beginning of a structure is
    a. medial
    b. lateral
    c. proximal
    d. distal

15. The term that refers to the back is
    a. ventral
    b. dorsal
    c. cranial
    d. caudal

16. The term that means toward the tail is
    a. ventral
    b. dorsal
    c. cranial
    d. caudal

17. The term that means toward the head is
    a. ventral
    b. dorsal
    c. cranial
    d. caudal

18. The term that refers to the belly or underside of a body is
    a. ventral
    b. dorsal
    c. cranial
    d. caudal

19. Which type of tissue covers internal and external body surfaces?
    a. adipose
    b. epithelial
    c. connective
    d. cytoplasm

20. The term for a hole or hollow space in the body that contains and protects internal organs is
    a. abdomen
    b. peritoneal
    c. cavity
    d. membrane

Matching

Match the number in Column I with its prefix in Column II. Each number may have more than one correct answer.

Column I	Column II
1. _____ one	a. hept-
2. _____ two	b. mono-
3. _____ three	c. tri-
4. _____ four	d. deka-
5. _____ five	e. duo-
6. _____ six	f. uni-
7. _____ seven	g. penta-
8. _____ eight	h. octo-
9. _____ nine	i. quadri-, quadro-
10. _____ ten	j. tetra-
	k. sex-
	l. nonus-
	m. deca-
	n. quinqu-

Match the pathology term in Column I with its definition in Column II.

Column I	Column II
11. _____ anaplasia	a. abnormal growth or development of an organ or a tissue or a cell
12. _____ aplasia	b. a change in the structure of cells and their orientation to each other
13. _____ dysplasia	c. an abnormal increase in the number of normal cells in normal arrangement in an organ, a tissue, or a cell
14. _____ hyperplasia	d. incomplete or less than normal development of an organ, a tissue, or a cell
15. _____ hypoplasia	e. lack of development of an organ, a tissue, or a cell

Fill in the Blanks

1. The _____ is also known as the navel.

2. _____ glands secrete chemical substances directly into the bloodstream.

3. A(n) _____ is any new growth of tissue in which multiplication of cells is uncontrolled, more rapid than normal, and progressive.

4. A(n) _____ is a deviation from what is regarded as normal.

5. The _____ cavity contains the heart and lungs.

6. The caudal surface of the front paw, foot, or hoof is the _____ surface.

7. The shoulder is _____ to the pelvis.

8. A(n) _____ is the basic structural unit of the body.

9. The stomach is located _____ to the heart.

10. _____ is the palmar or plantar movement of joint angles.

11. Another term for groin is _____.

12. The _____ is a layer of the peritoneum that suspends parts of the intestine in the abdominal cavity.

13. _____ is the suffix for formative material of cells.

14. Not malignant is _____.

15. The five combining forms for uterus are _____, _____, _____, _____, and _____.

16. The plane that divides the animal into equal right and left halves is the _____.

17. The plane that divides the animal into cranial and caudal parts is the _____.

18. The study of body structure is called _____.

19. The study of body function is called _____.

20. The aspect of the tooth of the mandible that faces the tongue is called the _____.

21. The aspect of the tooth that faces the cheek is called the _____.

22. Movement toward the midline is known as _____.

23. Movement away from the midline is known as _____.

24. Groups of specialized cells that secrete material used elsewhere in the body are known as _____.

25. A part of the body that performs a special function or functions is known as a(n) _____.

CROSSWORD PUZZLES

Directional Terms and Planes of the Body

Supply the correct term in the appropriate space for the definition listed.

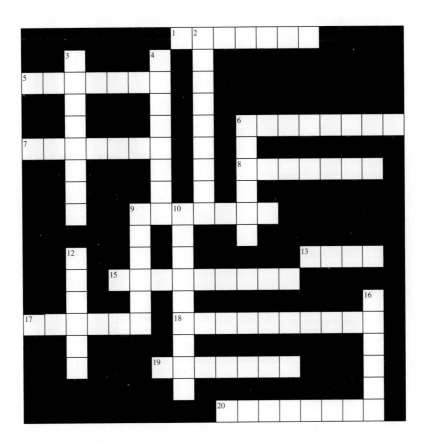

Across

1 pertaining to the belly
5 to reduce the angle between two bones
6 situated nearest the midline
7 bottom of the rear foot, hoof, or paw
8 pertaining to towards the side
9 pertaining to towards the head
13 positioned away from the surface
15 to move towards midline
17 situated farthest from midline
18 plane dividing the body into cranial
   and caudal portions
19 pertaining to the nose end of the head
20 plane dividing the body into left
   and right portions

Down

2 to increase the angle between two bones
3 pertaining to the head
4 below or lowermost
6 bottom of the front foot, hoof, or paw
9 pertaining to towards the tail
10 to move away from midline
12 pertaining to the back
16 pertaining to towards midline

Organ Combining Forms

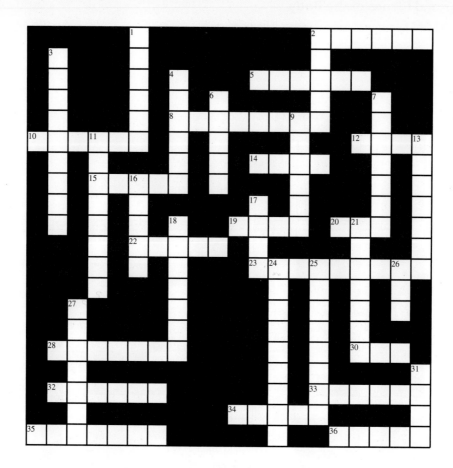

Across

2 kidney
5 lungs
8 pharynx
10 spleen
12 blood
14 nose or nares
15 bone
19 vein
20 ear
22 spinal cord
23 eye
28 urethra
30 mouth
32 small intestine
33 lymph vessel, fluid, or node
34 uterus
35 cartilage
36 urinary bladder

Down

1 joint
2 nerves
3 brain
4 liver
6 fascia
7 ureters
9 stomach
11 esophagus
13 testes
16 thymus
17 tendon
18 skin
21 trachea
24 pancreas
25 tonsil
26 muscle
27 larynx
31 large intestine

LABEL THE DIAGRAMS

For Figures 2–16, 2–17, 2–18, 2–19, 2–20, and 2–21, follow the instructions provided in the captions.

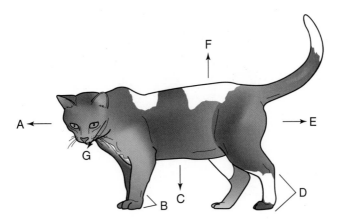

Figure 2–16 Label the arrows with the proper directional term.

Figure 2–17 Label the arrows with the proper directional term.

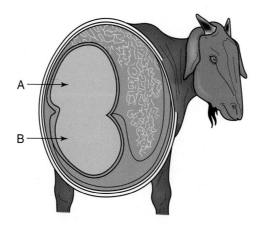

Figure 2–18   Label the sacs of the rumen. Through which plane is this goat sectioned?

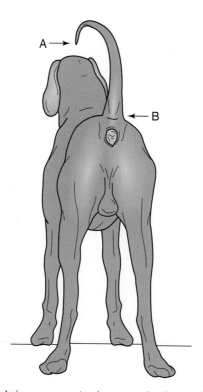

Figure 2–19   Is point A the more proximal or more distal end of the tail? Is point B the more proximal or more distal end of the tail?

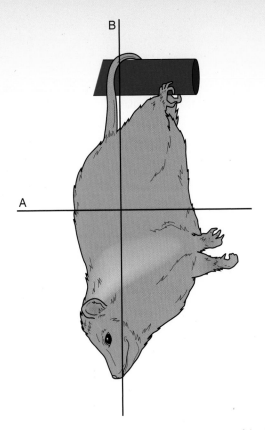

Figure 2–20   What plane of the body is plane A? What plane of the body is plane B?

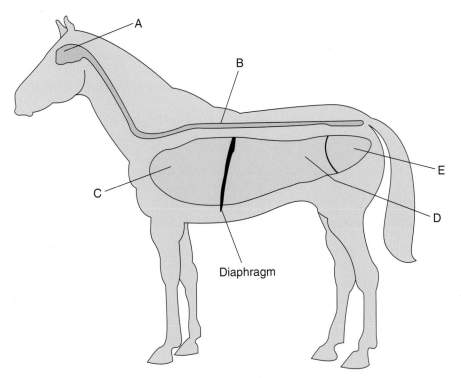

Figure 2–21   Label the body cavities in this drawing using the terminology on page 20 as a guide.

CASE STUDY

After reading the case study, define the list of terms.

 A 7-year-old male Siberian husky presented to the clinic with a cough that has become more severe in the past few weeks. Today the dog collapsed while playing fetch and was rushed to the veterinary clinic. Once the dog was stabilized, thoracic radiographs were taken and a tumor was seen in the cranial thoracic area. There also was hypertrophy of the right side of the heart. The veterinarian was concerned that the dog may have a malignant tumor and requested more tests.

Define the terms using the word parts.

1. thoracic   _____

2. tumor   _____

3. cranial   _____

4. hypertrophy   _____

5. malignant   _____

CHAPTER 3

MEAT AND BONES

Objectives

Upon completion of this chapter, the reader should be able to

- Identify and describe the major structures and functions of the musculoskeletal system
- Describe bone anatomy terms
- Differentiate between the axial and appendicular skeletons
- Recognize, define, spell, and pronounce terms related to the diagnosis, pathology, and treatment of the musculoskeletal system
- Construct musculoskeletal terms from word parts

FUNCTIONS OF THE SKELETAL SYSTEM

The **musculoskeletal** (muhs-kyoo-lō-skehl-eh-tahl) **system** consists of two systems that work together to support the body and allow movement of the animal—the skeletal system and the muscular system. The skeletal system forms the framework that supports and protects an animal's body. Within bone is the red bone marrow, which functions to form red blood cells, white blood cells, and clotting cells. Joints aid in the movement of the body. Cartilage protects the ends of bones where they contact each other. Cartilage is also found in the ear and nose. The muscular system is covered later in this chapter.

STRUCTURES OF THE SKELETAL SYSTEM

The skeletal system consists of bones, cartilage, joints, ligaments, and tendons.

Make the Connection

The skeleton is made up of various forms of connective tissues. Connective tissue is a type of tissue in which the proportion of cells to extracellular matrix is small. Connective tissue binds together and supports various structures of the body. Bone, tendons, ligaments, and cartilage are all connective tissues associated with the skeletal system.

Bone

Bone, a form of connective tissue, is one of the hardest tissues in the body. Embryonically, the skeleton is made of cartilage and fibrous membranes that harden into bone before birth.

Ossification (ohs-ih-fih-kā-shuhn), the formation of bone from fibrous tissue, continues until maturity, which varies with species. Normal bone goes through a continuous process of building up and breaking down throughout an animal's life. This process allows bone to heal and repair itself. Bone growth

Descriptive Word Parts for the Skeletal System

epi- = above, **physis** = growth, **dia-** = between, **peri-** = surrounding, **oste/o-** = bone, **-um** = structure, **endo-** = within or inner, **meta-** = beyond

Table 3–1 Terminology Applied to Bone

Types of Bone

cortical bone (kōr-tih-kahl)	hard, dense, strong bone that forms the outer layer of of bone; also called compact bone
	cortex = bark or shell in Latin
cancellous bone (kahn-sehl-uhs)	lighter, less strong bone that is found in the ends and inner portions of long bones; also called spongy bone
	cancellous = latticework in Latin

Bone Anatomy Terms

epiphysis (eh-pihf-ih-sihs)	wide end of a long bone, which is covered with articular cartilage and is composed of cancellous bone
	proximal epiphysis = located nearest the midline of the body
	distal epiphysis = located farthest away from the midline of the body
diaphysis (dī-ahf-ih-sihs)	shaft of a long bone that is composed mainly of compact bone
physis (fī-sihs)	cartilage segment of long bone that involves growth of the bone; also called the **growth plate** or **epiphyseal cartilage** (Figure 3–3)
metaphysis (meh-tahf-ih-sihs)	wider part of long bone shaft located adjacent to the physis; in adult animals, it is considered part of the epiphysis
periosteum (pehr-ē-ohs-tē-uhm)	tough, fibrous tissue that forms the outer covering of bone
endosteum (ehn-dohs-tē-uhm)	tough, fibrous tissue that forms the lining of the medullary cavity

Bone Classification

long bones	bones consisting of shaft, two ends, and a marrow cavity (i.e., femur)
short bones	cube-shaped bones with no marrow cavity (i.e., carpal bones)
flat bones	thin, flat bones (i.e., pelvis)
pneumatic bones	sinus-containing bones (i.e., frontal bone)
irregular bones	unpaired bones (i.e., vertebrae)
sesamoid bones	small bones embedded in a tendon (i.e., patella) (the only exception is the distal sesamoid of the horse)

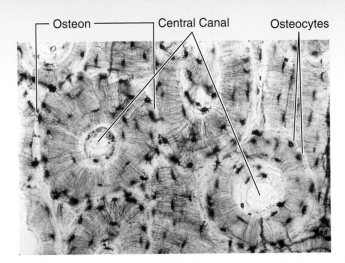

Figure 3–1 Microscopic structure of compact bone, showing osteons with a central canal. The bone matrix is deposited in a circular arrangement. Darkly stained osteocytes are visible in the matrix. (Photomicrograph courtesy of William J. Bacha, PhD, and Linda M. Bacha, MS, VMD.)

is balanced between the actions of **osteoblasts** (ohs-tē-ō-blahsts) and **osteoclasts** (ohs-tē-ō-klahsts). Osteoblasts (**oste/o** = bone, **-blasts** = immature) are immature bone cells that produce bony tissue, and osteoclasts (**oste/o** = bone, **-clasts** = break) are phagocytic cells that eat away bony tissue from the medullary cavity of bone (Figure 3–1). When osteoblasts mature, they become **osteocytes** (ohs-tē-ō-sītz). The combining forms for bone are **oste/o, oss/e,** and **oss/i.**

Red bone marrow, located in cancellous bone, is **hematopoietic** (hēm-ah-tō-poy-eht-ihck). The combining form **hemat/o** means blood, and the suffix **-poietic** means pertaining to formation. Thus, red bone marrow produces red blood cells, white blood cells, and clotting cells. The **medullary** (mehd-yoo-lahr-ē) **cavity** of bone, or the inner space of bone, contains yellow bone marrow. In adult animals, yellow bone marrow replaces red bone marrow. Yellow bone marrow is composed mainly of fat cells and serves as a fat storage area.

Bone is divided into different categories based on bone types, bone shapes, and bone functions (Table 3–1 and Figure 3–2).

Cartilage

Cartilage (kahr-tih-lihdj) is another form of connective tissue that is more elastic than bone. The elasticity of cartilage makes it useful in the more flexible portions of the skeleton. **Articular** (ahr-tihck-yoo-lahr) **cartilage,** a specific type of cartilage, covers the joint surfaces of bone. The **meniscus** (meh-nihs-kuhs) is a curved fibrous cartilage found in some joints, such as the canine stifle, that cushions forces applied to the joint. The combining form for cartilage is **chondr/o.**

Soft versus Hard

Bone diseases can cause abnormal changes. Bones can become softer than normal or harder than normal. To describe these changes, the suffixes **-malacia** (abnormal softening) and **-sclerosis** (abnormal hardening) are used.

Joints

Joints or **articulations** (ahr-tihck-yoo-lā-shuhns) are connections between bones. *Articulate* means to join in a way that allows motion between the parts. The combining form for joint is **arthr/o.** The different types of joints are based on their function and degree of movement.

Joints are classified based on their degree of movement (Figure 3–4). **Synarthroses** (sihn-ahrth-rō-sēz) allow no movement, **amphiarthroses** (ahm-fih-ahrthr-ō-sēz) allow slight movement, and **diarthroses** (dī-ahrth-rō-sēz) allow free movement.

Synarthroses are immovable joints usually united with fibrous connective tissue. An example of a synarthrosis is a suture. A **suture** (soo-chuhr) is a jagged line where bones join and form a nonmovable joint. Sutures typically are found in the skull. A **fontanelle** (fohn-tah-nehl) is a soft spot remaining at the junction of sutures that usually closes after birth.

Amphiarthroses are semimovable joints. An example of an amphiarthrosis is a symphysis. A **symphysis** (sihm-fih-sihs) is a joint where two bones join and are held firmly together so that they function as one bone. Another term for *symphysis* is **cartilaginous joint.** The halves of the mandible fuse at a symphysis to form one bone. This fusion is the **mandibular symphysis.** The halves of the pelvis also fuse at a symphysis, which is called the **pubic symphysis.**

Diarthroses are freely movable joints. Examples of diarthroses are synovial joints. **Synovial** (sih-nō-vē-ahl) **joints** are further classified as ball-and-socket joints (also called **enarthrosis** (ehn-ahr-thrō-sihs) or spheroid joints), arthrodial (ahr-thrō-dē-ahl) or condyloid (kohn-dih-loyd) joints, trochoid (trō-koyd) or pivot (pih-voht) joints, ginglymus (jihn-glih-muhs) or hinge joints, and gliding joints. **Ball-and-socket joints** allow a wide range of motion in many directions, such as the hip and shoulder joints. **Arthrodial** or **condyloid joints** are joints with oval projections that fit into a socket, such as the carpal joints (where the radius meets the carpus). **Trochoid joints** include

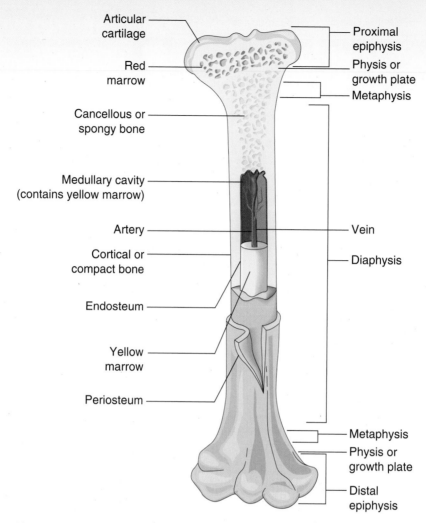

Articular cartilage
Red marrow
Cancellous or spongy bone
Medullary cavity (contains yellow marrow)
Artery
Cortical or compact bone
Endosteum
Yellow marrow
Periosteum

Proximal epiphysis
Physis or growth plate
Metaphysis
Vein
Diaphysis
Metaphysis
Physis or growth plate
Distal epiphysis

Figure 3–2   Anatomy of a long bone.

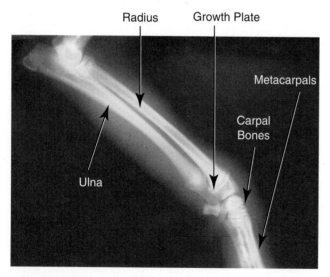

Radius   Growth Plate
Metacarpals
Carpal Bones
Ulna

Figure 3–3   Radiograph of the radius and ulna of a young dog. Note the growth plate (physis) visible in the radius and ulna.

pulley-shaped joints like the connection between the atlas to the axis. **Hinge joints** allow motion in one plane or direction, such as canine stifle and elbow joints. **Gliding joints** move or glide over each other, as in the radioulnar joint or the articulating process between successive vertebrae. Primates have an additional joint called the saddle joint. The only **saddle joint** is located in the carpometacarpal joint of the thumb. This saddle joint allows primates to flex, extend, abduct, adduct, and circumduct the thumb.

Ligaments and Tendons

A **ligament** (lihg-ah-mehnt) is a band of fibrous connective tissue that connects one bone to another bone. **Ligament/o** is the combining form for ligament. A ligament is different from a tendon. A **tendon** (tehn-dohn) is a band of fibrous connective tissue that connects muscle to bone. The combining forms for tendon are **ten/o, tend/o,** and **tendin/o.**

Bursa

A **bursa** (bər-sah) is a fibrous sac that acts as a cushion to ease movement in areas of friction. Within the shoulder joint is a bursa where a tendon passes over bone. The combining form for bursa is **burs/o.** More than one bursa is **bursae** (bər-sā).

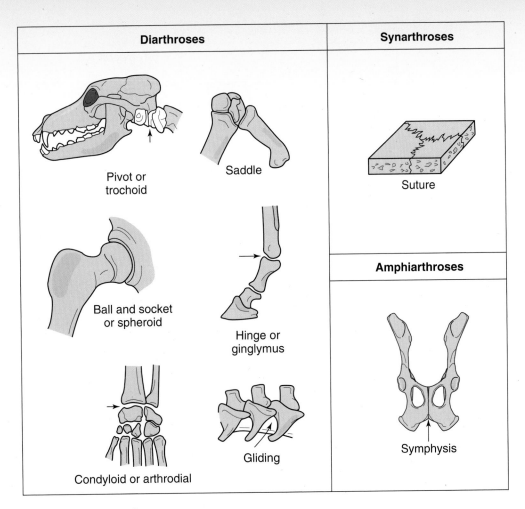

Figure 3–4 Types of joints.

A way to remember that a tendon connects a muscle to bone is that both tendon and muscle have the same number of letters or that the Achilles tendon attaches the calf muscle to a bone.

Synovial Membrane and Fluid

Bursae and synovial joints have an inner lining called the **synovial** (sih-nō-vē-ahl) **membrane.** The synovial membrane secretes synovial fluid, which acts as a lubricant to make joint movement smooth. **Synovi/o** is the combining form for synovial membrane and synovial fluid.

BONING UP

The skeleton is descriptively divided into two parts: the axial skeleton and the appendicular skeleton. The **axial** (ahcks-ē-ahl) **skeleton** is the framework of the body that includes the skull, auditory ossicles, hyoid bones, vertebral column, ribs, and sternum. The **appendicular** (ahp-ehn-dihck-yoo-lahr) **skeleton** is the framework of the body that consists of the extremities, shoulder, and pelvic girdle. *Append* means to add or hang, so think of the appendages or extremities as structures that hang from the axial skeleton.

The Axial Skeleton

The axial skeleton is composed of bones that lie around the body's center of gravity.

Take It From The Top

The **cranium** (krā-nē-uhm) is the portion of the skull that encloses the brain. The combining form **crani/o** means skull. The cranium consists of the following bones (Figure 3–5):

- **frontal** (frohn-tahl) = forms the roof of the cranial cavity or "front" or cranial portion of the skull. In some species, the horn, or cornual (kohrn-yoo-ahl) process, arises from the frontal bone (Figure 3–6).
- **parietal** (pah-rī-ih-tahl) = paired bones that form the roof of the caudal cranial cavity.

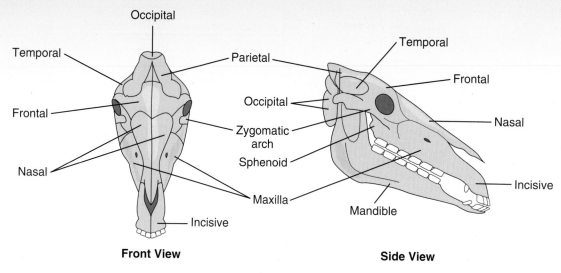

Figure 3–5 Selected bones of the skull and face.

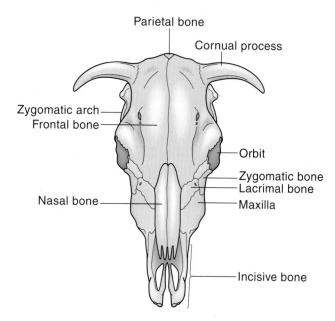

Figure 3–6 Skull of an ox, dorsal view.

Skull shapes in dogs can vary by breed. Examples of skull shapes in dogs include the following:

- **Brachycephalic** (brā-kē-seh-fahl-ihck) dogs have short, wide heads, as do pugs and Pekingese.
- **Dolichocephalic** (dō-lih-kō-seh-fahl-ihck) dogs have narrow, long heads, as do collies and greyhounds.
- **Mesocephalic** (mehs-ō-seh-fahl-ihck) dogs have average width to their heads, as do Labrador retrievers. Also called **mesaticephalic** (mehs-āt-ih-seh-fahl-ihck).

- **occipital** (ohck-sihp-ih-tahl) = forms the caudal aspect of the cranial cavity where the foramen magnum, or opening for the spinal cord, is located. **Foramen** (fō-rā-mehn) is an opening in bone through which tissue passes. **Magnum** (māg-nuhm) means large.
- **temporal** (tehm-pohr-ahl) = paired bones that form the sides and base of the cranium.
- **sphenoid** (sfeh-noyd) = paired bones that form part of the base of the skull and parts of the floor and sides of the bony eye socket.
- **ethmoid** (ehth-moyd) = forms the rostral part of the cranial cavity.
- **incisive** (ihn-sīs-ihv) = forms the rostral part of the hard palate and lower edge of nares.

- **pterygoid** (tahr-ih-goyd) = forms the lateral wall of the nasopharynx.

In addition to bones, the skull also has air- or fluid-filled spaces. These air- or fluid-filled spaces are called **sinuses** (sīn-uhs-ehz).

Let's Face It

The bones of the face consist of the following:

- **zygomatic** (zī-gō-mah-tihck) = projections from the temporal and frontal bones to form the cheekbone.
- **maxilla** (mahck-sih-lah) = forms the upper jaw.
- **mandible** (mahn-dih-buhl) = forms the lower jaw.
- **palatine** (pahl-ah-tihn) = forms part of the hard palate.

- **lacrimal** (lahck-rih-mahl) = forms the medial part of the orbit.
- **incisive** (ihn-sī-sihv) = forms the rostral part of the hard palate and lower edge of nares.
- **nasal** (nā-sahl) = forms the bridge of the nose.
- **vomer** (vō-mər) = forms the base of the nasal septum. The **nasal septum** (nā-sahl sehp-tuhm) is the cartilaginous structure that divides the two nasal cavities.
- **hyoid** (hī-oyd) = bone suspended between the mandible and the laryngopharynx.

Back To Basics

The **vertebral** (vər-teh-brahl) **column** (also called the **spinal column** and **backbone**) supports the head and body and protects the spinal cord. The vertebral column consists of individual bones called **vertebra** (vər-teh-brah). The combining forms for vertebra are **spondyl/o** and **vertebr/o.** More than one vertebra are called **vertebrae** (vər-teh-brā).

Vertebrae are divided into parts, and the parts may vary depending on the location of the vertebra and its function (Figure 3–7). The **body** is the solid portion ventral to the spinal cord. The **arch** is the dorsal part of the vertebra that surrounds the spinal cord. The **lamina** (lahm-ih-nah) is the left or right dorsal half of the arch. Processes project from the vertebrae. The term *process* means projection. A **spinous process** is a single projection from the dorsal part of the vertebral arch. **Transverse processes** project laterally from the right and left sides of the vertebral arch. **Articular processes** are paired cranial and caudal projections located on the dorsum of the vertebral arch.

Foramen (fō-rā-mehn) means opening. The opening in the middle of the vertebra through which the spinal cord passes is the **vertebral foramen.** The vertebrae are separated and cushioned from each other by cartilage discs called **intervertebral discs.**

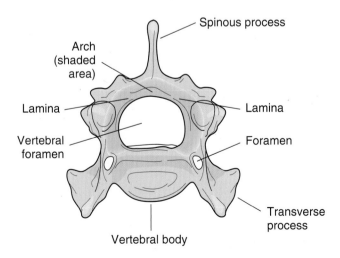

Figure 3–7 Parts of a vertebra.

The vertebral formulas for different species are as follows:

- dogs and cats: C = 7, T = 13, L = 7, S = 3, Cy = 6–23
- equine: C = 7, T = 18, L = 6 (or L = 5 in some Arabians), S = 5, Cy = 15–21
- bovine: C = 7, T = 13, L = 6, S = 5, Cy = 18–20
- pigs: C = 7, T = 14–15, L = 6 – 7, S = 4, Cy = 20–23
- sheep and goats: C = 7, T = 13, L = 6–7, S = 4, Cy = 16–18
- chicken: C = 14, T = 7, LS = 14, Cy = 6 (lumbar and sacral vertebrae are fused)

Vertebrae are organized and named by region. The regions are identified in Table 3–2 and Figures 3–8 and 3–9.

In addition, the first two vertebrae have individual names. C1 (or cervical vertebra one) is called the **atlas,** and C2 (or cervical vertebra two) is called the **axis.** (Remember that they follow alphabetical order.)

Stick to Your Ribs

Ribs are paired bones that attach to thoracic vertebrae (Figure 3–10). The combining form for rib is **cost/o.** Ribs are sometimes called **costals.**

The **sternum** (stər-nuhm), or breastbone, forms the midline ventral portion of the rib cage. The sternum is divided into three parts: the manubrium, body, and xiphoid process. The **manubrium** (mah-nū-brē-uhm) is the cranial portion of the sternum. The **body** of the sternum is the middle portion. The caudal portion of the sternum is known as the **xiphoid** (zī-foyd) **process.**

The ribs, sternum, and thoracic vertebrae make up the boundaries of the thoracic cavity. The **thoracic cavity, or rib cage,** protects the heart and lungs (Figure 3–11).

The Appendicular Skeleton

The appendicular skeleton is composed of the bones of the limbs.

From the Front

The bones of the front limb from proximal to distal consist of the scapula, clavicle, humerus, radius, ulna, carpus, metacarpals, and phalanges.

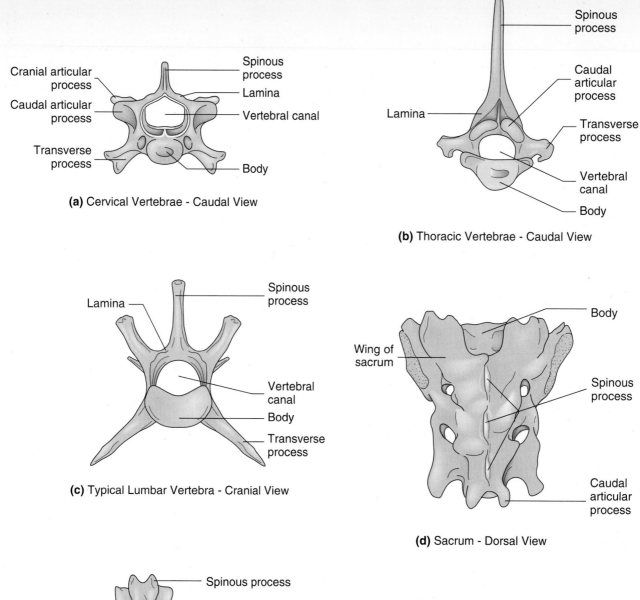

(a) Cervical Vertebrae - Caudal View

(b) Thoracic Vertebrae - Caudal View

(c) Typical Lumbar Vertebra - Cranial View

(d) Sacrum - Dorsal View

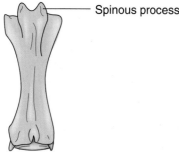

(e) Coccygeal Vertebrae - Dorsal View

Figure 3–8   Comparison of the structure of cervical, thoracic, lumbar, sacral, and coccygeal vertebrae: (a) caudal view of a cervical vertebra, (b) caudal view of a thoracic vertebra, (c) caudal view of a lumbar vertebra, (d) dorsal view of fused sacral vertebrae, and (e) dorsal view of a coccygeal vertebra.

Table 3–2 Vertebral Regions

Cervical (sihr-vih-kahl)	Thoracic (thō-rahs-ihck)	Lumbar (luhm-bahr)	Sacral (sā-krahl)	Coccygeal (kohck-sih-jē-ahl) (also called caudal)
Neck area "C"	Chest area "T"	Loin area "L"	Sacrum area "S"	Tail area "Cy" or "Cd"

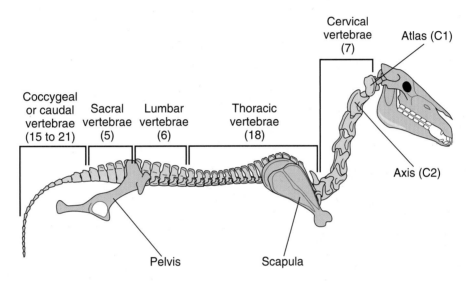

Figure 3–9 Vertebral column of a horse.

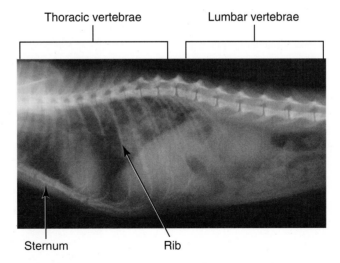

Figure 3–10 Radiograph of a cat showing the thoracic and lumbar spine. Ribs and sternum also are visible.

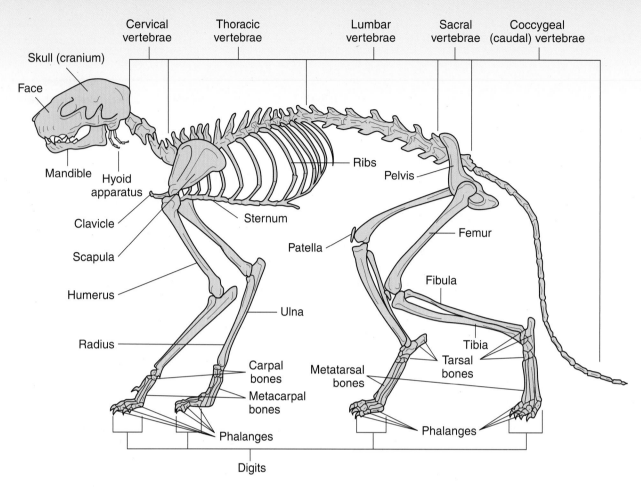

Figure 3–11 Cat skeleton.

The **scapula** (skahp-yoo-lah), or **shoulder blade,** is a large triangular bone on the side of the thorax (Figure 3–12). The **clavicle** (klahv-ih-kuhl), or **collarbone,** is a slender bone that connects the sternum to the scapula. Some animal species have only a **vestigial** (vehs-tihj-ē-ahl), or **rudimentary,** clavicle, whereas other animal species, such as swine, ruminants, and equine, do not have a clavicle.

The **humerus** (hū-mər-uhs) is the long bone of the proximal front limb. The humerus is sometimes called the **brachium.** The radius and ulna are the two bones of the forearm or distal front limb. This region is called the **antebrachium. Ante-** means before. The **radius** (rā-dē-uhs) is the cranial bone of the front limb, and the **ulna** (uhl-nah) is the caudal bone of the front limb. The ulna has a proximal projection called the **olecranon** (ō-lehck-rah-nohn) that forms the point of the elbow. Some species have a fused radius and ulna.

The **carpal** (kahr-pahl) **bones** are irregularly shaped bones in the area known as the **wrist** in people. In small animals, this joint is called the **carpus,** and in large animals, this joint is called the **knee.** The **metacarpals** (meht-ah-kahr-pahlz) are bones found distal to the carpus (**meta-** = beyond). The metacarpals are identified by numbers from medial to lateral.

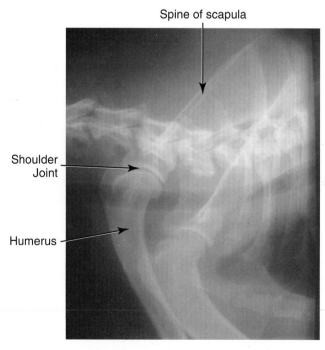

Figure 3–12 Radiograph of the scapula and shoulder (scapulohumeral) joint of a dog.

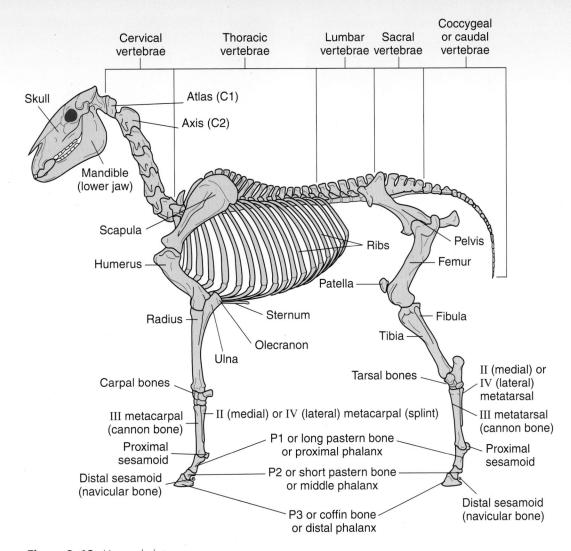

Cervical vertebrae | Thoracic vertebrae | Lumbar vertebrae | Sacral vertebrae | Coccygeal or caudal vertebrae

Skull
Atlas (C1)
Axis (C2)
Mandible (lower jaw)
Scapula
Ribs
Pelvis
Humerus
Femur
Patella
Radius
Sternum
Fibula
Olecranon
Tibia
Ulna
Tarsal bones
II (medial) or IV (lateral) metatarsal
Carpal bones
III metacarpal (cannon bone)
II (medial) or IV (lateral) metacarpal (splint)
III metatarsal (cannon bone)
Proximal sesamoid
P1 or long pastern bone or proximal phalanx
Proximal sesamoid
Distal sesamoid (navicular bone)
P2 or short pastern bone or middle phalanx
Distal sesamoid (navicular bone)
P3 or coffin bone or distal phalanx

Figure 3–13 Horse skeleton.

In some species such as the horse, certain metacarpals do not articulate with the phalanges. In the horse, metacarpals (and metatarsals) II and IV do not articulate with the phalanges and are commonly called **splint bones.** Splint bones are attached by an **interosseous** (ihn-tər-ohs-ē-uhs) ligament to the large third metacarpal (or metatarsal) bone, which is commonly called the **cannon bone** (Figure 3–13).

The **phalanges** (fā-lahn-jēz) are the bones of the digit. One bone of the digit is called a **phalanx** (Figures 3–14a and b). Phalanges are numbered from proximal to distal. Most digits have three phalanges, but the most medial phalanx (digit I) has only two phalanges. **Digits** are the bones analogous to the human finger and vary in number in animals (Figure 3–15). Digit I of dogs is commonly called the **dewclaw** and may be removed shortly after birth. **Ungulates** (uhng-yoo-lātz), or animals with hooves, also have digits that are numbered in the same fashion. Animals with a cloven hoof, or split hoof, have digits III and IV, and digits II and V are vestigial. The vestigial digits of cloven-hoofed animals are also called

dewclaws. Cloven-hoofed animals, such as ruminants and swine, have three phalanges in their digits, with the distal-most phalanx (P3) encased in the hoof. Equine species have one digit (digit III). Within that digit are three phalanges. In livestock, the joints between the phalanges or between the phalanges and other bones have common names. The joint between metacarpal (metatarsal) III and the digit is the **fetlock joint.** The joint between P1 and P2 is known as the **pastern joint.** The joint between P2 and P3 is known as the **coffin joint.** The phalangeal bones also have common names in livestock. P1 is the **long pastern bone,** P2 is the **short pastern bone,** and P3 is the **coffin bone.** (See Chapter 4 for an illustration.)

Phalanx 3 (P3) also may be called a **claw** in nonhooved animals. The combining form for claw is **onych/o.** In cats, a surgical procedure to remove the claws is commonly called a **declaw;** the medical term is **onychectomy** (ohn-ih-kehk-tō-mē).

Sesamoid (sehs-ah-moyd) **bones** are small nodular bones embedded in a tendon or joint capsule. There are multiple sesamoid

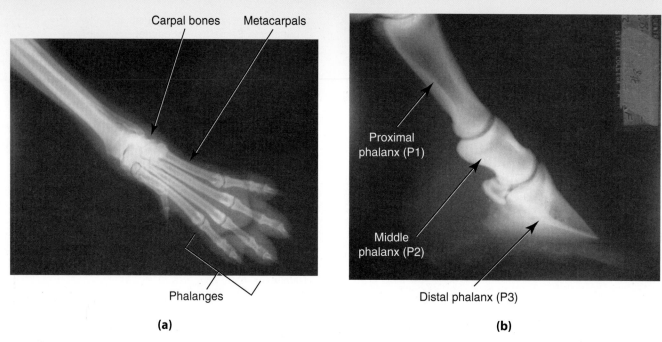

Figure 3–14 (a) Radiograph of the carpal bones, metacarpals, and phalanges of a dog. (b) Radiograph of the foot of a horse. The phalanges of a horse's foot are numbered proximal to distal. The proximal phalanx is commonly known as P1, or the long pastern bone. The middle phalanx is commonly known as P2, or the short pastern bone. The distal phalanx is commonly known as P3, or the coffin bone.

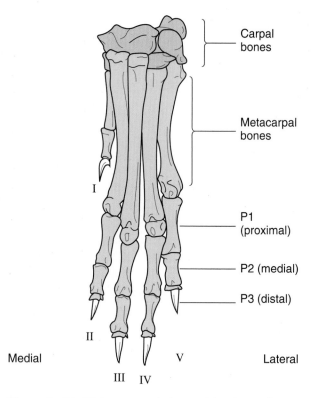

Figure 3–15 Digits versus phalanges. The digits of the hoof or paw are numbered medially to laterally. The phalanges are numbered from proximal to distal.

bones in animals. Some sesamoid bones also have a common name. The **navicular bone** of horses is the common name for the sesamoid bone located inside the hoof on the palmar or plantar surface of P3.

To the Back

The bones of the rear limb include the pelvis, femur, tibia, fibula, tarsals, metatarsals, and phalanges.

The **pelvis** (pehl-vihs), or **hip,** consists of three pairs of bones: ilium, ischium, and pubis. The **ilium** (ihl-ē-uhm) is the largest pair that is blade-shaped. The ilium articulates with the sacrum to form the **sacroiliac** (sā-krō-ihl-ē-ahck) **joint.** The **ischium** (ihs-kē-uhm) is the caudal pair of bones. The **pubis** (pehw-bihs) is the ventral pair of bones that are fused on midline by a cartilaginous joint called the **pubic symphysis** (pehw-bihck sihm-fih-sihs). The **acetabulum** (ahs-eh-tahb-yoo-luhm) is the large socket of the pelvic bone that forms where the three bones meet. The acetabulum forms the ball-and-socket joint with the femur (Figure 3–16).

The **femur** (fē-muhr), or **thigh bone,** is the proximal long bone of the rear leg. The head of the femur articulates proximally with the acetabulum. The **femoral** (fehm-ohr-ahl) **head,** or head of the femur, is connected to a narrow area, which is called the **femoral neck.** Other structures found on

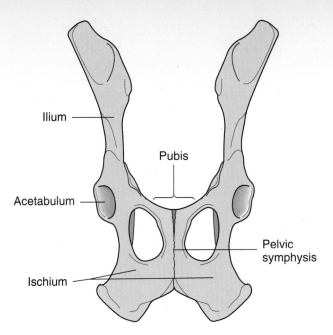

Ilium

Pubis

Acetabulum

Pelvic symphysis

Ischium

Figure 3–16 Parts of the pelvis.

Sound Alikes

Ileum and *ilium* are pronounced the same, yet they have different meanings. Ileum is the distal or aboral (the end opposite the mouth) part of the small intestine, and ilium is part of the pelvic bone. One way to keep these spellings straight is to remember that ileum has an *e* in it, as in *eating* and *enter/o*, which involve the digestive tract. Ilium and pelvis both have an *i* in them.

One way to remember the order of the ilium and ischium is that they follow alphabetical order from cranial to caudal. In cattle, the points of the ilium and ischium are called **hooks** and **pins,** respectively. They too follow alphabetical order from cranial to caudal.

Knee Deep in Trouble

The term *knee* can be a confusing term in veterinary medicine. Laypeople may use the term *knee* to refer to the stifle joint of dogs and cats; however, in large animals, *knee* is used to describe the carpal joint. Most people in veterinary medicine use the term *stifle* for the joint located in the rear leg between the femur and tibia/fibula and reserve *knee* for the carpal joint in large animals.

---

the femur are the **trochanters** (trō-kahn-tehrs), which means large, flat, broad projections on a bone, and **condyles** (kohn-dilz), which means rounded projection (Figure 3–17).

The **patella** (pah-tehl-ah) is a large sesamoid bone in the rear limb. In people, it is called the kneecap and the joint is known as the knee. *Knee* is not a good term to use to describe the joint between the femur and tibia in animals because in large animals, *knee* is commonly used to describe the carpus. In animals, the joint that houses the patella is called the **stifle** (stī-fuhl) **joint.** Another sesamoid bone in the rear limb of some animals is the **popliteal** (pohp-liht-ē-ahl). The popliteal sesamoid is located on the caudal surface of the stifle.

The tibia and fibula are the distal long bones of the rear limb. The **tibia** (tihb-ē-ah) is the larger and more weight-bearing bone of the two. The **fibula** (fihb-yoo-lah) is a long, slender bone. Some animals do not have a fibula that extends to the distal end, whereas other animals have the tibia and fibula fused. The area of the rear limb between the stifle and hock is called the **crus** (kruhs).

The tarsal bones are irregularly shaped bones found in the area known as the ankle in people. In small animals, this joint is called the **tarsus** (tahr-suhs), and in large animals, it is called the **hock** (hohck). One of the tarsal bones is the **talus** (tahl-uhs). The talus is the shorter, medial tarsal bone located in the proximal row of tarsal bones. Talus and tarsus both begin with *t* and sound similar, which makes associating them together easy. The long, lateral tarsal bone located in the proximal row of tarsal bones is the **calcaneus** (kahl-kā-nē-uhs). Calcaneus and carpus both begin with *c* and sound similar, but they are not located in the same area. Calcaneus was named because it reminded someone of a piece of chalk, which consists mainly

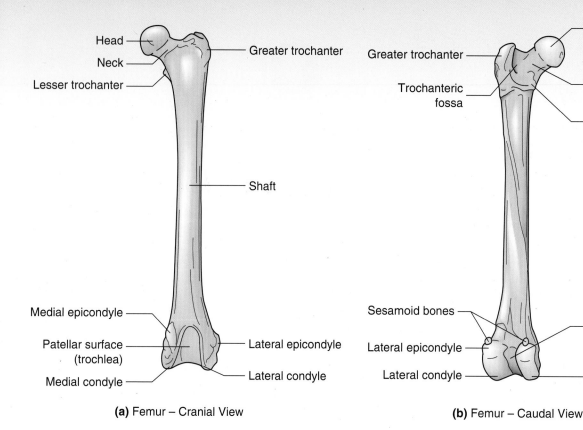

Figure 3–17   Femur of the cat: (a) cranial view (b) caudal view.

of calcium. The metatarsals are bones found distal to the tarsus (**meta-** = beyond). The metatarsals are numbered and have similar names as the metacarpals.

The phalanges are the bones of the digit (both front and rear limbs). The terminology used for the phalanges in the front limb is also used for the rear limb.

STRUCTURAL SUPPORT

Bones are not structurally smooth and often have bumps or grooves or ridges (Figures 3–18a, b, and c). All of these structures have a medical term that describes them. Knowing what these descriptive terms mean can make learning bone parts easier.

aperture (ahp-ər-chər) = opening.
canal (kahn-ahl) = tunnel.
condyle (kohn-dil) = rounded projection (that articulates with another bone).
crest (krehst) = high projection or border projection.
crista (krihs-tah) = ridge.
dens (dehnz) = toothlike structure.
eminence (ehm-ih-nehns) = surface projection.
facet (fahs-eht) = smooth area.
foramen (fō-rā-mehn) = hole.
fossa (fohs-ah) = trench or hollow depressed area.
fovea (fō-vē-ah) = small pit.
head = major protrusion.
lamina (lahm-ih-nah) = thin, flat plate.

line = low projection or ridge.
malleolus (mah-lē-ō-luhs) = rounded projection (distal end of tibia and fibula).
meatus (mē-ā-tuhs) = passage or opening.
process (proh-sehs) = projection.
protuberance (prō-too-bər-ahns) = projecting part.
ramus (rā-muhs) = branch or smaller structure given off by a larger structure.
sinus (sīn-uhs) = space or cavity.
spine (spīn) = sharp projection.
sulcus (suhl-kuhs) = groove.
suture (soo-chuhr) = seam.
trochanter (trō-kahn-tehr) = broad, flat projection (on femur).
trochlea (trōck-lē-ah) = pulley-shaped structure in which other structures pass or articulate.
tubercle (too-behr-kuhl) = small, rounded surface projection.
tuberosity (too-beh-rohs-ih-tē) = projecting part.

TEST ME: SKELETAL SYSTEM

Diagnostic procedures performed on the skeletal system include the following:

- **arthrocentesis** (ahr-thrō-sehn-tē-sihs) = surgical puncture of a joint to remove fluid for analysis.
- **arthrography** (ahr-throhg-rah-fē) = injection of a joint with contrast material for radiographic examination.

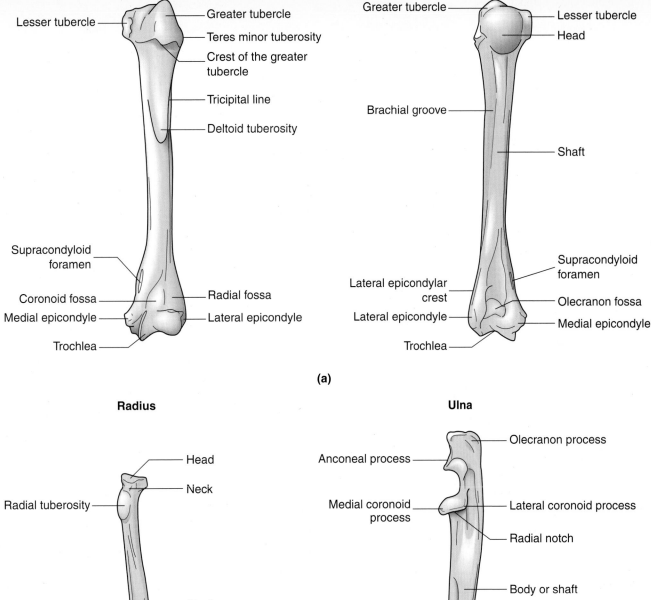

Humerus – Cranial View

Lesser tubercle
Greater tubercle
Teres minor tuberosity
Crest of the greater tubercle
Tricipital line
Deltoid tuberosity
Supracondyloid foramen
Coronoid fossa
Radial fossa
Medial epicondyle
Lateral epicondyle
Trochlea

Humerus – Caudal View

Greater tubercle
Lesser tubercle
Head
Brachial groove
Shaft
Supracondyloid foramen
Lateral epicondylar crest
Olecranon fossa
Lateral epicondyle
Medial epicondyle
Trochlea

(a)

Radius

Head
Neck
Radial tuberosity
Shaft
Medial styloid process

Ulna

Olecranon process
Anconeal process
Medial coronoid process
Lateral coronoid process
Radial notch
Body or shaft
Articular circumterence
Lateral styloid process

(b)

Figure 3–18 (a) Cranial and caudal view of the humerus of the cat; (b) radius and ulna of the cat; and (c) ventral view of the tibia and fibula of the cat.

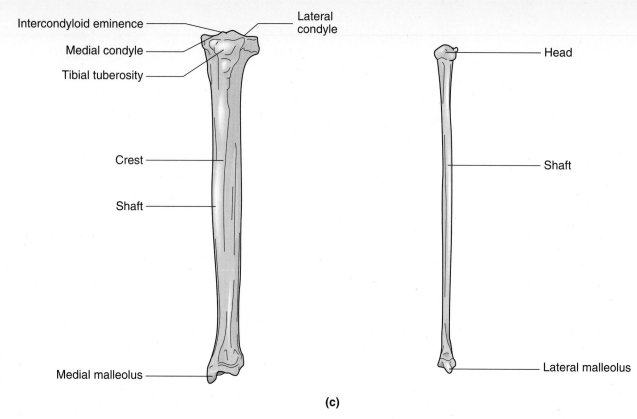

Tibia – Ventral View

Intercondyloid eminence
Lateral condyle
Medial condyle
Tibial tuberosity
Crest
Shaft
Medial malleolus

Fibula – Ventral View

Head
Shaft
Lateral malleolus

(c)

Figure 3–18 *(continued)*

- **arthroscopy** (ahr-throhs-kō-pē) = visual examination of the joint using a fiberoptic scope (called an arthroscope when used in the joint).
- **radiology** (rā-dē-ohl-ō-jē) = study of internal body structures after exposure to ionizing radiation; used to detect fractures and diseases of bones (Figure 3–19).

PATHOLOGY: SKELETAL SYSTEM

Pathologic conditions of the skeletal system include the following:

- **ankylosis** (ahng-kih-lō-sihs) = loss of joint mobility caused by disease, injury, or surgery.
- **arthralgia** (ahr-thrahl-jē-ah) = joint pain.
- **arthritis** (ahr-thrī-tihs) = inflammatory condition of joints.
- **arthrodynia** (ahr-thrō-dihn-ē-ah) = joint pain.
- **arthropathy** (ahr-throhp-ah-thē) = joint disease.
- **bursitis** (bər-sī-tihs) = inflammation of the bursa.
- **chondromalacia** (kohn-drō-mah-lā-shē-ah) = abnormal cartilage softening.
- **chondropathy** (kohn-drohp-ah-thē) = cartilage disease.
- **discospondylitis** (dihs-kō-spohn-dih-lī-tihs) = inflammation of the intervertebral disc and vertebrae.

- **exostosis** (ehck-sohs-tō-sihs) = benign growth on the bone surface.
- **gouty arthritis** (gow-tē ahr-thrī-tihs) or **gout** = joint inflammation associated with the formation of uric acid crystals in the joint (seen more commonly in birds).
- **hip dysplasia** (dihs-plā-zē-ah) = abnormal development of the pelvic joint causing the head of the femur and the acetabulum not to be aligned properly; most commonly seen in large breed dogs (Figure 3–19).
- **intervertebral disc disease** (ihn-tər-vər-tē-brahl dihsk dih-zēz) = rupture or protrusion of the cushioning disc found between the vertebrae that results in pressure on the spinal cord or spinal nerve roots; also called **herniated disc, ruptured disc,** or **IVDD** (Figure 3–20).
- **kyphosis** (kī-fō-sihs) = dorsal curvature of the spine; also called hunchback.
- **Legg-Calvé-Perthes disease** (lehg cah-veh pər-thehz dih-zēz) = idiopathic necrosis of the femoral head and neck of small breed dogs; also called avascular necrosis of the femoral head and neck.
- **lordosis** (lōr-dō-sihs) = position in which the vertebral column is abnormally curved ventrally; seen in cats in heat; commonly called **swayback.**

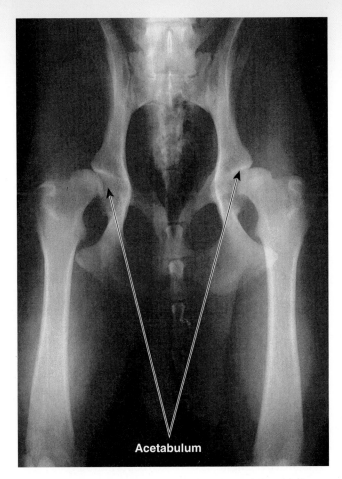

Acetabulum

Figure 3–19 Radiograph of the canine hip with hip dysplasia. Note the shallow acetabulum present in this dog's pelvis. (Courtesy of Lodi Veterinary Hospital, Lodi, Wisconsin.)

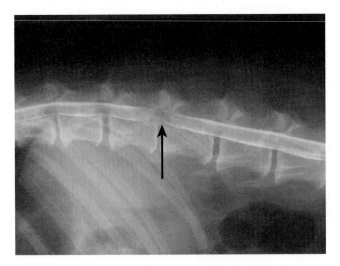

Figure 3–20 Myelogram showing intervertebral disc disease. Dye injected into the spinal column shows compression of the spinal cord. (Courtesy of David Sweet, VMD.)

- **luxation** (luhck-sā-shuhn) = dislocation or displacement of a bone from its joint.
- **myeloma** (mī-eh-lō-mah) = tumor composed of cells derived from hematopoietic tissues of bone marrow.

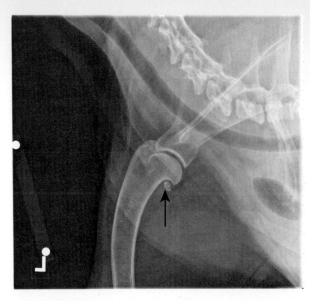

Figure 3–21 Radiograph of dog's shoulder with osteophyte at proximal end of humerus. (Courtesy of Amy Lang, University of Wisconsin Veterinary Teaching Hospital.)

- **ostealgia** (ohs-tē-ahl-jē-ah) = bone pain.
- **osteitis** (ohs-tī-tihs) = inflammation of bone.
- **osteoarthritis** (ohs-tē-ō-ahr-thrī-tihs) = degenerative joint disease commonly associated with aging or wear and tear on the joints; also called **degenerative joint disease,** or **DJD.**
- **osteochondrosis** (ohs-tē-ō-kohn-drō-sihs) = degeneration or necrosis of bone and cartilage followed by regeneration or recalcification.
- **osteochondrosis dissecans** (ohs-tē-ō-kohn-drō-sihs dehs-ih-kahns) = degeneration or necrosis of bone and cartilage followed by regeneration or recalcification with dissecting flap of articular cartilage and some inflammatory joint changes; detached pieces of articular cartilage are called **joint mice** or **osteophytes** (ohs-tē-ō-fitz) (Figure 3–21).
- **osteomalacia** (ohs-tē-ō-mah-lā-shē-ah) = abnormal softening of bone.
- **osteomyelitis** (ohs-tē-ō-mī-eh-lī-tihs) = inflammation of bone and bone marrow.
- **osteonecrosis** (ohs-tē-ō-neh-krō-sihs) = death of bone tissue.
- **osteoporosis** (ohs-tē-ō-pō-rō-sihs) = abnormal condition of marked loss of bone density and an increase in bone porosity.
- **osteosclerosis** (ohs-tē-ō-skleh-rō-sihs) = abnormal hardening of bone.
- **periostitis** (pehr-ē-ohs-tī-tihs) = inflammation of the fibrous tissue that forms the outermost covering of bone.
- **rheumatoid arthritis** (roo-mah-toyd ahr-thrī-tihs) = autoimmune disorder of the connective tissues and joints.
- **sequestrum** (sē-kwehs-truhm) = piece of dead bone that is partially or fully detached from the adjacent healthy bone (Figure 3–22).

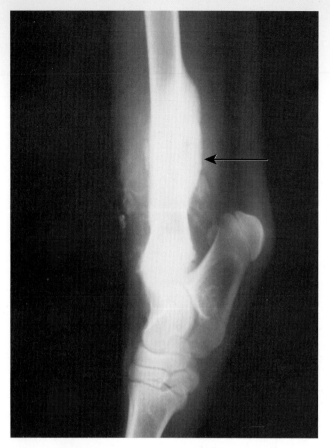

Figure 3–22 Radiograph of horse's leg with a sequestrum. (Courtesy of Amy Lang, University of Wisconsin Veterinary Teaching Hospital.)

- **spondylitis** (spohn-dih-lī-tihs) = inflammation of the vertebrae.
- **spondylosis** (spohn-dih-lō-sihs) = any degenerative disorder of the vertebrae.

- **spondylosis deformans** (spohn-dih-lō-sihs dē-fōr-mahnz) = chronic degeneration of the articular processes and the development of bony outgrowths around the ventral edge of the vertebrae (Figure 3–23).
- **subluxation** (suhb-luhck-sā-shuhn) = partial dislocation or displacement of a bone from its joint (Figure 3–24).
- **synovitis** (sihn-ō-vī-tihs) = inflammation of the synovial membrane of joints.

Fracture Terminology

See Figure 3–25.

- **avulsion** (ā-vuhl-shuhn) **fracture** = broken bone in which the site of muscle, tendon, or ligament insertion is detached by a forceful pull.
- **callus** (kahl-uhs) = bulging deposit around the area of a bone fracture that may eventually become bone.
- **closed fracture** = broken bone in which there is no open wound in the skin; also known as a **simple fracture.**
- **comminuted** (kohm-ih-noot-ehd) **fracture** = broken bone that is splintered or crushed into multiple pieces (Figure 3–26).
- **compression** (kohm-prehs-shuhn) **fracture** = broken bone produced when the bones are pressed together.
- **crepitation** (krehp-ih-tā-shuhn) = cracking sensation that is felt and heard when broken bones move together.
- **fracture** (frahck-shər) = broken bone.
- **greenstick fracture** = bone that is broken only on one side and the other side is bent; also called **incomplete fracture.**
- **immobilization** (ihm-mō-bihl-ih-zā-shuhn) = act of holding, suturing, or fastening a bone in a fixed position, usually with a bandage or cast.

Lumbar vertebrae

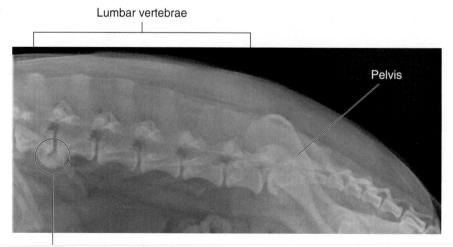

Pelvis

Bony outgrowth

Figure 3–23 Radiograph of the lumbar spine of a dog. This dog is showing an age-related change called spondylosis deformans. In this disease, bone spurs (bony outgrowths) are formed that eventually can bridge between vertebrae. (Courtesy of Eli Larson, DVM.)

Figure 3–24 Radiograph of atlas-axis subluxation in a dog. (Courtesy of Anne E. Chavet, DVM, Diplomate ACVIM-Neurology, University of Wisconsin School of Veterinary Medicine.)

In or out

Bones may abnormally bend in or bend out. Medical terms for this condition in bones are *varus* (vahr-uhs) and *valgus* (vahl-guhs). *Valgus* means bend out (think bend laterally; both *valgus* and *lateral* have an *l*), and *varus* means bend in.

- **manipulation** (mahn-ihp-yoo-lā-shuhn) = attempted realignment of the bone involved in a fracture or dislocation; also known as **reduction.**
- **oblique** (ō-blēck) **fracture** = broken bone that has an angular break diagonal to the long axis.
- **open fracture** = broken bone in which there is an open wound in the skin; also known as a **compound fracture.**

- **physeal** (fĭ-sē-ahl) **fracture** = bone that is broken at the epiphyseal line or growth plate; these fractures are further categorized as **Salter-Harris I–V fractures.**
- **spiral** (spī-rahl) **fracture** = broken bone in which the bone is twisted apart or spiraled apart.
- **transverse** (trahnz-vərs) **fracture** = broken bone that is broken at right angles to its axis or straight across the bone.

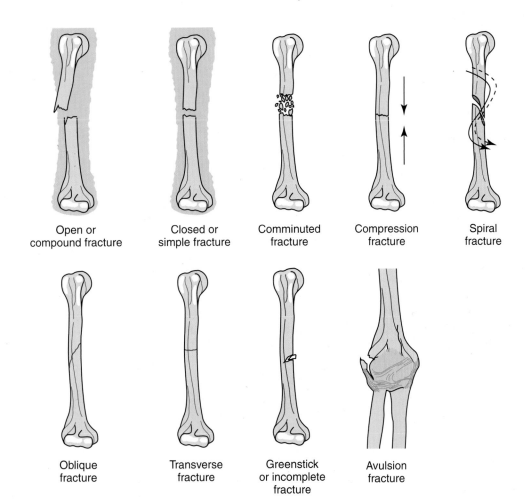

Figure 3–25 Fracture types.

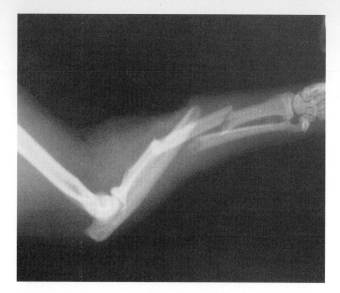

Figure 3–26 Comminuted fracture of the radius and ulna in a cat.

PROCEDURES: SKELETAL SYSTEM

Procedures performed on the skeletal system (Figure 3–27) include the following:

- **amputation** (ahmp-yoo-tā-shuhn) = removal of all or part of a body part.
- **arthrodesis** (ahr-thrō-dē-sihs) = fusion of a joint or the spinal vertebrae by surgical means.
- **chemonucleolysis** (kē-mō-nū-klē-ō-lī-sihs) = process of dissolving part of the center of an intervertebral disc by injecting a foreign substance.
- **craniotomy** (krā-nē-oht-ō-mē) = surgical incision or opening into the skull.
- **laminectomy** (lahm-ih-nehck-tō-mē) = surgical removal of the dorsal arch of a vertebra.

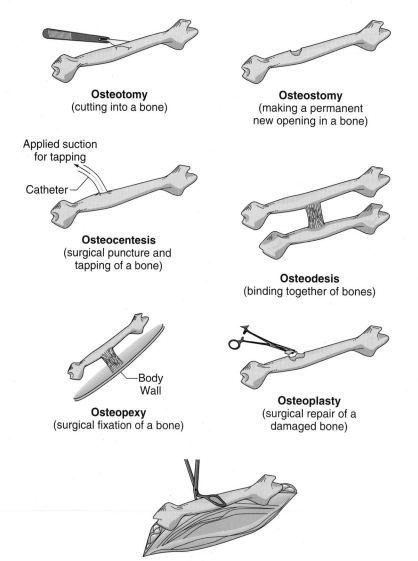

Osteotomy
(cutting into a bone)

Osteostomy
(making a permanent new opening in a bone)

Applied suction for tapping
Catheter
Osteocentesis
(surgical puncture and tapping of a bone)

Osteodesis
(binding together of bones)

Body Wall
Osteopexy
(surgical fixation of a bone)

Osteoplasty
(surgical repair of a damaged bone)

Ostectomy
(removal of a bone)

Figure 3–27 Surgical procedures of bone.

- **onychectomy** (ohn-ih-kehk-tō-mē) = surgical removal of a claw.
- **ostectomy** (ohs-tehck-tō-mē) = surgical removal of bone.
- **osteocentesis** (ohs-tē-ō-sehn-tē-sihs) = surgical puncture of a bone.
- **osteodesis** (ohs-tē-ō-dē-sihs) = fusion of bones.
- **osteopexy** (ohs-tē-ō-pehck-sē) = surgical fixation of a bone to the body wall.
- **osteoplasty** (ohs-tē-ō-plahs-tē) = surgical repair of bone.
- **osteostomy** (ohs-tē-ohs-tō-mē) = surgical creation of a permanent new opening in bone.
- **osteotomy** (ohs-tē-oht-ō-mē) = surgical incision or sectioning of bone.
- **trephination** (treh-fih-nā-shuhn) = process of cutting a hole into a bone using a **trephine** (trē-fin) (circular sawlike instrument used to remove bone or tissue).

FUNCTIONS OF THE MUSCULAR SYSTEM

Muscles are organs that contract to produce movement. Muscles make movement possible. One type of movement is **ambulation** (ahm-bū-lā-shuhn), or walking, running, or otherwise moving from one place to another. Another type of movement is contraction of organs or tissues that result in normal functioning of the body. For example, contraction of sections of the gastrointestinal tract allows food to move through the digestive system, and contraction of vessels allows movement of fluids such as blood. Movement also results in heat generation to keep the body warm.

STRUCTURES OF THE MUSCULAR SYSTEM

The muscular system is composed of specialized cells called muscle fibers whose predominant function is contractibility.

Muscle Fibers

Muscles are made up of long, slender cells called muscle fibers. Each muscle consists of a group of muscle fibers encased in a fibrous sheath. The combining form for muscle is **my/o;** the combining forms for fibrous tissue are **fibr/o** and **fibros/o.**

Three types of muscle cells are based on their appearance and function. The three types are skeletal, smooth, and cardiac (Table 3–3 and Figure 3–28).

Making Another Connection

Like the skeletal system, the muscular system also contains various forms of connective tissue. The connective tissues that support the muscular system are described next.

Fascia

Fascia (fahsh-ē-ah) is a sheet of fibrous connective tissue that covers, supports, and separates muscles. The combining forms for fascia are **fasci/o** and **fasc/i.**

Tendons

A **tendon** (tehn-dohn) is a narrow band of connective tissue that attaches muscle to bone. The combining forms for tendon are **tend/o, tendin/o,** and **ten/o.** Remember to make the distinction between a tendon and a ligament (Figure 3–29).

Table 3-3 Muscle Types

Muscle Type	Description	Microscopic Appearance	Function
skeletal	**striated** (strī-āt-ehd) **voluntary**	long, cylindrical, multinucleated cells with dark and light bands to create a striated or striped look	attach bones to the body and make motion possible
smooth	**nonstriated unstriated involuntary visceral**	spindle-shaped without stripes or striations	produce slow contractions to allow unconscious functioning of internal organs
cardiac	**striated involuntary**	elongated branched cells that lie parallel to each other and have dark and light bands	involuntary contraction of heart muscle

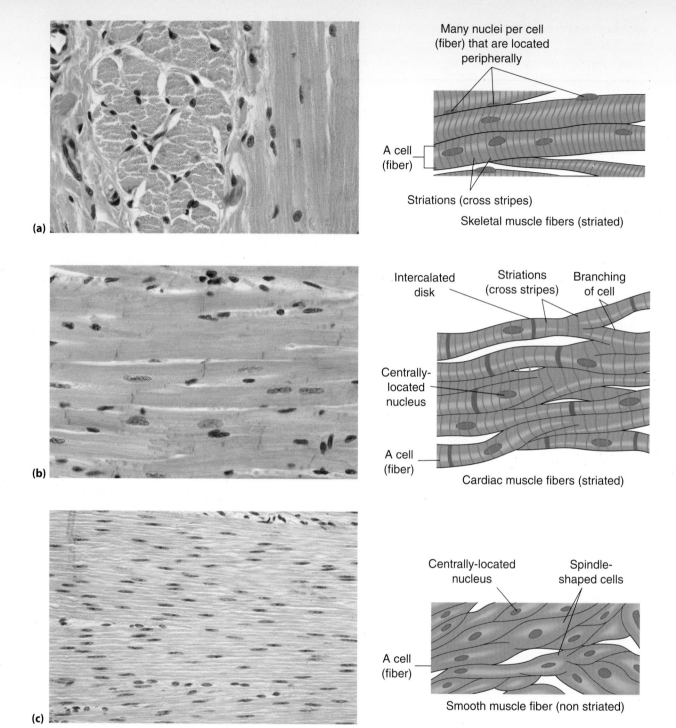

Figure 3–28 The three types of muscle tissues. (a) Skeletal muscle from the tongue of a cat. The photomicrograph shows cells cut along their length and in cross section. (b) Cardiac muscle from the heart of a goat. Note the intercalated disks that connect the cells. These structures allow the cells to act together with an organized contraction. (c) Smooth muscle from the colon of a horse. Note that smooth muscle lacks the striations found in cardiac and skeletal muscle. (Photomicrographs courtesy of William J. Bacha, PhD, and Linda M. Bacha, MS, VMD.)

Tendons connect muscles to bones or other structures. One example is the linea alba. *Linea alba* means white line in Latin. The **linea alba** is a fibrous band of connective tissue on the ventral abdominal wall that is the median attachment of the abdominal muscles (Figure 3–30).

Aponeurosis

An **aponeurosis** (ahp-ō-nū-rō-sihs) is a fibrous sheet that provides attachment to muscular fibers and is a means of origin or insertion of a flat muscle. The combining form for

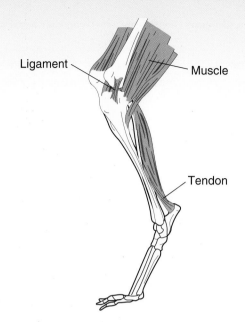

Figure 3–29 Tendons versus ligaments. Tendons are dense fibrous connective tissues that connect muscle to bone; ligaments are dense fibrous connective tissues that connect bone to bone.

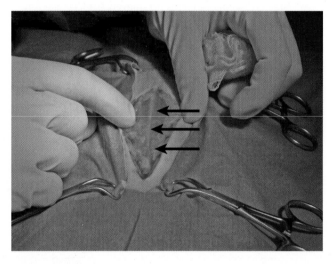

Figure 3–30 The linea alba of a cat. (Courtesy of Kelly Gilligan, DVM.)

aponeurosis is **aponeur/o,** and the plural form of aponeurosis is **aponeuroses.**

SHOW SOME MUSCLE

One of the functions of muscle is to allow movement. The combining form **kinesi/o** and the suffix **-kinesis** mean movement. **Kinesiology** (kih-nē-sē-ohl-ō-jē) is the study of movement.

Some muscles are arranged in pairs that work opposite or against each other. One muscle pair may produce movement in one direction, whereas another muscle pair produces movement in the opposite direction. Muscles that work against or opposite each other are called **antagonistic** (ahn-tā-gohn-ihs-tihck). **Anti-** means against; **agon** means struggle. Other muscles of the body are arranged to work with another muscle. **Synergists** (sihn-ər-jihsts) are muscles that contract at the same time as another muscle to help movement or support movement (synergists are also called agonistic (ā-gohn-ihs-tihck). **Syn-** means together; **erg** means work. Antagonistic muscles work by producing contraction of one pair of muscles while the other pair relaxes. **Contraction** (kohn-trahck-shuhn) means tightening. **Relaxation** (rē-lahk-sā-shuhn) means lessening of tension. During contraction, the muscle becomes shorter and thicker. During relaxation, the muscle returns to its original shape. Muscles are signaled to contract or relax by nerve impulses. A **neuromuscular** (nū-rō-muhs-kū-lahr) **junction** is the point at which nerve endings come in contact with the muscle cells.

Tonus (tō-nuhs), or muscle tone, is balanced muscle tension. The combining form for tone, tension, or stretching is **ton/o.**

WHAT'S IN A NAME?

At first glance, the names of muscles and the task of learning them may seem impossible. Dividing muscle names into their basic components or taking a closer look at how the names were derived may help in learning their names and functions.

Beginning and Ending

Muscles are formed by where they begin and where they end. Terms used to denote these two locations are **muscle origin** (ōr-ih-jihn) and **muscle insertion** (ihn-sihr-shuhn), respectively. Muscle origin is the place where a muscle begins, or originates, and is the more fixed attachment or the end of the muscle closest to the midline. Muscle insertion is the place where a muscle ends, or inserts, and is the more movable end or portion of the muscle farthest from the midline. Muscles may be named according to where they originate and end. Brachioradialis muscles are connected to the brachium (humerus) and to the radius.

How Do They Move?

Muscles move in a variety of ranges. **Range of motion** is a term used to describe the types of muscle movements. Range of motion is sometimes abbreviated ROM. Muscles may be named for the manner in which they move, as follows:

- **abductor** (ahb-duhck-tər) = muscle that moves a part away from the midline.
- **adductor** (ahd-duhck-tər) = muscle that moves a part toward the midline.

- **flexor** (flehck-sər) = muscle that bends a limb at its joint or decreases the joint angle.
- **extensor** (ehcks-tehn-sər) = muscle that straightens a limb or increases the joint angle.
- **levator** (lē-vā-tər) = muscle that raises or elevates a part.
- **depressor** (dē-prehs-sər) = muscle that lowers or depresses a part.
- **rotator** (rō-tā-tər) = muscle that turns a body part on its axis.
- **supinator** (soo-pih-nā-tər) = muscle that rotates the palmar or plantar surface upward.
- **pronator** (prō-nā-tər) = muscle that rotates the palmar or plantar surface downward.

Where Are They?

Muscles also are named for their location on the body or organ they are near. **Pectoral muscles** are located on the chest (**pector** = chest). Muscles may also be named for their location in relation to something else. **Epaxial** (ehp-ahcks-ē-ahl) **muscles** are located above the pelvic axis (**epi-** = above, **axis** = line about which rotation occurs), **intercostal muscles** are located between the ribs (**inter-** = between, **cost/o** = rib), **infraspinatus muscles** are located beneath the spine of the scapula (**infra-** = beneath or below), and **supraspinatus muscles** are located above the spine of the scapula (**supra-** = above). Muscle names also may indicate their location within a group, such as **inferior** (below or deep), **medius** (middle), and **superior** (above). Other terms indicating depth of muscles are **externus** (outer) and **internus** (inner). **Orbicularis** are muscles surrounding another structure.

Which Way Do They Go?

Muscles may also be named according to the direction of the muscle fibers.

- **Rectus** (rehck-tuhs) means straight. Rectus muscles align with the vertical axis of the body.
- **Oblique** (ō-blēck) means slanted. Oblique muscles slant outward away from the midline.
- **Transverse** means crosswise. Transverse muscles form crosswise to the midline.
- **Sphincter** means tight band. Sphincter muscles are ring-like and constrict the opening of a passageway.

How Many Parts Are There?

Some muscles are named for the number of divisions they have.

- **Biceps** (bī-sehpz) generally have two divisions (heads); **bi-** means two.
- **Triceps** (trī-sehpz) generally have three divisions (heads); **tri-** means three.

- **Quadriceps** (kwohd-rih-sehps) generally have four divisions (heads); **quadri-** means four.

Some muscles are not paired or divided. **Azygous** (ah-zī-guhs) means not paired (**a-** means without; **zygoto** means joined).

How Big Is It?

Muscles may also be named for their size. Muscles may be small (**minimus**) or large (**maximus** or **vastus**), broad (**latissimus**) or narrow (**longissimus** or **gracilis**). **Major** and **minor** also are used to describe larger and smaller parts, respectively.

How Is It Shaped?

Some muscles are shaped like familiar objects and have been named accordingly.

- **Deltoid** (dehl-toyd) muscles look like the Greek letter delta (Δ).
- **Quadratus** (kwohd-rā-tuhs) muscles are square or four-sided.
- **Rhomboideus** (rohm-boy-dē-uhs) muscles are diamond-shaped. (Rhomboid is a four-sided figure that may have unequal adjoining sides but equal opposite sides.)
- **Scalenus** (skā-lehn-uhs) muscles are unequally three-sided. (*Skalenos* is Greek for uneven.)
- **Serratus** (sihr-ā-tuhs) muscles are saw-toothed. (*Serratus* is Latin for notched.)
- **Teres** (tər-ēz) muscles are cylindrical. (*Teres* is Latin for smooth and round or cylindrical.)

No Rules

Sometimes muscles are named for what they look like or how they relate to something else. **Sartorius muscle** (one muscle of the thigh area) is named because this muscle flexes and adducts the leg of a human to that position assumed by a tailor sitting cross-legged at work (*sartorius* means tailor). The **gemellus** is named because it is a twinned muscle (*gemellus* means twin). The **gastrocnemius muscle** is the leg muscle that resembles the shape of the stomach (**gastr/o** means stomach, **kneme** means leg).

TEST ME: MUSCULAR SYSTEM

A diagnostic procedure performed on the muscular system is as follows:

- **electromyography** (ē-lehck-trō-mī-ohg-rah-fē) = process of recording the electrical activity of the muscle cells near the recording electrodes; abbreviated EMG. An **electromyogram** (ē-lehck-trō-mī-ō-grahm) is the record of the strength of muscle contraction caused by electrical stimulation.

PATHOLOGY: MUSCULAR SYSTEM

Pathologic conditions of the muscular system include the following:

- **adhesion** (ahd-hē-shuhn) = band of fibers that hold structures together in an abnormal fashion.
- **ataxia** (ā-tahck-sē-ah) = lack of voluntary control of muscle movement; "wobbliness."
- **atonic** (ā-tohn-ihck) = lacking muscle control.
- **dystrophy** (dihs-trō-fē) = defective growth.
- **fasciitis** (fahs-ē-ī-tihs) = inflammation of the sheet of fibrous connective tissue that covers, supports, and separates muscles (fascia).
- **fibroma** (fi-brō-mah) = tumor composed of fully developed connective tissue; also called **fibroid** (fi-broyd).
- **hernia** (hər-nē-ah) = protrusion of a body part through tissues that normally contain it.
- **laxity** (lahcks-ih-tē) = looseness.
- **leiomyositis** (lī-ō-mī-ō-sī-tihs) = inflammation of smooth tissue.
- **myasthenia** (mī-ahs-thē-nē-ah) = muscle weakness.
- **myoclonus** (mī-ō-klō-nuhs) = spasm of muscle.
- **myopathy** (mī-ohp-ah-thē) = abnormal condition or disease of muscle.
- **myositis** (mī-ō-sī-tihs) = inflammation of voluntary muscles.
- **myotonia** (mī-ō-tō-nē-ah) = delayed relaxation of a muscle after contraction.
- **tendinitis** (tehn-dih-nī-tihs) = inflammation of the band of fibrous connective tissue that connects muscle to bone.
- **tetany** (teht-ahn-ē) = muscle spasms or twitching.

PROCEDURES: MUSCULAR SYSTEM

Procedures performed on the muscular system include the following:

- **myectomy** (mī-ehck-tō-mē) = surgical removal of muscle or part of a muscle.
- **myoplasty** (mī-ō-plahs-tē) = surgical repair of muscle.
- **myotomy** (mī-oht-ō-mē) = surgical incision into a muscle.
- **tenectomy** (teh-nehck-tō-mē) = surgical removal of a part of a tendon (fibrous connective tissue that connects muscle to bone).
- **tenotomy** (teh-noht-ō-mē) = surgical division of a tendon (fibrous connective tissue that connects muscle to bone).

REVIEW EXERCISES

Multiple Choice

Choose the correct answer.

1. A common name for the tarsus is the
   - a. elbow
   - b. calcaneus
   - c. hock
   - d. wrist

2. The _____ joints are the freely movable joints of the body.
   - a. suture
   - b. synovial
   - c. symphysis
   - d. cartilaginous

3. The correct order of the vertebral segments is
   - a. cervical, thoracic, lumbar, sacral, and coccygeal
   - b. cervical, lumbar, thoracic, coccygeal, and sacral
   - c. thoracic, lumbar, cervical, sacral, and coccygeal
   - d. thoracic, cervical, lumbar, sacral, and coccygeal

4. A _____ is a fibrous band of connective tissue that connects bone to bone.
   - a. fascia
   - b. tendon
   - c. synovial membrane
   - d. ligament

5. The acetabulum is the
   - a. patella
   - b. cannon bone
   - c. large socket in the pelvic bone
   - d. crest of the scapula

6. The three parts of the pelvis are
   - a. ileum, pubis, and acetabulum
   - b. ilium, pubis, and sacrum
   - c. ilium, sacrum, and coccyx
   - d. ilium, ischium, and pubis

7. The digits contain bones that are called

   a. carpals

   b. phalanges

   c. tarsals

   d. tarsus

8. Components of the axial skeleton include

   a. scapula, humerus, radius, ulna, and carpus

   b. skull, auditory ossicles, hyoid, vertebrae, ribs, and sternum

   c. pelvic girdle, femur, tibia, fibula, and tarsus

   d. scapula, pelvis, humerus, femur, tibia, fibula, radius, and ulna

9. Another term for growth plate is

   a. physis

   b. shaft

   c. diaphysis

   d. trophic

10. Bones of the front limb include

   a. humerus, tibia, fibula, tarsal, metatarsal, and phalanges

   b. humerus, radius, ulna, carpal, metacarpal, and phalanges

   c. femur, tibia, fibula, tarsal, metatarsal, and phalanges

   d. radius, humerus, ulna, carpal, metatarsal, and phalanges

11. *Rectus* means

   a. ringlike

   b. straight

   c. angled

   d. rotating

12. Muscles may be classified as

   a. voluntary

   b. involuntary

   c. cardiac

   d. all of the above

13. A term for when a muscle becomes shorter and thicker is

   a. relaxation

   b. contraction

   c. rotation

   d. depression

14. Levator muscles _____ a body part.

   a. decrease the angle of

   b. increase the angle of

   c. raise

   d. depress

15. A fibrous band of connective tissue that connects muscle to bone is

   a. cartilage

   b. tendon

   c. ligament

   d. aponeurosis

16. Looseness is called

   a. laxity

   b. rigidity

   c. spasm

   d. tonus

17. Protrusion of a body part through tissues that normally contain it is called a

   a. projection

   b. hernia

   c. prominence

   d. myotonia

18. A muscle that forms a tight band is called a(n)

   a. purse-string

   b. sartorius

   c. sphincter

   d. oblique

19. Surgical removal of a muscle or part of a muscle is called

   a. myositis

   b. myotomy

   c. myectomy

   d. myostomy

20. Abnormal condition or disease of muscle is called

   a. myodynia

   b. myography

   c. myasthenia

   d. myopathy

21. Which term describes the shaft of a long bone?

    a. diaphysis
    b. epiphysis
    c. endosteum
    d. periosteum

22. The tarsal bones are found in the

    a. digits
    b. wrist
    c. stifle
    d. hock

23. The manubrium is the

    a. lower jaw
    b. cranial portion of the sternum
    c. upper jaw
    d. caudal portion of the sternum

24. Which term describes the freely movable joints of the body?

    a. synarthroses
    b. amphiarthroses
    c. diarthroses
    d. articulations

25. The opening in a bone through which blood vessels, nerves, and ligaments pass is a(n)

    a. fontanel
    b. foramen
    c. meatus
    d. lamina

26. The acetabulum is commonly called the

    a. collar bone
    b. patella
    c. hip socket
    d. knee

27. Muscles located above the pelvic axis are called

    a. spinatus muscles
    b. orbicularis muscles
    c. epaxial muscles
    d. inferior muscles

28. Muscles under voluntary control are known as

    a. involuntary
    b. nonstriated
    c. skeletal
    d. visceral

29. Minimus is to maximus as

    a. longissimus is to gracilis
    b. biceps is to triceps
    c. oblique is to sphincter
    d. minor is to major

30. A muscle that bends a limb at its joint or decreases the joint angle is called a(n)

    a. flexor
    b. extensor
    c. supinator
    d. pronator

Matching

Match the bone or joint in Column I with its common name in Column II.

Column I   **Column II**

1. _____ P1   a. carpus in large animals

2. _____ P2   b. hock

3. _____ P3   c. coffin bone

4. _____ tarsus   d. short pastern

5. _____ splint bone   e. long pastern

6. _____ fetlock joint   f. metacarpal/metatarsal III in equine and metacarpal/metatarsal III and IV in ruminants

7. _____ pastern joint

8. _____ coffin joint   g. collarbone

9. _____ knee   h. metacarpo-/metatarsophalangeal joint of equine and ruminants

10. _____ stifle   i. metacarpal/metatarsal II and IV in equine

11. _____ clavicle   j. connection between phalanx I and II in equine and ruminants

12. _____ cannon bone   k. distal interphalangeal joint of phalanx II and III in equine and ruminants

13. _____ dewclaw   l. variable digit depending on species; digit I in dogs, digits II and V in ruminants

14. _____ sternum

m. synovial joint located between the femur and tibia

n. breastbone

Match the bone in Column I with the area it is located in Column II.

Column I   **Column II**

15. _____ humerus   a. distal front limb

16. _____ fibula   b. proximal front limb

17. _____ tibia   c. proximal hind limb

18. _____ ulna   d. distal hind limb

19. _____ femur   e. joint in front limb

20. _____ tarsus   f. oint in hind limb

21. _____ radius   g. distal part of front and hind limbs

22. _____ carpus

23. _____ metacarpal

24. _____ metatarsal

25. _____ phalanx

Fill in the Blanks

1. _____ and _____ are terms used for displacement of a bone from its joint.

2. The _____ is the tough, fibrous tissue that forms the outermost covering of bone.

3. A(n) _____ is a curved fibrous cartilage found in some synovial joints.

4. Connections between two bones are called _____ or _____.

5. The caudal portion of the sternum is called the _____.

6. A(n) _____ is removal of all or part of a limb or body part.

7. A(n) _____ is a piece of dead bone that is partially or fully detached from the surrounding healthy bone.

8. Inward curvature of a bone is called _____.

9. Visual examination of the internal structure of a joint using a fiberoptic instrument is _____.

10. _____ is loss of mobility of a joint.

11. _____ is abnormal softening of cartilage.

12. A muscle that straightens a limb at a joint is called a(n) _____.

13. Straightening of a limb beyond its normal limits is called _____.

14. A(n) _____ is a band of fibers that holds structures together in an abnormal fashion.

15. Dogs with short, wide skulls are said to be _____.

16. Involuntary muscle is also called _____, _____, or _____.

17. Surgical removal of a claw is _____.

18. A(n) _____ is a broken bone in which there is an open wound in the skin.

19. The _____ is the fibrous band of connective tissue on the ventral abdominal wall that is the median attachment of the abdominal muscles.

20. A(n) _____ is the place where muscle ends that is the more movable end or portion away from the midline.

21. Inflammation of a tendon is called _____.

22. The opposite of extension is _____.

23. The opposite of contraction is _____.

24. A muscle that lowers or depresses a part is called a(n) _____.

25. The opposite of inferior is _____.

26. The term for crosswise is _____.

27. The term for slanted is _____.

28. The crackling sensation that is felt and heard when broken bones move together is called _____.

29. Broken bones that are splintered or crushed into multiple pieces are called _____.

30. Abnormal development of the pelvic joint causing the head of the femur and the acetabulum not to be aligned properly is called _____.

True or False

1. Arthrodesis is fusion of a joint or the spinal vertebrae by surgical means.
2. A craniotomy is a surgical incision into a joint.
3. An osteotomy is the surgical removal of a bone.
4. Ataxia is lack of voluntary control of muscle movement.
5. An adhesion is a band of fibers that holds structures together in an abnormal fashion.

CROSSWORD PUZZLE

Supply the correct term in the appropriate space for the definition listed.

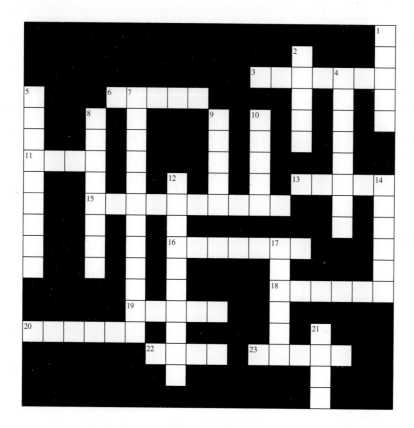

Across

3 projection
6 sharp projection
11 low projection or ridge
13 space or cavity
15 projecting part
16 rounded projection (that articulates with another bone)
18 passage or opening
19 tunnel
20 seam
22 toothlike structure
23 smooth area

Down

1 trench or hollow depressed
2 small pit
4 surface projection
5 rounded projection (distal end of tibia and fibula)
7 projecting part
8 opening
9 branch or smaller structure given off by a larger structure
10 high projection or border projection
12 broad, flat projection (on femur)
14 groove
17 thin, flat plate
21 major protrusion

WORD SEARCH

Define the following terms; then find each term in the puzzle.

```
G S N E S I T I S O Y M X O N
P T S I G R E N Y S I O I E O
S S E C O R P D I O H P I X I
I E L T S I N O G A T N A D T
Y S S C T M I A N I G I R O A
H A Y G O L O E T S O I O O T
P M S T B O E T A L U G N U O
O O G T N O I D S O O I H T R
R I Q U A D R U P E D I P G G
T D O G M Y O T O M Y I D O X
S T E O I T C I N O T A N O H
Y M N U U M T M A N D I B L E
D M V A L G U S M H H P O S A
M A X I L L A I H O I G O O A
D O L I C H O C E P H A L I C
```

defective growth   _____

bent outward   _____

things that work together _____

place where muscle begins _____

narrow, long skull   _____

surgical incision into a muscle _____

inflammation of voluntary muscle _____

lack of muscle control   _____

circular movement around an axis _____

small bone embedded in tendon _____

four-footed animal   _____

study of bone   _____

upper bone of jaw   _____

lower bone of jaw   _____

animal with hooves   _____

caudal part of sternum   _____

things that work opposite each other _____

LABEL THE DIAGRAMS

Label the Diagrams in Figures 3–31 and 3–32.

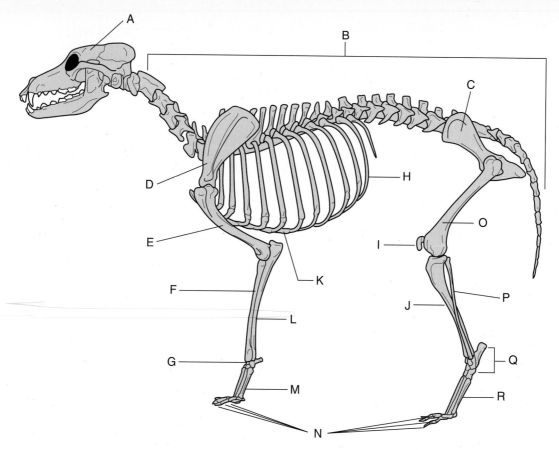

Figure 3–31   Dog skeleton. Label the parts of the dog skeleton indicated by the letters.

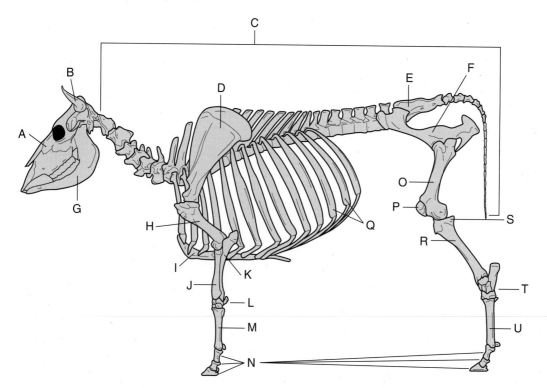

Figure 3–32   Bovine skeleton. Label the parts of the bovine skeleton indicated by the letters.

CHAPTER 4

HEAD TO TOE (AND ALL PARTS IN BETWEEN)

Objectives

Upon completion of this chapter, the reader should be able to

- Identify and describe the body parts of various species
- Recognize, define, spell, and pronounce anatomical terms

TWO WORDS, SAME MEANING

Medical terminology is a specific language that health care professionals (both human and veterinary) use to describe conditions in a concise manner. Laypeople also have a language that is used to describe anatomy and medical conditions in a concise manner. In veterinary medicine, many different terms are used to describe the anatomy and diseases of many different species. In this chapter, the anatomical lay terms that many people use in the veterinary community are described. The lay terms for diseases are covered in the chapters on individual species.

COMMON ANATOMICAL TERMS FOR EQUINE SPECIES

See Figures 4–1a and b.

barrel (ba-rəl): capacity of the chest or trunk.

bars (bahrz): raised V-shaped structure on distal surface of hoof.

cannon (kahn-nohn) **bone:** third metacarpal (metatarsal) of the horse; also called the shin bone.

cheek (chēk): fleshy portion of either side of the face; forms the sides of the mouth and continues rostrally to the lips.

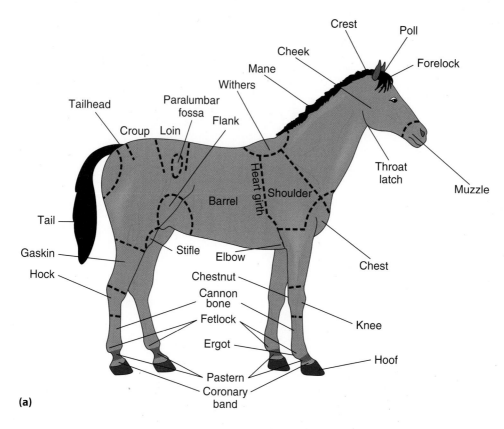

(a)

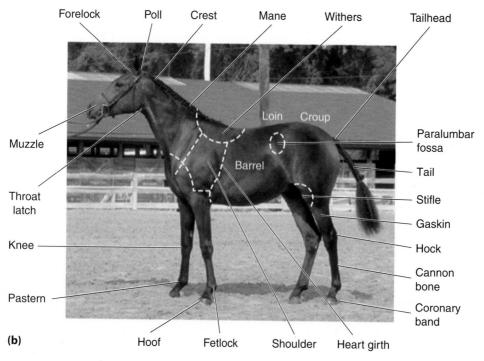

(b)

Figure 4–1   Anatomical parts of a horse [(b) Courtesy of the USDA by Bill Tarpenning.]

chest (chehst): part of the body between the neck and abdomen; the thorax.

chestnuts (chehs-nuhtz): horny, irregular growths on the medial surface of the equine leg; in the front legs, the chestnuts are just above the knee; in the rear legs, the chestnuts are near the hock.

coffin (kawf-ihn) **joint:** distal interphalangeal joint (joint between the short pastern and coffin bones [phalanges II and III, respectively]) in ungulates (Figures 4–2a and b).

corners (kawr-nrz): third incisors of equine.

coronary (kohr-ō-nār-ē) **band:** junction between the skin and the horn of the hoof; also called the **coronet** (kōr-oh-neht).

crest (krehst): root of the mane.

croup (kroop): muscular area around and above the tail base.

cutters (kuht-ərz): second incisors of equine.

dock (dohck): solid part of the equine tail.

elbow (ehl-bō): forelimb joint formed by distal humerus, proximal radius, and proximal ulna.

ergot (ahr-goht or ər-goht): small keratinized (kehr-ə-tə-nīzd) mass of horn in a small bunch of hair on the palmar or plantar aspects of the equine fetlock.

fetlock (feht-lohck): area of the limb between the pastern and the cannon.

fetlock (feht-lohck) **joint:** metacarpophalangeal and metatarsophalangeal joint (joint between the cannon bone and long pastern bone [phalanx I]) in ungulates.

flank (flânk): side of the body between the ribs and ilium.

forelock (fōr-lohck): in maned animals, the most cranial part of the mane hanging down between the ears and onto the forehead.

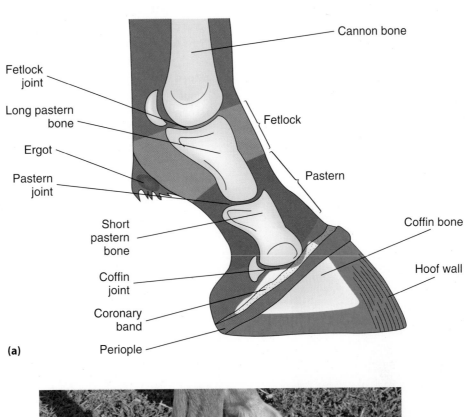

(a)

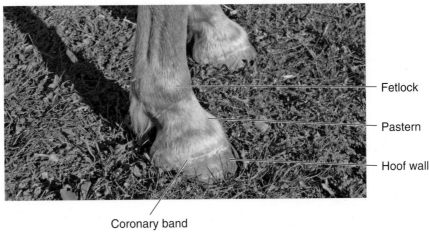

(b)

Figure 4–2 Anatomical parts of a horse's foot

frog (frohg): V-shaped pad of soft horn between the bars on the sole of the equine hoof.

gaskin (gahs-kihn): muscular portion of the hindlimb between the stifle and hock; also called the **crus.**

heart girth (hahrt gərth): circumference of the chest just caudal to the shoulders and cranial to the back.

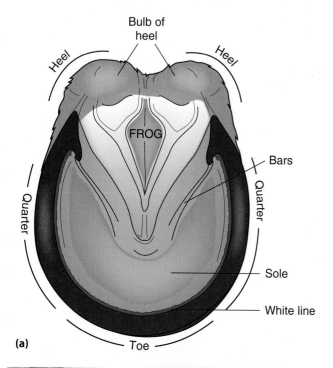

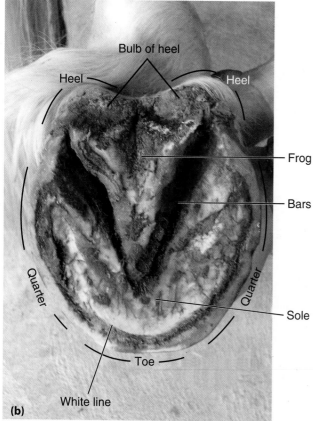

(b)

Figure 4–3 Anatomical parts of a horse's hoof [(b) Courtesy of Laura Lien, CVT, BS.]

heel: caudal region of the hoof that has an area of soft tissue called the bulb.

hock (hohck): tarsal joint.

hoof (hoof): hard covering of the digit in ungulates (Figures 4–3a and b).

hoof wall (hoof wahl): hard, horny outer layer of the covering of the digit in ungulates.

knee (nē): carpus in ungulates (an ungulate is an animal with hooves).

loin (loyn): lumbar region of the back between the thorax and pelvis.

mane (mān): region of long, coarse hair at the dorsal border of the neck and terminating at the poll.

muzzle (muh-zuhl): two nostrils (including the skin and fascia) and the muscles of the upper and lower lip.

nippers (nihp-pərz): central incisors of equine.

paralumbar fossa (pahr-ah-luhm-bahr fohs-ah): hollow area of the flank whose boundaries are the transverse processes of the lumbar vertebrae (dorsally), the last rib (cranially), and the thigh muscles (caudally).

pastern (pahs-tərn): area of the limb between the fetlock and hoof.

pastern (pahs-tərn) **joint:** proximal interphalangeal joint (joint between the long and short pastern bones [phalanges I and II, respectively]) in ungulates.

poll (pōl): top of the head; occiput; nuchal crest.

quarter: lateral or medial side of the hoof.

shoulder (shōl-dər): region around the large joint between the humerus and scapula.

sole (sōl): palmar or plantar surface of the hoof; irregular crescent-shaped bottom of hoof.

stifle (stī-fuhl) **joint:** femorotibial and femoropatellar joint in quadrupeds.

tail (tāl): caudal part of the vertebral column extending beyond the trunk.

tail head (tāl hehd): base of the tail where it connects to the body.

teat (tēt): nipple of mammary gland.

toe: cranial side of the hoof.

udder (uh-dər): mammary gland.

white line: fusion between the wall and sole of the hoof.

withers (wih-thərz): region over the dorsum where the neck joins the thorax and where the dorsal margins of the scapula lie just below the skin.

COMMON ANATOMICAL TERMS FOR CATTLE

See Figures 4–4a, b, and c.

brisket (brihs-kiht): mass of connective tissue, muscle, and fat covering the cranioventral part of the ruminant chest between the forelegs.

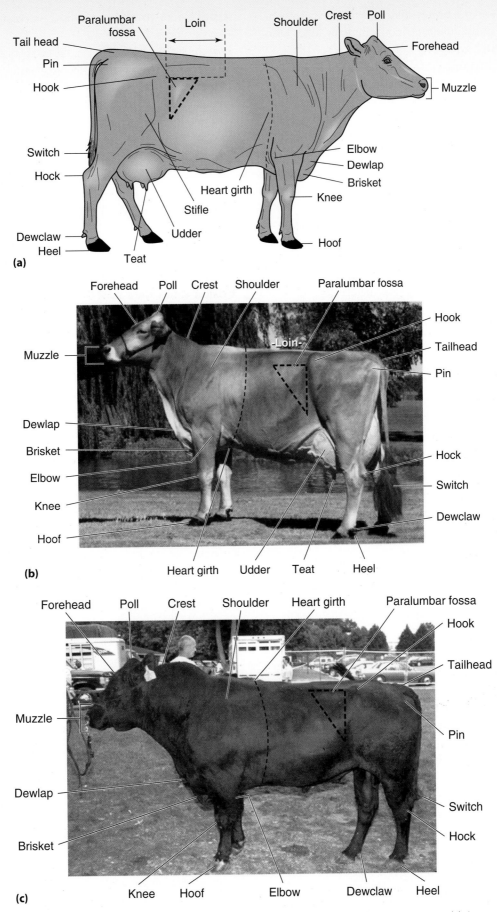

Figure 4–4   Anatomical parts of a cow [(b) Courtesy of the Brown Swiss Cattle Breeders' Association. (c) Anatomical parts of a bull Courtesy of the USDA by Bob Bill Tarpenning.]

cannon (kahn-nohn) **bone:** third and fourth metacarpal (metatarsal) of the ruminant (not commonly used); also called the **shin bone.**

coffin (kawf-ihn) **joint:** distal interphalangeal joint (joint between the short pastern and coffin bones [phalanges II and III, respectively]) in ungulates.

crest (krehst): dorsal margin of the neck.

dewclaw (doo-klaw): accessory claw of the ruminant foot that projects caudally from the fetlock.

dewlap (doo-lahp): loose skin under the throat and neck, which may become pendulous in some breeds.

dock (dohck): solid part of the tail.

elbow (ehl-bō): forelimb joint formed by distal humerus, proximal radius, and proximal ulna.

fetlock (feht-lohck) **joint:** metacarpophalangeal and metatarsophalangeal joint (joint between the cannon bone and the long pastern bone [phalanx I]) in ungulates.

flank (flānk): side of the body between the ribs and ilium.

forearm (fōr-ahrm): part of the foreleg supported by the radius and ulna, between the elbow and knee.

forehead (fōr-hehd): region of the head between the eyes and ears.

heart girth (hahrt gərth): circumference of the chest just caudal to the shoulders and cranial to the back.

heel (hēl): caudal region of the hoof that has an area of soft tissue called the bulb.

hock (hohck): tarsal joint.

hoof (hoof): hard covering of the digit in ungulates.

hoof wall (hoof wahl): hard, horny outer layer of the covering of the digit in ungulates.

hooks (hookz): protrusion of the wing of the ilium on the dorsolateral area of ruminants.

knee (nē): carpus in ungulates.

loin (loyn): lumbar region of the back, between the thorax and pelvis.

muzzle (muh-zuhl): two nostrils (including the skin and fascia) and the muscles of the upper and lower lip.

paralumbar fossa (pahr-ah-luhm-bahr fohs-ah): hollow area of the flank whose boundaries are the transverse processes of the lumbar vertebrae (dorsally), the last rib (cranially), and the thigh muscles (caudally).

pastern (pahs-tərn) **joint:** proximal interphalangeal joint (joint between the long and short pastern bones [phalanges I and II, respectively]) in ungulates.

pedal (pē-dahl): pertaining to the foot.

pins (pihnz): protrusion of the ischium bones just lateral to the base of the tail in ruminants.

poll (pōl): top of the head; occiput; nuchal crest.

quarter: one of the four glands in the cow's udder.

shoulder (shōl-dər): region around the large joint between the humerus and scapula.

sole (sōl): palmar or plantar surface of the hoof; bottom of hoof.

stifle (stī-fuhl) **joint:** femorotibial and femoropatellar joint in quadrupeds.

switch (swihtch): tuft of hair at the end of the tail.

tail (tāl): caudal part of the vertebral column extending beyond the trunk.

tail head (tāl hehd): base of the tail where it connects to the body.

teat (tēt): nipple of mammary gland.

toe (tō): cranial end of the hoof.

udder (uh-dər): mammary gland.

COMMON ANATOMICAL TERMS FOR GOATS

See Figures 4–5 a and b.

brisket (brihs-kiht): mass of connective tissue, muscle, and fat covering the cranioventral part of the ruminant chest between the forelegs.

cannon (kahn-nohn) **bone:** third and fourth metacarpal (metatarsal) of the ruminant (not commonly used); also called the **shin bone.**

chine (chīn): thoracic region of the back.

coffin (kawf-ihn) **joint:** distal interphalangeal joint (joint between the short pastern and coffin bones [phalanges II and III, respectively]) in ungulates.

crest (krehst): dorsal margin of the neck.

dewclaw (doo-klaw): accessory claw of the ruminant foot that projects caudally from the fetlock.

elbow (ehl-bō): forelimb joint formed by distal humerus, proximal radius, and proximal ulna.

fetlock (feht-lohck) **joint:** metacarpophalangeal and metatarsophalangeal joint (joint between the cannon bone and the long pastern bone [phalanx I]) in ungulates.

flank (flānk): side of the body between the ribs and ilium.

forearm (fōr-ahrm): part of the foreleg supported by the radius and ulna, between the elbow and knee.

forehead (fōr-hehd): region of the head between the eyes and ears.

heart girth (hahrt gərth): circumference of the chest just caudal to the shoulders and cranial to the back.

heel (hēl): caudal region of the hoof that has an area of soft tissue called the bulb.

hock (hohck): tarsal joint.

hoof (hoof): hard covering of the digit in ungulates.

hoof wall (hoof wahl): hard, horny outer layer of the covering of the digit in ungulates.

hooks (hookz): protrusion of the wing of the ilium on the dorsolateral area of ruminants.

horn butt (hōrn buht): poll region between the eyes and ears of previous horn growth.

knee (nē): carpus in ungulates.

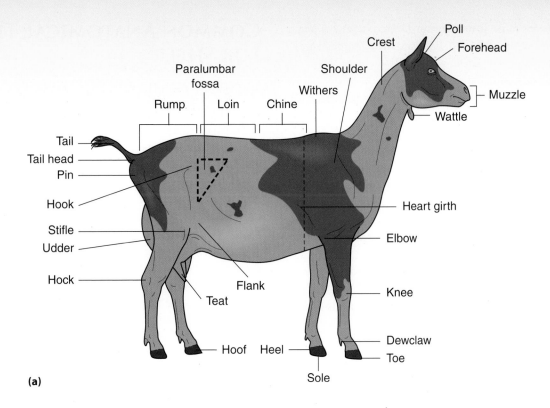

(a)

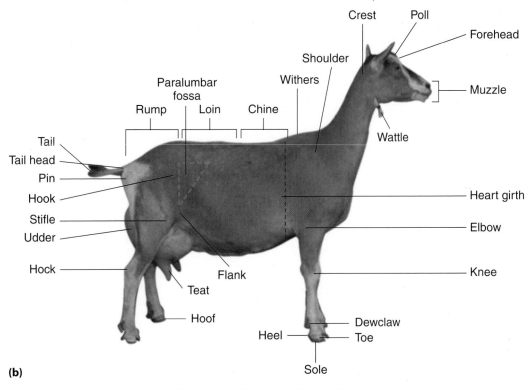

(b)

Figure 4–5 Anatomical parts of a goat [(b) Courtesy of iStock Photo.]

loin (loyn): lumbar region of the back, between the thorax and pelvis.

muzzle (muh-zuhl): two nostrils (including the skin and fascia) and the muscles of the upper and lower lip.

paralumbar fossa (pahr-ah-luhm-bahr fohs-ah): hollow area of the flank whose boundaries are the transverse processes of the lumbar vertebrae (dorsally), the last rib (cranially), and the thigh muscles (caudally).

pastern (pahs-tərn) **joint:** proximal interphalangeal joint (joint between the long and short pastern bones [phalanges I and II, respectively]) in ungulates.

pedal (pē-dahl): pertaining to the foot.

pins (pihnz): protrusion of the ischium bones just lateral to the base of the tail in ruminants.

poll (pōl): top of the head; occiput; nuchal crest.

rump (ruhmp): sacral to tailhead region of the back.

shoulder (shōl-dər): region around the large joint between the humerus and scapula.

sole (sōl): palmar or plantar surface of the hoof; bottom of hoof.

stifle (stī-fuhl) **joint:** femorotibial and femoropatellar joint in quadrupeds.

tail (tāl): caudal part of the vertebral column extending beyond the trunk.

tail head (tāl hehd): base of the tail where it connects to the body.

teat (tēt): nipple of mammary gland.

toe (tō): cranial end of the hoof.

udder (uh-dər): mammary gland.

wattle (waht-tuhl): appendages suspended from the head (usually under the chin).

withers (wih-thərz): region over the dorsum where the neck joins the thorax and where the dorsal margins of the scapula lie just below the skin.

COMMON ANATOMICAL TERMS FOR SHEEP

See Figures 4–6a and b.

brisket (brihs-kiht): mass of connective tissue, muscle, and fat covering the cranioventral part of the ruminant chest between the forelegs.

cannon (kahn-nohn) **bone:** third and fourth metacarpal (metatarsal) of the ruminant (not commonly used); also called the **shin bone.**

coffin (kawf-ihn) **joint:** distal interphalangeal joint (joint between the short pastern and coffin bones [phalanges II and III, respectively]) in ungulates.

crest (krehst): dorsal margin of the neck.

dewclaw (doo-klaw): accessory claw of the ruminant foot that projects caudally from the fetlock.

dock (dohck): solid part of the tail.

elbow (ehl-bō): forelimb joint formed by distal humerus, proximal radius, and proximal ulna.

fetlock (feht-lohck) **joint:** metacarpophalangeal and metatarsophalangeal joint (joint between the cannon bone and the long pastern bone [phalanx I]) in ungulates.

flank (flānk): side of the body between the ribs and ilium.

forearm (fōr-ahrm): part of the foreleg supported by the radius and ulna, between the elbow and knee.

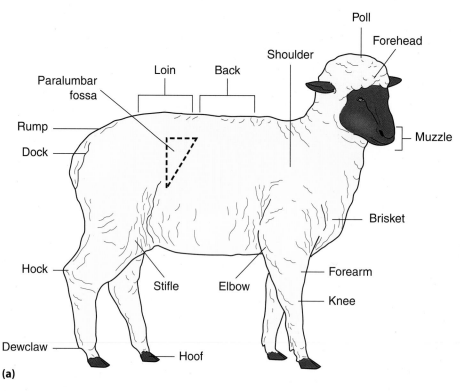

(a)

Figure 4–6 Anatomical parts of a sheep [(b) Courtesy of the American Hampshire Sheep Association.]

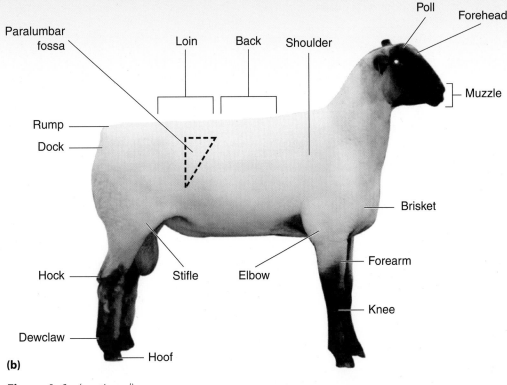

Figure 4–6 (continued)

forehead (fōr-hehd): region of the head between the eyes and ears.

heart girth (hahrt gərth): circumference of the chest just caudal to the shoulders and cranial to the back.

heel (hēl): caudal region of the hoof that has an area of soft tissue called the bulb.

hock (hohck): tarsal joint.

hoof (hoof): hard covering of the digit in ungulates.

hoof wall (hoof wahl): hard, horny outer layer of the covering of the digit in ungulates.

knee (nē): carpus in ungulates.

loin (loyn): lumbar region of the back, between the thorax and pelvis.

muzzle (muh-zuhl): two nostrils (including the skin and fascia) and the muscles of the upper and lower lip.

paralumbar fossa (pahr-ah-luhm-bahr fohs-ah): hollow area of the flank whose boundaries are the transverse processes of the lumbar vertebrae (dorsally), the last rib (cranially), and the thigh muscles (caudally).

pastern (pahs-tərn) joint: proximal interphalangeal joint (joint between the long and short pastern bones [phalanges I and II, respectively]) in ungulates.

pedal (pē-dahl): pertaining to the foot.

poll (pōl): top of the head; occiput; nuchal crest.

rump (ruhmp): sacral to tailhead region of the back.

shoulder (shōl-dər): region around the large joint between the humerus and scapula.

sole (sōl): palmar or plantar surface of the hoof; bottom of hoof.

stifle (stī-fuhl) joint: femorotibial and femoropatellar joint in quadrupeds.

tail head (tāl hehd): base of the tail where it connects to the body.

teat (tēt): nipple of mammary gland.

toe (tō): cranial end of the hoof.

udder (uh-dər): mammary gland.

COMMON ANATOMICAL TERMS FOR SWINE

See Figures 4–7a and b.

coffin (kawf-ihn) joint: joint between the short pastern and coffin bones (phalanges II and III, respectively) in ungulates.

dewclaw (doo-klaw): accessory claw of the porcine foot that projects caudally from the fetlock.

elbow (ehl-bō): forelimb joint formed by distal humerus, proximal radius, and proximal ulna.

fetlock (feht-lohck) joint: metacarpophalangeal and metatarsophalangeal joint in ungulates.

flank (flānk): side of the body between the ribs and ilium.

ham (hahm): musculature of the upper thigh.

hock (hohck): tarsal joint.

hoof (hoof): hard covering of the digit in ungulates.

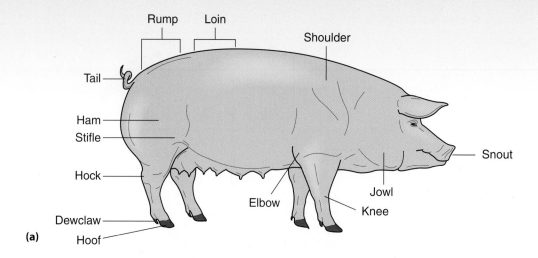

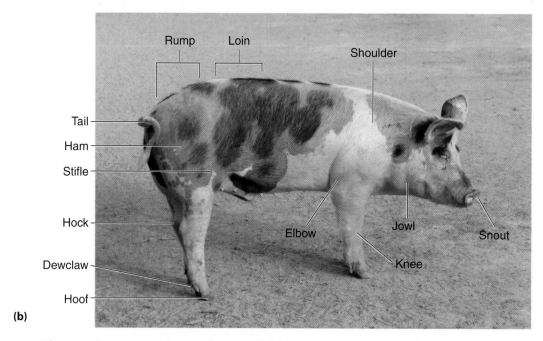

Figure 4–7   Anatomical parts of a swine [(b) Courtesy of iStock photo.]

hoof wall (hoof wahl): hard, horny outer layer of the covering of the digit in ungulates.

jowl (jowl): external throat, especially when fat or loose skin is present.

knee (nē): carpus in ungulates.

loin (loyn): lumbar region of the back, between the thorax and pelvis.

pastern (pahs-tərn) **joint:** joint between the long and short pastern bones (phalanges I and II, respectively) in ungulates.

rump (ruhmp): sacral to tailhead region of the back.

shoulder (shōl-dər): region around the large joint between the humerus and scapula.

snout (snowt): upper lip and apex of the nose of swine.

stifle (stī-fuhl) **joint:** femorotibial and femoropatellar joint in quadrupeds.

tail (tāl): caudal part of the vertebral column extending beyond the trunk.

COMMON ANATOMICAL TERMS FOR DOGS AND CATS

See Figures 4–8a and b and 4–9a and b.

cheek (chēk): fleshy portion of either side of the face; forms the sides of the mouth and continues rostrally to the lips.

chest (chehst): part of the body between the neck and abdomen; the thorax.

chin (chihn): rostroventral protrusion of the mandible.

dewclaw (doo-klaw): rudimentary first digit of dogs and cats.

elbow (ehl-bō): forelimb joint formed by distal humerus, proximal radius, and proximal ulna.

flank (flānk): side of the body between the ribs and ilium.

forehead (fōr-hehd): region of the head between the eyes and ears.

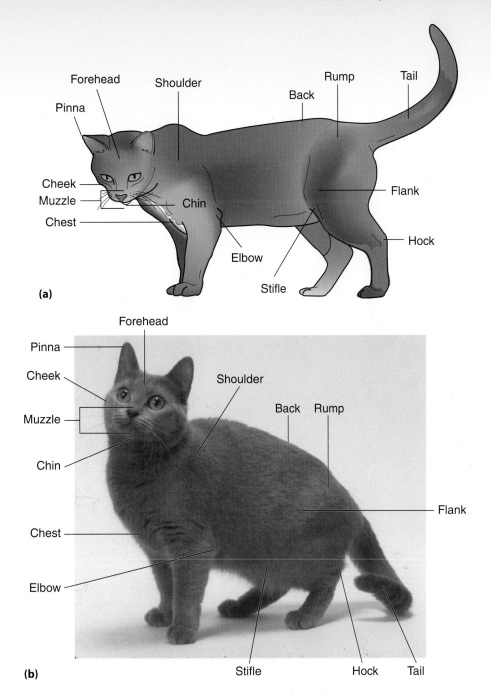

Figure 4–8 Anatomical parts of a cat [(b) Photo by Isabelle Francais.]

hock (hohck): tarsal joint; **tarsus** also is used for this joint.

muzzle (muh-zuhl): two nostrils (including the skin and fascia) and the muscles of the upper and lower lip.

pinna (pihn-ah): projecting part of the ear lying outside the head; the **auricle.**

rump (ruhmp): sacral to tailhead region of the back; also called the **croup.**

shoulder (shōl-dər): region around the large joint between the humerus and scapula.

stifle (stī-fuhl) **joint:** femorotibial and femoropatellar joint in quadrupeds.

tail (tāl): caudal part of the vertebral column extending beyond the trunk.

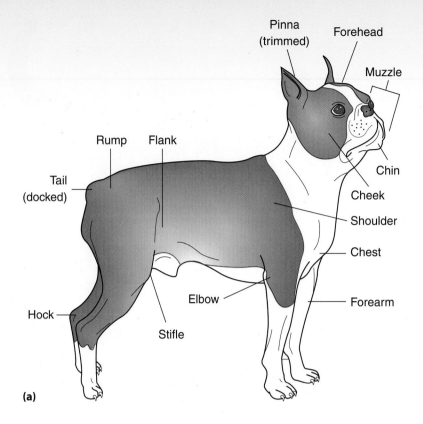

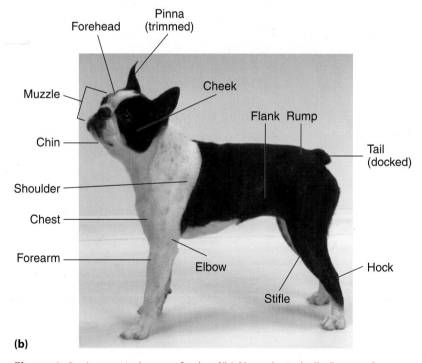

Figure 4–9   Anatomical parts of a dog [(b) Photo by Isabelle Francais.]

REVIEW EXERCISES

Multiple Choice

Choose the correct answer.

1. Another term for the distal interphalangeal joint in ungulates is the
   a. coffin joint
   b. fetlock joint
   c. pastern joint
   d. coronary band

2. The V-shaped pad of soft horn between the bars on the sole of the equine hoof is known as the
   a. sole
   b. white line
   c. ergot
   d. frog

3. The anatomical term for the top of the head is the
   a. crest
   b. poll
   c. sole
   d. forehead

4. The common name for the tarsal joint in animals is the
   a. loin
   b. stifle
   c. hock
   d. wrist

5. The side of the body between the ribs and ilium is called the
   a. rump
   b. flank
   c. loin
   d. ham

6. The upper lip and apex of the nose of swine is called the
   a. jowl
   b. ham
   c. snout
   d. pinna

7. The two nostrils and the muscles of the upper and lower lip are called the
   a. snout
   b. muzzle
   c. cheek
   d. crest

8. The proximal interphalangeal joint in ungulates is called the
   a. pastern joint
   b. coffin joint
   c. fetlock joint
   d. stifle joint

9. The protrusions of the ischium bones just lateral to the base of the tail in ruminants are known as
   a. tailheads
   b. hooks
   c. pins
   d. docks

10. The protrusions of the wing of the ilium on the dorsolateral area of ruminants are known as
    a. tailheads
    b. hooks
    c. pins
    d. docks

11. The hollow area of the flank is called the
    a. pedal fossa
    b. paralumbar fossa
    c. rump fossa
    d. tail fossa

12. The mass of connective tissue, muscle, and fat covering the cranioventral part of the ruminant chest is the
    a. brisket
    b. crest
    c. flank
    d. loin

13. The rudimentary first digit of dogs and cats is the
    a. claw
    b. digit
    c. dewclaw
    d. declaw

14. The "knee" in people is known as what in animals?
    a. hock joint
    b. pastern joint
    c. coffin joint
    d. stifle joint

15. The auricle is also known as the

    a. cheek
    b. forehead
    c. chin
    d. pinna

16. In swine, the external throat, especially when fat or loose skin is present, is called the

    a. snout
    b. ham
    c. jowl
    d. loin

17. In equine, the region over the dorsum where the neck joins the thorax and where the dorsal margins of the scapula lie is called the

    a. shoulder
    b. withers
    c. flank
    d. loin

18. Ilium is to ischium as

    a. hooks is to pins
    b. pins is to hooks
    c. pastern is to coffin
    d. coffin is to pastern

19. The lumbar region of the back is called the

    a. rump
    b. tailhead
    c. loin
    d. flank

20. The lateral or medial side of the hoof is the

    a. frog
    b. coronet
    c. heel
    d. quarter

Matching

Match the anatomical term in Column I with its definition in Column II.

Column I

1. _____ mane
2. _____ cannon bone
3. _____ knee
4. _____ stifle joint
5. _____ sole
6. _____ cutters
7. _____ corners
8. _____ dock
9. _____ dewclaw
10. _____ barrel

Column II

a. carpus in ungulates
b. capacity of the chest or trunk
c. amputation of the tail; solid part of the equine tail
d. region of long, coarse hair at the dorsal border of the neck
e. rudimentary first digit of dogs and cats
f. third metacarpal or metatarsal bone of the horse
g. femorotibial and femoropatellar joint in quadrupeds
h. bottom of the hoof
i. second incisors of equine
j. third incisors of equine

CROSSWORD PUZZLE

Supply the correct anatomical term in the appropriate space for the definition listed.

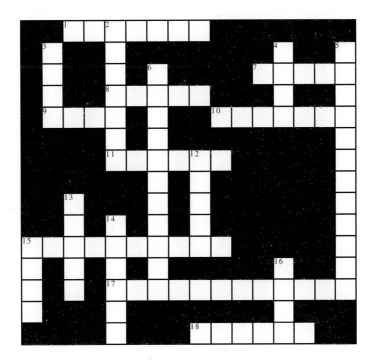

Across

1 Rudimentary first digit of dogs and cats
7 Projecting part of the ear lying outside the head; the auricle
8 Protrusion of the wing of the ilium on the dorsolateral area of ruminants
9 The carpus in ungulates (an ungulate is an animal with hooves)
10 Mass of connective tissue, muscle, and fat covering the cranioventral part of the ruminant chest between the forelegs
11 Tuft of hair at the end of the tail
15 Circumference of the chest just caudal to the shoulders and cranial to the back
17 Metacarpo-phalangeal and metatarso-phalangeal joint in ungulates (2 wds.)
18 The two nostrils (including the skin and fasica) and the muscles of the upper and lower lip

Down

2 Region in equine over the dorsum where the neck joins the thorax and where the dorsal margins of the scapula lie just below the skin
3 Amputation of the tail
4 Protrusion of the ischium bones just lateral to the base of the tail in ruminants
5 Joint between the long and short pastern bones (phalanx I and II respectively) in ungulates (2 wds.)
6 Joint between the short pastern and coffin bones (phalanx II and III respectively) in ungulates (2 wds.)
12 The root of the mane
13 Side of the body between the ribs and ilium
14 Femorotibial and femoropatellar joint in quadrupeds
15 Tarsal joint
16 Top of the head; the occiput

WORD SEARCHES

Find the following anatomical terms in the puzzle below. (Make sure you understand what the terms mean as you find them.)

Dog and Cat Word Search

```
W L L E C L N W I C E A I L M
L N W L E C C I L L W W U W L
E K A Z W O T C Z C L L L H H
A L T W N S L Z E H I C Z T L
T N I A I H U L L M F W I W E
L I W L L M U I L W L L T W S
O L U L C Z S U I M I U L L W
T N O L L U L C F I Z L Z C A
W Z L W N N I H C Z T I M K U
A I K S T I F L E J O I N T W
L L N T I T A C A T L T A W Z
C K C O H A N A N E W I L N W
W Z L C N I N O I L I W C P U
E E Z H T L I U C N F D H L W
D L H T K C P N A T Z F W S C
```

Hock   chin
stifle joint   dewclaw
pinna   tail
muzzle

Ruminant Word Search

```
T A W A T T L E R T L O T U L
R D R O A D C H K P T L U W K
O N I D N R P T N L T R L T S
T P T L R A B S A N S D L K L
L T P E D A L P L A O A R O O
E N D P S P T D F D A T O L I
N P D T N A L W O G E W T R T
O U I N A C I O A E L W T D K
B I C N N P C R K O L C L L C
N U L U S T D L I I A R T A L
O D T R L T N N W R S P R N P
N D A E P O U P T E K S I R B
N E S P C U O I L C R B G A S
A R T I C E S O C S K O O H T
C H T R I G T R A E H W L T L
```

cannon bone

udder

wattle

heart girth

pins

hooks

loin

flank

pedal

brisket

dewlap

Equine Word Search

```
O T O T E N A H E E S U E N L
G N A N T S N G E N T N N N A
A S R E H U N G G E U T H T S
I U L E N G A T O E N N P N O
L L O P H S S O R P T I K I A
U R E H T T U G F T S O S O N
T G A S K I N R S U E J U J W
T T S M S R H E N I H N G N I
N O E O H E E E E C I N R T
A G U A U A C N E N A F N E H
C P E G T G H U P E K F M T E
E E N T E U O S O U N O A S R
L A R S H E U T P E E C N A S
H N I R R I G R E G E S E P A
G E S K N N S O L E E A E K H
```

knee   frog
poll   gaskin
mane   withers
pastern joint   chestnuts
coffin joint   ergot
sole

Swine Word Search

```
O O L P O O O L N T O S O O H
L R H L O F M L O R M T O O O
H O R T S F R O L H H O M L O
A A L H H H A O O N M P T F
M M O O N M L O O H R O U A U
T T O T M O B W S R T U O N S
O P T A L M L H O U S H L L S
S H S H M R L T O M H L O A O
O O P O O U U L O L M S O F R
A O N O R U F H M R L L O S O
O L A O J O L O S L E L H U O
J J S P O P H S U S O L O H L
O J S O O M S M H O O U B S S
W O L S H U M J A M P O H O O
L M F U O R O O T H H S H O W
```

rump	snout
ham	hoof
jowl	elbow

LABEL THE DIAGRAMS

Label Diagrams (Figures 4–10, 4–11, 4–12, 4–13, 4–14, and 4–15)

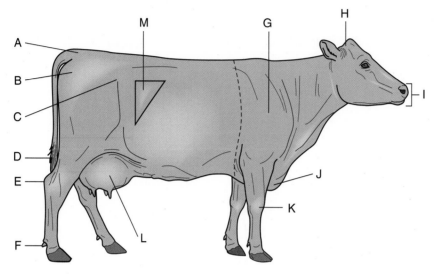

Figure 4–10 Identify the parts of this cow.

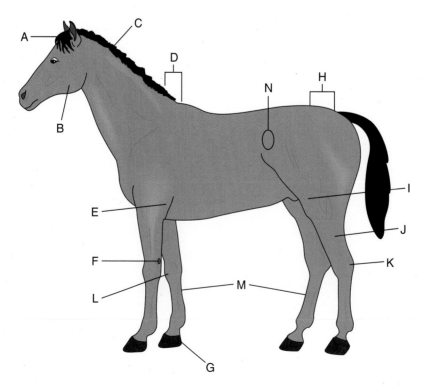

Figure 4–11 Identify the parts of this horse.

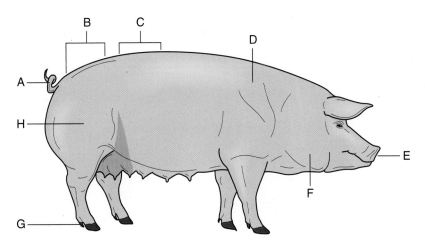

Figure 4–12 Identify the parts of this swine.

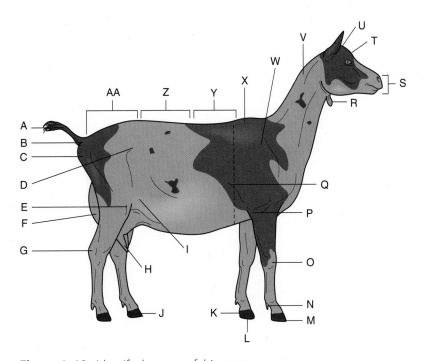

Figure 4–13 Identify the parts of this goat.

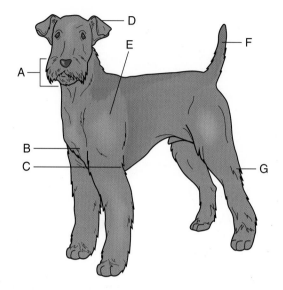

Figure 4–14   Identify the parts of this dog.

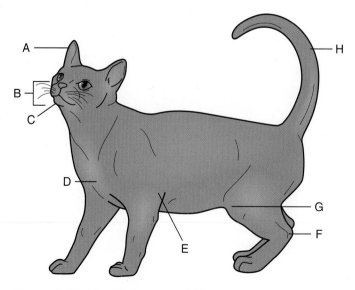

Figure 4–15   Identify the parts of this cat.

Label the anatomical parts of a horse's foot.

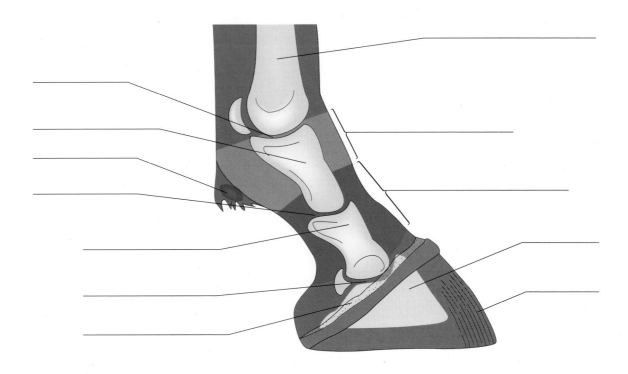

1. cannon bone

2. coffin bone

3. coffin joint

4. coronary band

5. ergot

6. fetlock

7. fetlock joint

8. hoof wall

9. long pastern bone

10. pastern

11. pastern joint

12. short pastern bone

CHAPTER 5

WHAT IS IN A NAME?

Objectives

Upon completion of this chapter, the reader should be able to

- Identify and recognize common terms used for animals
- Define common terms used to denote sex and age of animals
- Define common terms used to denote birthing and grouping of animals

WHAT IS YOUR NAME?

Laypeople and professionals use terms to describe in one word the status of an animal. The term may relate to the sexual status of an animal (intact or sexually functional, or altered or sexually nonfunctional) or the age status of an animal. Terms have also been derived to denote the process of giving birth and the grouping of animals. The following lists provide the terms used to describe animals.

canine (kā-nīn) = dog

dog/stud = intact (not sexually altered) male dog.
bitch = intact female dog.
whelp (wehlp) or pup = young dog.
whelping (wehl-pihng) = giving birth to whelps.
pack = group of dogs.
litter = multiple offspring born during same labor.

feline (fē-līn) = cat

tom = intact male cat.
queen = intact female cat.
kitten = young cat.
queening = giving birth to kittens.

lagomorph (lāg-ō-mōrf) = rabbit

buck = intact male rabbit.
doe = intact female rabbit.
lapin (lahp-ihn) = neutered male rabbit.
kit = young (blind, deaf) rabbit.
kindling (kihnd-lihng) = giving birth to rabbits.
herd = group of rabbits.

ferret (fehr-reht)

hob = intact male ferret.
jill = intact female ferret.
gib (gihb) = neutered male ferret.
sprite (sprīt) = spayed female ferret.
kit = young ferret.
kindling = giving birth to ferrets.

cavy (kā-vē) = guinea pig

boar = intact male guinea pig.
sow = intact female guinea pig.
pup = young guinea pig.

murine (moo-rēn) = mouse or rat

sire (sī-ər) = intact male mouse or rat.
dam (dahm) = intact female mouse or rat.
pup = young mouse or rat.

psittacine (siht-ah-sēn) = parrot includes other birds with bills for cracking seeds)

cock = intact male parrot.
hen = intact female parrot.
chick = young parrot.
flock = group of parrots (also called a **company**).

turkey = one kind of poultry

tom = intact male turkey.
hen = intact female turkey.
poult (pōlt) = young turkey.
flock = group of turkeys.
clutch (kluhtch) = group of eggs.

chicken = one kind of poultry

rooster = sexually mature male chicken; also called cock.
hen = intact female chicken.
capon (kā-pohn) = young castrated male chicken or domestic fowl.
cockerel (kohck-ər-ehl) = immature male chicken.
pullet (puhl-eht) = immature female chicken.
poult = young chicken.
chick = very young chicken.
flock = group of chickens.

goose = one kind of anserine (ahn-sehr-ihn)

gander = intact male goose.
goose = intact female goose.
gosling = young goose.
gaggle = group of geese.

Gender versus sex

Veterinary professionals talk only about the sex of an animal. Gender is used to denote whether words pertaining to a noun are masculine, feminine, or neutral. Gender also can be used to denote social constructs, such as the gender or social roles of men and women.

Dam and *sire* are terms used to denote female parent and male parent, respectively, for many species. When animals are bred, these terms may be used instead of the ones in the lists here. Used correctly, these mean that the male and female have mated and produced an embryo or fetus. **Dam** is a female parent; **sire** is a male parent.

duck = one type of anserine

drake = intact male duck.
duck = intact female duck.
duckling = young duck.
flock = group of ducks.

ratite (rah-tīt) = large, flightless bird (ostrich, emu, and rhea)

rooster = intact male ratite.
hen = intact female ratite.
hatchling = ostrich up to 2 days old.
chick = young ratite (usually less than 6 months of age).
yearling = ostrich from 6 months to 12 months of age; rhea from 6 months to 18 months of age.
flock = group of ratites.

porcine (poor-sīn) = pig = swine

boar (bōr) = intact male pig.
sow = intact female pig.
barrow (bār-ō) = male pig castrated when young.
stag = male pig castrated after maturity.
gilt (gihlt) = young female pig that has not farrowed.
pig or piglet = young pig; old term is *shoat.*
farrowing (fār-ō-ihng) = giving birth to pigs.
herd = group of pigs.

There are unique names for groups of birds that are particular to one type of bird. Examples include the following:

- flock, flight, pod, or volery of birds
- pandemonium of parrots
- murder of crows
- dole or flight of doves
- brace, bunch, flock, or team of ducks
- aerie or convocation of eagles
- gaggle of geese
- colony of gulls
- cast of hawks
- charm of hummingbirds
- band of blue jays
- watch of nightingales
- ostentation of peacocks
- bevy or covey of quails

equine (ē-kwīn) = horse, pony, donkey, and mule

stallion (stahl-yuhn) = intact male equine = 4 years old.
colt (kōlt) = intact male equine = 4 years old.
mare (mār) = intact female equine = 4 years old.
filly (fihl-ē) = intact female equine = 4 years old.
gelding (gehld-ihng) = castrated male equine.
ridgeling (rihdj-lihng) or **rig** = cryptorchid equine (one or both testicles have not descended from the abdomen).
foal = young equine of either sex.
weanling = young equine =1 year old.
yearling = young equine between 1 and 2 years old.
foaling = giving birth to equine.
herd = group of equine.
band = group of horses consisting of one mature stallion, his breeding mares, and the immature male and female offspring of his mares.
brood mare (bruhd mār) = breeding female equine.
maiden mare (mā-dehn mār) = female equine never bred.
barren mare (bār-ehn mār) = intact female horse that was not bred or did not conceive the previous season = open mare.
wet mare = intact female horse that has foaled during the current breeding season.
agalactic mare (ā-gahl-ahck-tihck mār) = intact female horse not producing milk.
pony = equine between 8.2 and 14.2 hands when mature (not a young horse).

donkey = ass = burro

jack or **jack ass** = intact male donkey.
jenny = intact female donkey.

What is a mule?

Mule is a general term that applies to the hybrid crossing of equines. *Mule* also is used to denote the offspring of a jack (male donkey) and a mare (female horse). **Hinny** (hihn-ē) is used to denote the offspring of a stallion (male horse) and a jenny (female donkey). Think mule:mare to remember the lineage of this hybrid. Both mules and hinnies are sterile.

ovine (ō-vīn) = sheep

ram = intact male sheep.
ewe (yoo) = intact female sheep.
wether (weh-thər) = castrated male sheep.
lamb = young sheep.
hothouse lamb = young sheep less than 3 months of age.
spring lamb = young sheep 3–7 months of age.
yearling = sheep 1–2 years of age (used when marketing sheep).
lambing = giving birth to sheep.
flock = group of sheep.

caprine (kahp-rīn) = goat

buck = intact male goat.
doe = intact female goat.
wether = castrated male goat.
kid = young goat.
kidding = giving birth to goats.
freshening = giving birth to dairy animals; also called freshen.
herd = group of goats.

camelid = llama, alpaca, guanaco, vicuna

bull = intact male llama (also called a stallion).
cow = intact female llama.
gelding = castrated male llama.
cria (krē-ah) = young llama.

bovine = cattle

bull = intact male bovine.
jumper bull = intact male bovine that has just reached maturity and is used for breeding.
cow = intact female bovine that has given birth.
steer = male bovine castrated when young.
stag = male bovine castrated after maturity.
heifer (hehf-ər) = young female bovine that has not given birth.
calf = young bovine.
calving = giving birth to cattle.
freshening = giving birth to dairy animals; also called freshen.
herd = group of cattle.
springing heifer = young female pregnant with her first calf = first calf heifer.
freemartin (frē-mahr-tihn) = sexually imperfect, usually sterile female calf born as a twin with a male calf.
gomer bull = bull used to detect female bovines in heat; bull may have penis surgically deviated to the side, may be treated with androgens, or may be vasectomized so as not to impregnate female; also called **teaser bull.**

cervidae = deer, elk, moose, caribou

buck = intact male deer.
bull = intact male elk, moose, caribou.
doe = intact female deer.
cow = intact female elk, moose, caribou.
fawn = young deer.
heifer = young female elk, moose, caribou.
herd = group of cervidae.

What is a bellwether?

A *wether* is a neutered sheep or goat. *Bell* is a ringing device. Originally, *bellwether* was used to describe the practice of putting a bell on the lead wether of a flock or herd.

The symbols denoting male ♂ and female ♀ originally stood for Mars (the god of war) and Venus (the goddess of love), respectively. *Mars* and *male* both begin with *m*.

The term **stud** is commonly used in reference to a male animal used for breeding. However, the term *stud* actually refers to the facility or farm where breeding animals are kept.

REVIEW EXERCISES

Multiple Choice

Choose the correct answer.

1. A neutered male sheep or goat is a(n)

    a. bull
    b. ovine
    c. wether
    d. caprine

2. A sexually imperfect, usually sterile female calf born as a twin with a male calf is a

    a. heifer
    b. freemartin
    c. gilt
    d. filly

3. A cross between a stallion and a jenny is a

    a. donkey
    b. mule
    c. jenny
    d. hinny

4. Cow is to mare as steer is to

    a. stallion
    b. ridgling
    c. colt
    d. gelding

5. Parrots are in a group of birds called

    a. amazon
    b. psittacine
    c. lagomorph
    d. murine

6. In canines, the act of giving birth is

    a. whelping
    b. pupping
    c. packing
    d. gestation

7. Male and female ferrets are called

    a. jacks and jills, respectively
    b. kits and jills, respectively
    c. hobs and jills, respectively
    d. jacks and kits, respectively

8. A young llama is called a

    a. calf
    b. cria
    c. clutch
    d. colt

9. *Freshening* is a term that means

    a. cleaning an animal to make it smell fresh
    b. giving birth to a dairy animal
    c. the act of mating in cattle
    d. removing the horns of a bovine to enhance mating

10. Giving birth to swine is called

    a. barrowing
    b. queening
    c. farrowing
    d. tupping

11. Tom is to queen as hob is to

    a. doe
    b. bitch
    c. jill
    d. cow

12. An intact male rabbit is known as a

    a. stud
    b. stallion
    c. buck
    d. gander

13. A young female bovine that has not given birth is called a

    a. cow
    b. stag
    c. calf
    d. heifer

14. A female equine that has never been bred is known as a

    a. ridgeling mare
    b. brood mare
    c. maiden mare
    d. barren mare

15. A group of eggs is known as a

    a. flock
    b. clutch
    c. poult
    d. herd

16. The symbol for female is

    a. f
    b. I
    c. o
    d. ♀

17. A ram is to a wether as a bull is to a

    a. barrow
    b. steer
    c. gelding
    d. gib

18. A mule is an offspring of a

    a. male donkey and a female pony
    b. male pony and a female horse
    c. male horse and a female donkey
    d. male donkey and a female horse

19. Giving birth to rabbits is known as

    a. whelping
    b. farrowing
    c. freshening
    d. kindling

20. A group of dogs is known as a

    a. pack
    b. herd
    c. flock
    d. gaggle

Matching

Match the species common name in Column I with its taxonomic name in Column II.

Column I	Column II
1. _____ cattle	a. canine
2. _____ cat	b. bovine
3. _____ pig	c. equine
4. _____ parrot	d. feline
5. _____ rat	e. caprine
6. _____ dog	f. ovine
7. _____ sheep	g. porcine
8. _____ mouse	h. psittacine
9. _____ donkey	i. murine
10. _____ horse	
11. _____ goat	

Match the female name in Column I with its species in Column II.

Column I	Column II
12. _____ bitch	a. horse
13. _____ queen	b. bovine
14. _____ dam	c. goat
15. _____ hen	d. dog

16. _____ ewe   e. mouse or rat

17. _____ doe   f. cat

18. _____ goose   g. sheep

19. _____ jill   h. goose

20. _____ mare   i. chicken

21. _____ cow   j. ferret

Fill in the Blanks

1. A young dog is called a(n) _____ or _____ .

2. A young cat is called a(n) _____ .

3. A young horse is called a(n) _____ .

4. A young bovine is called a(n) _____ .

5. A young goat is called a(n) _____ .

6. A young sheep is called a(n) _____ .

7. A young swine is called a(n) _____ or _____ .

8. Young rabbits or ferrets are called _____ .

9. Young mice or rats are called _____ .

10. A young llama is called a(n) _____ .

11. A group of rabbits is called a(n) _____ .

12. A group of cattle is called a(n) _____ .

13. A group of turkeys is called a(n) _____ .

14. A group of sheep is called a(n) _____ .

15. A group of goats is called a(n) _____ .

16. An ostrich up to 2 days old is called a(n) _____ .

17. A ratite less than 6 months of age is called a(n) _____ .

18. A young deer is called a(n) _____ .

19. A young castrated male chicken is called a(n) _____ .

20. A group of eggs is called a(n) _____ .

CROSSWORD PUZZLE

Supply the correct term in the appropriate space for the definition listed.

Across

1 male rabbit or goat
6 male horse over 4 yrs old
7 4 yr old intact male horse
9 female rabbit or goat
11 female cat
12 canine (pl.)
14 young goat
15 young horse
16 castrated rabbit
18 feline (pl.)
20 group of bovine
21 young ferret or rabbit
22 caprine (pl.)
24 group of geese
25 female dog

Down

1 male swine castrated
   when young
4 male swine
5 female caprine
7 immature male chicken
8 male turkey
10 immature female chicken
13 castrated llama
14 giving birth to a ferret or rabbit
17 young mouse or rat
18 group of eggs
19 ovine
23 male ovine

CHAPTER 6

GUT INSTINCTS

Objectives

Upon completion of this chapter, the reader should be able to

- Identify and describe the major structures and functions of the digestive tract
- Distinguish between monogastric and ruminant digestive system anatomy and physiology
- Describe the processes of digestion, absorption, and metabolism
- Recognize, define, spell, and pronounce terms related to the diagnosis, pathology, and treatment of the digestive system

FUNCTIONS OF THE DIGESTIVE SYSTEM

The **digestive** (dī-jehs-tihv) **system, alimentary** (ăl-ih-mehn-tahr-ē) **system, gastrointestinal** (gahs-trō-ihn-tehst-ihn-ahl) **system,** and **GI system** are terms used to describe the body system that is a long, muscular tube that begins at the mouth and ends at the anus. The digestive system is responsible for the intake and digestion of food and water, the absorption of nutrients, and the elimination of solid waste products. The combining form for nourishment is **aliment/o.**

STRUCTURES OF THE DIGESTIVE SYSTEM

The major structures of the digestive system include the oral cavity, pharynx, esophagus, stomach, and small and large intestines. The liver, gallbladder, and pancreas are organs associated with the digestive system.

Mouth

The mechanical and chemical process of digestion begins in the mouth. The mouth contains the major structures of the oral cavity. The **oral** (ōr-ahl) **cavity** contains the lips and cheeks, hard and soft palates, salivary glands, tongue, teeth, and periodontium. The combining forms **or/o** and **stomat/o** mean mouth.

The maxilla and mandible are bones that are the boundaries of the oral cavity. The combining form for jaw is **gnath/o.** An animal that has **prognathia** (prohg-nah- thē-ah) has an elongated mandible, or a mandible that is overshot. Prognathia is sometimes called sow mouth. An animal that has **brachygnathia** (brahck-ē-nah-thē-ah) has a shortened mandible, or a mandible that is undershot. Brachygnathia is sometimes called parrot mouth.

The lips form the opening to the oral cavity. The term **labia** (lā-bē-ah) is the medical term for lips. A single lip is a **labium** (lā-bē-uhm). The combining forms for lips are **cheil/o** and **labi/o.** The cheeks form the walls of the oral cavity. The combining form for cheek is **bucc/o.** The term **buccal** (buhk-ahl or būk-ahl) means pertaining to or directed toward the cheek.

The **palate** (pahl-aht) forms the roof of the mouth. The palate consists of two parts: the hard and soft palates. The **hard palate** forms the bony rostral portion of the palate that is covered with specialized mucous membrane. This specialized mucous membrane contains irregular folds called **rugae** (roo-gā) (Figure 6–1). Rugae also are found in the stomach. **Rug/o** is the combining form for wrinkle or fold. The **soft palate** forms the flexible caudal portion of the palate. The soft palate is involved in closing off the nasal passage during swallowing so that food does not move into the nostrils. The combining form **palat/o** means palate.

The **tongue** (tuhng) is a movable muscular organ in the oral cavity used for tasting and processing food, grooming, and articulating sound. The tongue moves food during chewing and swallowing. The dorsum of the tongue has **papillae** (pah-pihl- à), or elevations, and the ventral surface of the tongue is highly vascular (Figure 6–2). The types of papillae located on the dorsum of the tongue may appear threadlike, or **filiform** (fihl-ih-fōrm); mushroomlike, or **fungiform** (fuhn-jih-fōrm); or cup-shaped, or **vallate** (vahl-āt). Taste buds are located in the fungiform and vallate papillae. The tongue is connected to the ventral surface of the oral cavity by a band of connective

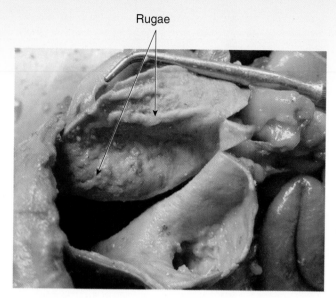

Figure 6–1 Rugae of the stomach are visible on this preserved cat.

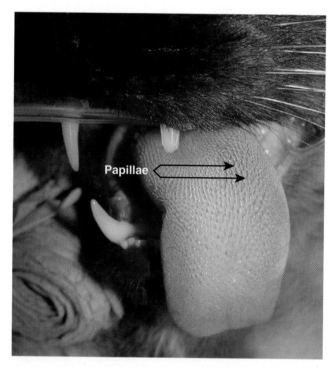

Figure 6–2 Papillae are found on the dorsum of the tongue. (Courtesy of Teri Raffel, CVT.)

tissue called the **frenulum** (frehn-yoo-luhm). The combining forms for tongue are **gloss/o** and **lingu/o.** The **lingual surface** of the cheek is the side adjacent to the tongue.

The combining forms **dent/o, dent/i,** and **odont/o** are used to refer to teeth. **Dentition** (dehn-tihsh-uhn) refers to the teeth as a whole, that is, the teeth arranged in the maxillary (upper) and mandibular (lower) arcades. The primary dentition or **deciduous** (deh-sihd-yoo-uhs) **dentition** is the temporary set of teeth that erupt in young animals and are replaced at or near maturity. **Decidu/o** is the combining form for shedding. The **permanent dentition** is the set of teeth designed to last the lifetime of an animal. Sometimes a deciduous tooth of brachydontic animals is not shed at the appropriate time, and both the deciduous and permanent teeth are situated beside each other. The period in which both deciduous and permanent teeth are present is the mixed dentition. The deciduous tooth that has not been shed is called a **retained deciduous tooth** and may be extracted (removed) professionally.

The four types of teeth have different functions (Figure 6–3):

- **incisor** (ihn-sīz-ōr) = front tooth used for cutting; an incision is a cut; abbreviated I.
- **canine** (kā-nīn) = long, pointed bonelike tooth located between the incisors and premolars; also called **fang** and **cuspid** (*cusp* means having a tapering projection; *cuspid* means having one point); abbreviated C.
- **premolar** (prē-mō-lahr) = cheek tooth found between the canine teeth and molars; also called **bicuspids** because they have two points; abbreviated P.
- **molar** (mō-lahr) = most caudally located permanent cheek tooth used for grinding; *molar* comes from the Latin term *to grind*; abbreviated M.

When the number and type of teeth found in an animal are written, a shorthand method is used. This method is called the dental formula. The **dental formula** of an animal represents the type of tooth and the number of each tooth type found in that species. In the dental formula, only one side of the mouth is represented and is preceded by a 2 to indicate that this arrangement is the same on both sides. For example, the dental formula for the adult dog is 2 (I 3/3, C 1/1, P 4/4, M 2/3). This means that an adult dog has 3 upper incisors, 1 upper canine, 4 upper premolars, and 2 upper molars on the left and right sides of its mouth. Similarly, an adult dog has 3 lower incisors, 1 lower canine, 4 lower premolars, and 3 lower molars on the left and right sides of its mouth. Adding all of the numbers together gives the total dentition, which in the adult dog is 42 (Figure 6–4 and Table 6–1).

Teeth also can be identified using the triadan (trī-ə-dahn) system. In this numbering system, each tooth has a three-digit number. The first digit (the quadrant number) represents the quadrant in which the tooth is located. The quadrant numbers start on the animal's upper right side (right side of the maxilla) and continue clockwise to the lower right side (right side of the mandible). The second and third digits represent individual teeth. The numbering of individual teeth starts at rostral midline and increases in number from rostral to caudal. For example, all central incisor teeth in dogs and cats are numbered 01; therefore,

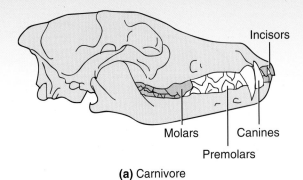

(a) Carnivore

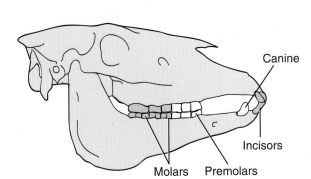

(b) Herbivore

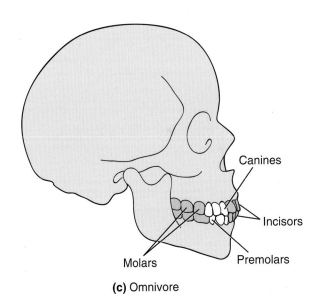

(c) Omnivore

Figure 6–3 Comparison of skull and teeth. (a) Carnivores (e.g., cats) have pointed canines and incisors for obtaining and tearing flesh. Sharp molars and premolars are essential for shearing flesh. (b) Herbivores (e.g., horses) have teeth adapted to biting off plant material and grinding the food into smaller pieces. (c) Omnivores (e.g., humans) have teeth adapted for eating a variety of foods.

the central incisor on the right side of the maxilla is 101, the central incisor on the left side of the maxilla is 201, the central incisor on the left side of the mandible is 301, and the central incisor on the right side of the mandible is 401. Likewise, all

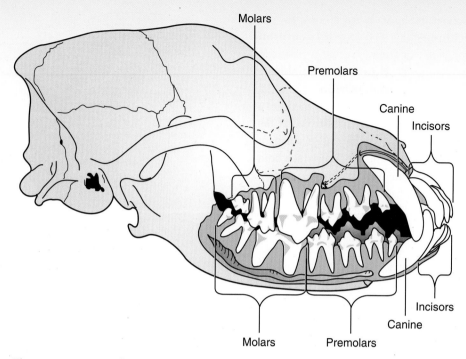

Figure 6-4   Types of teeth found in an adult dog.

Table 6-1   Dental Formulas of Select Adult Animals

Animal	Formula	Total Number of Adult Teeth
Dog	2(I 3/3, C 1/1, P 4/4, M 2/3)	42
Cat	2(I 3/3, C 1/1, P 3/2, M 1/1)	30
Bovine, sheep, goat	2(I 0/4, C 0/0, P 3/3, M 3/3) or 2(I 0/3, C 0/1, P 3/3, M 3/3)	32
Horse	2(I 3/3, C 1/1, P 3-4/3, M 3/3)	40 or 42
Pig	2(I 3/3, C 1/1, P 4/4, M 3/3)	44
Llama (female)	2(I 1/3, C 1/1, P 2/1, M 3/2)	28
(male)	2(I 1/3, C 1/1, P 2/1, M 3/3)	30
Rabbit	2(I 2/1, C 0/0, P 3/2, M 3/3)	28
Ferret	2(I 3/3, C 2/2, P 4/3, M 1/2)	40
Chinchilla	2(I 1/1, C 0/0, P 1/1, M 3/3)	20
Gerbil	2(I 1/1, C 0/0, P 0/0, M 3/3)	16
Hamster	2(I 1/1, C 0/0, P 0/0, M 3/3)	16
Rat and mouse	2(I 1/1, C 0/0, P 0/0, M 3/3)	16
Guinea pig	2(I 1/1, C 0/0, P 1/1, M 3/3)	20

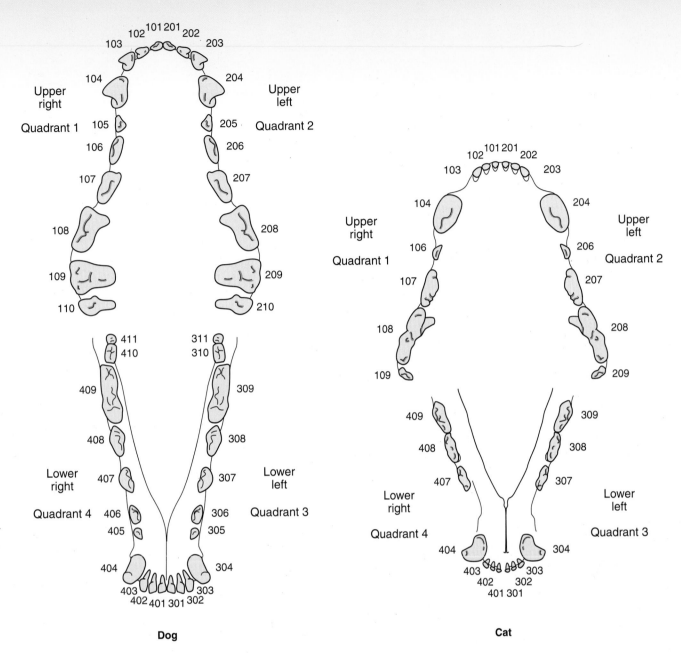

Figure 6–5 Triadan system of tooth identification.

canine teeth in dogs and cats are numbered 04; therefore, the canine tooth on the right side of the maxilla is 104, the canine tooth on the left side of the maxilla is 204, the canine tooth on the left side of the mandible is 304, and the canine tooth on the right side of the mandible is 404 (Figure 6–5).

The anatomy of the tooth consists of the enamel (located in the crown) or cementum (located in the root), dentin, and pulp (Figure 6–6). The **enamel** (ē-nahm-ahl) is the hard white substance covering the dentin of the crown of the tooth. **Cementum** (sē-mehn-tuhm) is the bonelike connective tissue that covers the root of the tooth. The **dentin** (dehn-tihn) is the connective tissue surrounding the tooth pulp. The tooth **pulp** consists of nerves, blood vessels, and loose connective tissue. The hole at the tip of the root where nerves and blood vessels enter the tooth is the **apical foramen** (ā-pih-kahl fō-rā-mehn). The **periodontia** (pehr-ē-ō-dohn-shē-ah) are the structures that support the teeth. Teeth are situated in sockets or saclike dilations called **alveoli** (ahl-vē-ō-lī). A single socket is called an alveolus (ahl-vē-ō-luhs). The thin layer of compact bone that forms the tooth socket is **alveolar bone** (ahl-vē-ō-lahr bōn). Alveolar bone surrounds the roots of the teeth. The fibrous structure that holds the tooth in the alveolus is the **periodontal** (pehr-ē-ō-dohn-tahl) **ligament.** The periodontal ligament contains collagen fibers that are anchored to the cementum of the tooth and to the alveolar bone.

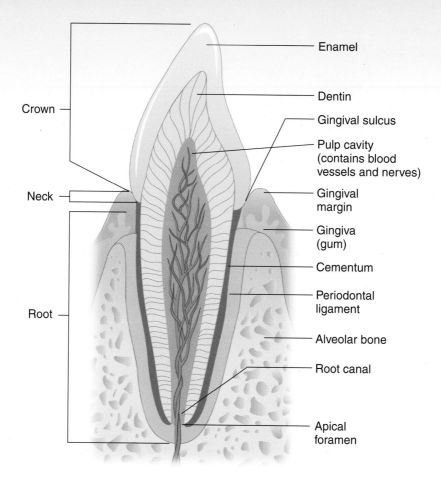

Crown

Neck

Root

Enamel

Dentin

Gingival sulcus

Pulp cavity
(contains blood
vessels and nerves)

Gingival
margin

Gingiva
(gum)

Cementum

Periodontal
ligament

Alveolar bone

Root canal

Apical
foramen

Figure 6–6 Structure of the tooth in a dog or cat.

Occasionally, teeth are referred to by terms other than those seen in dental formulas. Examples of lay terms for teeth include the following:

- **cheek teeth** premolars and molars
- **needle teeth** deciduous canines and third incisor of pigs
- **wolf teeth** rudimentary premolar 1 in horses
- **milk teeth** first set of teeth
- **tusks** permanent canine teeth of pigs
- **carnassial** (kahr-nā-zē-ahl) **tooth** large, shearing cheek tooth; upper P4 and lower M1 in dogs; upper P3 and lower M1 in cats
- **fighting teeth** set of six teeth in llamas that include upper vestigial incisors and upper and lower canines on each side

The **gingiva** (jihn-jih-vah) is the mucous membrane that surrounds the teeth and forms the mouth lining. The gingiva is sometimes called the gums. The combining form **gingiv/o** means gums. The space that surrounds the tooth is the **gingival sulcus** (jihn-jih-vahl suhl-kuhs). The gingival sulcus is located between the tooth and gingival margin.

Salivary glands (sahl-ih-vahr-ē glahndz) are a group of cells located in the oral cavity that secrete a clear substance containing digestive enzymes (saliva). The **saliva** (sah-lī-vah) moistens food, begins the digestive process by aiding in bolus formation and some digestive enzyme activity (amylase in some animal species), and cleanses the mouth. Different salivary glands are named for the location in which they are found: the **mandibular** (mahn-dihb-yoo-lahr) **salivary glands** are found near the lower jaw, the **sublingual** (suhb-lihn-gwahl) **salivary glands** are found under the tongue, the **zygomatic** (zī-gō-mah-tihck) **salivary glands** are found medial to the zygomatic arch, and the **parotid** (pah-roht-ihd) **salivary glands** are found near the ear. **Para-** is the prefix for near, and the combining form **ot/o** means ear. The combining forms for salivary glands are **sialaden/o** and **sial/o** (which also means saliva) (Figure 6–7).

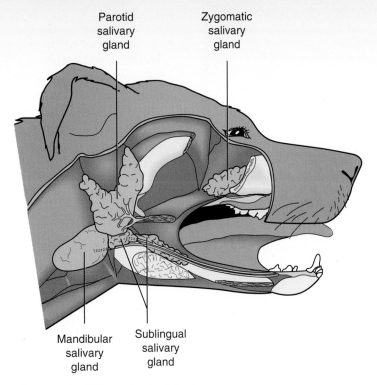

Parotid salivary gland

Zygomatic salivary gland

Mandibular salivary gland

Sublingual salivary gland

Figure 6–7   Salivary glands of the dog.

Animals may be grouped based on the types of teeth they have. Examples include the following:

- **selenodont** (sē-lēn-ō-dohnt) = animals with teeth that have crescents on their grinding surfaces (e.g., ruminants)
- **lophodont** (lō-fō-dohnt) = animals with teeth that have ridged occlusal surfaces (e.g., equine)
- **bunodont** (boon-ō-dohnt) = animals with teeth that have worn, rounded surfaces (e.g., swine)
- **hypsodont** (hihps-ō-dohnt) = animals with continuously erupting teeth (e.g., cheek teeth of ruminants)
- **pleurodont** (pluhr-ō-dohnt) = animals with teeth attached by one side on the inner jaw surface (e.g., lizards)
- **brachydont** (brā-kē-dohnt) = animals with permanently rooted teeth (e.g., carnivores)

Branches of dentistry

Veterinary dental specialties consist of the following:

- **endodontics** (ehn-dō-dohn-tihks) = branch of dentistry that involves treatment of diseases that affect the tooth pulp
- **exodontics** (ehcks-ō-dohn-tihks) = branch of dentistry that involves extraction of teeth and related procedures
- **oral surgery** = branch of surgery that involves surgical correction of the jaw, gums, and inside of the mouth
- **orthodontics** (ohrth-ō-dohn-tihks) = branch of dentistry that involves the guidance and correction of malocclusion
- **periodontics** (pehr-ē-ō-dohn-tihks) = branch of dentistry that studies and treats the diseases of tooth-supporting structures

Hard to swallow

Chewing, also called **mastication** (mahs-tih-kā-shuhn), makes food easier to swallow and can increase the surface area of food particles or ingesta. **Ingesta** (ihn-jehst-ah) is the material taken in orally. This increased surface area increases the contact between digestive enzymes and the food and may speed up the breakdown of food. The first digestive enzymes with which food comes in contact are found in saliva. Sometimes animals salivate or drool when they smell food. **Hypersalivation** (hī-pər-sahl-ih-vā-shuhn) is excessive production of saliva. Hypersalivation also is called **ptyalism** (tī-uh-lihz-uhm) and **hypersialosis** (hī-pər-sī-ahl-ō-sihs). The combining forms for saliva are **sial/o** and **ptyal/o.**

The process of swallowing is called **deglutition** (dē-gloo-tih-shuhn). The combining form **phag/o** means eating or ingestion. Swallowed food passes from the mouth to the pharynx and then to the esophagus.

Pharynx

The **pharynx** (făr-ihncks) is the cavity in the caudal oral cavity that joins the respiratory and gastrointestinal systems. The pharynx also is called the **throat.** The combining form for pharynx is **pharyng/o.** The pharynx is covered in Chapter 9.

Gullet

The **esophagus** (ē-sohf-ah-guhs) is a collapsible, muscular tube that leads from the oral cavity to the stomach. The esophagus is located dorsal to the trachea. The combining form **esophag/o** means esophagus. The esophagus enters the stomach through an opening that is surrounded by a **sphincter.** A sphincter (sfihngk-tər) is a ringlike muscle that constricts an opening.

Stomach

After the esophagus, the remaining organs of digestion are located in the abdominal cavity (Figure 6–8). The **abdomen** is the cavity located between the diaphragm and pelvis. The combining forms for abdomen are **abdomin/o** and **celi/o. Lapar/o** is the combining form for abdomen and flank. The **peritoneum** (pehr-ih-tō-nē-uhm) is the membrane lining that covers the abdominal and pelvic cavities and some of the organs in that area. The layer of the peritoneum that lines the abdominal and pelvic cavities is called the **parietal peritoneum,** and the layer of the peritoneum that covers the abdominal organs is called the **visceral peritoneum.**

Food enters the stomach from the esophagus, where it is stored, and the act of digestion begins. The stomach is connected to other visceral organs by a fold of peritoneum called the **lesser omentum** (ō-mehn-tuhm) and to the dorsal abdominal wall by another fold of peritoneum called the **greater omentum** (Figure 6–9). *Omentum* is Latin for apron.

The combining form for stomach is **gastr/o.** Animals can be classified as **monogastric** (mohn-ō-gahs-trihck) or **ruminant** (roo-mihn-ehnt). Monogastric animals have one true, or glandular, stomach. The **glandular stomach** is the one that produces secretions for digestion. Ruminants also have one true, or glandular, stomach (the abomasum), but they also have three forestomachs (the rumen, reticulum, and omasum). The forestomachs of ruminants are actually outpouchings of the esophagus. Because all animals have only one true or glandular stomach, monogastric animals often are called simple nonruminant animals. The parts of the true stomach are as follows (Figure 6–10):

- **cardia** (kahr-dē-ah) = entrance area located nearest the esophagus.
- **fundus** (fuhn-duhs) = base of an organ, which is the cranial, rounded part.
- **body** (boh-dē) = main portion of an organ, which is the rounded base or bottom; also called the **corpus** (kōr-puhs).
- **antrum** (ahn-truhm) = caudal part, which is the constricted part of the stomach that joins the pylorus.
- **pylorus** (pī-lōr-uhs) = narrow passage between the stomach and the duodenum. The combining form **pylor/o** means gatekeeper and refers to the narrow passage between the stomach and duodenum. The **pyloric sphincter** is the muscle ring that controls the flow of material from the stomach to the duodenum of the small intestine.
- **rugae** (roo-gā) = folds present in the mucosa of the stomach. Rugae contain glands that produce gastric juices that aid in digestion, and the mucus forms a protective coating for the stomach lining.

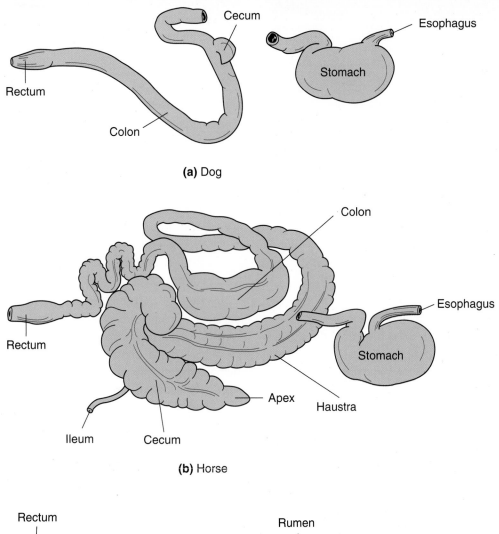

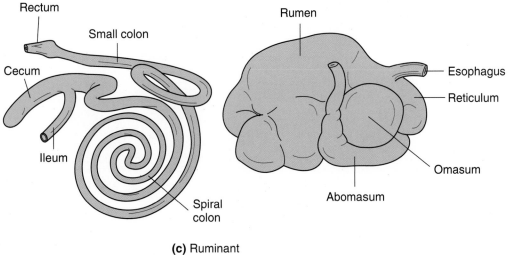

Figure 6–8   Gastrointestinal tracts. (Small intestinal segments were omitted for clarity.) (a) Dog. (b) Horse. (c) Ruminant.

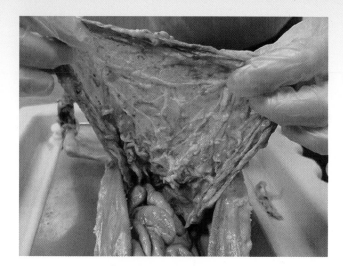

Figure 6–9 Omentum in a preserved cat.

Ruminants

Ruminants are animals that can **regurgitate** (rē-guhr-jih-tāt) and **remasticate** (rē-mahs-tih-kāt) their food. The ruminant stomach is adapted for fermentation of ingested food by bacterial and protozoan microorganisms. Normal microorganisms residing in the gastrointestinal tract are called **intestinal flora** (ihn-tehs-tih-nahl flō-rah). These microbes produce enzymes that can digest plant cells through fermentation. Fermentation is aided by regurgitation (return of undigested material from the rumen to the mouth) and remastication (rechewing). Regurgitation and remastication provide finely chopped material with a greater

surface area to the stomach. Regurgitated food particles, fiber, rumen fluid, and rumen microorganisms are called **cud** (kuhd).

The ruminant stomach is divided into the following four parts (Figures 6–11 and 6–12):

- **rumen** (roo-mehn) = largest compartment of the ruminant stomach that serves as a fermentation vat; also called the **paunch.** The rumen is divided into a ventral sac and a dorsal sac.
- **reticulum** (re-tihck-yoo-luhm) = most cranial compartment of the ruminant stomach; also called the honeycomb because it is lined with a mucous membrane that contains numerous intersecting ridges.
- **omasum** (ō-mā-suhm) = third compartment of the ruminant stomach. The omasum has short, blunt papillae that grind food before it enters the abomasum. Omasal contractions also squeeze fluid out of the food bolus.
- **abomasum** (ahb-ō-mā-suhm) = fourth compartment of the ruminant stomach. Also called the true stomach. The abomasum is the glandular portion that secretes digestive enzymes.

The layout of the ruminant stomach depends on the age of the animal. In adult ruminants, the rumen is the largest compartment and occupies a prominent portion of the left side of the animal. For the most part, the abomasum is on the right side of the animal. In young ruminants, the abomasum is the largest compartment. Forestomach development is associated with roughage intake, and calves are fed only milk for a period of time after birth.

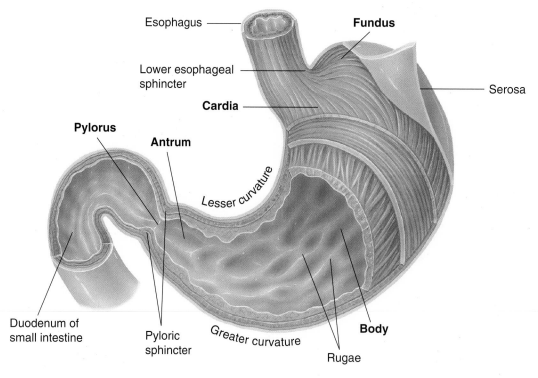

Figure 6–10 Structures of the stomach.

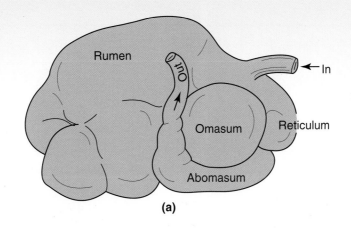

(a)

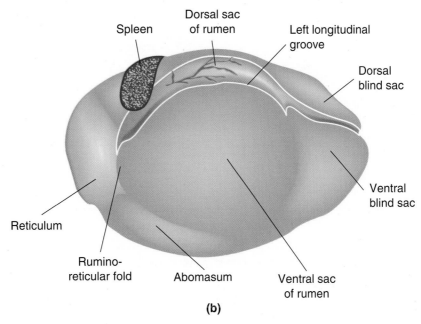

(b)

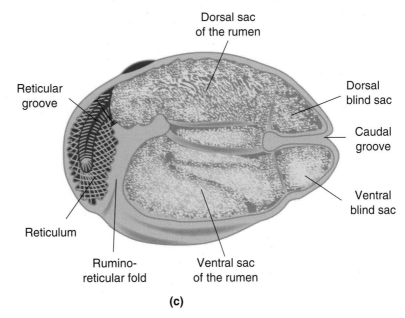

(c)

Figure 6–11   (a) Parts of the ruminant stomach. (b) Left side of ruminant stomach. (c) Interior view of the rumen and reticulum.

Figure 6–12 Linings of the four stomachs of cattle. (a) Rumen. Note the small fingerlike projections, the rumen papillae, that increase the surface area for absorption. (b) Reticulum. The distinct lining of the reticulum looks like a honeycomb. (c) Omasum. The lining has large sheets with ingesta between layers. (d) Abomasum. This smooth glandular lining appears very similar to the monogastric stomach.

In adult ruminants, food enters the rumen. A contraction transfers the rumen contents into the reticulum. The foodstuff is then regurgitated or directed toward the caudal part of the rumen or the omasum. The plies of the omasum grind the food, and water is removed. Food enters the abomasum, or true stomach, which is similar to the glandular stomach of the monogastric. In young ruminants, a reticular groove shuttles milk from the esophagus to the abomasum.

Small Intestine

The small intestine, or small bowel, extends from the pylorus to the proximal part of the large intestine. The small intestine is attached to the dorsal abdominal wall by a fold of the peritoneum called the **mesentery** (mehs-ehn-tehr-ē) (Figure 6–13). Digestion of food and absorption takes place in the small intestine for animals not needing extensive fermentation of their ingested

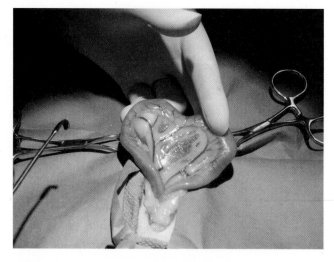

Figure 6–13 The mesentery is a fold of peritoneum that attaches the small intestine to the dorsal abdominal wall.

Rumination (roo-mehn-ā-shuhn) is the process of bringing up food material from the stomach to the mouth for further chewing. Rumination is a cycle of four phases: regurgitation, remastication, resalivation, and redeglutition.

Step 1: Regurgitation. The animal takes a deep breath (glottis is closed); the thoracic cavity enlarges; intrapleural pressure decreases; the cardia opens; and because of the low pressure in the esophagus, the rumen content is aspirated into the esophagus. Reverse peristalsis occurs, and the food bolus quickly enters the mouth.

Steps 2 and 3: Remastication and resalivation. Liquid is squeezed out of bolus, and the liquid is reswallowed. Remastication and resalivation occur together, and the animal may chew its cud 100 or more times before swallowing.

Step 4: Redeglutition. The bolus is reswallowed, and the next rumination cycle begins. Rumination usually occurs when the animal is quiet. The time spent ruminating each day varies with species and diet. Smaller particles (such as ground feed) take less rumination time.

A **herbivore** (hərb-ih-vōr) is an animal that is able to sustain life by eating only plants.

An **omnivore** (ohm-nih-vōr) is an animal that sustains life by eating plant and animal tissue.

A **carnivore** (kahr-nih-vōr) is an animal that is able to sustain life by eating only animal tissue. A carnivore may eat a plant, but that does not make it an omnivore. A cat is a carnivore, but cats may occasionally eat grass. This does not make them omnivores.

food. The combining form **enter/o** means small intestine. **Gastroenterology** (gahs-trō-ehn-tər-ohl-ō-jē) is the study of the stomach and small intestine.

The small intestine has the following three segments (Figure 6–14):

- **duodenum** (doo-ō-də-nuhm or doo-wahd-nuhm) = proximal or first portion of the small intestine. The proximal portion also is known as the most oral portion. The duodenum is the segment of the small intestine located nearest the mouth. The combining forms **duoden/i** and **duoden/o** mean duodenum.
- **jejunum** (jeh-joo-nuhm) = middle portion of the small intestine. The combining form **jejun/o** means jejunum.
- **ileum** (ihl-ē-uhm) = distal or last portion of the small intestine. The distal portion also is known as the most **aboral** (ahb-ōr-ahl) portion. The ileum is the segment of the small intestine located furthest from the mouth. The combining form **ile/o** means ileum. (Remember that *ileum* is spelled with an *e*, as in *eating* or *entero*. The ilium is part of the hip bone.)

Once food is digested in the small intestine, it forms a milky fluid. This milky fluid is called **chyle** (kī-uhl). Chyle is absorbed through the intestinal wall and travels via the thoracic duct, where it is passed into veins.

Large Intestine

The large intestine, or large bowel, extends from the ileum to the anus. The large intestine consists of the **cecum** (sē-kuhm), **colon** (kō-lihn), **rectum** (rehck-tuhm), and **anus** (ā-nuhs). Development of the large intestine varies between species. Fermentation occurs to some extent in the large intestine of all animals but is a more consuming process in herbivorous animals. In ruminants, the forestomach constitutes a fermentation vat (hence, they are called foregut fermenters) and in nonruminant herbivores (e.g., rabbits and horses), the cecum and colon provide fermentation (hence, they are called hindgut fermenters).

Food enters a pouch, called the cecum, from the ileum. The cecum may be poorly developed, as in the dog, or large, as in the horse, or double-pouched, as in birds. **Cec/o** is the combining form for cecum.

The colon continues from the cecum to its termination at the rectum. The colon consists of ascending, transverse, and descending portions. All animals have a transverse and descending portion. The arrangement from the cecum to the transverse colon varies between species. Dogs and cats have an ascending colon, pigs and ruminants have spiral colons, and horses have large colons. **Col/o** is the combining form for colon. The cecum and colon of pigs and horses are sacculated; these sacculations are called **haustra** (haw-strah) (Figure 6–15). Haustra act as buckets and prolong retention of material so that the microbes have more time for digestion. Haustra are formed because of the longitudinal smooth muscle bands, or **teniae** (tehn-ē-ā), in the cecal wall.

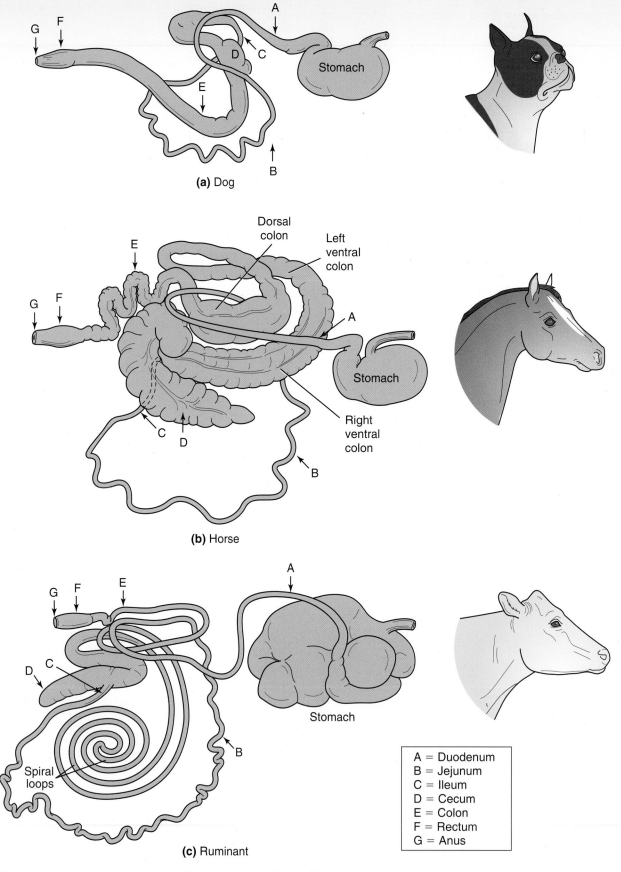

(a) Dog

(b) Horse

Dorsal colon

Left ventral colon

Right ventral colon

Stomach

(c) Ruminant

Spiral loops

Stomach

A = Duodenum
B = Jejunum
C = Ileum
D = Cecum
E = Colon
F = Rectum
G = Anus

Figure 6–14 Gastrointestinal tracts of various animals. (a) Dog. (b) Horse. (c) Ruminant.

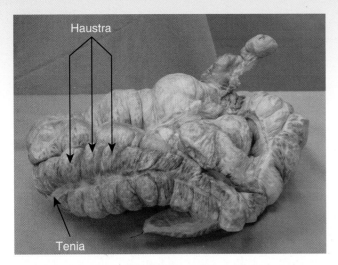

Figure 6–15 Haustra and teniae of the equine large intestine.

The colon

The colon is divided into three parts: ascending, transverse, and descending. **Ascending** is the part that progresses upward, or cranially. *Ascend* means to move up. **Transverse** is the part that travels across. **Trans-** is the prefix for across. **Descending** is the part that progresses downward, or caudally. To descend a flight of stairs is to move down. Bends or curves are called **flexures** (flehck-shərz); therefore, the **pelvic flexure** is a bend in the colon near the pelvis, and the **diaphragmatic flexure** is a bend in the colon near the diaphragm.

The rectum is the caudal portion of the large intestine. The combining form **rect/o** means rectum. The anus is the caudal opening of the gastrointestinal tract. The anus is controlled by two anal sphincter muscles that tighten or relax to allow or control defecation. The combining form for anus is **an/o.** **Anorectal** (ā-nō-rehck-tahl) is a term that means pertaining to the anus and rectum. The combining form **proct/o** refers to the anus and rectum collectively.

The anal canal is a short terminal portion of the digestive tract. Dogs and cats have a pair of pouches in the skin between the internal and external anal sphincters. These pouches are called **anal sacs.** Anal sacs are lined with microscopic anal glands that secrete a foul-smelling fluid. Normally, the anal sacs

are compressed during defecation. The fluid may be related to social recognition in dogs and cats.

Accessories

Accessory organs aid the digestive tract in different ways. The accessory organs of the digestive tract include the salivary glands (covered previously), the liver, the gallbladder, and the pancreas (Figure 6–16).

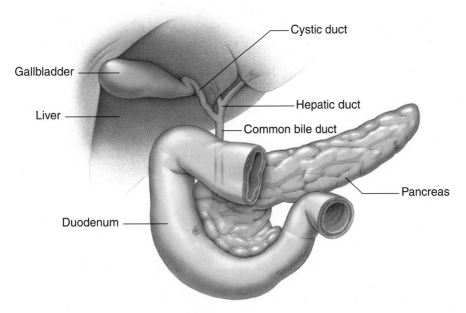

Figure 6–16 Accessory organs of digestion.

Liver

The liver is located caudal to the diaphragm and has several important functions. The combining form for liver is **hepat/o.** The liver removes excess **glucose** (gloo-kohs) from the bloodstream and stores it as **glycogen** (glī-koh-jehn). When blood sugar is low, a condition called **hypoglycemia** (hī-pō-glī-sē-mē-ah) results, and the liver converts glycogen back into glucose and releases it. The liver also destroys old erythrocytes; removes toxins from the blood; produces some blood proteins; and stores iron and vitamins A, B$_{12}$, and D. Liver cells are called **hepatocytes** (heh-paht-ō-sītz). The liver also contains **sinusoids** (sīn-yoo-soydz), or channels. The hepatocytes make up the liver **parenchyma** (pahr-ehnk-ih-mah), or the functional elements of a tissue or organ.

The digestive function of the liver involves the production of **bile** (bīl). The term **biliary** (bihl-ē-ār-ē) means pertaining to bile. Bile travels down the hepatic duct to the cystic duct, which leads to the gallbladder (in species that have a gallbladder). Bile alkalinizes the small intestine, and bile salts play a part in fat digestion. Fat digestion is called **emulsification** (ē-muhl-sih-fih-kā-shuhn). **Bilirubin** (bihl-ē-roo-bihn) is a pigment produced from the destruction of hemoglobin that is released by the liver in bile.

Gallbladder

The gallbladder is a sac embedded in the liver that stores bile for later use. When bile is needed, the gallbladder contracts, which forces bile out of the cystic duct into the common bile duct. The common bile duct carries bile into the duodenum. The rat and horse do not have gallbladders but have a continuous flow of bile into the duodenum. The combining form **cyst/o** means cyst, sac of fluid, or urinary bladder. The combining form **chol/e** means bile or gall. The combining form **doch/o** means receptacle. Together these combining forms are used to refer to the bile and its associated structures. **Cholecystic** (kō-lē-sihst-ihck) means pertaining to the gallbladder. **Choledochus** (kō-lehd-uh-kuhs) means common bile duct.

Pancreas

The **pancreas** (pahn-krē-ahs) is an elongated gland located near the cranial portion of the duodenum. The main pancreatic duct enters the duodenum close to the common bile duct. (In species such as sheep and goats, it empties directly into the common bile duct.)

Pancreat/o is the combining form for pancreas. The pancreas is an organ that has both exocrine and endocrine functions. The endocrine functions are covered in Chapter 11. The exocrine function of the pancreas involves the production of pancreatic juices, which are filled with digestive enzymes. **Trypsin** (trihp-sihn) is an enzyme that digests protein, **lipase** (lī-pās) is an enzyme that digests fat, and **amylase** (ahm-ih-lās) is an enzyme that digests starch. All are produced in the pancreas.

DIGESTION

Digestion (dī-jehst-shuhn) is the process of breaking down foods into nutrients that the body can use. **Enzymes** (ehn-zīmz) are substances that chemically change another substance. Digestive enzymes are responsible for the chemical changes that break foods into smaller forms for the body to use. Enzymes typically end with **-ase.** For example, proteases are enzymes that work on proteins, and lipases are enzymes that work on fats. One enzyme that does not end in *-ase* is pepsin, which is an enzyme that digests protein.

Metabolism (meh-tahb-ō-lihzm) is the processes involved in the body's use of nutrients. The prefix **meta-** means change or beyond. **Anabolism** (ah-nahb-ō-lihzm) is the building of body cells and substances. **Catabolism** (kah-tahb-ō-lihzm) is the breaking down of body cells and substances.

Absorption (ahb-sōrp-shuhn) is the process of taking digested nutrients into the circulatory system. A **nutrient** (nū-trē-ehnt) is a substance that is necessary for normal functioning of the body. Absorption occurs mainly in the small intestine. The small intestine contains tiny hairlike projections called **villi** (vihl-ī). A single projection is called a **villus** (vihl-uhs).

The scoop on poop

The following terms are used for gastrointestinal waste:

- feces
- dung
- manure
- stool
- bowel movement
- excrement
- poop

Terms used to describe gastrointestinal waste include the following:

- solid
- loose
- soft
- watery
- mucoid
- bloody

The combining form **vill/i** means tuft of hair. Because the small intestine has villi, or projections, it also has blind sacs, or valleys. These blind sacs are called **crypts** (krihptz). An intestinal crypt is a valley of the intestinal mucous membrane lining the small intestine.

Path of Digestion

- **Prehension** (prē-hehn-shuhn), or grasping of food, involves collecting food in the oral cavity. The material taken orally is known as **ingesta** (ihn-jehst-ah).
- **Mastication** breaks food into smaller pieces and mixes the ingesta with saliva.
- **Deglutition** moves chewed ingesta into the pharynx and into the esophagus. (The epiglottis closes off the entrance to the trachea and allows food to move into the esophagus.)
- **Peristalsis** (pehr-ih-stahl-sihs) is the series of wavelike contractions of smooth muscles (Figure 6–17). (The suffix **-stalsis** means contraction.) Food moves down the esophagus by gravity and peristalsis.

Ingesta moves into the stomach. In ruminants, it enters the rumen, reticulum, and omasum before entering the true stomach, or abomasum. The true glandular stomach contains **hydrochloric** (hī-drō-klōr-ihck) **acid** and the enzymes **protease** (prō-tē-ās), **pepsin** (pehp-sihn), and **lipase** (lī-pās). The muscular action of the stomach mixes the ingesta with the gastric juices to convert the food to chyme. **Chyme** (kīm) is the semifluid mass of partly digested food that passes from the stomach.

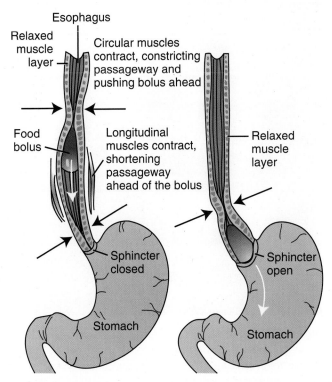

Figure 6–17 Peristalsis actively propels ingesta through the intestinal tract.

Esophagus

Relaxed muscle layer

Circular muscles contract, constricting passageway and pushing bolus ahead

Food bolus

Longitudinal muscles contract, shortening passageway ahead of the bolus

Relaxed muscle layer

Sphincter closed

Sphincter open

Stomach

Stomach

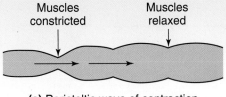

Muscles constricted

Muscles relaxed

(a) Peristaltic wave of contraction

(b) Segmentation

Figure 6–18 Peristalsis versus segmentation. (a) Peristalsis propels food through the digestive system. (b) Segmentation helps break down and mix food through cement-mixer-type action.

Chyme passes from the stomach into the duodenum. Food moves through the small intestine by peristaltic action and **segmentation** (sehg-mehn-tā-shuhn). Peristalsis consists of contractile waves that propel ingesta caudally; segmentation mixes and thus delays movement of ingesta (Figure 6–18). Digestion is completed in the duodenum after chyme has mixed with bile and pancreatic secretions.

Digested food is absorbed in the small intestine. Another term for absorption is **assimilation** (ah-sih-mih-lā-shuhn).

The large intestine receives waste products of digestion and in some species is responsible for fermentation of plant material and vitamin synthesis. Excess water is absorbed from the waste, and solid feces are formed. **Defecation** (dehf-eh-kā-shuhn) is the emptying of the bowels.

TEST ME: DIGESTIVE SYSTEM

Diagnostic tests performed on the digestive system include the following:

- **ballottement** (bahl-oht-mehnt) = diagnostic technique of hitting or tapping the wall of a fluid-filled structure to bounce a solid structure against a wall; used for pregnancy diagnosis and determination of abdominal contents.
- **barium** (bār-ē-uhm) = contrast material used for radiographic studies. To evaluate the gastrointestinal tract, barium sulfate may be given orally (and the resulting test is called a barium swallow or upper GI or barium sulfate may be given rectally (and the resulting test is called a barium enema or lower GI). An **enema** (ehn-ah-mah) is introduction of fluid into the rectum (Figure 6–19).
- **biopsy** (bī-ohp-sē) = removal of tissue to examine. Biopsies can be **incisional** (ihn-sih-shuhn-ahl), in which part of the tissue is removed and examined, or they can be **excisional** (ehcks-sih-shuhn-ahl), in which the entire tissue is removed.

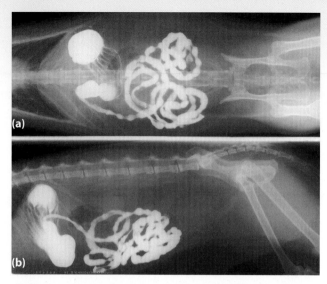

Figure 6–19 Radiograph of a cat during a barium series. The bright white areas are barium in the gastrointestinal tract. (a) Ventrodorsal view. (b) Lateral view.

- **blood tests** = determination of blood parameters used to detect some diseases of the gastrointestinal tract. For example, bile acids are used to assess liver disease, and elevated amylase levels may indicate pancreatitis. Sometimes it is important to know whether the blood sample was taken as a preprandial or postprandial sample. **Preprandial** (prē-prahn-dē-ahl) is before eating, and **postprandial** (pōst-prahn-dē-ahl) is after eating.
- **colonoscopy** (kō-luhn-ohs-kō-pē) = endoscopic visual examination of the inner surface of the colon; the scope is passed from the rectum through the colon. An **endoscope** (ehn-dō-skōp) is a tubelike instrument with lights and refracting mirrors that is used to examine the body or organs internally (Figure 6–20).
- **esophagoscopy** (ē-sohf-ah-gohs-kō-pē) = endoscopic visual examination of the esophagus; the scope is passed from the oral cavity through the esophagus.
- **fecal examinations** = various procedures used to detect parasitic diseases of animals (Figures 6–21 and 6–22). Specialized fecal tests also can identify bacteria, isolate viruses, or demonstrate abnormal substances present in the stool.
- **gastroscopy** (gahs-trohs-kō-pē) = endoscopic visual examination of the inner surface of the stomach; the scope is passed from the oral cavity through the stomach.
- **hemoccult** (hēm-ō-kuhlt) = test for hidden blood in the stool. **Occult** means hidden.
- **radiography** (rā-dē-ohg-rah-fē) = imaging of internal structures is created by the exposure of sensitized film to X-rays. Radiographs of the gastrointestinal system demonstrate foreign bodies, torsions, organ distention or enlargement, and some masses.
- **ultrasound** (uhl-trah-sownd) = imaging of internal body structures by recording echoes of sound waves.

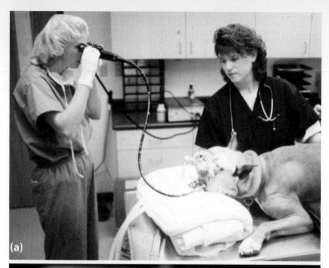

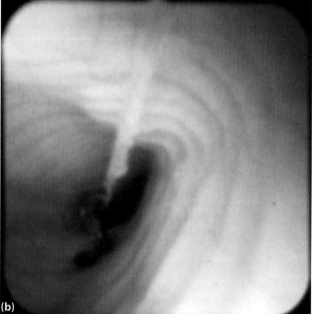

Figure 6–20 Endoscopy. (a) An endoscope is passed through the esophagus (esophagoscopy) to view parts of the gastrointestinal tract. (b) View of the intestine from an endoscope. A string foreign body is identified. (c) View of the intestine during surgery to remove the string foreign body. [(a) Courtesy of Lodi Veterinary Hospital, SC; (b) Mark Jackson, DVM, PhD; (c) Eli Larson, DVM.]

PATHOLOGY: DIGESTIVE SYSTEM

Pathologic conditions of the digestive system include the following:

- **achalasia** (ahk-ah-lahz-ah) = inability to relax the smooth muscle of the gastrointestinal tract.
- **adontia** (ā-dohnt-shah) = absence of teeth.
- **aerophagia** (âr-ō-fā-jē-ah) = swallowing of air.
- **anal sacculitis** (ā-nahl sahck-yoo-lī-tihs) = inflammation of the pouches located around the anus. The term **inspissation** (ihn-spihs-sā-shuhn) is the process of rendering dry or thick by evaporation and is used to describe the anal sac fluid in animals with anal sacculitis.
- **anorexia** (ahn-ō-rehck-sē-ah) = lack or loss of appetite.
- **ascariasis** (ahs-kah-rī-ah-sihs) = parasitic infestation with roundworms of the genus *Ascaris* (Figure 6–23).

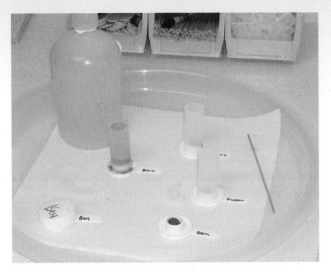

Figure 6–21 A fecal flotation is a basic test that checks for intestinal parasites (usually by identification of the parasite eggs, or oocysts). Flotation methods are based on the use of flotation fluid with a specific gravity suitable to float the eggs, or oocysts.

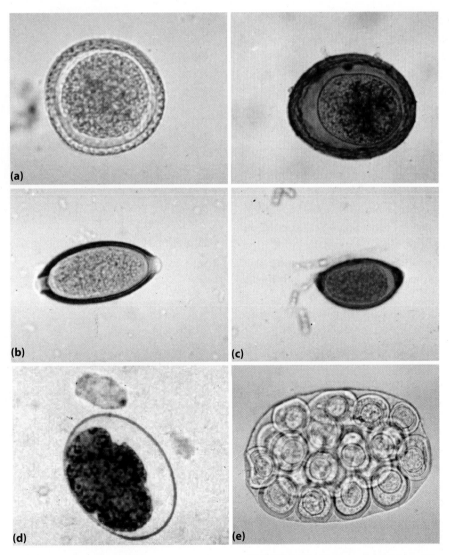

Figure 6–22 Parasite eggs commonly found on fecal flotation. (a) Roundworm eggs (*Toxocara canis*, left; *Toxocaris leonina*, right). (b) Whipworm, left (*Trichuris vulpis*). (c) Lungworm, right (*Capillara aerophila*). (d) Hookworm (*Ancylostoma caninum*). (e) Tapeworm (*Dipylidium caninum*).

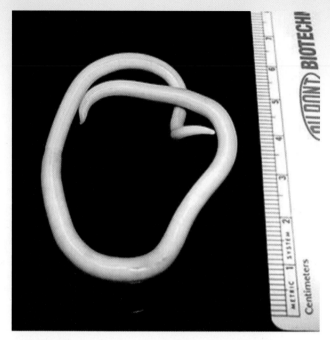

Figure 6–23 Ascariasis is a parasitic infestation with roundworms. (Courtesy of North Dakota Department of Education.)

- **ascites** (ah-sī-tēz) = abnormal accumulation of fluid in the abdomen.
- **atresia** (ah-trēz-ah) = occlusion or absence of normal body opening or tubular organ.
- **bloat** (blōt) = accumulation of gas in the digestive tract. In monogastric animals, bloat is accumulation of gas in the stomach. In ruminants, bloat is accumulation of gas in the rumen, abomasum, or cecum. In ruminants, gas accumulation in the rumen is also called **ruminal tympany** (tihm-pahn-ē).
- **borborygmus** (bohr-bō-rihg-muhs) = gas movement in the gastrointestinal tract that produces a rumbling noise.
- **bruxism** (bruhck-sihzm) = involuntary grinding of the teeth.
- **cachexia** (kah-kehcks-ē-ah) = general ill health and malnutrition; used in describing the condition of cancer patients.
- **cholecystitis** (kō-lē-sihs-tī-tihs) = inflammation of the gallbladder.
- **cirrhosis** (sihr-rō-sihs) = degenerative disease that disturbs the structure and function of the liver.
- **colic** (kohl-ihck) = severe abdominal pain.
- **colitis** (kō-lī-tihs) = inflammation of the colon.
- **constipation** (kohn-stah-pā-shuhn) = condition of prolonged gastrointestinal transit time, making the stool hard, dry, and difficult to pass.
- **coprophagia** (kō-prō-fā-jē-ah or kohp-rohf-ah-jē-ah) = ingestion of fecal material. The combining form **copr/o** means feces.

- **cribbing** (krihb-ihng) = vice of equine in which an object is grasped between the teeth, pressure is applied, and air is inhaled. See Figure 19–10.
- **dehydration** (dē-hī-drā-shuhn) = condition of excessive loss of body water or fluid.
- **dental calculus** (dehn-tahl kahl-kyoo-luhs) = abnormal mineralized deposit that forms on teeth (Figure 6–24). Calculus is mineral deposit. Dental calculus also is called **tartar** (tahr-tahr).
- **dental caries** (dehn-tahl kār-ēz) = decay and decalcification of teeth, producing a hole in the tooth.
- **diarrhea** (dī-ah-rē-ah) = abnormal frequency and liquidity of fecal material (Figure 6–25).
- **displaced abomasum** (dihs-plāsd ahb-ō-mā-suhm) = disease of ruminants in which the fourth stomach compartment becomes trapped under the rumen; also called DA. It is denoted LDA (left displaced abomasum)

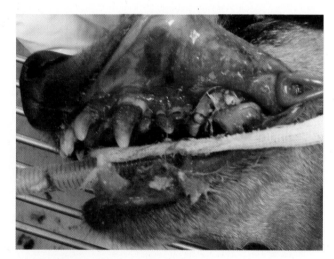

Figure 6–24 Dental calculus and gingivitis in a dog. (Courtesy of Eli Larson, DVM.)

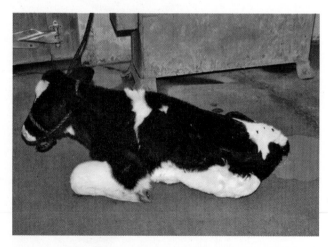

Figure 6–25 Bovine with diarrhea caused by *Samonella* bacteria. (Courtesy of Dr. Ramos-Vara, College of Veterinary Medicine. Michigan State University.)

or RDA (right displaced abomasum) depending on its location. LDAs are more common.

- **diverticulitis** (dī-vər-tihck-yoo-lī-tihs) = inflammation of a pouch or pouches occurring in the wall of a tubular organ. A **diverticulum** (dī-vər-tihck-yoo-luhm) is a pouch occurring on the wall of a tubular organ; diverticula are pouches occurring on the wall of a tubular organ.
- **dyschezia** (dihs-kē-zē-ah) = difficulty defecating. *Chezein* is Greek for stool. The prefix **dys-** means difficult.
- **dysentery** (dihs-ehn-tər-ē) = number of disorders marked by inflammation of the intestine, abdominal pain, and diarrhea.
- **dysphagia** (dihs-fā-jē-ah) = difficulty swallowing or eating. The suffix **-phagia** means eating or swallowing.
- **emaciation** (ē-mā-sē-ā-shuhn) = marked wasting or excessive leanness (Figure 6–26).
- **emesis** (ehm-eh-sihs) = forcible expulsion of stomach contents through the mouth; also known as vomiting. The material vomited is known as **vomitus** (voh-mih-tus). When an animal is vomiting, the recommendation is not to give it anything orally. The term for orally is **per os,** which is abbreviated **PO.** If the desire is to give nothing orally, the abbreviation is **NPO.**
- **enteritis** (ehn-tər-ī-tihs) = inflammation of the small intestine.
- **enterocolitis** (ehn-tehr-ō-kō-lī-tihs) = inflammation of the small intestine and large intestine.
- **epulis** (ehp-uhl-uhs) = benign tumor arising from periodontal mucous membranes.
- **eructation** (ē-ruhck-tā-shuhn) = belching or raising gas orally from the stomach.
- **esophageal reflux** (ē-sohf-ah-jē-ahl rē-fluhcks) = return of stomach contents into the esophagus; also called gastroesophageal reflux disease, or GERD.

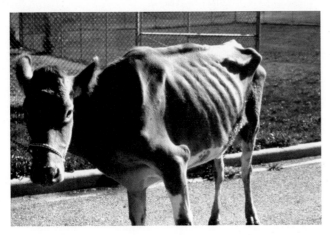

Figure 6–26 Emaciation in a Jersey cow with Johne's disease. (Courtesy of Michael T. Collins, DVM, PhD, University of Wisconsin School of Veterinary Medicine.)

- **eviscerate** (ē-vihs-ər-āt) = remove or expose internal organs. Evisceration is used to describe the exposure of internal organs after unsuccessful surgical closure of the abdomen (or another area containing organs).
- **exocrine pancreatic insufficiency** (ehck-sō-krihn pahn-krē-ah-tihck ihn-suh-fih-shehn-sē) = metabolic disease in which the pancreas does not secrete adequate amounts of digestive enzymes and is associated with weight loss, fatty stools, and borborygmus; abbreviated EPI.
- **fecalith** (fēck-ah-lihth) = stonelike fecal mass. The suffix **-lithiasis** (lih-thī-ah-sihs) means presence of stones. **Coprolith** (kō-prō-lihth) is another name for a fecalith.
- **flatulence** (flaht-yoo-lehns) = excessive gas formation in the gastrointestinal tract.
- **gastric dilatation** (gahs-trihck dihl-ah-tā-shuhn) = condition usually seen in deep-chested canines in which the stomach fills with air and expands. Dilatation is stretching beyond normal.
- **gastric dilatation volvulus** (gahs-trihck dihl-ah-tā-shuhn vohl-vū-luhs) = condition usually seen in deep-chested canines in which the stomach fills with air, expands, and twists on itself; also called GDV (Figures 6–27 and 6–28).
- **gastritis** (gahs-trī-tihs) = inflammation of the stomach.
- **gastroenteritis** (gahs-trō-ehn-tehr-ī-tihs) = inflammation of the stomach and small intestine. Note that anatomically the stomach occurs first, followed by the small intestine. This order is reflected in the order of the medical terms as well.
- **gingival hyperplasia** (jihn-jih-vahl hī-pər-plā-zē-ah) = overgrowth of the gingiva characterized by firm, nonpainful swellings associated with the gingiva.
- **gingivitis** (jihn-jih-vī-tihs) = inflammation of the gums (Figure 6–24).
- **glossitis** (glohs-ī-tihs) = inflammation of the tongue.
- **hematemesis** (hēm-ah-tehm-eh-sihs) = vomiting blood.
- **hematochezia** (hēm-aht-ō-kē-zē-ah) = passage of bloody stool.
- **hemoperitoneum** (hēm-ō-pehr-ih-tō-nē-uhm) = blood in the peritoneum.
- **hepatitis** (hehp-ah-tī-tihs) = inflammation of the liver.
- **hepatoma** (heh-pah-tō-mah) = tumor of the liver.
- **hepatomegaly** (hehp-ah-tō-mehg-ah-lē) = abnormal enlargement of the liver.
- **hiatal hernia** (hī-ā-tahl hər-nē-ah) = protrusion of part of the stomach through the esophageal opening in the diaphragm.
- **hydrops** (hī-drohps) = abnormal accumulation of fluid in tissues or a body cavity; also called **dropsy** (drohp-sē).
- **hyperglycemia** (hī-pər-glī-sē-mē-ah) = elevated blood sugar levels.
- **hypoglycemia** (hī-pō-glī-sē-mē-ah) = lower-than-normal blood sugar levels.

Gastric Dilatation Volvulus

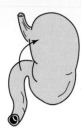
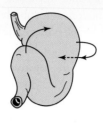

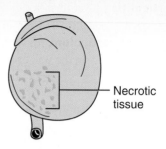

Pyloric antrum
is displaced
aborally

Pylorus crosses
midline, passes
underneath
distended oral
part of stomach

Fundus
moves ventrally
and becomes
located in
ventral abdomen

Gastric dilatation
displaces greater
curvature ventrally

— Necrotic
tissue

Figure 6–27 Gastric dilatation volvulus formation. *Gastric dilatation volvulus* is a term commonly used to describe a condition in deep-chested dogs in which the stomach fills with air, expands, and twists on itself. Because the stomach is "attached" at only two points, it should be more accurately described as a torsion.

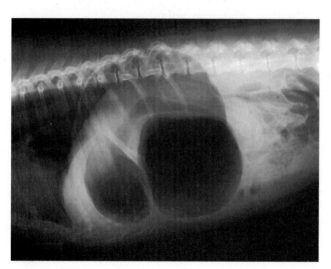

Figure 6–28 Radiograph of a dog with GDV. Note the large gas-distended stomach.

- **ileitis** (ihl-ē-ī-tihs) = inflammation of the ileum.
- **ileus** (ihl-ē-uhs) = stoppage of intestinal peristalsis.
- **impaction** (ihm-pahck-shuhn) = obstruction of an area, usually when feed is too dry.
- **inappetence** (ihn-ahp-eh-tehns) = lack of desire to eat.
- **incontinence** (ihn-kohn-tihn-ehns) = inability to control. A descriptive term usually is applied in front of it; for example, fecal incontinence is the inability to control bowel movements.
- **inguinal hernia** (ihng-gwih-nahl hər-nē-ah) = protrusion of bowel through the inguinal canal; protrusion is seen in the groin.
- **intussusception** (ihn-tuhs-suhs-sehp-shuhn) = telescoping of one part of the intestine into an adjacent part (Figure 6–29).

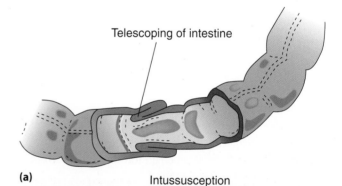

Telescoping of intestine

(a)   Intussusception

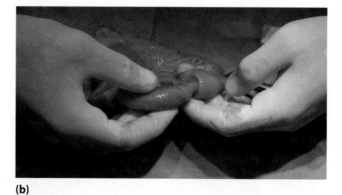

(b)

Figure 6–29 Intussusception is the telescoping of the bowel onto itself.

- **jaundice** (jawn-dihs) = yellow discoloration of the skin and mucous membranes caused by elevated bilirubin levels, also called **icterus** (ihck-tər-uhs) (Figure 6–30).
- **lethargy** (lehth-ahr-jē) = condition of drowsiness or indifference.
- **malabsorption** (mahl-ahb-sōrp-shuhn) = impaired uptake of nutrients from the intestine.

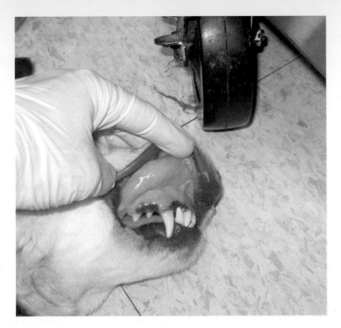

Figure 6–30 Jaundice is the yellow discoloration of the skin and mucous membranes. (Courtesy of Kim Kruse Sprecher, CVT.)

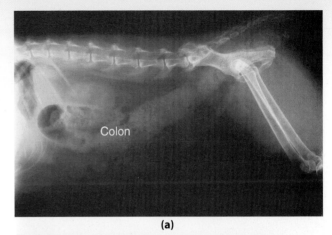

(a)

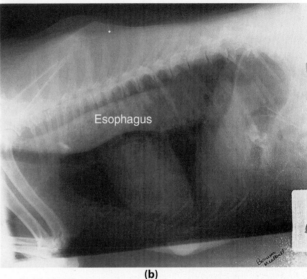

(b)

Figure 6–31 Contrast radiography. (a) Megacolon is diagnosed in the dog using contrast radiography. (b) Megaesophagus is diagnosed in the dog using contrast radiography. (Courtesy of Lodi Veterinary Hospital, SC.)

- **malocclusion** (mahl-ō-kloo-shuhn) = abnormal contact between the teeth. The prefix **mal-** means bad, and **occlusion** means any contact between the chewing surfaces of the teeth.
- **megacolon** (mehg-ah-kō-lihn) = abnormally large colon (Figure 6–31a).
- **megaesophagus** (mehg-ah-ē-sohf-ah-guhs) = abnormally large esophagus (Figure 6–31b).
- **melena** (meh-lē-nah) = black stools containing digested blood. Melena suggests a bleeding problem in the upper gastrointestinal tract.
- **nausea** (naw-sē-ah) = stomach upset or sensation of urge to vomit; difficult to use descriptively in animals.
- **obstruction** (ohb-struhck-shuhn) = complete stoppage or impairment to passage. Obstructions usually are preceded by a term that describes its location, as in **intestinal obstruction** (Figure 6–32). When the obstruction is not complete, it is called a **partial obstruction.**
- **oronasal fistula** (ohr-ō-nā-zahl fihs-tyoo-lah) = abnormal opening between the nasal cavity and the oral cavity. Oronasal fistulas may be congenital, traumatic, or associated with dental disease.
- **palatoschisis** (pahl-ah-tohs-kih-sihs) = congenital fissure of the roof of the mouth that may involve the upper lip, hard palate, and soft palate; more commonly called a **cleft palate.**
- **perforating ulcer** (pər-fohr-āt-ihng uhl-sihr) = erosion through the entire thickness of a surface.
- **periapical abscess** (pehr-ē-ā-pih-cahl ahb-sehsz) = inflammation of tissues and collection of pus surrounding the apical portion of a tooth root due to pulpal disease.

- **periodontitis** (pehr-ē-ō-dohn-tī-tihs) = inflammation of the tissue surrounding and supporting the teeth; also called **periodontal disease. Inflammation** is a localized protective response elicited by injury or destruction of tissue. The signs of inflammation are heat, redness, pain, swelling, and loss of function.
- **pica** (pī-kah) = eating and licking abnormal substances or a depraved appetite.
- **plaque** (plahck) = small, differentiated area on a body surface. In the gastrointestinal system, it is used to refer to the mixed colony of bacteria, leukocytes, and salivary products that adhere to the tooth enamel; also called dental plaque.
- **polydipsia** (pohl-ē-dihp-sē-ah) = excessive thirst or drinking.
- **polyp** (pohl-uhp) = small growth on a mucous membrane.
- **polyphagia** (pohl-ē-fā-jē-ah) = excessive eating or swallowing. The prefix **poly-** means many or much.

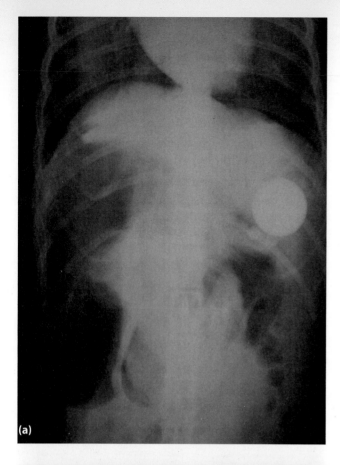

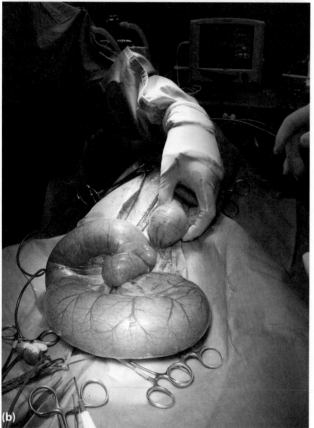

Figure 6–32 (a) Radiograph of a dog with an intestinal obstruction (golf ball). (b) Surgical view of the dog with an intestinal obstruction. (Courtesy of Eli Larson, DVM.)

- **prolapse** (prō-lahpz) = protrusion of viscera. A descriptive term usually precedes the term *prolapse*. For example, a rectal prolapse is protrusion of the rectum through the anus.
- **quidding** (kwihd-ihng) = condition in which food is taken into the mouth and chewed but falls from the mouth.
- **regurgitation** (rē-gərj-ih-tā-shuhn) = return of swallowed food into the oral cavity; a passive event compared with the force involved with vomiting.
- **salivary mucocele** (sahl-ih-vahr-ē myoo-kō-sēl) = collection of saliva that has leaked from a damaged salivary gland or duct and is surrounded by granulation tissue (Figure 6–33).
- **scours** (skowrz) = diarrhea in livestock.
- **shunt** (shuhnt) = to bypass or divert. In a **portosystemic** (poor-tō-sihs-tehm-ihck) **shunt,** blood vessels bypass the liver and the blood is not detoxified properly.
- **stenosis** (steh-nō-sihs) = narrowing of an opening. The term *stenosis* usually is used with a descriptive term in front of it. For example, a pyloric stenosis is narrowing of the pylorus as it leads into the duodenum.
- **stomatitis** (stō-mah-tī-tihs) = inflammation of the mouth.
- **tenesmus** (teh-nehz-muhs) = painful, ineffective defecation. Tenesmus also means painful, ineffective urination but is rarely used in this context.
- **torsion** (tōr-shuhn) = axial twist; twist around the long axis of gut.
- **trichobezoar** (trī-kō-bē-zōr) = hairball. The combining form **trich/o** means hair.
- **ulcer** (uhl-sihr) = erosion of tissue.
- **volvulus** (vohl-vū-luhs) = twisting on itself (end-to-end twist); twist around long axis of mesentery (Figure 6–34).

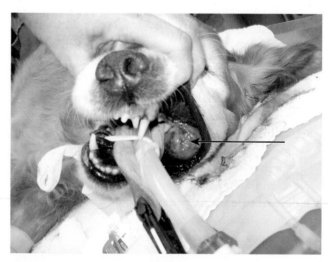

Figure 6–33 Salivary mucocele in a dog. (Courtesy of Kimberly Kruse Sprecher, CVT.)

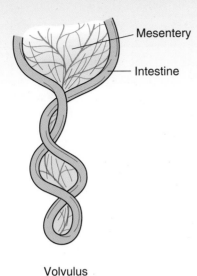

Volvulus

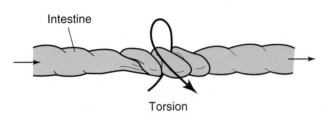

Torsion

Figure 6–34 Different types of twists. Volvulus is a twist around the long axis of the mesentery; torsion is a twist around the long axis of the gut.

PROCEDURES: DIGESTIVE SYSTEM

Procedures performed on the digestive system include the following:

- **abdominocentesis** (ahb-dohm-ihn-ō-sehn-tē-sihs) = surgical puncture to remove fluid from the abdomen.
- **abomasopexy** (ahb-ō-mahs-ō-pehcks-ē) = surgical fixation of the abomasum of ruminants to the abdominal wall (Figure 6–35). The suffix **-pexy** means to surgically fix something to a body surface.
- **anastomosis** (ah-nahs-tō-mō-sihs) = surgical connection between two tubular or hollow structures.
- **anoplasty** (ah-nō-plahs-tē) = surgical repair of the anus.
- **antidiarrheal** (ahn-tih-dī-ə-rē-ahl) = substance that prevents frequent and extremely liquid stool.
- **bolus** (bō-luhs) = rounded mass of food or large pharmaceutical preparation or to give something rapidly.
- **cholecystectomy** (kō-lē-sihs-ehck-tō-mē) = surgical removal of the gallbladder.
- **colectomy** (kō-lehck-tō-mē) = surgical removal of the colon.
- **colostomy** (kō-lahs-tō-mē) = surgical production of an artificial opening between the colon and the body surface.

- **colotomy** (kō-loht-ō-mē) = surgical incision into the colon.
- **crown** (krown) = restoration of teeth using materials that are cemented into place; used to cap or completely cover a tooth; also called a **cap.**
- **drench** (drehnch) = to give medication in liquid form by mouth and forcing the animal to drink.
- **emetic** (ē-meh-tihck) = producing vomiting. An **antiemetic** (ahn-tih-ē-meh-tihck) prevents vomiting.
- **enterostomy** (ehn-tər-ohs-tō-mē) = surgical production of an artificial opening between the small intestine and the abdominal wall.
- **esophagoplasty** (ē-sohf-ah-gō-plahs-tē) = surgical repair of the esophagus.
- **extraction** (ehcks-trahck-shuhn) = removal; used to describe surgical removal of a tooth.

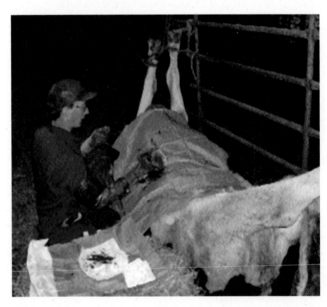

Figure 6–35 A displaced abomasum is repaired surgically with a procedure called an abomasopexy

Figure 6–36 A rumen fistula is an artificial opening created between the rumen and the body surface. (Courtesy of Laura Lien, CVT, BS.)

- **fistula** (fihs-tyoo-lah) = abnormal passage from an internal organ to the body surface or between two internal organs. A ruminant that has an artificial opening created between the rumen and the body surface has a rumen fistula (Figure 6–36). This also is called a **rumenostomy** (roo-mehn-ah-stō-mē). A **perianal fistula** (pehr-ih-ā-nahl fihsh-too-lah) is an abnormal passage around the caudal opening of the gastrointestinal tract. **Perianal** means around the anus.
- **float** (flōt) = instrument used to file or rasp an equine's premolar or molar teeth (Figure 6–37); also used to describe the procedure of filing equine teeth.
- **gastrectomy** (gahs-trehck-tō-mē) = surgical removal of all or part of the stomach. To clarify the extent of the excision, the term **partial gastrectomy** is used to denote surgical removal of part of the stomach.
- **gastroduodenostomy** (gahs-trō-doo-ō-deh-nohs-tōm-ē) = removal of part of the stomach and duodenum and making a connection between them.
- **gastropexy** (gahs-trō-pehcks-ē) = surgical fixation of the stomach to the abdominal wall.
- **gastrostomy** (gahs-trohs-tō-mē) = surgical production of an artificial opening between the stomach and abdominal wall. The suffix **-stomy** means surgical production of an opening between an organ and a body surface. The opening created during this procedure is a **stoma** (stō-mah). **Effluent** (ehf-floo-ehnt) means discharge and an effluent flow from the stoma created by a -stomy surgery (e.g., gastrostomy).
- **gastrotomy** (gahs-troht-ō-mē) = surgical incision into the stomach (Figure 6–38).
- **gavage** (gah-vahzh) = forced feeding or irrigation through a tube passed into the stomach.
- **gingivectomy** (jihn-jih-vehck-tō-mē) = surgical removal of the gum tissue.
- **hepatotomy** (hehp-ah-toht-ō-mē) = surgical incision into the liver.
- **ileectomy** (ihl-ē-ehck-tō-mē) = surgical removal of the ileum.
- **ileostomy** (ihl-ē-ohs-tō-mē) = surgical production of an artificial opening between the ileum and abdominal wall.

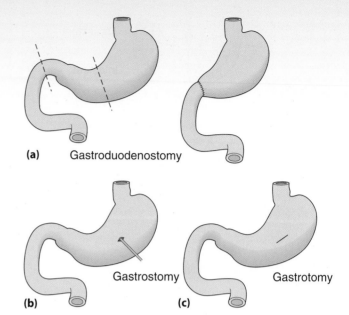

(a)   Gastroduodenostomy

(b)   Gastrostomy   **(c)**   Gastrotomy

Figure 6–38   (a) Gastroduodenostomy, (b) Gastrostomy, and (c) Gastrotomy.

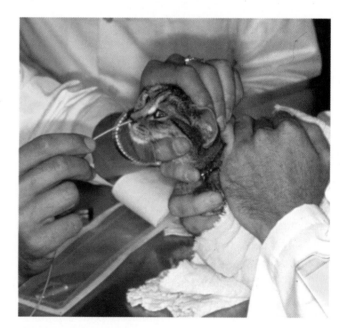

Figure 6–39   Nasogastric tube placement in a cat. (Courtesy of Mark Jackson, DVM, PhD.)

- **laparotomy** (lahp-ah-roht-ō-mē) = surgical incision into the abdomen; **lapar/o** is the combining form for abdomen or flank.
- **nasogastric intubation** (nā-zō-gahs-trihck ihn-toob-ā-shuhn) = placement of a tube through the nose into the stomach (Figure 6–39).
- **orogastric intubation** (ōr-ō-gahs-trihck ihn-too-bā-shuhn) = passage of a tube from the mouth to the stomach; also called a stomach tube. *Orogastric* means pertaining to the mouth and stomach.
- **palatoplasty** (pahl-ah-tō-plahs-tē) = surgical repair of a cleft palate.

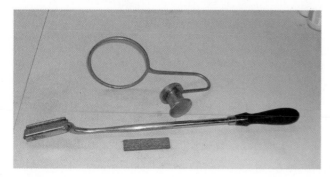

Figure 6–37   A tooth float used to file the teeth of horses. A gag is used to hold the mouth open for the procedure.

- **trocarization** (trō-kahr-ih-zā-shuhn) = insertion of a pointed instrument (trocar) into a body cavity or an organ. The trocar usually is inside a cannula so that once the trocar penetrates the membrane, it can be withdrawn and the cannula remains in place. Trocar-ization usually is performed for acute cases of bloat to relieve pressure. When trocarization is performed for treatment of ruminal bloat, it may be called ruminal **paracentesis** (pahr-ah-sehn-tē-sihs).

REVIEW EXERCISES

Multiple Choice

Choose the correct answer.

1. Mixing of ingesta in the intestine is called
   a. propulsion
   b. peristalsis
   c. segmentation
   d. separation

2. Abnormal accumulation of fluid in the abdomen is called
   a. ascites
   b. effusion
   c. icterus
   d. bloat

3. Telescoping of one part of the intestine into an adjacent part is called
   a. volvulus
   b. diverticulum
   c. parenchyma
   d. intussusception

4. The small intestine is attached to the dorsal abdominal wall by the
   a. peritoneum
   b. emesis
   c. mesentery
   d. omentum

5. Eating and licking of abnormal substances is called
   a. coprophagy
   b. pica
   c. dysphagia
   d. polyphagia

6. Inflammation of the mouth is
   a. stomatitis
   b. orititis
   c. dentitis
   d. osititis

7. Straining, painful defecation is called
   a. strangstolia
   b. colostrangia
   c. tenesmus
   d. epulis

8. A tumor of the liver is a
   a. hematoma
   b. hemoma
   c. hepatoma
   d. hemotoma

9. Marked wasting or excessive leanness is
   a. evaluation
   b. elimination
   c. emesis
   d. emaciation

10. Forced feeding or irrigation through a tube passed into the stomach is called
    a. gavage
    b. drench
    c. bolus
    d. cachexia

11. The combining form for the first part of the large intestine is
    a. ile/o
    b. cec/o
    c. duoden/o
    d. jejun/o

12. The muscular, wavelike movement used to transport food through the digestive system is
    a. mastication
    b. peristalsis
    c. anastomosis
    d. regurgitation

13. The part of the tooth that contains a rich supply of nerves and blood vessels is the

    a. enamel
    b. dentin
    c. pulp
    d. cementum

14. *Buccal* means

    a. pertaining to the cheek
    b. pertaining to the tongue
    c. pertaining to the throat
    d. pertaining to the palate

15. *Stomat/o* means

    a. mouth
    b. cheek
    c. stomach
    d. intestine

16. The term for erosion of tissue is

    a. melena
    b. plaque
    c. shunt
    d. ulcer

17. The narrow passage between the stomach and the duodenum is the

    a. ileum
    b. pylorus
    c. peritoneum
    d. mesentery

18. Incontinence means

    a. prolonged transit time
    b. dribbling urine
    c. loose stool
    d. inability to control

19. A drug used to prevent vomiting is known as an

    a. emetic
    b. atresic
    c. antiemetic
    d. antiemaciation

20. The formation of a new opening from the large intestine to the surface of the body is known as a(n)

    a. enterostomy
    b. colostomy
    c. colectomy
    d. enterotomy

Matching

Match the term in Column I with the definition in Column II.

Column I

1. _____ chyme
2. _____ bile
3. _____ glycogen
4. _____ rugae
5. _____ mastication
6. _____ aerophagia
7. _____ trypsin
8. _____ amylase
9. _____ villus
10. _____ assimilation

Column II

a. absorption

b. tiny hairlike projection on the small intestine

c. semifluid mass of partly digested food

d. substance made by the liver that helps with fat digestion

e. folds present in the mucosa of the stomach

f. stored energy present in the liver

g. swallowing of air

h. breaking food into smaller pieces and mixing with saliva

i. enzyme that digests starch

j. enzyme that digests protein

Fill in the Blanks

1. Gloss/o and lingu/o mean _____ .

2. Or/o and stomat/o mean _____ .

3. Cheil/o and labi/o mean _____ .

4. Abdomin/o, celi/o, and lapar/o mean _____ .

5. Dent/o, dent/i, and odont/o mean _____ .

Spelling

Circle the term that is spelled correctly.

1. twisting of the intestine:	vulvulus	volvulus	volvolus
2. inability to relax the smooth muscles of the digestive tract:	achalsia	achalasia	achaelasia
3. yellow discoloration of the skin and mucous membranes:	jaundise	jawndise	jaundice
4. remove or expose internal organs:	eviserate	eviscerate	evicerate
5. abnormal accumulation of fluid in the abdomen:	ascites	asites	ascetes

CROSSWORD PUZZLES

Disease Terms Puzzle

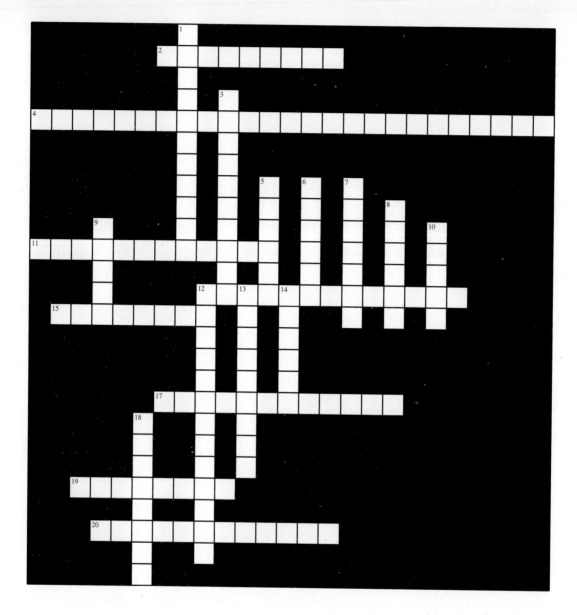

Across

2 abnormally large portion of the large intestine located between the cecum and rectum

4 condition usually seen in deep-chested canines in which the stomach fills with air, expands, and twists on itself

11 inability to control

12 abnormally large tube that connects the oral cavity to the stomach

15 general ill health and malnutrition

17 process of rendering dry or thick by evaporation

19 stonelike fecal mass

20 hairball

Down

1 condition of excessive loss of body water or fluid

3 excessive gas formation in the GI tract

5 black stools containing digested blood

6 diarrhea in livestock

7 abnormal accumulation of fluid in tissues or a body cavity

8 another term for dental calculus

9 severe abdominal pain

10 rounded mass of food, large pharmaceutical preparation, or to give something rapidly

12 impaired uptake of nutrients from the intestine

13 inflammation of the tongue

14 benign tumor arising from periodontal mucous membranes

18 excessive salivation

Digestive Organs Puzzle

Across

1 throat
4 caudal portion of the small intestine
6 caudal portion of the large intestine
8 organ that produces bile
11 one of the combining forms for mouth
14 cheek teeth
15 cranial portion of the small intestine

Down

1 gland that secretes digestive juices as well as hormones
2 combining form for nourishment
3 entero is the combining form for this GI organ
5 collapsible muscular tube that leads from the oral cavity to the stomach
6 most cranial compartment of the ruminant stomach
7 one of the combining forms for tongue
9 organ that stores bile
10 part of large intestine located between the cecum and rectum
12 middle portion of the small intestine
13 combining form for stomach
14 roof of the mouth

WORD SEARCH

Define the following terms; then find each term in the puzzle.

```
R T Y T N N D L T Y U B T Y I M A S N N O O
Y A M A O A C A T E E H Y M Y C I I T T E N
O E O N A N A P O Y S I S O M O T S A N A A
A Y N E M Y O A A T E H M T T E T E T E B S
D T G O I T T R O T E S L S T N A T N N D O
O T N T E N T O Y O A P Y O E O I N O O A G
A O A T O O O T T B T T A L O I O E O D M A
T N N H O N M O O N Y N T O Y S S C G T Y S
E D T O C O O M N N T B I C E B I O P S Y T
O N O E T O A Y Y I O Y O N O C D N R E S R
S O I S O S M O D N M E M E T I C I Y T S I
O T M E O O D I S M T A A O A G N M H N C C
L C S P B R A O O D N A O A N Y H O T T N I
E M E O M R M I N T T A M G M T E D S N T N
M X Y C R O O N T Y S A O R C O O B T T E T
Y S C H T T T Y E E O C A G A P O A O O B U
O Y E I A Y I N E T O S A C A N E M E N T B
T A S O O E E N T H S C Y G I O Y T O A A A
L X N T T M S A L E M N T E T A O N T N G T
H T D M A A N O M C N E X T R A C T I O N I
A M O A N C Y N S G A S T R E C T O M Y N O
A N O P L A S T Y O T N O N A B S I I T M N
```

surgical connection between two tubular or hollow organs

_____

surgical incision into the abdomen

_____

surgical removal of all or part of the stomach

_____

produces vomiting

_____

to remove (teeth)

_____

surgical repair of the anus

_____

surgical production of an artificial opening between the portion of the large intestine between the cecum and rectum and the body surface

_____

surgical fixation in ruminants of the fourth stomach compartment to the abdominal wall

_____

removing tissue to examine
substance that prevents frequent and extremely liquid stool

_____

LABEL THE DIAGRAMS

Label the diagram in Figure 6–40.

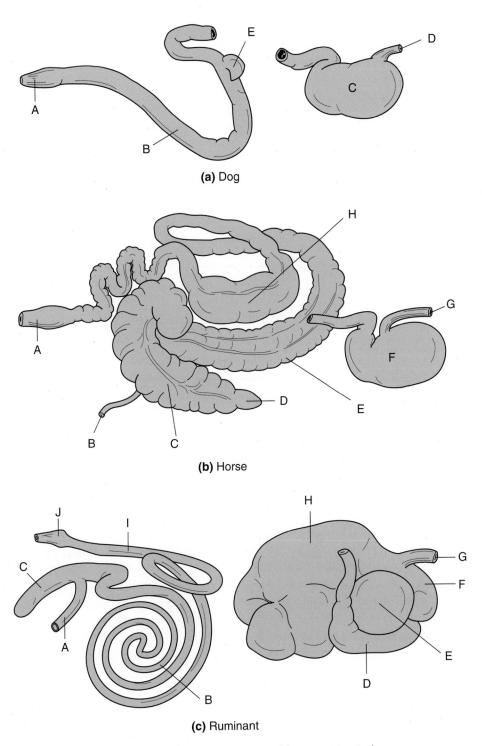

(a) Dog

(b) Horse

(c) Ruminant

Figure 6–40 Gastrointestinal tracts. Label the parts of the gastrointestinal tracts.

CASE STUDIES

Using terms presented in this chapter, the appendices, and a dictionary, define the underlined terms in each case study.

A 3-yr-old F/S New Zealand white rabbit is presented for a few months hx of quidding and ptyalism. The problem has become progressively worse, and although the rabbit has a good appetite, it has difficulty prehending food. On PE, it is noted that there is severe overgrowth of the upper and lower incisors. This problem in rabbits is most often caused by malocclusion of the dental arcades. The premolars and molars are found to be normal. Conservative treatment includes periodic trimming of the overgrown teeth. A more permanent solution involves the extraction of the incisors. Under general anesthesia, all incisors are removed. Analgesics are provided for the first few days postoperatively because rabbits are easily stressed by pain. Once healed, the rabbit returned to its normal eating habits.

1. yr _____

2. F/S _____

3. hx _____

4. quidding _____

5. ptyalism _____

6. prehending _____

7. incisors _____

8. malocclusion _____

9. dental arcades _____

10. premolars _____

11. molars _____

12. extraction _____

13. anesthesia _____

14. analgesics _____

15. postoperatively _____

An 8-wk-old M coonhound is presented to the clinic with an acute history of emesis, hemorrhagic diarrhea, lethargy, and anorexia. The pup was not vaccinated and was healthy until yesterday. Upon PE, it was noted that the pup was pyrexic, dehydrated, and lethargic. Heart and lungs ausculted normally. Stool was collected for parasitic examination, and blood was collected for a CBC and chemistry panel. The stool was negative for parasites, the blood count revealed lymphopenia, and the chemistry panel was normal except for indications of dehydration. A dx of canine parvoviral enteritis was suspected because of the lymphopenia and clinical signs, so virus isolation was performed on a stool sample. Pending virus isolation results, the pup was hospitalized and isolated, IV fluids were administered, and antibiotics were given to prevent a secondary septicemia. Twelve hours after hospitalization the pup expired. A necropsy was done, and the intestines demonstrated loss of intestinal villi and crypt necrosis. The virus isolation test was positive for canine parvovirus infection. The facility was thoroughly disinfected, and the owners were advised to disinfect their facility and vaccinate any future pups.

16. wk _____

17. acute _____

18. emesis _____

19. hemorrhagic _____

20. diarrhea   _____

21. lethargy   _____

22. anorexia   _____

23. pyrexic   _____

24. dehydrated   _____

25. ausculted   _____

26. stool   _____

27. lymphopenia   _____

28. dx   _____

29. canine   _____

30. enteritis   _____

31. IV   _____

32. secondary   _____

33. septicemia   _____

34. expired   _____

35. necropsy   _____

36. intestinal villi _____

37. crypt   _____

38. disinfected   _____

CHAPTER 7

[NULL AND VOID]

Objectives

Upon completion of this chapter, the reader should be able to

- Identify the major organs and tissues of the urinary system
- Describe the major functions of the urinary system
- Recognize, define, spell, and pronounce terms relating to diagnosis, pathology, and treatment of the urinary system

FUNCTIONS OF THE URINARY SYSTEM

The urinary system's main responsibility is the removal of wastes from the body. The **urinary** (yoo-rih-nār-ē) **system** removes wastes from the body by constantly filtering blood. The major waste product of protein metabolism is **urea** (yoo-rē-ah), which is filtered by the kidney and used in some diagnostic tests to determine the health status of the kidney.

In addition to filtering wastes, the urinary system also maintains proper balance of water, electrolytes, and acids in body fluids and removes excess fluids from the body. Maintaining a proper balance of water, electrolytes, and acids allows the body to have a stable internal environment. This stable internal environment is called **homeostasis** (hō-mē-ō-stā-sihs), and it involves continually adjusting to conditions to maintain a relatively constant internal environment.

One part of the urinary system, the kidney, also produces hormones and affects the secretory rate of other hormones.

Urin/o and **ur/o** are combining forms meaning urine or pertaining to the urinary organs.

STRUCTURES OF THE URINARY SYSTEM

The structures of the normal urinary system include a pair of kidneys, a pair of ureters, a single urinary bladder, and a single urethra (Figure 7–1).

> Urine is formed in the kidneys, flows through the ureters to the urinary bladder, is stored in the urinary bladder, and flows through the urethra and outside the body.

Kidney

Kidneys are located **retroperitoneally** (reh-trō-peh-rih-tō-nē-ah-lē), which means that they are located behind the lining of the abdominal cavity or outside the peritoneal cavity. One kidney sits on each side of the vertebral column below the diaphragm. **Ren/o** (Latin for kidney) and **nephr/o** (Greek for kidney) are combining forms for kidney. Ren/o (the Latin form) is used

as an adjective, as in *renal pelvis* and *renal disease*. Nephr/o (the Greek term) tends to be used to describe pathologic conditions and surgical procedures, as in *nephritis* and *nephrectomy*.

Blood flows into each kidney through the renal artery and leaves the kidney via the renal vein. Filtration of waste products by the kidney depends on this blood flow; therefore, blood pressure can affect the rate at which filtration takes place.

Each kidney has two layers that surround the renal pelvis. The outer layer of the kidney is known as the **cortex** (kōr-tehckz). **Cortic/o** means outer region and is used to describe the outer region of many organs. The **medulla** (meh-doo-lah) of the kidney is the inner layer. **Medull/o** means middle or inner portion and is used to describe the middle or inner region of many organs. The cortex contains the majority of the nephron, and the medulla contains most of the collecting tubules.

The **nephron** (nehf-rohn) is the functional unit of the kidney. The nephron consists of the glomerulus, Bowman's capsule, a proximal convoluted tubule, a loop of Henle, a distal convoluted tubule, and a collecting duct. The nephrons form urine by the processes of filtration, reabsorption, and secretion. Filtration occurs in the glomerulus; reabsorption occurs in the proximal convoluted tubule, loop of Henle, and collecting tubule; and secretion occurs in the distal convoluted tubule (Table 7–1).

The **glomerulus** (glō-mər-yoo-luhs) is a cluster of capillaries surrounded by Bowman's capsule. The combining form for

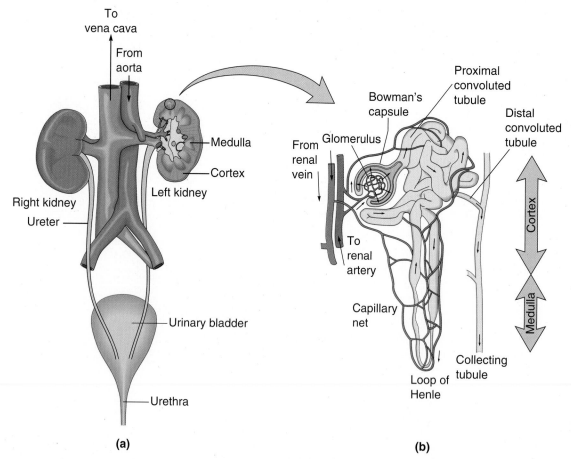

Figure 7–1   (a) Structures of the urinary tract. (b) Structures of a nephron.

Table 7–1 Parts of the Nephron

glomerulus	cluster of capillaries that filter blood
Bowman's capsule	cup-shaped structure that contains the glomerulus
proximal convoluted tubules	hollow tubes located between Bowman's capsule and loops of Henle that are involved in reabsorption
loop of Henle	U-shaped turn in the convoluted tubule of the kidney located between the proximal and distal convoluted tubules that is involved in reabsorption; has ascending and descending loop
distal convoluted tubules	hollow tubes located between the loops of Henle and the collecting tubules that are involved in secretion
collecting tubules	hollow tubes that carry urine from the cortex to the renal pelvis

glomerulus is **glomerul/o,** which means to wind into a ball, which is what the glomerulus looks like microscopically. The plural form of glomerulus is **glomeruli** (glō-mər-yoo-lī).

The **renal pelvis** is the area of the kidney where the nephrons collect before entering the ureters. The combining form for renal pelvis is **pyel/o.** The renal pelvis collects urine from the **calyces** (kā-lah-sēz), which are irregular cuplike spaces that collect urine from the kidney. **Calyx** (kā-lihcks) is the singular form of *calyces.*

Each kidney has a concave depression called the **hilus** (hī-luhs) that serves as the point of attachment of the renal blood vessels, nerves, and ureter (Figures 7–2a and b). The hilus is located on the medial surface of the kidney and gives some kidneys their bean shape. Not all species have kidneys that are bean-shaped. Cattle have lobulated kidneys (and no renal pelvis), and the right kidney of horses is heart-shaped (Figure 7–3).

Descriptive structural terms of the kidney

- **distal** = farthest from midline
- **proximal** = closest to midline
- **convoluted** (kohn-vō-lūt-ehd) = rolled or coiled
- **cortex** = outer
- **medulla** = middle or inner
- **hilus** = point of attachment or depression
- **calyx** or **calix** = cuplike organ
- **ascending** (ā-sehnd-ihng) = moving upward or cranially
- **descending** (dē-sehnd-ihng) = moving downward or caudally

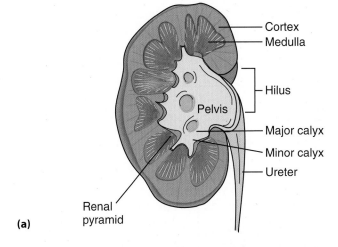

(a)

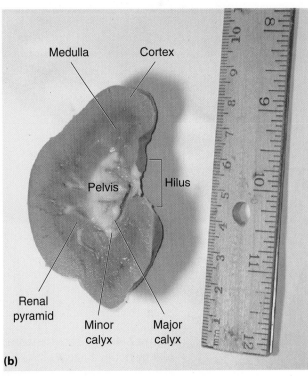

(b)

Figure 7–2   (a) Structures of the kidney. (b) Sectioned sheep kidney.

Figure 7–3   External appearance of a bovine kidney.

Ureters

The **ureters** (yoo-rē-tərz) are a pair of narrow tubes that carry urine from the kidneys to the urinary bladder. The combining form for ureter is **ureter/o.** The ureters enter the urinary bladder at the **trigone** (trīg-ōn). *Trigone* comes from the Greek term meaning triangle. The trigone of the urinary bladder is a triangular portion at the base of that organ where the three angles are marked by the two ureteral openings and one urethral opening.

Urinary Bladder

The **urinary bladder** (yoo-rihn-ār-ē blah-dər) is a singular hollow muscular organ that holds urine. **Cyst/o** is the combining form for urinary bladder. The urinary bladder is very elastic, and its shape and size depend on the amount of urine it is holding. The flow of urine into the urinary bladder enters from the ureters at such an angle that it serves as a natural valve to control backflow. The flow of urine out of the urinary bladder to the urethra is controlled by **sphincters** (sfihngk-tərz). Sphincters are ringlike muscles that close a passageway (Figure 7–4).

Five words, one meaning?

The terms **urination** (yoo-rih-nā-shuhn), **excretion** (ehcks-krē-shuhn), **voiding** (voy-dihng), **elimination** (eh-li-mə-nā-shuhn), and **micturition** (mihck-too-rih-shuhn) have been used to describe the act of excreting urine. Actually, the terms *excretion, elimination,* and *voiding* mean elimination of a substance and can be used in other body systems. The term *urination* means the elimination of urine from the body. The term *micturition* means the elimination of urine from the body; however, it implies voluntary control of the sphincter muscles of the urinary tract. This voluntary control of voiding urine is learned and implies a more intelligent form of animal life.

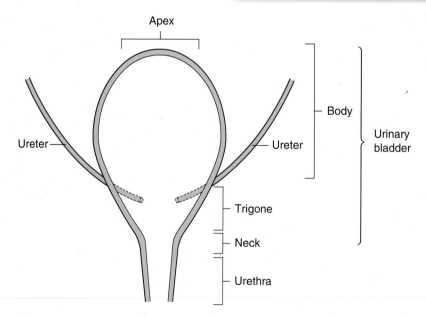

Figure 7–4   Divisions of the urinary bladder. The divisions of the urinary bladder include the apex (cranial free end), the body (central main part), the trigone (triangular portion where ureters enter the urinary bladder), and the neck (constricted portion that joins the urethra).

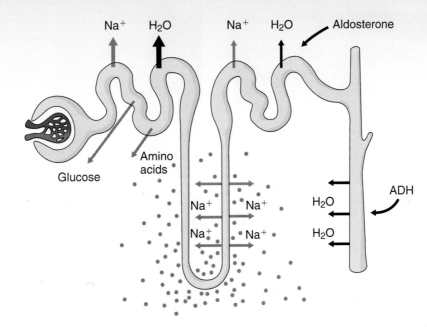

Figure 7–5 The effects of the hormones aldosterone and ADH on the nephron.

Urinary system hormones

Hormones that affect or are produced by the urinary system include the following:

- **erythropoietin** (ē-rihth-rō-poy-ē-tihn) = hormone produced by the kidney that stimulates red blood cell production in the bone marrow. Also pronounced ē-rihth-rō-pō-ih-tihn.
- **antidiuretic** (ahn-tih-dī-yoo-reht-ihck) = hormone released by the posterior pituitary gland that suppresses urine formation by reabsorbing more water; abbreviated as **ADH.**
- **aldosterone** (ahl-dah-stər-own) = hormone secreted by the adrenal cortex that regulates electrolyte balance via the reabsorption of sodium (Figure 7–5).

Urethra

The **urethra** (yoo-rē-thrah) is a tube extending from the urinary bladder to the outside of the body. **Urethr/o** is the combining form for urethra. The external opening of the urethra is the **urethral meatus** (yoo-rē-thrahl mē-ā-tuhs), or urinary meatus. The combining form **meat/o** means opening.

In females, the only function of the urethra is to transport urine from the urinary bladder to the outside of the body. In males, the urethra transports urine from the urinary bladder and reproductive fluids from the reproductive organs out of the body.

URINE

The end product of renal filtration of wastes is urine. The process of urine production is **uropoiesis** (yoo-rō-pō-ē-sihs). The suffix **-poiesis** means formation. Normal urine of most species is clear and pale yellow in color. The suffix for color is **-chrome.** However, some species may have normal urine that is **turbid** (tər-bihd), or cloudy, and may be brown, white, or another color. Sometimes the color of urine depends on the diet (as in rabbits) or the reproductive cycle. Urine color also can reflect hydration status. In dehydrated animals, urine is more concentrated and therefore a deeper shade of yellow. Urine also has a pH, which depends on species and diet. Herbivores tend to have basic urine (a higher pH), whereas carnivores tend to have acidic urine (a lower pH). Through dietary management, urine pH can be manipulated to treat or prevent disease.

TEST ME: URINARY SYSTEM

Diagnostic procedures performed on the urinary system include the following:

- **cystocentesis** (sihs-tō-sehn-tē-sihs) = surgical puncture of the urinary bladder, usually to collect urine. A cystocentesis usually is performed with a needle and syringe (Figure 7–6).

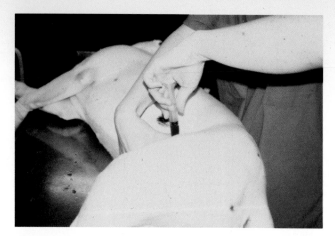

Figure 7–6 Cystocentesis in a dog. (Courtesy of Lodi Veterinary Hospital.)

- **cystography** (sihs-tohg-rah-fē) = radiographic study of the urinary bladder after contrast material has been placed in the urinary bladder via a urethral catheter. The contrast material used in the urinary bladder is water-soluble. Cystography can be single-contrast, when one contrast material is used, or double-contrast, when more than one contrast material is used. Double-contrast cystography is a radiographic study of the urinary bladder after air and contrast material have been placed in the urinary bladder via a urethral catheter. A **cystogram** (sihs-tō-grahm) is the radiographic film of the urinary bladder after contrast material has been placed in the urinary bladder via a urethral catheter. **Retrograde** (reh-trō-grād) means going backward and can be used to describe the path that contrast material takes. If the contrast material goes in reverse order of how urine normally flows in the body, it is referred to as retrograde.
- **cystoscopy** (sihs-tohs-kō-pē) = visual examination of the urinary bladder using a fiberoptic instrument. A **cystoscope** (sihs-toh-skōp) is the fiberoptic instrument used to access the interior of the urinary bladder.
- **intravenous pyelogram** (ihn-trah-vē-nuhs pī-eh-lō-grahm) = radiographic study of the kidney (especially the renal pelvis) and ureters in which a dye is injected into a vein to define structures more clearly. The urinary bladder also may be visualized better with an intravenous pyelogram. It is abbreviated IVP (Figure 7–7).
- **pneumocystography** (nū-mō-sihs-tohg-rah-fē) = radiographic study of the urinary bladder after air has been placed in the bladder via a urethral catheter.
- **radiography** (rā-dē-ohg-rah-fē) = imaging of internal structures that is created by exposing specialized film to X-rays. A **scout film** is a plain X-ray made without the use of contrast material (Figure 7–8).

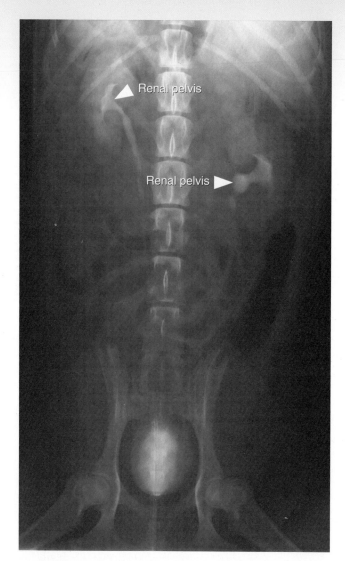

Figure 7–7 IVP in a dog. Contrast material outlines the renal pelvis in this radiograph. Some contrast material also is present in the urinary bladder. (Courtesy of University of Wisconsin Veterinary Teaching Hospital—Radiology.)

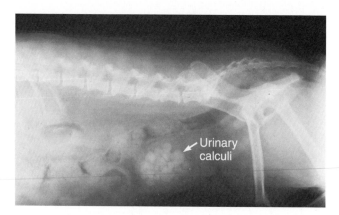

Figure 7–8 Radiograph of the urinary bladder. Urinary calculi in the canine urinary bladder are seen on this lateral scout radiograph. (Courtesy of Lodi Veterinary Hospital.)

- **retrograde pyelogram** (reh-trō-grād pī-eh-lō-grahm) = radiographic study of the kidney and ureters in which a contrast material is placed directly in the urinary bladder.
- **urinalysis** (yoo-rih-nahl-ih-sihs) = examination of urine components. It is abbreviated UA. Urinalyses can tell us about pH (hydrogen ion concentration that indicates acidity or alkalinity), leukocytes, erythrocytes, protein, glucose, specific gravity (measurement that reflects the amount of wastes, minerals, and solids in urine), and other factors (Table 7–2, Table 7–3, and Figure 7–9).
- **urinary catheterization** (kahth-eh-tər-ih-zā-shuhn) = insertion of a tube through the urethra into the urinary bladder (usually to collect urine). A **catheter** (kahth-eh-tər) is the hollow tube that is inserted into a body cavity to inject or remove fluid (Figure 7–10).

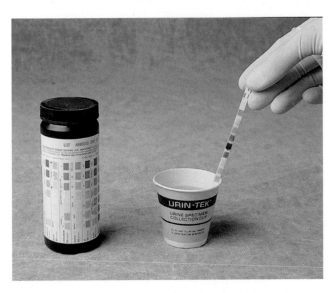

Figure 7–9 Urinalysis. Chemical properties of urine, such as pH, glucose, ketones, and bilirubin, are tested with a dipstick.

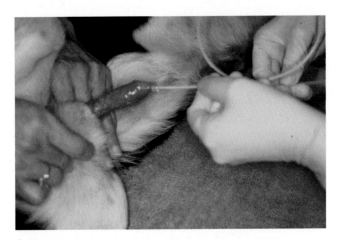

Figure 7–10 Urinary catheterization in a male dog. (Courtesy of Teri Raffel, CVT.)

Table 7–2 Descriptive Terms for Urine and Urination

albuminuria (ahl-bū-mihn-yoo-rē-ah)	presence of the major blood protein in urine
anuria (ah-nū-rē-ah)	complete suppression of urine production
bacteriuria (bahck-tē-rē-yoo-rē-ah)	presence of bacteria in urine
crystalluria (krihs-tahl-yoo-rē-ah)	urine with naturally produced angular solid of definitive form (crystals)
dysuria (dihs-yoo-rē-ah)	difficult or painful urination
glucosuria (gloo-kohs-yoo-rē-ah)	glucose (sugar) in urine
glycosuria (glī-kohs-yoo-rē-ah)	glucose (sugar) in urine
hematuria (hēm-ah-toor-ē-ah)	blood in urine
ketonuria (kē-tō-nū-rē-ah)	presence of ketones in urine (ketones are produced during increased fat metabolism)
nocturia (nohck-too-rē-ah)	excessive urination at night
oliguria (ohl-ih-goo-rē-ah)	scanty or little urine
pollakiuria (pōl-lahck-ē-yoo-rē-ah)	frequent urination
polyuria (pohl-ē-yoo-rē-ah)	excessive urination
proteinuria (prō-tēn-yoo-rē-ah)	presence of proteins in urine
pyuria (pī-yoo-rē-ah)	pus in urine
stranguria (strahng-yoo-rē-ah)	slow or painful urination

Table 7–3   Urine Test Strips: Common Tests and Significance

Test	Significance
Urobilinogen	This test is used more commonly in human medicine to evaluate liver disease or the breakdown of red blood cells. This test is less useful in veterinary medicine. (The test strips used are designed for human urianalysis.)
Glucose	This test is used to screen for diabetes mellitus. Diabetics have elevated blood sugar. The kidneys are unable to conserve all the sugar once the level becomes too high. The test is also used to monitor diabetics once treatment has been started.
Ketones	In dogs and cats, the presence of ketones is typical of an animal with uncontrolled diabetes mellitus. Ketones result when metabolism is shifted from carbohydrates to lipids.
Bilirubin	Aged red blood cells are removed from the circulation in organs such as the spleen. Bilirubin is formed in the breakdown process of hemoglobin. (Normally bilirubin is cleared from the blood by the liver and excreted in bile). Bilirubin is found in the urine in liver disease or excessive blood cell breakdown.
Protein	Proteins are large molecules that are not normally filtered into the urine. Protein in the urine can be present with a disease of the glomerulus (making it leaky) or with inflammation of the urinary tract (such as a bladder infection). A small amount of protein may normally be detected in very concentrated urine.
Blood	The test strip detects occult blood (i.e., blood that cannot be visibly seen in the urine). Blood can be present in diseases that cause inflammation of the urinary tract, much like protein. Bladder infections, stones, tumors, and trauma (e.g., hit by car) can all cause blood to be present in the urine. Bleeding disorders may also cause the test to be positive.
pH	Urine pH is influenced by diet and disease states in the body. Acidic pH is typical in animals with a meat diet or with acidosis (the kidney attempting to rid the body of excess acid). Basic or alkaline pH is typical in animals with a cereal grain diet, some urinary tract infections, and alkalosis in the body.

PATHOLOGY: URINARY SYSTEM

Pathologic conditions of the urinary system include the following:

- **azotemia** (ā-zō-tē-mē-ah) = presence of urea or other nitrogenous elements in the blood.
- **calculus** (kahl-kyoo-luhs) = abnormal mineral deposit. Urine may contain crystals that remain in solution in urine. Stones, calculi, and liths are formed when crystals precipitate and form solids. **Lith/o** is the combining form for stone or calculus, and **-lith** is the suffix for stone or calculus. When used in relationship to the urinary system, *calculus* is modified to *urinary calculus*, *lith* is modified to *urolith*, and *stones* is modified to *urinary stones* or *kidney stones* to clarify which system is involved.
- **casts** (kahstz) = fibrous or protein materials found in the urine with renal disease or another abnormality (Figure 7–11).

- **crystals** (krihs-tahlz) = naturally produced angular solid of definitive form (Figure 7–11).
- **cystalgia** (sihs-tahl-jē-ah) = urinary bladder pain = **cystodynia** (sihs-tō-dihn-ē-ah).
- **cystitis** (sihs-tī-tihs) = inflammation of the urinary bladder.
- **cystocele** (sihs-tō-sēl) = displacement of the urinary bladder through the vaginal wall.
- **epispadias** (ehp-ih-spā-dē-uhs) = abnormal condition in which the urethra opens on the dorsum of the penis.
- **feline lower urinary tract disease** = common disease of cats in which cystitis, urethritis, and crystalluria are found; formerly called feline urologic syndrome (FUS). In male cats, urethral obstruction is commonly associated with this disease, abbreviated FLUTD.
- **glomerulonephritis** (glō-mər-yoo-lō-nehf-rī-tihs) = inflammation of the kidney involving the glomeruli.

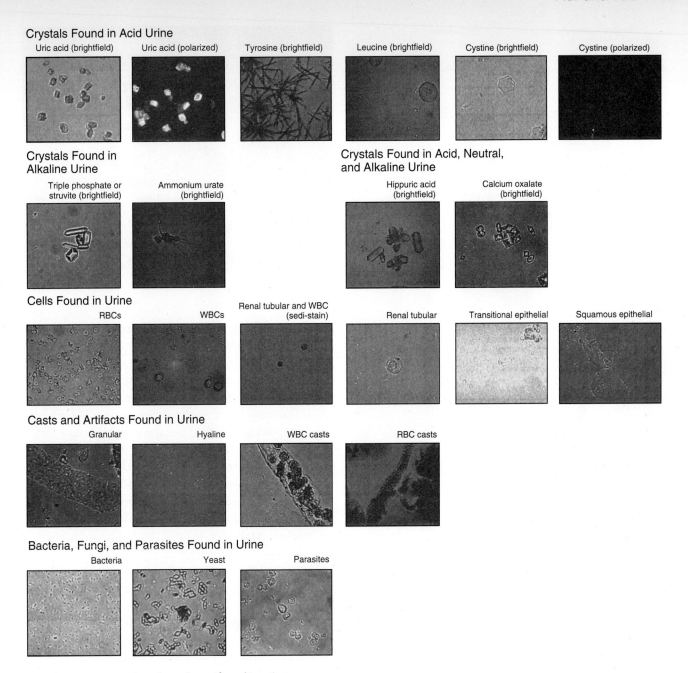

Crystals Found in Acid Urine

Uric acid (brightfield) | Uric acid (polarized) | Tyrosine (brightfield) | Leucine (brightfield) | Cystine (brightfield) | Cystine (polarized)

Crystals Found in Alkaline Urine

Triple phosphate or struvite (brightfield) | Ammonium urate (brightfield)

Crystals Found in Acid, Neutral, and Alkaline Urine

Hippuric acid (brightfield) | Calcium oxalate (brightfield)

Cells Found in Urine

RBCs | WBCs | Renal tubular and WBC (sedi-stain) | Renal tubular | Transitional epithelial | Squamous epithelial

Casts and Artifacts Found in Urine

Granular | Hyaline | WBC casts | RBC casts

Bacteria, Fungi, and Parasites Found in Urine

Bacteria | Yeast | Parasites

Figure 7–11 Crystals, cells, and casts found in urine.

- **hydronephrosis** (hī-drō-nehf-rō-sihs) = dilation of the renal pelvis as a result of an obstruction to urine flow (Figure 7–12).
- **hydroureter** (hī-drō-yoo-rē-tər) = distention of the ureter with urine caused by any blockage.
- **hypospadias** (hī-pō-spā-dē-uhs) = abnormal condition in which the urethra opens on the ventral surface of the penis.
- **inappropriate urination** (ihn-ah-prō-prē-ət yoo-rihn-ā-shuhn) = eliminating urine at the wrong time or in the wrong place.
- **incontinence** (ihn-kohn-tih-nehns) = inability to control excretory functions. The term *urinary* is applied in front of this term to refer to the inability to control urine.

- **interstitial cystitis** (ihn-tər-stihsh-ahl sihs-tī-tihs) = inflammation within the wall of the urinary bladder.
- **-lithiasis** (lih-thī-ah-sihs) = suffix meaning the presence of stones or calculi, as in nephrolithiasis (disorder characterized by the presence of kidney stones), urolithiasis (disorder characterized by the presence of urinary bladder stones), and ureterolithiasis (disorder characterized by the presence of stones in the ureters).
- **nephrectasis** (neh-frehck-tah-sihs) = distention of the kidneys. **Distention** means enlargement, and the suffix **-ectasis** means distention or stretching.
- **nephritis** (neh-frī-tihs) = inflammation of the kidneys.
- **nephrolith** (nehf-rō-lihth) = kidney stone or renal calculus.

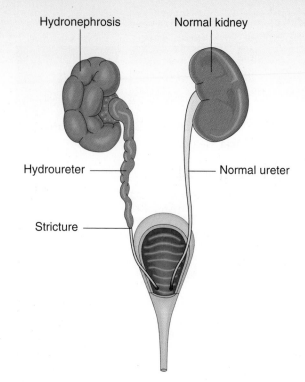

Figure 7–12 Hydroureter and hydronephrosis caused by urethral stricture.

Figure 7–13 Uroliths.

- **nephromalacia** (nehf-rō-mah-lā-shē-ah) = abnormal softening of the kidney.
- **nephropathy** (neh-frohp-ah-thē) = disease of the kidneys.
- **nephrosclerosis** (nehf-rō-skleh-rō-sihs) = abnormal hardening of the kidney.
- **nephrosis** (neh-frō-sihs) = abnormal condition of the kidney(s).
- **prerenal** (prē-rē-nahl) and **postrenal** (pōst-rē-nahl) = before and after the kidney, respectively. These terms are used to describe other pathologic conditions, such as prerenal azotemia and postrenal obstruction.
- **prolapse** (prō-lahps) = downward displacement of a body organ. The suffix **-ptosis** means drooping or dropping down, and **nephroptosis** (neh-frohp-tō-sihs) may be used to describe a prolapsed kidney.
- **pyelitis** (pī-eh-lī-tihs) = inflammation of the renal pelvis.
- **pyelonephritis** (pī-eh-lō-nehf-rī-tihs) = inflammation of the renal pelvis and kidney.
- **renal failure** (rē-nahl fāl-yər) = inability of the kidneys to function. Renal failure may be **acute** (ā-kūt) or **chronic** (krohn-ihck). Acute means occurring suddenly or over a short period. Acute renal failure (ARF) is the sudden onset of the inability of the kidneys to function. ARF may be caused by a **nephrotoxin** (nehf-rō-tohcks-ihn), which is a poison having destructive effects on the kidneys. Chronic means having a longer onset. Chronic renal failure (CRF) is the progressive onset of the inability of the kidneys to function. Signs

of renal failure may include **polyuria** and **polydipsia** (pohl-ē-dihp-sē-ah); polyuria/polydipsia is abbreviated PU/PD. Polyuria is elevated urine production, and polydipsia is excessive thirst or drinking. Animals with renal failure may undergo **diuresis** (dī-yoo-rē-sihs). Diuresis is the elevated excretion of urine. Diuresis may be produced by fluid therapy or drug therapy. Drugs that increase urine production are **diuretics.**
- **renal infarction** (rē-nahl ihn-fahrck-shuhn) = obstruction of blood flow to the kidney(s).
- **uremia** (yoo-rē-mē-ah) = waste products in the blood. Uremia is seen with many types of kidney disease.
- **ureterectasis** (yoo-rē-tər-ehck-tah-sihs) = distention of the ureter.
- **ureterolith** (yoo-rē-tər-ō-lihth) = stone in the urethra.
- **urethritis** (yoo-rē-thrī-tihs) = inflammation of the urethra.
- **urethrostenosis** (yoo-rē-thrō-steh-nō-sihs) = stricture of the urethra. A **stricture** (strihck-shər) is an abnormal band of tissue narrowing a passage.
- **urinary retention** (yoo-rih-nār-ē rē-tehn-shuhn) = inability to completely empty the urinary bladder.
- **urinary tract infection** (yoo-rihn-ār-ē trahckt ihn-fehck-shuhn) = invasion of microorganisms in the urinary system, which results in local cellular injury; abbreviated UTI.
- **urolith** (yoo-rō-lihth) = urinary bladder stone; also called **cystolith** (sihs-tō-lihth) (Figure 7–13).

PROCEDURES: URINARY SYSTEM

Procedures performed on the urinary system include the following:

- **cystectomy** (sihs-tehck-tō-mē) = surgical removal of all or part of the urinary bladder.

- **cystopexy** (sihs-tō-pehck-sē) = surgical fixation of the urinary bladder to the abdominal wall.
- **cystoplasty** (sihs-tō-plahs-tē) = surgical repair of the urinary bladder.
- **cystostomy** (sihs-tohs-tō-mē) = surgical creation of a new opening between the skin and urinary bladder.
- **cystotomy** (sihs-tah-tō-mē) = surgical incision into the urinary bladder.
- **dialysis** (dī-ahl-ih-sihs) = procedure to remove blood waste products when the kidneys are no longer functioning. **Peritoneal dialysis** is the removal of blood waste products by fluid exchange through the peritoneal cavity; **hemodialysis** (hēm-ō-dī-ahl-ih-sihs) is the removal of blood waste products by filtering blood through a machine.
- **lithotripsy** (lihth-ō-trihp-sē) = destruction of stone using ultrasonic waves traveling through water (the suffix **-tripsy** means to crush).
- **nephrectomy** (neh-frehck-tō-mē) = surgical removal of a kidney.
- **nephropexy** (nehf-rō-pehcks-sē) = surgical fixation of a kidney to the abdominal wall.
- **ureterectomy** (yoo-rē-tər-ehck-tō-mē) = surgical removal of the ureter.
- **ureteroplasty** (yoo-rē-tər-ō-plahs-tē) = surgical repair of the ureter.
- **urethrostomy** (yoo-re-throhs-tō-mē) = surgical creation of a permanent opening between the urethra and the skin. **Perineal urethrostomy** (pər-ih-nē-ahl yoo-rē-throhs-tō-mē) is the surgical creation of a permanent opening between the urethra and the skin between the anus and scrotum.

REVIEW EXERCISES

Multiple Choice

Choose the correct answer.

1. The combining forms for kidney are

    a. ren/o and ureter/o
    b. ren/o and nephr/o
    c. ren/o and cyst/o
    d. ren/o and periren/o

2. Inflammation of the kidney is

    a. nephrosis
    b. nephroptosis
    c. nephritis
    d. nephropathy

3. Insertion of a hollow tube through the urethra into the urinary bladder is called

    a. cystocentesis
    b. cystogram
    c. urinary injection
    d. urinary catheterization

4. The hormone produced by the kidney that stimulates red blood cell production in the bone marrow is

    a. ADH
    b. erythropoietin
    c. aldosterone
    d. renerythogenin

5. Retrograde means

    a. going backward
    b. going forward
    c. going sideways
    d. repeating

6. Examination of the components of urine is a

    a. urinoscopy
    b. cystoscopy
    c. urinalysis
    d. cystolysis

7. Inflammation of the urinary bladder is

    a. cystitis
    b. urolithiasis
    c. urology
    d. uritis

8. UTI is the abbreviation for

    a. urinary treatment for infection
    b. urinary tract infection
    c. urinary tract inflammation
    d. urinary trigone infarct

9. Stable internal environment is

    a. stricture

    b. status

    c. homeostasis

    d. isostatic

10. Diuretics are chemical substances that

    a. cause painful urination

    b. cause complete cessation of urine

    c. cause nighttime urination

    d. cause an increase in urine production

11. The triangular part of the urinary bladder is the

    a. calyx

    b. hilus

    c. medulla

    d. trigone

12. The term for production of urine is

    a. urinogenesis

    b. uropoiesis

    c. turbidity

    d. renogenesis

13. Inability to control excretory functions is

    a. inappropriate urination

    b. urinary retention

    c. incontinence

    d. urethritis

14. The medical term for excessive urination is

    a. pyuria

    b. polyuria

    c. polydipsia

    d. pollakiuria

15. Oliguria means

    a. scanty or little urine

    b. blood in urine

    c. frequent urination

    d. excessive urination

16. The term for frequent urination is

    a. scanty or little urine

    b. dysuria

    c. stranguria

    d. pollakiuria

17. A surgical incision into the urinary bladder is known as

    a. cystectomy

    b. cystotomy

    c. cystopexy

    d. cystostomy

18. The presence of urea or other nitrogenous elements in the blood is called

    a. uremia

    b. diuresis

    c. azotemia

    d. proteinuria

19. Obstruction of blood flow to the kidney(s) is

    a. dialysis

    b. azotemia

    c. prerenal infarction

    d. renal infarction

20. The term that means no urine production is

    a. diuresis

    b. anuria

    c. nocturia

    d. hematuria

Matching

Match the term in Column I with the definition in Column II.

Column I	Column II
1. _____ bacteriuria	a. complete suppression of urine production
2. _____ glycosuria or glucosuria	b. difficult or painful urination
3. _____ nocturia	c. scanty or little urine
4. _____ proteinuria	d. excessive urination at night
5. _____ anuria	e. slow or painful urination
6. _____ oliguria	f. blood in urine
7. _____ albuminuria	g. frequent urination
8. _____ stranguria	h. presence of bacteria in urine
9. _____ polyuria	i. glucose (sugar) in urine
10. _____ pyuria	j. presence of ketones in urine
11. _____ pollakiuria	k. presence of proteins in urine
12. _____ ketonuria	l. presence of the major blood protein in urine
13. _____ crystalluria	m. pus in urine
14. _____ dysuria	n. increased urination
15. _____ hematuria	o. crystals in urine
16. _____ diuresis	p. pertaining to the outer portion of an organ
17. _____ calculus	q. stone
18. _____ cortical	r. inflammation of the urinary bladder
19. _____ erythropoietin	s. elevated urine excretion
20. _____ cystitis	t. hormone produced by the kidney that stimulates red blood cell production

Fill in the Blanks

1. Urin/o and ur/o mean _____.

2. Ren/o and nephr/o mean _____.

3. Glycosuria and glucosuria mean _____.

4. Cystolith and urolith mean _____.

5. Excretion, elimination, and voiding mean _____.

Spelling

Circle the term that is spelled correctly.

1. examination of urine components:	urinanalysis	urinalysis	urinealisis
2. functional unit of the kidney:	nephron	nefron	nephrone
3. constant internal environment:	homostasis	homeostasis	homeostatis
4. excessive thirst:	polydipsia	polydypsia	polydyspsia
5. abnormal mineral deposit:	calkulus	kalkulus	calculus

CROSSWORD PUZZLES

Urinary System Combining Forms Puzzle

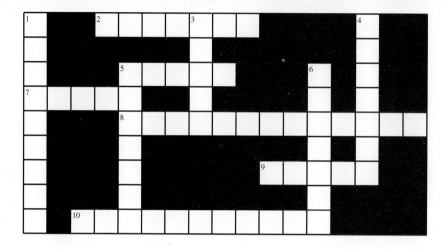

Across

2 combining form for inner or middle
5 combining form for urinary bladder
7 combining form for opening
8 combining forms for kidney
9 combining form for renal pelvis
10 combining forms for urine/urinary organs

Down

1 combining form for cluster of capillaries
3 combining form for stone/calculus
4 combining form for the tube that takes urine from the urinary bladder to the outside
5 combining form for outer
6 combining form for paired narrow tubes that are located between the kidneys and urinary bladder

Disease Terms Puzzle

Across

1 surgical fixation of the kidney to the abdominal wall
4 surgical removal of a kidney
8 surgical puncture of the urinary bladder (to remove fluid)
10 study of the urinary system
12 urinary bladder stone or calculus
13 inflammation of the kidney involving the cluster of capillaries
14 inability to control the urine

Down

2 inflammation of the renal pelvis and kidney
3 surgical repair of the hollow tube that connects the urinary bladder to the outside of the body
4 abnormal hardening of the kidney
5 functional unit of the kidney
6 surgical creation of a permanent opening between the hollow tube that connects the urinary bladder to the outside of the body and the skin between the anus and scrotum
7 surgical incision into the urinary bladder
9 abnormal band of tissue narrowing a passage
11 voluntary control of urination

LABEL THE DIAGRAMS

Label the parts of the urinary system in Figure 7–14. Provide the combining forms for the parts labeled.

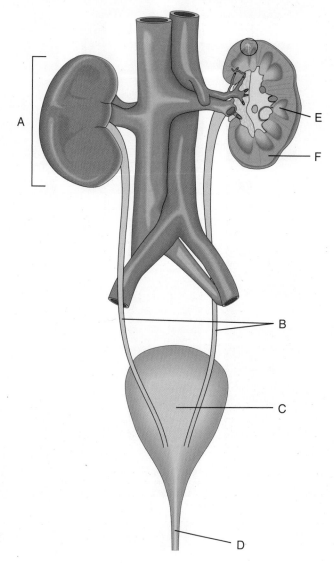

Figure 7–14   Urinary system structures.

CASE STUDIES

Supply the proper abbreviation or medical term for the lay terms and definitions underlined.

A 5-year-old male neutered springer spaniel was presented with difficulty urinating and blood in the urine. The veterinarian performed examination of the urine by breaking it into its components by inserting a needle in the urinary bladder and withdrawing urine.

The test revealed large numbers of erythrocytes and struvite crystals. Suspecting more than inflammation of the urinary bladder, the veterinarian took an X-ray of the dog's urinary bladder, and urinary bladder stones were detected (Figure 7–15). An incision into the urinary bladder was performed to remove the urinary stones, and the dog recovered uneventfully.

1. year   _____

2. male neutered   _____

3. difficulty urinating   _____

4. blood in the urine   _____

5. examination of the urine by breaking it into its components   _____

6. inserting a needle in the urinary bladder and withdrawing urine _____

7. erythrocytes   _____

8. inflammation of the urinary bladder   _____

9. X-ray   _____

10. urinary bladder stones   _____

11. incision into the urinary bladder   _____

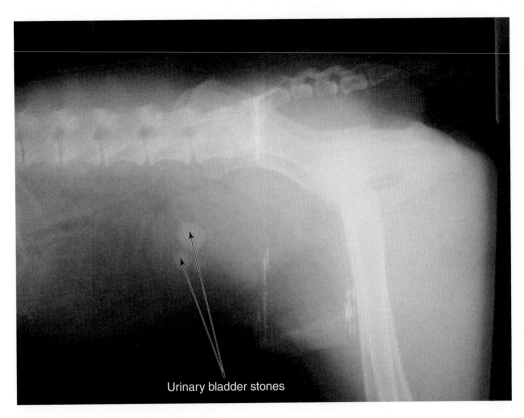

Urinary bladder stones

Figure 7–15   Radiograph showing bladder stones.

Define the underlined terms in each case study.

A 2-yr-old <u>M/N</u> <u>DLH</u> cat was presented to the clinic for crying when urinating in the litter box. The owner stated that the cat had been <u>inappropriately urinating</u> on the carpet for the past few days but otherwise seemed normal. Upon examination, the cat was about 7% <u>dehydrated</u>, T = 102° F, HR = 180 bpm, and RR = panting. Heart and lungs <u>auscultated</u> normally. Abdominal palpation revealed <u>cystomegaly</u>. The <u>urethra</u> was red and swollen. When the cat was put in a litter box, he was straining but was unable to <u>urinate</u>. The cat was <u>dx'd</u> with <u>urethral obstruction</u>. The owner was informed of the emergency status of relieving the obstruction, and the cat was hospitalized. The cat was <u>anesthetized</u> using <u>inhalant anesthesia</u> via a mask, and a <u>urinary catheter</u> was passed and sutured to the <u>perineum</u>. <u>IV</u> fluids were administered to reverse the cat's dehydration. Urine was collected via the urinary catheter, and a <u>UA</u> was performed. The UA revealed a <u>pH</u> = 7.0, large amounts of blood in the urine, a large number of <u>leukocytes</u> in the urine, a <u>specific gravity</u> of 1.040, and large amounts of <u>struvite crystals</u> and bacteria in the urine. In addition to the urethral obstruction, the cat also had <u>cystitis</u> and was treated with antibiotics and a urinary acidifying diet. The cat was hospitalized until the urinary catheter was removed. The cat was discharged with its medication, and the owner was advised to have a recheck UA when the medication was completed.

12. M/N _____

13. DLH _____

14. inappropriately urinating _____

15. dehydrated _____

16. auscultated _____

17. cystomegaly _____

18. urethra _____

19. urinate _____

20. dx'd _____

21. urethral obstruction _____

22. anesthetized _____

23. inhalant anesthesia _____

24. urinary catheter _____

25. perineum _____

26. IV _____

27. UA _____

28. pH _____

29. leukocytes _____

30. specific gravity _____

31. struvite crystals _____

32. cystitis _____

A 2-yr-old milking <u>cow</u> was showing a decrease in milk production and weight loss over the past few weeks. She was <u>off feed</u> and restless according to the farmer. Upon arrival at the farm, the veterinarian noted that the cow was switching its tail, vital signs were normal, and the cow was <u>polyuric</u>. <u>Hematuria</u> was noted, and urine was collected for analysis. <u>Rectal palpation</u> revealed <u>nephro-megaly</u>. A tentative <u>diagnosis</u> of <u>pyelonephritis</u> was made. Antibiotics were given to this cow, the cow's milk was pulled for antibiotic withdrawal times, and the farmer was advised to monitor the other cows for similar abnormalities. The veterinarian returned to the clinic and performed a <u>UA</u>, which showed a large amount of blood in the urine and rod-shaped bacteria on microscopic examination. The sample was sent in for bacterial <u>culture and sensitivity</u> to determine whether the cow had a *Corynebacterium renale* infection. *C. renale* infections may indicate a herd with subclinical infection, which would affect the treatment and hygiene practices on this farm.

33. cow _____

34. off feed _____

35. polyuric _____

36. hematuria _____

37. rectal palpation _____

38. nephromegaly _____

39. diagnosis _____

40. pyelonephritis _____

41. UA _____

42. culture and sensitivity _____

CHAPTER 8

[HAVE A HEART]

Objectives

Upon completion of this chapter, the reader should be able to

- Identify the structures of the cardiovascular system
- Differentiate between the types of blood vessels and describe their functions
- Describe the functions of the cardiovascular system
- Describe the flow of blood throughout the body
- Recognize, define, spell, and pronounce the terms related to the diagnosis, pathology, and treatment of the cardiovascular system

FUNCTIONS OF THE CARDIOVASCULAR SYSTEM

The cardiovascular system delivers oxygen, nutrients, and hormones to the various body tissues and transports waste products to the appropriate waste removal systems. Occasionally, the cardiovascular system is called the circulatory system; however, the circulatory system is divided into systemic circulation (blood flow to all parts of the body except the lungs) and pulmonary circulation (blood flow out of the heart through the lungs and back to the heart).

STRUCTURES OF THE CARDIOVASCULAR SYSTEM

Cardiovascular (kahr-dē-ō-vahs-kyoo-lər) means pertaining to the heart and vessels (blood vessels in this context). There are three major parts of this system: the heart, the blood vessels, and the blood. Blood is discussed in Chapter 15.

The Heart

The heart is a hollow muscular organ that provides the power to move blood through the body. The combining form for heart is **cardi/o.** The heart is located inside the **thoracic** (thō-rahs-ihck) **cavity,** or **chest cavity.** The heart lies between the lungs in a cavity called the **mediastinum** (mē-dē-ahs-tī-nuhm). The mediastinum also contains the large blood vessels, trachea, esophagus, lymph nodes, and other structures.

The Pericardium

Surrounding the heart is a double-walled membrane called the **pericardium** (pehr-ih-kahr-dē-uhm) (Figure 8–1). The two layers of the pericardium are the fibrous and serous. The **fibrous pericardium** is the tough external layer. The **serous layer** is the inner layer and is divided into two parts: the **parietal** (pahr-ī-ih-tahl) **layer** and the **visceral** (vihs-ər-ahl) **layer.** The parietal layer (in Latin, *parietal* means belonging to the wall) is the serous layer that lines the fibrous pericardium. The visceral layer (*viscus* is Latin for an organ) is the serous layer that lines the heart. The visceral layer also is called the **epicardium** (ehp-ih-kahr-dē-uhm). Between the two serous layers of the pericardium is a space called the **pericardial** (pehr-ih-kahr-dē-ahl) **space. Pericardial fluid** is a liquid in the pericardial space. Pericardial fluid prevents friction between the heart and the pericardium when the heart beats.

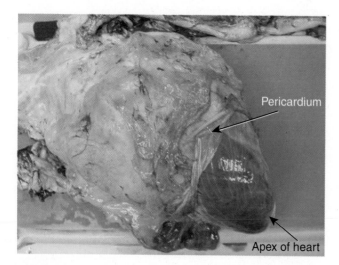

Figure 8–1 The pericardium is a double-walled membrane that surrounds the heart.

Layers of the Heart Walls

The heart wall is made up of the following three layers (Figure 8–2):

- **epicardium** (ehp-ih-kahr-dē-uhm) = external layer of the heart; also part of the serous layer of the pericardium. The prefix **epi-** means upper.
- **myocardium** (mī-ō-kahr-dē-uhm) = middle and thickest layer of the heart; the actual heart muscle. The combining form **my/o** means muscle.
- **endocardium** (ehn-dō-kahr-dē-uhm) = inner layer of the heart; lines the heart chambers and valves. The prefix **endo-** means within.

Blood Supply

Heart tissue beats constantly and must have a continuous supply of oxygen and nutrients and the prompt removal of waste. The problem is that it cannot fulfill these needs from the blood it is pumping. These needs are supplied by its own arteries, and its wastes are removed by its own veins. The arteries that serve the heart are known as the **coronary arteries** (kōr-oh-nār-ē ahr-tər-ēz) because they resemble a crown. **Coron/o** is the combining form meaning crown. The **coronary veins** remove waste products from the myocardium.

If the blood supply to the heart is disrupted, the myocardium cannot function. Disruption of blood to the myocardium may be caused by **coronary occlusion** (ō-kloo-shuhn). *Occlusion* means blockage. Coronary occlusion may lead to **ischemia** (ihs-kē-mē-ah), which is deficiency in the blood supply to an area. Ischemia can lead to **necrosis** (neh-krō-sihs), which is tissue death. The area of necrosis due to ischemia is called an **infarct** (ihnfahrckt) or **infarction** (ihn-fahrck-shuhn). An infarct is a localized area of necrosis caused by an interrupted blood supply.

Chambers

The heart is divided into right and left sides. The right and left sides are further subdivided into chambers. Mammalian and avian hearts are four-chambered; most reptile hearts have three chambers. The craniodorsal chambers of the heart are called the **atria** (ā-trē-ah); the singular form of *atria* is **atrium** (ā-trē-uhm). All vessels coming into the heart enter here. **Atri/o** is the combining form for atria. The left and right atria are separated by the **interatrial septum** (ihn-tər-ā-trē-ahl sehp-tuhm). A septum is a separating wall or partition.

The caudoventral chambers of the heart are **ventricles** (vehn-trih-kuhlz). **Ventricul/o** is the combining form for ventricle. The ventricles are separated from each other by the **interventricular septum.** The interventricular septum in reptiles is not complete, so the ventricles are open to each other and count as only one heart chamber. The ventricles are the pumping chambers of the heart, and all vessels leave the heart via the ventricles.

The narrow tip of the heart is called the **apex** (ā-pehcks) or cardiac apex (Figure 8–3).

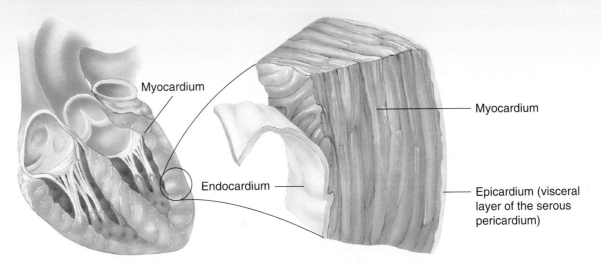

Figure 8-2 Layers of the heart.

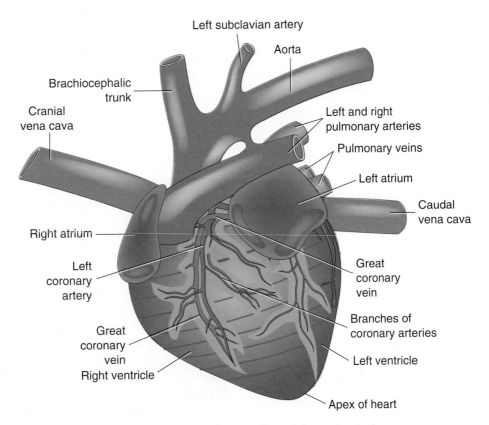

Figure 8-3 External heart structures of a canine heart (left lateral aspect).

Valves

Blood flow through the heart is controlled by valves (Figure 8–4 and Table 8–1). A **valve** (vahlv) is a membranous fold. The combining form for valve is **valv/o** or **valvul/o**. The four heart valves are as follows:

- **right atrioventricular valve** (ā-trē-ō-vehn-trihc-kyoo-lahr vahlv), or right AV valve. This valve controls the opening between the right atrium and right ventricle

(Figure 8–5). It also is called the **tricuspid valve** (trī-kuhs-pihd vahlv) because it has three points, or cusps (tri- = three; cusps = points).

- **pulmonary semilunar valve** (puhl-mah-nār-ē sehm-ē-loo-nahr vahlv) or pulmonary valve. This valve is located between the right ventricle and the pulmonary artery and controls blood entering the lungs. *Semilunar* means half-moon, and this valve is shaped like a half-moon.

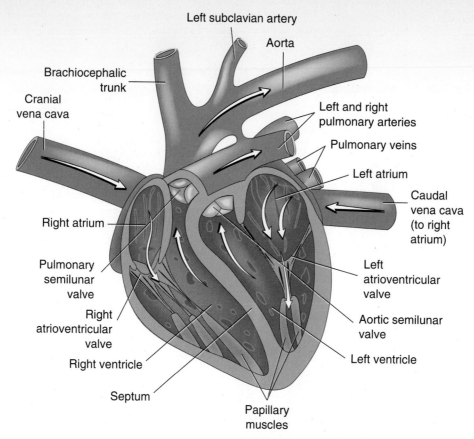

Figure 8–4 Internal heart structures of a canine heart (left lateral aspect).

Table 8–1 Blood Flow through the Heart

1. The right atrium receives blood from all tissues, except the lungs, through the cranial and caudal venae cavae. Blood flows from the right atrium through the tricuspid valve into the right ventricle. (This is systemic circulation.)

2. The right ventricle pumps blood through the pulmonary semilunar valve and into the pulmonary artery, which carries it to the lungs. (This is pulmonary circulation.)

3. The left atrium receives oxygenated blood from the lungs through the four pulmonary veins. The blood flows through the mitral valve into the left ventricle. (This is pulmonary circulation.)

4. The left ventricle receives blood from the left atrium. From the left ventricle, blood goes out through the aortic semilunar valve into the aorta and is pumped to all parts of the body except the lungs. (This is systemic circulation.)

5. Blood is returned by the venae cavae to the right atrium, and the cycle continues.

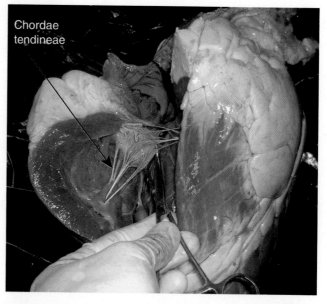

Figure 8–5 Atrioventricular valve. The hemostat is under the leave of the valve. AV valves attach to papillary muscles in the ventricles by connective tissue strings called chordae tendineae (kohr-dā tehn-dih-nā).

- **left atrioventricular valve** (ā-trē-ō-vehn-trihck-yoo-lahr vahlv), or left AV valve. This valve controls the opening between the left atrium and left ventricle. It also is called the **mitral valve** (mī-trahl vahlv) or **bicuspid** (bī-kuhs-pihd) because it has two points (bi- = two).
- **aortic semilunar valve** (ā-ōr-tihck sehm-ē-loo-nahr vahlv), or aortic valve. The aortic valve is located between the left ventricle and the aorta and controls blood entering the arterial system. It also is half-moon shaped.

Rhythm

A rhythm is the recurrence of an action or a function at regular intervals. The heart's contractions are supposed to be rhythmic. The rate and regularity of the heart rhythm, called the **heartbeat,** are modified by electrical impulses from nerves that stimulate the myocardium. The heartbeat or cardiac cycle is an alternating sequence of relaxation and contraction of the heart chambers. **Cardiac output** is the volume of blood pumped by the heart per unit time. To effectively pump blood throughout the body, contraction and relaxation of the heart must be synchronized accurately. These electrical impulses, also called the conduction system, are controlled by the sinoatrial node, atrioventricular node, bundle of His, and Purkinje fibers (Figure 8–6).

The **sinoatrial node** (sī-nō-ā-trē-ahl nōd), or SA node, is located in the wall of the right atrium near the entrance of the superior vena cava. The SA node, along with atypical cardiac muscle cells called **Purkinje** (pər-kihn-jē) **fibers,** establishes the basic rhythm of the heart and is called the pacemaker of the heart. Purkinje fibers are less developed in the atria and are usually associated with the ventricles.

Electrical impulses from the SA node start waves of muscle contractions in the heart. The impulse in the right atrium spreads over the muscles of both atria, causing them to contract simultaneously. This contraction forces blood into the ventricles. Atrial contraction is called **atrial systole** (sihs-stohl-ē). **Inotropy** (ihn-ō-trōp-ē) is the term meaning force of contraction.

The electrical impulses from the SA node continue to travel to the **atrioventricular node** (ā-trē-ō-vehn-trihck-yoo-lahr nōd), or AV node. The AV node is located in the interatrial septum.

The AV node conducts impulses more slowly than the SA node does. This slower conduction of the AV node causes a pause after atrial contraction to allow the ventricles to fill with blood. The AV node transmits the electrical impulses to the bundle of His (also called the AV bundle).

The **bundle of His** (hihs) is located in the interventricular septum. The bundle of His continues on through the ventricle as ventricular Purkinje fibers, which carry the impulse through the ventricular muscle, causing the ventricles to contract. Ventricular contraction is called **ventricular systole.** Ventricular contraction forces blood into the aorta and pulmonary arteries.

The normal heart rhythm is called the **sinus rhythm** because it starts in the sinoatrial node. If the SA node does

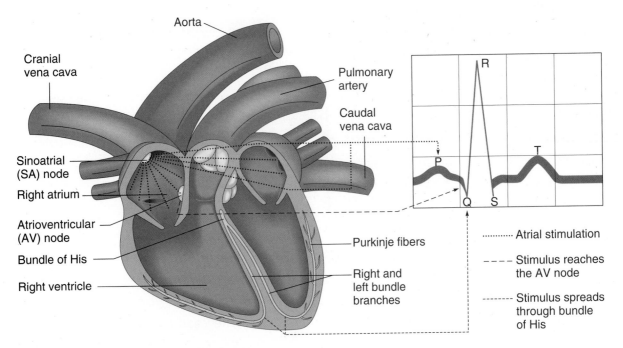

Figure 8–6 Conduction systems of the heart.

not function properly and is unable to send the impulse to the rest of the heart, other areas of the conduction system can take over and initiate a heartbeat. The resulting abnormal rhythm is called an **arrhythmia** (ā-rihth-mē-ah) or **dysrhythmia** (dihs-rihth-mē-ah) (Table 8–2). Antiarrhythmic drugs are substances that control heartbeat irregularities.

Electrical Activity

The electrical events in the conduction system can be visualized by wave movement on an **electrocardiogram** (ē-lehck-trō-kahr-dē-ō-grahm). An electrocardiogram, abbreviated ECG or EKG, is the record of the electrical activity of the myocardium. The ECG or EKG is a tracing that shows the changes in voltage and polarity (positive and negative) over time (Figure 8–7). The process of recording the electrical activity of the myocardium is **electrocardiography** (ē-lehck-trō-kahr-dē-ohg-rah-fē).

Table 8–2 Terms Relating to Rhythm

palpitation (pahl-pih-tā-shuhn) = heartbeat sensations that feel like pounding with or without irregularity in rhythm.

fibrillation (fih-brih-lā-shuhn) = rapid, random, and ineffective heart contractions.

flutter (fluht-tər) = cardiac arrhythmia in which atrial contractions are rapid but regular.

bradycardia (brā-dē-kahr-dē-ah) = abnormally slow heartbeat.

tachycardia (tahck-ē-kahr-dē-ah) = abnormally rapid heartbeat.

paroxysm (pahr-ohck-sihzm) = sudden convulsion or spasm.

normal sinus arrhythmia = irregular heart rhythm resulting from variation in vagal nerve tone as a result of respiration (a nonpathologic arrhythmia).

asystole (ā-sihs-tō-lē) = without contraction or lack of heart activity; flat line on an ECG.

syncope (sihn-kō-pē) = temporary suspension of respiration and circulation.

gallop (gahl-ohp) = low-frequency vibrations occurring during early diastole and late diastole.

What is systole?

Systole (sihs-tō-lē) is a term that generally means contraction, derived from the Greek word for drawing together (think systole = squeeze). However, the term *systole* is used to denote ventricular contraction. Remember that the atria and ventricles cannot contract together, although they follow in rapid succession.

Diastole (dī-ah-stō-lē) means expansion, derived from the Greek word for drawing apart (think diastole = dilate).

The term *diastole* is used to denote relaxation, or the time when the chambers are expanded. During diastole, the atria fill with blood. Then the atria contract, forcing blood into the ventricles, and the ventricles contract. If someone refers to systole and diastole without the modifiers *atrial* or *ventricular*, it is assumed that the person means ventricular contraction (systole) and ventricular relaxation (diastole).

Loading

The workload of the heart is divided into preload and afterload. **Preload** (prē-lōd) is the ventricular end-diastolic volume, or the volume of blood entering the right side of the heart.

Afterload (ahf-tər-lōd) is the impedance to ventricular emptying presented by aortic pressure.

Preload problems usually are associated with right-sided heart disease, whereas afterload usually is associated with left-sided heart disease. Drugs used to treat heart disease may alter preload or afterload.

Electrocardiography produces a tracing that represents the variations in electric potential caused by excitation of heart muscle and is detected at the body surface. These variations in electric potential are detected by conductors called **leads** (lēdz) (Figure 8–8).

Heart Sounds

Auscultation (aws-kuhl-tā-shuhn) is the act of listening to body sounds and usually involves the use of a stethoscope. A **stethoscope** (stehth-ō-skōp) is an instrument used to listen; however, when the term is broken down into its basic

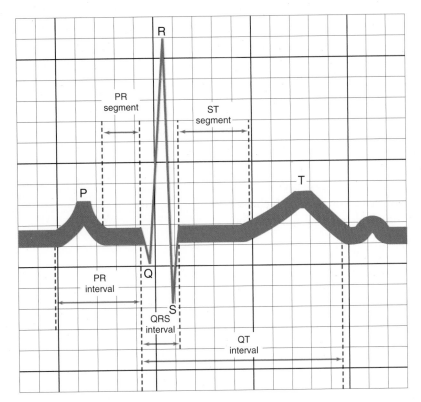

Figure 8–7   Anatomy of an electrocardiogram. The first deflection, the P wave, represents excitation (depolarization) of the atria. The PR interval represents conduction through the atrioventricular valve. The QRS complex results from excitation of the ventricles. The QT interval represents ventricular depolarization and repolarization. The ST segment represents the end of ventricular depolarization to the onset of ventricular repolarization. The T wave results from recovery (repolarization) of the ventricles.

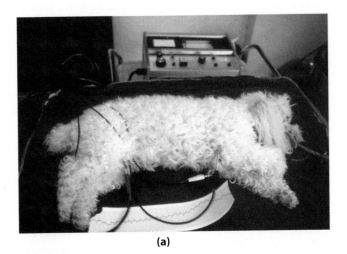

(a)

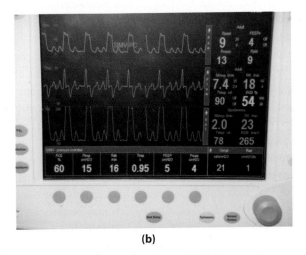

(b)

Figure 8–8   (a) Electrocardiography in a dog. (b) ECG monitor displaying the electrical activity of the heart. [(a) Courtesy of Lodi Veterinary Hospital. (b) Courtesy Kimberly Kruse Sprecher, CVT.]

components, **stetho-** means chest and **-scope** means instrument to visually examine or monitor. Traditionally, the stethoscope is considered to be a monitoring device of the chest area, but it also is used to auscultate other body parts.

When auscultating the heart, one hears a lubb-dubb sound. The lubb is the first sound heard (called the first heart sound) and is caused by closure of the AV valves. The dubb is the second sound heard (called the second heart sound) and is caused by closure of the semilunar valves. Systole, or ventricular contraction, occurs between the first and second heart sounds, whereas diastole, or ventricular relaxation, occurs between the second and first heart sounds.

A **heart murmur** (mər-mər) is an abnormal sound associated with the turbulent flow of blood. A murmur may be caused by a leak in a valve. A leak results in the inability of the valve to perform at the proper level, and this inability to perform at the proper level is called **insufficiency** (ihn-sah-fihsh-ehn-sē). Narrowing of a valve also may result in turbulent blood flow, causing a murmur.

Murmurs are described as systolic (the swooshing noise occurring between the first and second heart sounds) or diastolic (the swooshing noise occurring between the second and first heart sounds). Murmurs may be further described as **holosystolic** (hō-lō-sihs-stohl- ihck) or **pansystolic** (pahn-sihs-stohl-ihck), meaning that they occur during the entire ventricular contraction phase. The prefixes **holo-** and **pan-** mean all. These different phases in which a murmur occurs aid in identification of the cause of the murmur.

For example, murmurs heard during systole may be atrioventricular insufficiency or aortic or pulmonic valve **stenosis** (stehn-ō-sihs), or narrowing. Murmurs heard during diastole may be atrioventricular stenosis or aortic or pulmonic valve insufficiency.

In addition, murmurs can be described as crescendo and decrescendo. **Crescendo** (kreh-shehn-dō) **murmurs** are abnormal swooshing cardiac sounds that progressively increase in loudness, and **decrescendo** (deh-kreh-shehn-dō) **murmurs** progressively decrease in loudness.

The location of the murmur also is helpful in determining its cause. Where the murmur is heard the loudest is the point of maximal intensity (PMI). The PMI usually is located at the auscultation site of the defective valve. Occasionally, murmurs may result in vibrations felt on palpation of the chest. This vibration felt on palpation is called **thrill** (thrihl).

Clicks also may be heard during an examination. Clicks may be a sign of mitral insufficiency or may be of unknown origin. Other sounds that may be heard are split heart sounds (heartbeat sounds that are divided), crackles (which may be associated with movement or respiratory sounds), and rumbles (usually caused by shivering).

What are we hearing?

When auscultating the heart, there are various things on which to focus. One is heart rate, or the speed at which the heart is beating. Tachycardia and bradycardia can affect how blood is transported through the body. Abnormal sounds such as murmurs, clicks, and splits also should be assessed. Rhythm must be checked as well. Auscultation should be done in conjunction with pulse detection. The pulse is determined by feel and can be categorized as strong, weak, or thready. A pulse deficit occurs when the ventricles contract without enough force to propel blood to the periphery.

Blood Vessels

There are three major types of blood vessels: arteries, veins, and capillaries. The combining forms for vessel are **angi/o** and **vas/o.** These combining forms generally are used in reference to blood vessels but may be used to describe other types of vessels as well.

The **lumen** (loo-mehn) is the opening in a vessel through which fluid flows. The diameter of the lumen is affected by **constriction** (kohn-strihckt-shuhn), or narrowing of the vessel diameter, and **dilation** (dī-lā-shuhn), or widening of the vessel diameter. **Vasoconstrictors** (vahs-ō-kohn-strihck-tərz or vās-ō-kohn-strihck-tərz) are things that narrow a vessel's diameter; **vasodilators** (vahs-ō-dī-lāt-ərz or vās-ō-dī-lāt-ərz) are things that widen a vessel's diameter.

The **hilus** (hī-luhs) is the depression where vessels and nerves enter an organ.

The pumping action of the heart drives blood into the **arteries** (ahr-tər-ēz). An artery is a blood vessel that carries blood away from the heart. Blood in the arteries usually is oxygenated (the main exception is the pulmonary artery) and is bright red. The combining form for artery is **arteri/o.**

The **aorta** (ā-ōr-tah) is the main trunk of the arterial system that begins from the left ventricle of the heart. The combining form for aorta is **aort/o.** After leaving the left ventricle, the aorta arches dorsally and then progresses caudally. The aorta is located ventral to the vertebrae. The aorta branches into other arteries that supply many muscles and organs of the body. The branches from the aorta usually are named for the area in which they supply blood. For example, the **celiac** (sē-lē-ahck) **artery** supplies the liver, stomach, and spleen (**celi/o** is derived from the Greek term *koilia*, meaning belly); the **renal arteries** supply the kidneys; and the **ovarian** (or **testicular**) **arteries** supply the ovaries (or testicles). Occasionally, arteries are named for their location, as in the **subclavian** (suhb-klā-vē-ahn) **artery,** which is located under the collarbone (Figure 8–9).

The **arterioles** (ahr-tē-rē-ōlz) are smaller branches of arteries. The combining form **arter/i** means vessel that carries blood away from the heart, and the suffix **-ole** means small. Arterioles are smaller and thinner than arteries and carry blood to the capillaries.

The **capillaries** (kahp-ih-lār-ēz) are single-cell-thick vessels that connect the arterial and venous systems. Blood flows rapidly through arteries and veins; however, blood flow is slower through the capillaries due to their smaller diameter. This slower flow allows time for the diffusion of oxygen, nutrients, and waste products. Blood in the alveolar capillaries picks up oxygen and gives off carbon dioxide. In the rest of the body, oxygen diffuses (passes through) from the capillaries into tissue and carbon dioxide diffuses from tissue into the capillaries. Blood flow through tissues is called **perfusion** (pər-fū-shuhn). An indicator of perfusion is **capillary refill time,** or CRT. A CRT can be obtained by applying pressure to mucous membranes and timing how long it takes for the pink color to return (Figure 8–10).

Capillaries connect with **venules** (vehn-yoolz), which are tiny blood vessels that carry blood to the veins. **Veins** (vānz) form a low-pressure collecting system that returns blood to the heart. Veins have thinner walls and are less elastic than arteries, which have muscular walls to allow contraction and expansion to move blood throughout the body (Figure 8–11). Because the veins do not have muscular walls, contractions of the skeletal muscles cause the blood to flow through the veins toward the heart. Veins also have valves that permit blood flow toward the heart and prevent blood from flowing away from the heart (Figure 8–12). The combining forms for vein are **ven/o** and **phleb/o.**

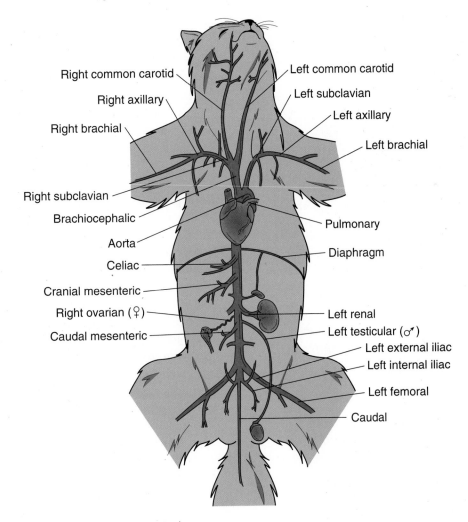

Figure 8–9   Major arteries in the cat.

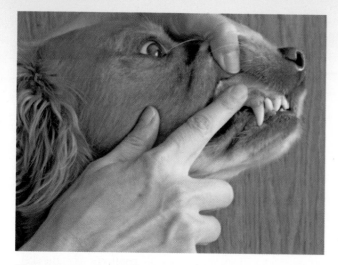

Figure 8–10 The dog's mucous membranes are checked for color and capillary refill time.

Vessels are used to distribute substances throughout the body. Many drugs are injected into vessels and transported to different areas of the body. Examples include the following:

- **intravenous** (ihn-trah-vē-nuhs) = within a vein. **Perivascular** (pehr-ih-vahs-kyoo-lahr), or around the vessels, is an undesired route of administration and usually is an error of intravenous injection.
- **intra-arterial** (ihn-trah-ahr-tēr-ē-ahl) = within an artery.

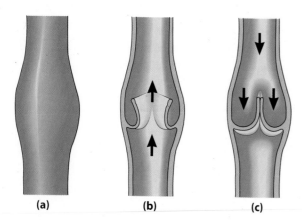

Figure 8–11 Cross section of a large vein (left) and artery (right). Compare the differences between the thin-walled vena cava on the left and the thick-walled aorta on the right.

(a)   (b)   (c)

Figure 8–12 Veins contain valves to prevent the backward flow of blood. (a) External view of the vein shows wider area of the valve. (b) Internal view with the valve open as blood flows through. (c) Internal view with the valve closed.

Similar to arteries, veins usually are named for the area from which they take blood away. The **jugular** (juhg-yoo-lahr) **vein** drains the head and neck area (*jugulum* is Latin for throat), **femoral veins** drain the legs, and **renal veins** drain the kidneys. An exception to this naming structure is the azygous vein. The **azygous** (ahz-ih-gihs) **vein** is a single vein that drains the chest wall and adjacent structures (the prefix **a-** means without, and the combining form **zygon** means yoke or pair) and is named based on the fact that it is not paired in the body (Figure 8–13).

Pressure

Blood pressure is the tension exerted by blood on the arterial walls. It is determined by the energy produced by the heart, the elasticity of the arterial walls, and the volume and **viscosity** (vihs-koh-siht-ē) (resistance to flow) of the blood. The **pulse** (puhlz) is the rhythmic expansion and contraction of an artery produced by pressure.

Blood pressure is measured by a **sphygmomanometer** (sfihg-mō-mah-nohm-eh-tər), which measures the amount of pressure exerted against the walls of the vessels.

Sphygm/o is the combining form for pulse, **man/o** is the combining form for pressure, and the suffix **-meter** means device. **Systolic** (sihs-stohl-ihck) **pressure** occurs when the ventricles contract and is highest toward the end of the stroke output of the left ventricle. **Diastolic** (dī-ah-stohl-ihck) **pressure** occurs when the ventricles relax and is lowest late in ventricular dilation.

The combining form **tensi/o** means pressure or tension and is used when describing blood pressure. **Hypertension** is high blood pressure (**hyper-** = excessive or above normal), and **hypotension** is low blood pressure (**hypo-** = deficient or less than normal). Drugs used to lower blood pressure are called **antihypertensives** (ahn-tih-hī-pər-tehns-ihvs).

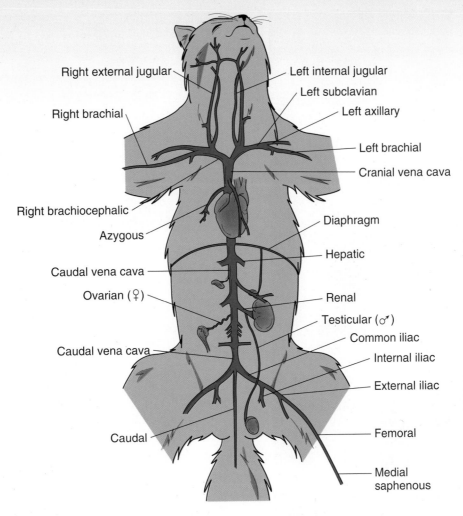

Figure 8–13 Major veins in the cat.

Right external jugular
Left internal jugular
Left subclavian
Right brachial
Left axillary
Left brachial
Cranial vena cava
Right brachiocephalic
Diaphragm
Azygous
Hepatic
Caudal vena cava
Ovarian (♀)
Renal
Testicular (♂)
Caudal vena cava
Common iliac
Internal iliac
External iliac
Femoral
Caudal
Medial saphenous

TEST ME: CARDIOVASCULAR SYSTEM

Diagnostic tests performed on the cardiovascular system include the following:

- **angiocardiography** (ahn-jē-ō-kahr-dē-ohg-rah-fē) = radiographic study of the blood vessels and heart using contrast material. The resulting film is an **angiocardiogram** (ahn-jē-ō-kahr-dē-ō-grahm).
- **angiography** (ahn-jē-ohg-rah-fē) = radiographic study of the blood vessels following injection of radiopaque material. An **angiogram** (ahn-jē-ō-grahm) is the film produced from this radiographic procedure (Figure 8–14).
- **cardiac catheterization** (kahr-dē-ahck kahth-eh-tər-ih-zā-shuhn) = radiographic study in which a catheter is passed into a blood vessel and is guided into the heart to detect pressures and patterns of blood flow.
- **echocardiography** (ehck-ō-kahr-dē-ohg-rah-fē) = process of evaluating the heart structures using sound waves. (**Ech/o** is a combing form for sound.) **Doppler**

(dohp-lər) **echocardiography** uses the differences in frequency between sound waves and their echoes to measure the velocity of a moving object (Figure 8–15).

- **electrocardiography** (ē-lehck-trō-kahr-dē-ohg-rahf-ē) = process of recording the electrical activity of the heart. An **electrocardiogram** (ē-lehck-trō-kahr-dē-ō-grahm) is the record of the electrical activity of the heart and is abbreviated ECG or EKG (Figure 8–16). The machine that records the electrical activity of the heart is an **electrocardiograph** (ē-lehck-trō-kahr-dē-ō-grahf).
- **Holter monitor** (hōl-tər mohn-ih-tər) = 24-hour ECG that records the heart rates and rhythms onto a specialized tape recorder.
- **radiography** (rā-dē-ohg-rah-fē) = procedure of imaging objects by exposing sensitized film to X-rays. The resulting film is called a **radiograph** (rā-dē-ō-grahf) or X-ray (Figure 8–17a). (This is an exception to the **-gram, -graph, -graphy** organization.)
- **tourniquet** (toor-nih-keht) = constricting band applied to a limb to control bleeding or to assist in drawing blood.

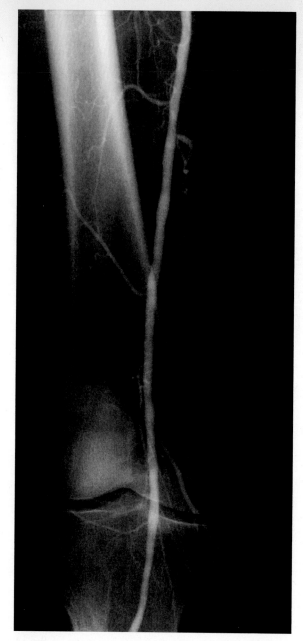

Figure 8–14   Angiogram of a dog showing the femoral arteries. The use of a contrast medium makes the arteries visible.

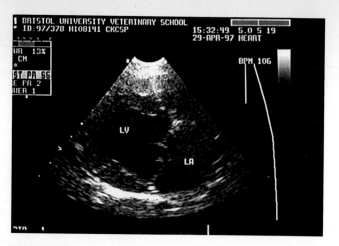

Figure 8–15   Echocardiogram of a canine heart. (Courtesy of Mark Jackson, DVM, PhD, Glasgow University.)

Doppler echocardiography is based on the Doppler effect. An example of the Doppler effect (differences in frequency between sound waves and their echoes to measure the velocity of a moving object) is a police siren. A police siren seems to have a higher pitch when the police car is approaching than after it passes. Similarly, the pitch of a train whistle is higher when the train is approaching than after it passes. Medical use of the Doppler effect enables detection of blood flow based on various pitches after the blood passes by a transducer. Christian Doppler, an American mathematician, was the first person to describe this effect.

PATHOLOGY: CARDIOVASCULAR SYSTEM

Pathologic conditions of the cardiovascular system include the following:

- **aneurysm** (ahn-yoo-rihzm) = localized balloonlike enlargement of an artery.
- **angiopathy** (ahn-jē-ohp-ah-thē) = disease of vessels.
- **aortic insufficiency** (ā-ōr-tihck ihn-sah-fihsh-ehns-ē) = inability of the aortic valve to perform at the proper levels, which results in blood flowing back into the left ventricle from the aorta.
- **atherosclerosis** (ahth-ər-ō-skleh-rō-sihs) = hardening and narrowing of the arteries. This may be caused by **plaque** (plahck), which is a patch or raised area. **Ather/o** is the combining form for plaque or fatty substance.
- **cardiac tamponade** (kahr-dē-ahck tahm-pō-nohd) = compression of the heart due to fluid or blood collection in the pericardial sac.
- **cardiomegaly** (kahr-dē-ō-mehg-ah-lē) = heart enlargement (Figure 8–17b).
- **cardiomyopathy** (kahr-dē-ō-mī-ohp-ah-thē) = disease of heart muscle. May be further classified as **hypertrophic** (hī-pər-trō-fihck), which is excessive growth of the left ventricle, or **dilated** (dī-lāt-ehd), which is characterized by a thin-walled left ventricle.

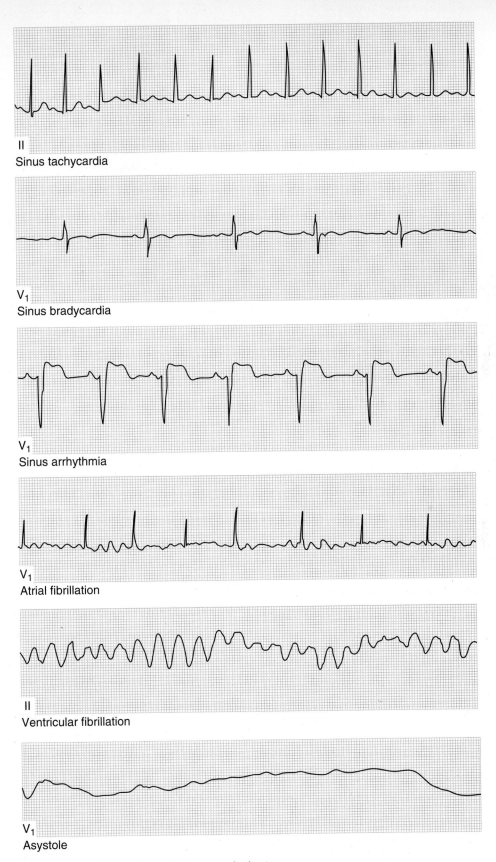

II
Sinus tachycardia

V₁
Sinus bradycardia

V₁
Sinus arrhythmia

V₁
Atrial fibrillation

II
Ventricular fibrillation

V₁
Asystole

Figure 8–16   ECG tracings of common arrhythmias.

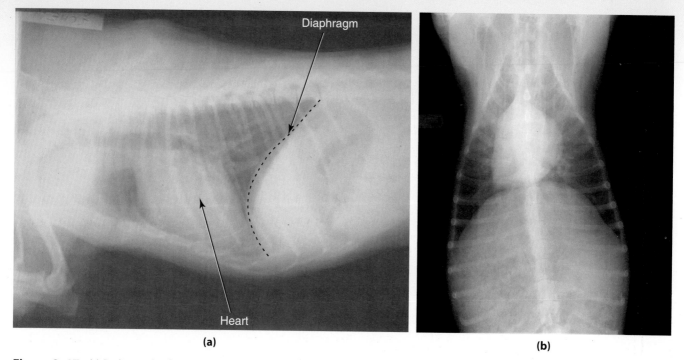

Figure 8–17 (a) Radiograph of a dog's chest showing the position of the heart in the thoracic cavity. (b) Radiograph of a dog's heart with cardiomegaly. Right-sided heart enlargement in a dog with heartworm disease gives the reverse D appearance of the heart in this radiograph. [(b) Courtesy of Country Hills Pet Hospital.]

Dilated cardiomyopathy also is known as **congestive** (kohn-jēhs-tihv) (Figures 8–18a and b).

- **carditis** (kahr-dī-tihs) = inflammation of the heart.
- **congestive** (kohn-jēhs-tihv) **heart failure** = syndrome that reflects insufficient cardiac output to meet the body's needs; abbreviated CHF. **Congestion** (kohn-jehs-chuhn), which is accumulation of fluid, and **edema** (eh-dē-mah), which is accumulation of fluid in the intercellular spaces, may be seen with CHF. **Ascites** (ah-sī-tēz) is fluid accumulation in the peritoneal cavity seen in dogs secondary to CHF and other diseases. **Pleural effusion** (ploor-ahl eh-fū-zhuhn) is abnormal fluid accumulation between the layers of the membrane encasing the lungs and is seen in cats secondary to CHF. Fluid accumulation can be relieved with the use of diuretics. **Diuretics** (dī-yoo-reht-ihcks) are substances that increase urine excretion.
- **cor pulmonale** (kōr puhl-mah-nahl-ē) = alterations in the structure or function of the right ventricle caused by pulmonary hypertension; also called pulmonary heart disease. **Cor** means heart, and **pulmon/o** is the combining form for lung.
- **dirofilariosis** (dī-rō-fihl-ahr-ē-ō-sihs) = heartworm infection; formerly called **dirofilariasis** (dī-rō-fihl-ahr-ē-ah-sihs). The scientific name of heartworm is *Dirofilaria immitis* (dī-rō-fihl-ahr-ē-ah ihm-ih-tihs), from which dirofilariosis is derived. Heartworm disease is found in dogs, cats, and ferrets (Figure 8–19).

Heartworms mature and breed in the larger blood vessels. Mature heartworms produce tiny larvae called **microfilariae** (mī-krō-fihl-ahr-ē-ah). Mature heartworms may obstruct blood flow through the heart and blood vessels. A dead heartworm can cause pulmonary embolism. Obstruction of blood flow from the vena cava caused by heavy heartworm infestation is called **caval** (kā-vahl) **syndrome.** Heartworm disease can be prevented by the use of **prophylactic** (prō-fih-lahck-tihck) medication. **Prophylaxis** (prō-fih-lahck-sihs) means prevention. If an animal has heartworm disease, treatment includes use of an **adulticide** (ah-duhlt-ih-sīd), or substance that kills mature or adult heartworms, and a **microfilaricide** (mī-krō-fihl-ahr-ih-sīd), or substance that kills larvae or juvenile heartworms.

- **embolus** (ehm-bō-luhs) = foreign object (e.g., a clot, air, or tissue) that is circulating in blood. An **embolism** (ehm-bō-lihzm) is blockage of a vessel by a foreign object.
- **endocarditis** (ehn-dō-kahr-dī-tihs) = inflammation of the endocardium and sometimes the heart valves (Figure 8–20). Endocarditis may be further modified, as in bacterial endocarditis.
- **heart block** = interference with the electrical conduction of the heart. Heart block may be partial or complete and is graded in degrees based on the characteristics of the block.

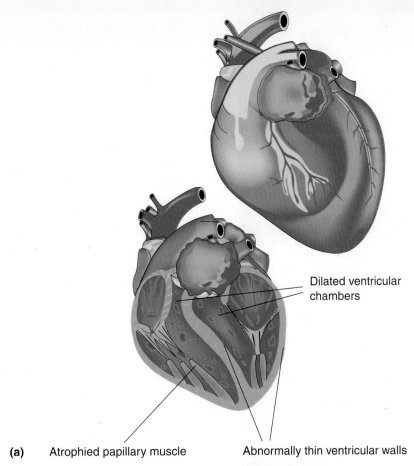

Dilated ventricular
chambers

(a)   Atrophied papillary muscle   Abnormally thin ventricular walls

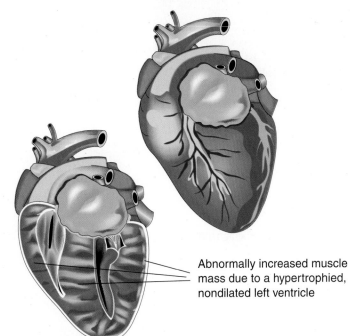

Abnormally increased muscle
mass due to a hypertrophied,
nondilated left ventricle

(b)

Figure 8–18   (a) Dilated cardiomyopathy. (b) Hypertrophic cardiomyopathy.

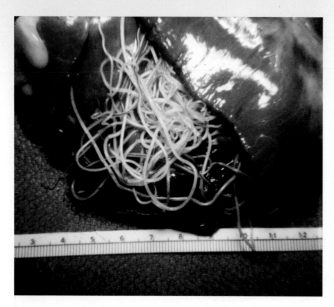

Figure 8–19 Heart with heartworms.

Figure 8–20 An AV valve with endocarditis. This valve would not create a tight seal, resulting in a murmur. (Courtesy of Dr. Arthur Hattel, Pennsylvania State University.)

- **hemangioma** (hē-mahn-jē-ō-mah) = benign tumor comprised of newly formed blood vessels.
- **hematoma** (hē-mah-tō-mah) = collection of blood.
- **hypercapnia** (hī-pər-kahp-nē-ah) = above-normal levels of carbon dioxide. Hypercapnia results in reduced levels of oxygen (**hypoxia** = hī-pohck-sē-ah) and may cause a bluish tinge to the skin and mucous membranes. This bluish tinge is called **cyanosis.** (The combining form **cyan/o** means blue.)
- **hypocapnia** (hī-pō-kahp-nē-ah) = below-normal levels of carbon dioxide.

- **hypoxia** (hī-pohck-sē-ah) = below-normal levels of oxygen.
- **infarct** (ihn-fahrckt) = localized area of necrosis caused by an interrupted blood supply.
- **ischemia** (ihs-kē-mē-ah) = deficiency in blood supply (the combining form **isch/o** means hold back) (Figure 8–21).
- **mitral valve insufficiency** (mī-trahl vahlv ihn-sah-fihsh-ehn-sē) = inability of the left atrioventricular valve to perform at the proper level; may be caused by fibrosis, endocarditis, or other conditions that occur in the mitral valve area.
- **mitral valve prolapse** (mī-trahl vahlv prō-lahps) = abnormal protrusion of the left atrioventricular valve that results in incomplete closure of the valve.
- **myocarditis** (mī-ō-kahr-dī-tihs) = inflammation of the myocardium.
- **occlusion** (ō-kloo-shuhn) = blockage in a vessel or passageway in the body.
- **patent ductus arteriosus** (pā-tehnt duhck-tuhs ahr-tē-rē-ō-sihs) = persistence of the fetal communication (ductus arteriosus) between the left pulmonary artery and aorta that should close shortly after birth; abbreviated PDA. (*Patent* means remaining open.) A PDA may cause overloading of the left ventricle, which may lead to left ventricular failure. A continuous heart murmur and enlarged heart are signs of a PDA.
- **pericarditis** (pehr-ih-kahr-dī-tihs) = inflammation of the pericardium.
- **pulmonic stenosis** (puhl-mah-nihck stehn-ō-sihs) = narrowing of the opening and valvular area between the pulmonary artery and right ventricle.
- **regurgitation** (rē-gərj-ih-tā-shuhn) = backflow; used to describe backflow of blood caused by imperfect closure of heart valves.

Figure 8–21 Ischemia to a section of the small intestine of a horse (dark area on right). (Courtesy of Dr. Arthur Hattel, Pennsylvania State University.)

- **shock** (shohck) = inadequate tissue perfusion. There are different types of shock, but one type occurs after cardiac arrest or cessation of heartbeat. Treatment of shock includes **resuscitation** (reh-suhs-ih-tā-shuhn), or the restoration of life. Resuscitative measures include fluid administration, cardiac massage, and artificial respiration. **Cardiopulmonary resuscitation** (kahr-dē-ō-puhl- mohn-ār-ē), or CPR, addresses only the cardiac and respiratory systems.
- **tetralogy of Fallot** (teht-rahl-ō-jē ohf fahl-ō) = congenital cyanotic cardiac condition that classically has four anatomical defects in the heart: pulmonary stenosis, ventricular septal defect, overriding aorta, and right ventricular hypertrophy.
- **thrombus** (throhm-buhs) = blood clot attached to the interior wall of a vein or artery. A **thrombosis** (throhm-bō-sihs) is an abnormal condition in which a blood clot develops in a blood vessel. Substances that prevent blood clotting are called **anticoagulants** (ahn-tih-kō-āg-yoo-lahnts).
- **vasculitis** (vahs-kyoo-lī-tihs) = inflammation of a blood or lymph vessel.
- **ventricular septal defect** (vehn-trihck-yoo-lahr sehp-tahl dē-fehckt) = opening in the wall dividing the right and left ventricles that may allow blood to shunt from the right ventricle to the left ventricle without becoming oxygenated. To **shunt** (shuhnt) means to bypass or divert. A shunt resulting from a ventricular septal defect would bypass the lungs. Ventricular septal defect is abbreviated VSD. A harsh holosystolic murmur usually is a sign of a VSD.

Terms used to describe disease

congenital (kohn-jehn-ih-tahl) present at birth

hereditary (hər-eh-dih-tahr-ē) genetically transmitted from parent to offspring

anomaly (ah-nohm-ah-lē) deviation from normal

idiopathic (ihd-ē-ō-pahth-ihck) of unknown cause

iatrogenic (ī-aht-rō-jehn-ihck) produced by treatment

PROCEDURES: CARDIOVASCULAR SYSTEM

Procedures performed on the cardiovascular system include the following:

- **angioplasty** (ahn-jē-ō-plahs-tē) = surgical repair of blood or lymph vessels. Angioplasties may be **transluminal** (trahnz-loo-mehn-ahl), which means the procedure is done through the opening of a vessel, or **percutaneous** (pehr-kyoo-tā-nē-uhs), which means the procedure is done through the skin.
- **angiorrhaphy** (ahn-jē-ōr-ah-fē) = suture of a vessel.
- **arteriectomy** (ahr-tē-rē-ehck-tō-mē) = surgical removal of part of a blood vessel that carries blood away from the heart.
- **arteriotomy** (ahr-tē-rē-oh-tō-mē) = incision of a blood vessel that carries blood away from the heart.
- **central venous pressure** (sehn-trahl vē-nuhs prehs-shər) = tension exerted by blood in the cranial vena cava; abbreviated CVP. CVP is monitored by catheterization of the cranial vena cava via the jugular vein. The catheter is connected to a fluid-filled column and a syringe or bag that serves as a fluid source.
- **defibrillation** (dē-fihb-rih-lā-shuhn) = use of electrical shock to restore the normal heart rhythm.
- **hemostasis** (hē-mō-stā-sihs) = control or stoppage of bleeding.
- **stent** (stehnt) = small expander implanted in a blood vessel to prevent it from collapsing. (A stent also is a device to hold tissue in place or to provide support for a graft.)
- **transfusion** (trahnz-fū-shuhn) = introduction of whole blood or blood components into the bloodstream of the recipient (Figure 8–22).
- **valvotomy** (vahl-vah-tō-mē) = surgical incision into a valve or membranous flap.

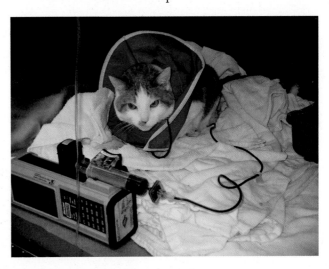

Figure 8–22 A cat receiving a transfusion. (Courtesy of Kimberly Kruse Sprecher, CVT.)

REVIEW EXERCISES

Multiple Choice

Choose the correct answer.

1. The right atrioventricular valve is also known as the

    a. mitral valve
    b. semilunar valve
    c. tricuspid valve
    d. bicuspid valve

2. The double-walled membranous sac enclosing the heart is the

    a. peritoneum
    b. pericardium
    c. perimyocardium
    d. pericardosis

3. A partition or wall separating something is called a

    a. septum
    b. valve
    c. lumen
    d. plaque

4. A localized area of necrosis caused by an interrupted blood supply is

    a. ischemia
    b. resuscitation
    c. pulse
    d. infarct

5. Introduction of whole blood or blood components into the bloodstream of the recipient is a(n)

    a. embolism
    b. thrombus
    c. transfusion
    d. stent

6. A bypass or diversion is called a

    a. preload
    b. shunt
    c. stent
    d. tourniquet

7. *Cor* means

    a. abnormality
    b. vessel
    c. heart
    d. valve

8. Disease of heart muscle is

    a. cardiopathy
    b. cor pulmonale
    c. cardiovalvopathy
    d. cardiomyopathy

9. Heart enlargement is

    a. cardiac swelling
    b. cardiac augmentation
    c. cardiac dilation
    d. cardiomegaly

10. Blood flow through tissue is

    a. ischemia
    b. infarct
    c. auscultation
    d. perfusion

11. The blood vessels that carry blood from the heart to the lungs are the

    a. pulmonary veins
    b. pulmonary arteries
    c. vena cava
    d. aorta

12. The contraction phase of the heartbeat is the

    a. septum
    b. diastole
    c. systole
    d. tachycardia

13. A disease produced by treatment is known as

    a. idiopathic
    b. iatrogenic
    c. congenital
    d. hereditary

14. The term for the external layer of the heart is the

    a. myocardium
    b. endocardium
    c. pericardium
    d. epicardium

15. The myocardium receives its blood supply from the

    a. aorta

    b. coronary arteries

    c. vena cava

    d. subclavian artery

16. Which heart valve is also known as the biscupid valve?

    a. aortic semilunar valve

    b. pulmonary semilunar valve

    c. mitral valve

    d. tricuspid valve

17. Which heart chamber pumps blood to the lungs?

    a. left atrium

    b. left ventricle

    c. right atrium

    d. right ventricle

18. Hemostasis means

    a. control of bleeding

    b. formation of new blood

    c. introduction of whole blood

    d. present at birth

19. A deficiency in blood supply is called

    a. infarct

    b. ischemia

    c. prophylaxis

    d. dilation

20. The opening in a vessel through which fluid flows is known as a

    a. hilum

    b. perfusion

    c. murmur

    d. lumen

Matching

Match the term in Column I with the definition in Column II.

Column I	Column II
1. _____ perfusion	a. vessels that return blood to the heart
2. _____ heart murmur	b. single-cell-thick vessels that connect the arterial and venous systems
3. _____ constriction	c. vessels that carry blood away from the heart
4. _____ vasculitis	d. blood flow through tissues
5. _____ dilation	e. prevention
6. _____ hematoma	f. abnormal sound associated with the turbulent flow of blood
7. _____ capillaries	g. narrowing of a vessel diameter
8. _____ veins	h. widening of a vessel diameter
9. _____ arteries	i. inflammation of a blood or lymph vessel
10. _____ prophylaxis	j. collection of blood

Fill in the Blanks

1. Angi/o and vas/o mean _____.

2. Ven/o and phleb/o mean _____.

3. Valv/o and valvul/o mean _____.

4. Cardi/o and cor mean _____.

5. Holo- and pan- mean _____.

Trace the flow of blood through the heart.

6. Blood enters the heart through two large veins called the _____ and _____, which empty blood into the _____.

7. Blood flows from the right atrium to the right ventricle through the _____, which also is called the _____.

8. After blood enters the right ventricle, it passes through the _____ and enters the lungs via the _____.

9. Blood becomes oxygenated in the lungs and returns to the heart via the _____.

10. Once blood enters the left atrium, it passes through the _____ or _____ on its way to the _____.

11. Blood passes through one last valve, the _____, before it enters the aorta and flows to various parts of the body.

Spelling

Circle the term that is spelled correctly.

1. bluish tinge to the skin and mucous membranes: cianosis   cyianosis   cyanosis

2. below-normal levels of carbon dioxide: hypokapnea   hypocapnia   hypocapnea

3. widening of a vessel: vasodialation   vasodilation   vasodilatation

4. relaxation of the heart: diastole   diastoly   diastooly

5. contraction of the ventricles: systole   sistolle   sistole

CROSSWORD PUZZLE

Cardiovascular Structure and Disease Terms Puzzle

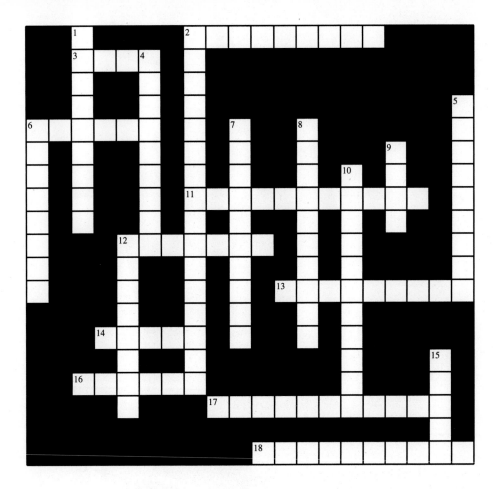

Across

2 to narrow
3 combining form for sound
6 vibration felt on palpation
11 abnormally rapid heart rate
12 deviation from normal
13 resistance to flow
14 craniodorsal chambers of the heart
16 to widen
17 above-normal levels of carbon dioxide
18 constricting band applied to limb to control bleeding or to assist in drawing blood

Down

1 caudoventral chamber of the heart
2 compression of the heart due to fluid or blood collection in the pericardial sac
4 blockage in a vessel
5 pertaining to the lungs
6 blood clot attached to the interior wall of a vessel
7 listening
8 external layer of the heart
9 blood vessel that returns blood to the heart
10 abnormally slow heart rate
12 without contraction
15 portion of the EKG that represents ventricular contraction

LABEL THE DIAGRAMS

Label the diagrams in Figures 8–23 and 8–24.

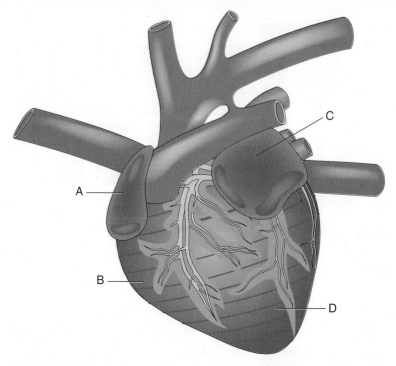

Figure 8–23   External heart structures. Label the external heart structures to which the lines are pointing.

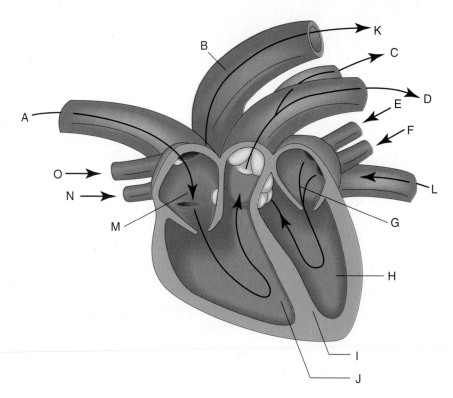

Figure 8–24   Internal heart structures. Label the internal heart structures to which the lines are pointing.

CASE STUDIES

Define the underlined terms in each case study.

A 5-yr-old M/N Doberman was presented with signs of lethargy, syncope, and cyanotic mucous membranes. After auscultating the heart, the veterinarian detected a cardiac arrhythmia and tachycardia. An ECG and radiograph of the heart were ordered. The radiograph revealed cardiomegaly, which helped to support the veterinarian's diagnosis of cardiomyopathy.

1. lethargy _____

2. syncope _____

3. cyanotic _____

4. cardiac arrhythmia _____

5. tachycardia _____

6. ECG _____

7. radiograph _____

8. cardiomegaly _____

9. diagnosis _____

10. cardiomyopathy _____

A farmer called the clinic because one of his cows was suddenly off feed and had not been producing as much milk as before. Upon arrival at the farm, the veterinarian noted that the cow was reluctant to move, had an arched back, and appeared tachypnic. PE revealed tachycardia, tachypnea, dyspnea, pyrexia, and abducted elbows. Auscultation of the thorax revealed muffled lung and heart sounds. The farmer was questioned as to his use of magnets to prevent metallic objects from staying in the rumen or reticulum. The owner did not use magnets as a prophylactic measure, so the veterinarian suspected acute traumatic reticuloperitonitis in this cow. Acute traumatic reticuloperitonitis is commonly called hardware disease and is seen when swallowed metallic objects fall into the reticulum of the ruminant stomach, pierce the reticulum wall, and contaminate the peritoneal cavity. Occasionally, the object punctures the diaphragm, enters the thoracic cavity, and punctures the pericardial sac, causing pericarditis. A magnet was placed in this cow via a balling gun (Figure 8–25), and antibiotics were initiated. If the cow does not improve, she may be sent to slaughter because of the high cost of treating this disease.

11. off feed _____

12. PE _____

13. tachycardia _____

14. tachypnea _____

15. dyspnea _____

16. pyrexia _____

17. thorax _____

18. rumen _____

19. prophylactic _____

20. acute _____

21. traumatic _____

22. reticuloperitonitis   _____

23. acute traumatic reticuloperitonitis _____

24. reticulum   _____

25. ruminant   _____

26. peritoneal cavity   _____

27. diaphragm   _____

28. pericardial sac   _____

29. pericarditis   _____

30. balling gun   _____

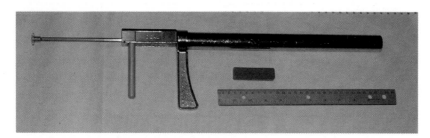

Figure 8–25 Balling gun used to administer oral medication to cattle. The rectangular gray magnet remains in the reticulum, where it can trap ingested metal.

CHAPTER 9

A BREATH OF FRESH AIR

Objectives

Upon completion of this chapter, the reader should be able to

- Identify and describe the major structures and functions of the respiratory system
- Recognize, define, spell, and pronounce terms related to the diagnosis, pathology, and treatment of the respiratory system

FUNCTIONS OF THE RESPIRATORY SYSTEM

The **respiratory** (rehs-pih-rah-tōr-ē) **system** is the body system that brings oxygen from the air into the body for delivery via the blood to the cells. Once the blood has delivered the oxygen to the cells, it picks up carbon dioxide and carries it back to the lungs, where this waste is expelled into the air. Carbon dioxide has acid properties and therefore also is involved in maintaining the body's acid–base status.

The term **respiration** means the diffusion of gases (oxygen and carbon dioxide) between the atmosphere and the cells of the body. The gas exchange between the blood and the cells is called internal or cellular respiration. External respiration is the absorption of atmospheric oxygen by the blood in the lungs and the diffusion of carbon dioxide from the blood in the lungs to atmospheric air.

Ventilation (vehn-tih-lā-shuhn) is a term that means the intake of fresh air. Ventilation is used to refer to breathing. Ventilation may be natural, as in normal breathing, or assisted, as in the use of a ventilator. Ventilators are devices that aid in breathing and therefore should not be called respirators.

STRUCTURES OF THE RESPIRATORY SYSTEM

The respiratory tract is routinely divided into the upper and lower respiratory tracts. The **upper respiratory tract** (uhp-pər rehs-pih-rah-tōr-ē trahckt) is the part of the respiratory system that consists of the nose, mouth, pharynx, epiglottis, and larynx. The **lower respiratory tract** (lō-ər rehs-pih-rah-tōr-ē trahckt) is the part of the respiratory system that consists of the trachea, bronchial tree, and lungs. The trachea sometimes is considered part of the upper respiratory tract (Figure 9–1).

Upper Respiratory Tract

The conducting passages of the upper respiratory tract consist of the nose, mouth, pharynx, epiglottis, and larynx. These structures open to the outside and are lined with mucous membranes that may or may not have cilia to help filter the air.

Nose

Air enters and exits the body through the **nose. Nas/o** and **rhin/o** are combining forms for nose. In animals, not all noses look alike. The rigidity of the nose in swine has led to its being called the **snout** (snowt) because it is so different from other species' noses. The nose consists of **nostrils,** or **nares** (nehr-ēz), which are the paired external openings of the respiratory tract.

Nares vary from species to species and may have the ability to open widely, as in equine, or may remain the same, as in the canine. In **endotherms** (ehn-dō-thərmz), or warm-blooded animals, the nasal passages contain **nasal turbinates** (tər-bih-nātz)

> **Rhin/o** is the Greek root for nose. Consider the term *rhinoceros*. Rhin/o means nose, and **cer/o** means horn; therefore, a rhinoceros is an animal with a horn on its nose. Where do you think a rhinovirus causes disease?

(Figure 9–2). The nasal turbinates, sometimes called conchae (kohn-kā), are scroll-like cartilages covered with highly vascular mucous membranes. The nasal turbinates warm, humidify, and filter inspired air. Two nasal turbinates, the dorsal and ventral, separate the nasal cavity into passages. Each passage is called a **meatus** (mē-ā-tuhs) (**meat/o** is the combining form for opening or passageway), and they are named based on their location: dorsal meatus, middle meatus, ventral meatus, and common meatus. A **nasogastric** (nā-zō-gahs-trihck) **tube** is a tube that passes through the nose down to the stomach. A nasogastric tube is placed through the ventral nasal meatus.

Air passes from the nose through the nasal cavity. The rostral part of the nostrils and nasal cavity are called the **vestibule** (vehs-tih-buhl). The nose is divided by a wall of cartilage called the nasal septum (nā-zahl sehp-tuhm). The combining form **sept/o** means partition.

The respiratory system is lined with **mucous membrane** (myoo-kuhs mehm-brān), which is a specialized form of

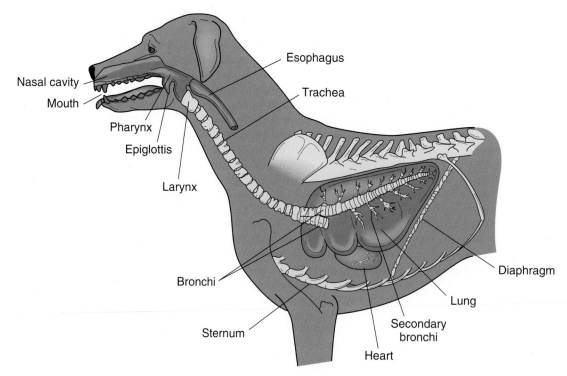

Figure 9–1   Structures of the respiratory system.

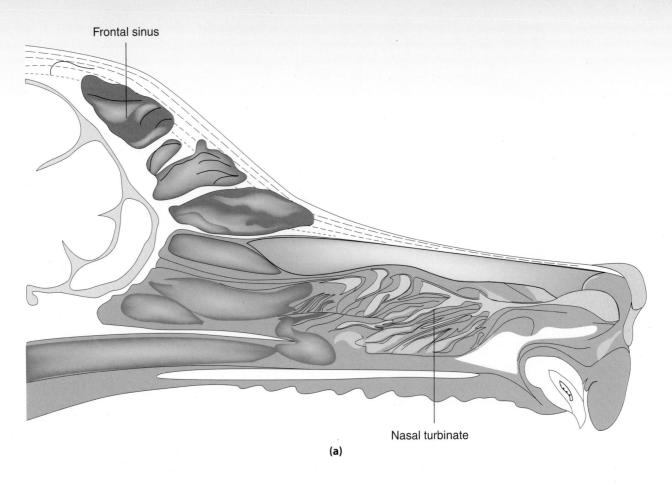

Frontal sinus

Nasal turbinate

(a)

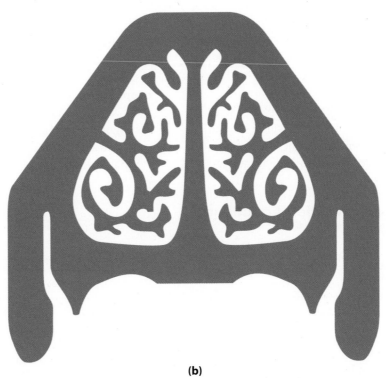

(b)

Figure 9–2   (a) Internal structures of the nasal cavity of a dog. (b) Cross section of the turbinates of a dog.

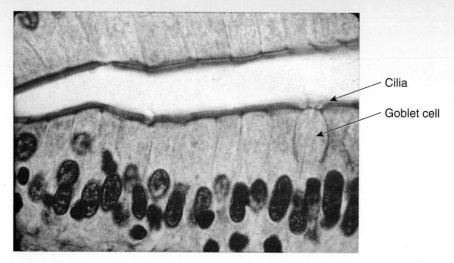

Cilia

Goblet cell

Figure 9–3   Respiratory epithelial tissue contains cilia and goblet cells.

epithelial tissue. The mucous membranes secrete **mucus** (myoo-kuhs). Mucus is a slimelike substance that is composed of glandular secretions, salts, cells, and leukocytes. Mucus helps to moisten, warm, and filter the air as it enters the nose. **Cilia** (sihl-ē-ah) are thin hairs located inside the nostrils (Figure 9–3). Cilia filter the air to remove **debris** (deh-brē). Debris is the remains of something destroyed or damaged.

The **olfactory** (ohl-fahck-tōr-ē) **receptors** are responsible for the sense of smell. The combining form **olfact/o** means smell. Olfactory receptors are nerve endings located in the mucous membranes of the nasal cavity.

The **tonsils** (tohn-sihlz) are lymphatic tissue that protects the nasal cavity and proximal (upper) throat. The combining form for tonsil is **tonsill/o.** The tonsils are discussed in Chapter 15.

Sinuses

A **sinus** (sīn-uhs) is an air-filled or fluid-filled space. A sinus in the respiratory system is an air-filled or fluid-filled space in bone. Sinuses have a mucous membrane lining. The functions of sinuses are to provide mucus, to make bone lighter, and to help to produce sound. The combining form for sinus is **sinus/o** (Table 9–1).

Pharynx

Air passes through the nasal cavity to the **pharynx** (far-ihnks). The pharynx is commonly called the throat. The pharynx is the common passageway for the upper respiratory and gastrointestinal tracts. The pharynx extends from the caudal part of the nasal passages and mouth to the larynx and connects the nasal passages to the larynx and the mouth to the esophagus. The combining form for throat is **pharyng/o.**

The pharynx has three divisions:

- **nasopharynx** (nā-zō-far-ihnckz) = portion of the throat posterior to the nasal cavity and above (dorsal to) the soft palate.
- **oropharynx** (ō-rō-far-ihnckz) = portion of the throat between the soft palate and epiglottis.
- **laryngopharynx** (lah-rihng-gō-far-ihnckz) = portion of the throat below the epiglottis that opens into the voice box and esophagus.

Table 9–1   Sinuses and Their Locations

Sinus	Species Found	Location
frontal (frohn-tahl)	all domestic species	dorsal part of skull between nasal cavity and orbit
maxillary (mahx-ihl-ār-ē)	all domestic species	maxilla with nasal cavity on each side (maxillary recess in carnivores)
sphenoid (sfehn-oyd)	feline, bovine, equine, swine	sphenoid bone; opens to nasal cavity
palatine (pahl-eh-tīn)	ruminants, equine	palatine bone; communicates with maxillary sinus
lacrimal (lahck-rih-mahl)	swine, ruminants	lacrimal bone
conchal (kohn-kahl)	swine, ruminants, equine	formed by enclosure of conchae

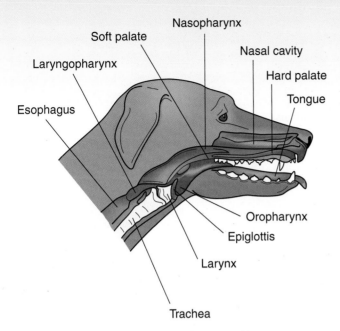

Figure 9–4 Structures of the nasal cavity and larynx.

The nasopharynx is the passageway for air entering through the nose, whereas the oropharynx and laryngopharynx are passageways for air entering through the nose and food entering through the mouth, respectively. During swallowing, the **soft palate** (pahl-aht) moves dorsally and caudally to close off the nasopharynx to prevent food from going into the nasal cavity.

Palate means roof of the mouth, and the combining form for palate is **palat/o.** The **epiglottis** (ehp-ih-gloht-ihs) acts like a lid and covers the larynx during swallowing. The covering of the larynx by the epiglottis does not allow food to enter the trachea and go into the lungs. The combining form for epiglottis is **epiglott/o** (Figure 9–4).

Larynx

The **larynx** (lār-ihnckz) is the part of the respiratory tract located between the pharynx and trachea (Figure 9–5). The larynx is commonly called the **voice box.** The larynx contains the **vocal cords** (vō-kahl kōrdz), which are paired membranous bands in the larynx that help produce sound. **Laryng/o** is the combining form for the voice box.

The vocal apparatus is found in the **glottis** (gloh-tihs). The glottis is the space between the vocal cords. Air passing through the glottis causes vibration of the vocal cords that produces sound. The combining form for glottis is **glott/o.** The vocal apparatus of avian species is the **syrinx** (sehr-ihnks), which is located between the trachea and bronchi.

Lower Respiratory Tract

The conducting passages of the lower respiratory tract consist of the trachea, bronchi, bronchioles, and alveoli. These structures enter or are found in the thoracic cavity.

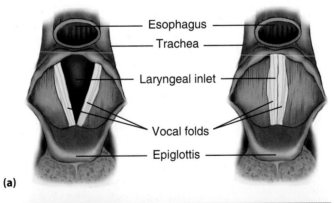

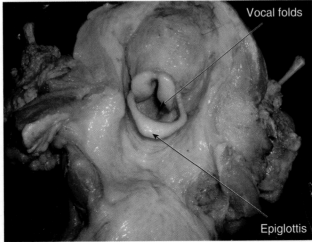

Figure 9–5 (a) Epiglottis and larynx. (b) Epiglottis and larynx of a bovine.

Trachea

Air passes from the larynx to the **trachea** (trā-kē-ah). The trachea is commonly called the **windpipe** and extends from the neck to the chest. The trachea attaches to the larynx in the neck and passes into the **thorax** (thōr-ahcks), or chest cavity, through the **thoracic** (thoh-rahs-ihck) **inlet.** The trachea is located ventral to the esophagus and is held open by a series of C-shaped cartilaginous rings. The open part of the Cs typically are along the dorsal aspect, which is adjacent to the esophagus. This allows easier expansion of the esophagus when the animal swallows. The trachea also is lined with cilia, which help filter debris. The combining form for the windpipe is **trache/o.**

Bronchi

The distal end of the trachea (bottom of the trachea) divides into two branches at the **tracheal bifurcation** (trāk-ē-ahl bī-fər-kā-shuhn). The branches from the trachea are called **bronchi** (brohng-kī). The combining form for bronchi is **bronch/o.** Each **bronchus** (brohng-kuhs) leads to a separate lung (right or left) and continues to divide. This continual division appears similar to a tree and its branches; therefore, the bronchi and its branches sometimes are called the **bronchial tree.** Each bronchus that leads to a separate lung is called a **principal** or **primary bronchus** (right principal or left principal bronchus). The principal bronchi divide into smaller branches called **secondary** (sehck-ohnd-ār-ē) or lobar (lō-bahr) **bronchi.** The secondary bronchi divide into **tertiary** (tər-shē-ār-ē) or **segmental** (sehg-mehn-tahl) **bronchi.** The tertiary bronchi are smaller units and are also called **bronchioles** (brohng-kē-ōlz) or **bronchiolus** (brohngk-ē-ō-luhs). The suffix **-ole** means small, indicating that bronchioles are smaller than bronchi. The combining form for bronchiole is **bronchiol/o.** Bronchioles contain no cartilage or glands. The bronchioles continue to divide. The terminal bronchioles are the last portion of a bronchiole that does not contain alveoli. The respiratory bronchioles are the final branches of the bronchioles. The respiratory bronchioles have alveolar outcroppings and branch into alveolar ducts (Figure 9–6).

Alveoli

Alveoli (ahl-vē-ō-lī) are air sacs in which most of the gas exchange occurs. An **alveolus** (ahl-vē-ō-luhs), which is Latin for small hollow thing, is a small grapelike cluster at the end of each bronchiole. The alveolus is connected to the bronchiole via an alveolar duct. The combining form **alveol/o** means small sac.

Alveoli have thin, flexible membrane walls that are surrounded by a network of microscopic capillaries. Gas exchange occurs across the alveolar membrane. Oxygen diffuses into the blood in alveolar capillaries and binds to the hemoglobin in erythrocytes. Carbon dioxide diffuses from the plasma

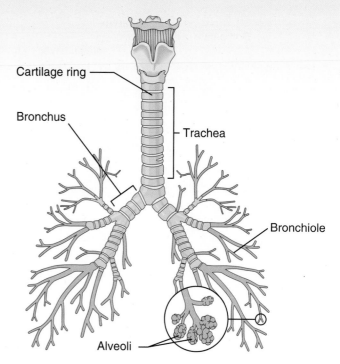

Figure 9–6 Lower respiratory tract.

across the alveolar membrane into the alveolus. When an animal exhales, much of this air is pushed out of the alveolus, back up through the respiratory tract, and out the nose or mouth.

Alveoli contain liquid that reduces alveolar surface tension. This liquid is called **surfactant** (sihr-fahck-tehnt). Surfactant prevents collapse of the alveoli during expiration.

Thorax

The thoracic cavity is contained within the ribs. The combining form **cost/o** means ribs. *Intercostal* means pertaining to between the ribs. The lungs are located in the thoracic cavity. The thoracic cavity also protects the lungs. The combining form **thorac/o** and the suffix **-thorax** mean chest cavity or chest.

The **lung** (luhng) is the main organ of respiration. There are two lungs (right and left) that are composed of divisions called **lobes** (lōbz). A lobe is a well-defined portion of an organ and is used in describing areas in the lung, liver, and other organs. The combining form **lob/o** means well-defined portion. The number and names of lung lobes vary between species. The combining forms **pneum/o, pneumon/o,** and **pneu** mean lung or air; the combining forms **pulm/o** and **pulmon/o** mean lung.

The term **parenchyma** (pahr-ehnk-ih-mah) refers to the functional elements of an organ, as opposed to its framework, or **stroma** (strō-mah). The functional elements of the lung are collectively called the lung parenchyma.

The region between the lungs is called the **mediastinum** (mē-dē-ahs-tī-nuhm). The mediastinum is the space between the lungs that houses the heart, aorta, lymph nodes, esophagus,

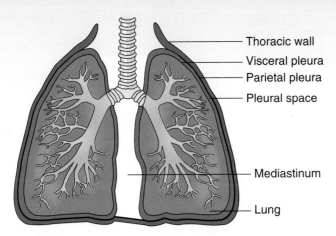

Figure 9–7 Respiratory structures of the thoracic cavity.

trachea, part of the bronchial tubes, nerves, thoracic duct, and thymus (Figure 9–7).

Pleura

Each lung is encased in a membranous sac called the **pleura** (ploor-ah). The combining form **pleur/o** means membrane surrounding the lung; the plural form of pleura is **pleurae.**

- **Parietal pleura** (pah-rī-eh-tahl ploor-ah) is the outer layer of the membrane lining the inner wall of the thoracic cavity (Figure 9–7).
- **Visceral pleura** (vihs-ər-ahl ploor-ah) is the inner layer of the membrane lining the outside of the lung.
- **Pleural** (ploor-ahl) **space** is the potential space between the parietal pleura and visceral pleura. The pleural space or pleural cavity contains a small amount of lubricating fluid called pleural fluid. Pleural fluid prevents friction when the membranes rub together during respiration and provides adhesive force to keep the lungs in contact with the chest wall as it expands during inspiration.

Diaphragm

The thoracic and peritoneal cavities are separated from each other by the **diaphragm** (dī-ah-frahm) (Figure 9–1). The prefix **dia-** means across, and the combining form **phragm/o** means wall. The diaphragm is a muscle, and contraction of the diaphragm causes air pressure in the lungs to drop below atmospheric pressure. This produces a vacuum in the thoracic cavity to draw in air. When the diaphragm relaxes, this negative pressure is no longer generated and air is forced out of the lung. The combining forms **diaphragmat/o** and **phren/o** mean diaphragm. A **diaphragmatic hernia** (dī-ah-frahg-mah-tihck hər-nē-ah) is an abnormal displacement of organs through the muscle separating the chest and abdomen (Figure 9–8). The phrenic nerve innervates the diaphragm.

> The combining form **phren/o** also is used to refer to the mind. The ancient Greeks once thought that the spleen and kidneys were the organs responsible for emotion, and because the diaphragm sits across the cranial portion of these organs, phren/o was used for both the diaphragm and the mind.

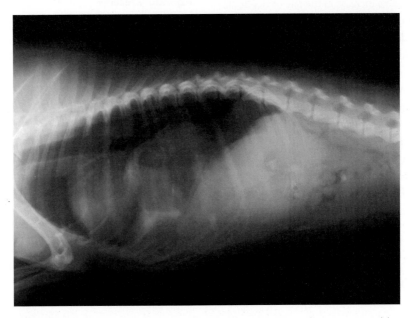

Figure 9–8 Radiograph of a diaphragmatic hernia. Intestinal organs are visible in the chest cavity.

BREATHING

Breathing is the inhalation and exhalation of air. **Inhalation** (ihn-hah-lā-shuhn) is the drawing in of breath. **Inspiration** is another term for the drawing in of breath. **Exhalation** (ehcks-hah-lā-shuhn) is the release of breath. **Expiration** is another term for the release of breath. *Spirare* in Latin means to breathe. The combining form **spir/o** also means breath or breathing. Most medical terms use the Greek word **-pnea** to refer to breathing.

- **apnea** (ahp-nē-ah) = absence of breathing.
- **dyspnea** (dihsp-nē-ah) = difficult or labored breathing.
- **bradypnea** (brād-ihp-nē-ah) = abnormally slow respiratory rates.
- **tachypnea** (tahck-ihp-nē-ah) = abnormally rapid respiratory rates.
- **hyperpnea** (hī-pərp-nē-ah) = abnormal increase in the rate and depth of respirations.
- **hypopnea** (hī-pōp-nē-ah) = abnormally slow or shallow respirations.
- **hyperventilation** (hī-pər-vehn-tih-lā-shuhn) = abnormally rapid deep breathing, which results in decreased levels of cellular carbon dioxide.
- **agonal** (āg-uh-nuhl) **breathing** = respirations near death or during extreme suffering.

Respiration involves the diffusion of oxygen (O_2) and carbon dioxide (CO_2). The combining forms **ox/i, ox/o,** and **ox/y** refer to O_2, and **capn/o** refers to CO_2.

Hypoxia (hī-pohck-sē-ah) refers to an inadequate supply of oxygen to tissue despite an adequate blood supply. **Hypercapnia** (hī-pər-kahp-nē-ah) refers to excessive amounts of carbon dioxide in the blood. **Hyperventilation** (hī-pər-vehn-tih-lā-shuhn), an abnormal increase in the rate or depth of breathing, may lead to **hypocapnia** (hī-pō-kahp-nē-ah), which is a decrease in the carbon dioxide levels in the blood.

When carbon dioxide dissolves in water, some of it reacts with the water to form carbonic acid (H_2CO_3). Because carbon dioxide breaks down into a weak acid, it affects the blood pH. An excessive amount of carbon dioxide in the blood due to decreased ventilation can lower the pH of blood; this is called **respiratory acidosis** (ah-sih-dō-sihs). If carbon dioxide levels are abnormally low due to increased ventilation, **respiratory alkalosis** (ahl-kah-lō-sihs) may result. Changes in blood pH also can result from metabolic factors (such as vomiting and renal disease) and are then called **metabolic acidosis** or **metabolic alkalosis** (Figure 9–9 and Table 9–2).

TEST ME: RESPIRATORY SYSTEM

Diagnostic procedures performed on the respiratory system include the following:

- **auscultation** (aws-kuhl-tā-shuhn) = act of listening. The respiratory tract is auscultated with a **stethoscope** (stehth-ō-skōp). Respiratory rhythm, rate, and sound are evaluated upon auscultation. Pathologic respiratory sounds are called **adventitious** (ahd-vehn-tish-uhs) **sounds.** Things to listen for include:
  - **bubbling** = sound of popping bubbles that suggests fluid accumulation.
  - **crepitation** (krehp-ih-tā-shuhn) = fine or coarse interrupted crackling noises coming from collapsed or fluid-filled alveoli during inspiration; also called **rales** (rahlz) or crackles.
  - **decreased lung sounds** = less or no sound of air movement, suggesting consolidation of lung tissue.
  - **respiratory rate** (RR) = number of respirations per minute. One inspiration and one expiration form a single respiration. RR varies with species.
  - **rhonchi** (rohn-kī) = abnormal, continuous, musical, high-pitched whistling sounds heard during inspiration; also called **wheezes** (wē-zehz).
  - **stridor** (strī-dōr) = snoring, squeaking, or whistling that suggests airway narrowing.
  - **vesicular** (vehs-ihck-yoo-lahr) **sounds** = sound resulting from air passing through small bronchi and alveoli.

Table 9–2   Lung Volume Terminology

tidal volume	amount of air exchanged during normal respiration (air inhaled and exhaled in one breath)
inspiratory reserve volume or **complemental air**	amount of air inspired over the tidal volume (extra amount that could be inhaled after normal inspiration)
expiratory reserve volume or **supplemental air**	amount of air expired over the tidal volume (extra amount that could be exhaled after normal expiration)
residual volume	air remaining in the lungs after a forced expiration (amount of air trapped in alveoli)
dead space	air in the pathway of the respiratory system (termed *dead* because this air is not currently participating in gas exchange)
minimal volume	amount of air left in alveoli after the lung collapses
vital capacity	largest amount of air that can be moved in the lung (tidal volume + inspiratory and expiratory reserve volumes)

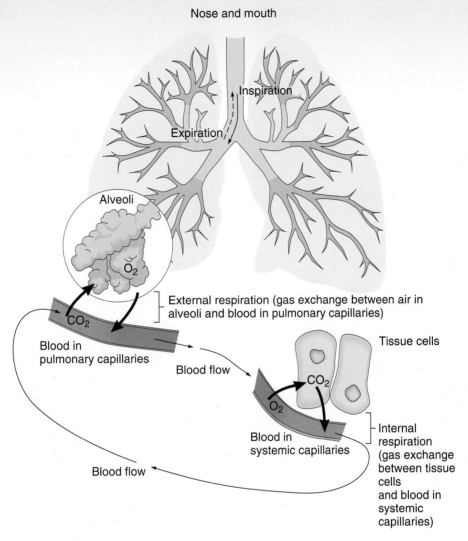

Figure 9–9 External and internal respiration.

- **bronchoalveolar lavage** (brohng-kō-ahl-vē-ō-lahr lah-vahj) = collection of fluid or mucus from the bronchi and/or alveoli via an endoscope or through an endotracheal tube inserted as far down the trachea caudally as possible before infusing fluid and aspirating a sample. Fluid may be used for cytologic examination.
- **bronchoscopy** (brohng-kohs-kō-pē) = visual examination of the bronchus. Bronchoscopy may be used to examine the bronchi for disease or foreign objects. A **bronchoscope** (brohng-kō-skōp) is an instrument used to visually examine the bronchus.
- **laryngoscopy** (lahr-ihng-gohs-kō-pē) = visual examination of the voice box. Laryngoscopy is used to examine the larynx for disease, tissue repair, or foreign objects. A **laryngoscope** (lahr-ihn-gō-skōp) is an instrument used to visually examine the voice box.
- **percussion** (pər-kuhsh-uhn) = diagnostic procedure used to determine density in which sound is produced by tapping various body surfaces with the finger or an

instrument. The sound produced over the chest where air is present differs from that of an area where fluid is present.
- **phlegm** (flehm) = thick mucus secreted by the respiratory lining. Mucus secretion from the lower respiratory tract is called **sputum** (spyoo-tuhm). Sputum can be used for cytologic examination.
- **radiography** (rā-dē-ohg-rah-fē) = image of internal structures created by exposure of sensitized film to X-rays (Figures 9–10a and b). Ultrasound does not work well for the respiratory system because the ultrasound beam cannot pass through a gas-containing structure to provide information about the internal structures.
- **spirometer** (spər-oh-mē-tər) = instrument used to measure air taken in and out of the lungs.
- **thoracocentesis** (thō-rah-kō-sehn-tē-sihs) = puncture of the chest wall with a needle to obtain fluid from the pleural cavity. This fluid may be used for cytologic

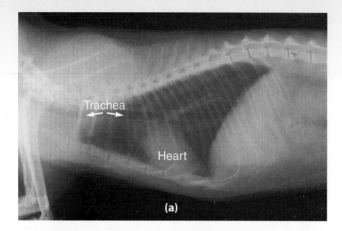

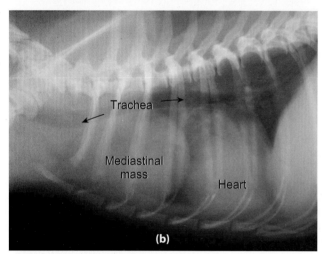

Figure 9–10 (a) Thoracic radiograph in a cat. This lateral thoracic radiograph reveals normal thoracic structures. (b) Thoracic radiograph of a dog with a mediastinal mass. [(a)Courtesy of Lodi Veterinary Hospital.]

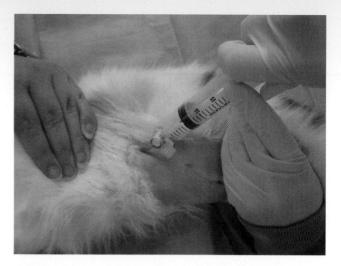

Figure 9–11 Thoracocentesis in a dog. Thoracocentesis is used to collect fluid or gas from the chest cavity. Fluid is collected ventrally, and gas is aspirated dorsally. (Courtesy of Kimberly Kruse Sprecher, CVT.)

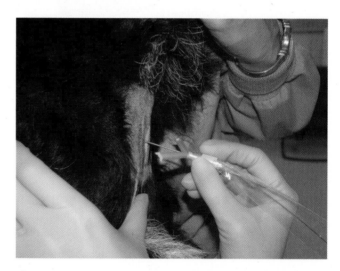

Figure 9–12 Transtracheal wash in a dog. A TTW is initiated with insertion of a catheter into the trachea of a dog. A syringe filled with saline is attached to the opposite end of the catheter so that a small volume of saline may be introduced into the trachea and then quickly aspirated along with mucus. This sample is then ready for cytologic and microbiologic assessment. (Courtesy of Teri Raffel, CVT.)

and microbiologic examination. Thoracocentesis also may be performed to drain pleural effusions or to reexpand a collapsed lung. Thoracocentesis also is called **thoracentesis** (thō-rah-sehn-tē-sihs) (Figure 9–11).

- **tracheal** (trā-kē-ahl) **wash** = collection of fluid or mucus from the trachea via an endotracheal tube to assess respiratory disease. Fluid may be used for cytologic and microbiologic examination.
- **transtracheal** (trahnz-trā-kē-ahl) **wash** = sterile collection of fluid or mucus from the trachea via a catheter inserted through the skin into the trachea to assess respiratory disease; abbreviated TTW. Fluid may be used for cytologic and microbiologic examination (Figure 9–12).
- **trephination** (trē-fin-ā-shuhn) = insertion of a hole-boring instrument (trephine) into a sinus to establish fluid drainage or to allow access to the roots of teeth.

PATHOLOGY: RESPIRATORY SYSTEM

Pathologic conditions of the respiratory tract include the following:

- **anoxia** (ā-nohck-sē-ah) = absence of oxygen (almost complete lack of oxygen).
- **asphyxiation** (ahs-fihck-sē-ā-shuhn) = interruption of breathing resulting in lack of oxygen; also called suffocation.

- **aspiration** (ahs-pih-rā-shuhn) = inhalation of a foreign substance into the upper respiratory tract.
- **asthma** (ahz-mah) = chronic allergic disorder.
- **atelectasis** (aht-eh-lehck-tah-sihs) = incomplete expansion of the alveoli; also may mean collapse of a lung.
- **bronchiectasis** (brohng-kē-ehck-tah-sihs) = dilation of the bronchi. Bronchiectasis may be a **sequela** (sē-kwehl-ah) of inflammation or obstruction. Sequela is a condition following as a consequence of a disease.
- **bronchitis** (brohng-kī-tihs) = inflammation of the bronchi. Bronchitis may be **acute** (ah-kūt), which means occurring over a short course with a sudden onset, or **chronic** (krohn-ihck), which means occurring over a long course with a progressive onset.
- **bronchopneumonia** (brohng-kō-nū-mō-nē-ah) = abnormal condition of the bronchi and lung.
- **chronic obstructive pulmonary disease** (krohn-ihck ohb-struhck-tihv puhl-mah-nār-ē dih-zēz) = general term for abnormal conditions in equine species in which expiratory flow is slowed; commonly called heaves and abbreviated COPD. Horses with heaves may have a heave line, which is increased abdominal musculature associated with increased expiratory effort in horses with COPD.
- **cyanosis** (sī-ah-nō-sihs) = abnormal condition of blue discoloration. Cyanosis is caused by inadequate oxygen levels.
- **diaphragmatic hernia** (dī-ah-frahg-mah-tihck hər-nē-ah) = abnormal opening in the diaphragm that allows part of the abdominal organs to migrate into the chest cavity (Figure 9–8).
- **emphysema** (ehm-fih-zē-mah) = chronic lung disease caused by enlargement of the alveoli or changes in the alveolar wall.
- **epistaxis** (ehp-ih-stahck-sihs) = nosebleed.
- **equine laryngeal hemiplegia** (ē-kwīn lahr-ihn-jē-ahl hehm-ih-plē-jē-ah) = disorder of horses that is characterized by abnormal inspiratory noise during exercise associated with degeneration of the left recurrent laryngeal nerve and atrophy of the laryngeal muscles; also called **left laryngeal hemiplegia** or **roaring.**
- **hemoptysis** (hē-mohp-tih-sihs) = spitting of blood from the lower respiratory tract.
- **hemothorax** (hē-mō-thō-rahcks) = accumulation of blood in the chest cavity.
- **inflammation** (ihn-flah-mā-shuhn) = localized protective response to destroy, dilute, or wall off injury; classic signs are heat, redness, swelling, pain, and loss of function.
- **laryngitis** (lahr-ihn-jī-tihs) = inflammation of the voice box.
- **laryngoplegia** (lahr-ihng-gō-plē-jē-ah) = paralysis of the voice box.
- **laryngospasm** (lah-rihng-ō-spahzm) = sudden fluttering or closure of the voice box.

- **pharyngitis** (făr-ihn-jī-tihs) = inflammation of the throat.
- **phonation** (fō-nā-shuhn) = act of producing sound. **Aphonation** (ā-foh-nā-shuhn) is the inability to produce sound.
- **pleural effusion** (ploor-ahl eh-fū-shuhn) = abnormal accumulation of fluid in the pleural space. **Effusion** (eh-fū-shuhn) is fluid escaping from blood or lymphatic vessels into tissues or spaces. A small amount of lubricating fluid in the pleural space is normal.
- **pleurisy** (ploor-ih-sē) = inflammation of the pleura; also called **pleuritis** (ploor-ī-tihs).
- **pleuropneumonia** (ploor-ō-nū-mō-nē-ah) = abnormal condition of the pleura and the lung (usually involves inflammation and congestion).
- **pneumonia** (nū-mō-nē-ah) = abnormal condition of the lung that usually involves inflammation and congestion of the lung (Figure 9–13). **Congestion** (kohn-jehs-zhuhn) is the abnormal accumulation of fluid. **Interstitial** (ihn-tər-stih-shahl) pertains to the area between the cells; **interstitial pneumonia** is an abnormal lung condition with increased fluid between the alveoli and a decrease in lung function.
- **pneumothorax** (nū-mō-thōr-ahckz) = abnormal accumulation of air or gas in the chest cavity (Figure 9–14).
- **polyp** (pohl-uhp) = growth or mass protruding from a mucous membrane (usually benign).
- **pulmonary edema** (puhl-mohn-ār-ē eh-dē-mah) = accumulation of fluid in the lung tissue. **Edema** (eh-dē-mah) is abnormally large amounts of fluid in the intercellular tissue spaces.
- **pulmonary fibrosis** (puhl-mō-nār-ē fī-brō-sihs) = abnormal formation of fibers in the alveolar walls.
- **pyothorax** (pī-ō-thō-rahcks) = accumulation of pus in the chest cavity. **Pus** is a fluid product of inflammation composed of leukocytes, **exudate** (ehcks-yoo-dāt) (high-protein fluid), and cell debris.

Figure 9–13 Heifer suffering from pneumonia. This heifer has a thick nasal discharge and labored breathing.

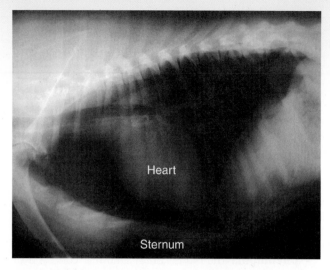

Figure 9–14 Pneumothorax in a dog. Lateral projection of a pneumothorax demonstrates dorsal displacement of the heart from the sternum. (Courtesy of Wisconsin Veterinary Teaching Hospital—Radiology.)

Figure 9–15 Young kitten with an upper respiratory tract infection. Note the nasal and ocular discharge.

- **rhinitis** (rī-nī-tihs) = inflammation of the nasal mucous membranes. Nasal discharge can be described by its appearance, such as mucopurulent (myoo-kō-pər-ū-lehnt), which means containing mucus and pus.
- **rhinopneumonitis** (rī-nō-nū-moh-nī-tihs) = inflammation of the nasal mucous membranes and lungs.
- **rhinorrhea** (rī-nō-rē-ah) = nasal discharge.
- **sinusitis** (sī-nuh-sī-tihs) = inflammation of a sinus.
- **snuffles** (snuhf-uhlz) = common term for upper respiratory disease of rabbits caused by *Pasteurella multocida.*
- **stenotic nares** (stehn-ah-tihck nār-ēz) = narrowed nostrils that reduce airway flow.
- **tracheitis** (trā-kē-ī-tihs) = inflammation of the windpipe.
- **tracheobronchitis** (trā-kē-ō-brohng-kī-tihs) = inflammation of the trachea and bronchi.
- **upper respiratory infection** = invasion of the nose, mouth, pharynx, epiglottis, or larynx (or trachea) by pathogenic organisms; abbreviated URI. Signs of URI include cough, nasal and ocular discharge, dyspnea, and respiratory noise (Figure 9–15).

A **cough** (kowf) is a sudden, noisy expulsion of air from the lungs. Coughs may be **paroxysmal** (pahr-ohck-sihz-mahl), which means spasmlike and sudden. **Tuss/i** is the combining form for cough.

PROCEDURES: RESPIRATORY SYSTEM

Procedures performed on the respiratory system include the following:

- **chest tube placement** = a chest tube is a hollow device inserted into the thoracic cavity to remove fluid or gas.

Drugs used on the respiratory system include **bronchoconstrictors** (brohng-kō-kohn-strihck-tərz), which are substances that narrow the openings into the lung, and **bronchodilators** (brohng-kō-dī-lā-tərz), which are substances that expand the openings into the lung. **Mucolytics** (mū-kō-lih-tihckz) are substances used to break down (**-lysis** means break down or separate) mucus. **Antitussives** (ahn-tih-tuhs-ihvz) are substances used to control or prevent coughing. The prefix **anti-** means against, and the combining form **-tussi** means cough.

Chest tubes are passed when animals are severely dyspneic because of pressure on the lungs (Figure 9–16).

- **endotracheal intubation** (ehn-dō-trā-kē-ahl ihn-too-bā-shuhn) = passage of a tube through the oral cavity, pharynx, and larynx into the windpipe. An endotracheal tube provides a **patent** (pā-tehnt) airway for administration of anesthetics or for critical care patients. *Patent* means open, unobstructed, or not closed.
- **laryngectomy** (lār-ihn-jehck-tō-mē) = surgical removal of the voice box.
- **laryngoplasty** (lah-rihng-ō-plahs-tē) = surgical repair of the voice box.

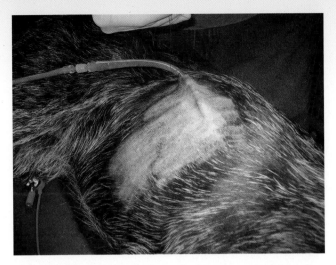

Figure 9–16 A chest tube is a hollow device inserted into the thoracic cavity to remove fluid or gas. This dog is having blood removed from the thoracic cavity via a chest tube. (Courtesy of Kimberly Kruse Sprecher, CVT.)

Figure 9–17 Tracheostomy tube in a dog. (Courtesy of Kimberly Kruse Sprecher, CVT.)

- **lobectomy** (lō-behck-tō-mē) = surgical removal of a lobe.
- **pharyngoplasty** (făr-rihng-ō-plahs-tē) = surgical repair of the throat.
- **pharyngostomy** (făr-ihng-ohs-tō-mē) = surgical creation of an opening into the throat. A **stoma** (stō-mah) is an opening on a body surface that may occur naturally or may be created surgically.
- **pharyngotomy** (făr-ihng-oht-ō-mē) = surgical incision into the throat.
- **pleurectomy** (ploor-ehck-tō-mē) = surgical removal of all or part of the pleura.

- **pneumonectomy** (nū-mō-nehck-tō-mē) = surgical removal of lung tissue.
- **sinusotomy** (sī-nuhs-oht-ō-mē) = surgical incision into a sinus.
- **thoracotomy** (thō-rah-koht-ō-mē) = surgical incision into the chest wall.
- **tracheoplasty** (trā-kē-ō-plahs-tē) = surgical repair of the windpipe.
- **tracheostomy** (trā-kē-ohs-tō-mē) = surgical creation of an opening into the windpipe (usually involves insertion and placement of a tube) (Figure 9–17).
- **tracheotomy** (trā-kē-oht-ō-mē) = surgical incision into the windpipe.

REVIEW EXERCISES

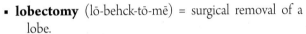

Multiple Choice
Choose the correct answer.

1. The wall that divides the nasal cavity is called the
   a. nasodivision
   b. nares
   c. nasal septum
   d. nasal meatus

2. Parts of the respiratory tract contain thin hairs called
   a. flagella
   b. naris
   c. surfactant
   d. cilia

3. An abnormal condition of blue discoloration is called
   a. bluing
   b. cyanosis
   c. xanthochromia
   d. erythemia

4. Inhaling a foreign substance into the upper respiratory tract is called
   a. asphyxiation
   b. effusion
   c. atelectasis
   d. aspiration

5. Hypoxia is

    a. below-normal levels of oxygen

    b. above-normal levels of oxygen

    c. below-normal levels of carbon dioxide

    d. below-normal levels of carbon dioxide and oxygen

6. Liquid that reduces alveolar surface tension is called

    a. surfactant

    b. mucus

    c. rhinorrhea

    d. mucorrhea

7. A condition following as a consequence of disease is a(n)

    a. chronic condition

    b. acute condition

    c. sequela

    d. consequensosis

8. A substance that works against, controls, or stops a cough is a(n)

    a. bronchoconstrictor

    b. bronchodilator

    c. mucolytic

    d. antitussive

9. Abnormal accumulation of blood in the pleural cavity is

    a. hemothorax

    b. hemoptysis

    c. hemopleuritis

    d. hemopneumonia

10. A growth or mass protruding from a mucous membrane is a

    a. nasogastric

    b. polyp

    c. bifurcation

    d. stridor

11. Tubes that bifurcate from the trachea are known as

    a. alveoli

    b. bronchi

    c. bronchioles

    d. nares

12. A nosebleed is known as

    a. pleurisy

    b. atelectasis

    c. aspiration

    d. epistaxis

13. Snoring, squeaking, or whistling that suggests airway narrowing is known as

    a. sequela

    b. polyp

    c. rhonchi

    d. stridor

14. Difficult breathing is known as

    a. dysphonia

    b. dyspnea

    c. dysphagia

    d. dyspepsia

15. Pus in the chest cavity is called

    a. pyothorax

    b. polyp

    c. hemiplegia

    d. hemoptysis

16. Tapping various body surfaces with the finger or an instrument to determine sound density is known as

    a. vesicular sounds

    b. crepitation

    c. percussion

    d. rales

17. The potential space between the parietal and visceral pleura is the

    a. parenchyma

    b. stroma

    c. diaphragmatic space

    d. pleural space

18. The alveoli are

    a. branches of the bronchial tree

    b. flexible air sacs where gas exchange occurs

    c. divisions of the lung

    d. sacs that surround the lung

19. The epiglottis

    a. acts as a lid over the entrance to the esophagus

    b. acts as a lid over the entrance to the trachea

    c. is commonly known as the Adam's apple

    d. is also known as the voice box

20. Which term means an abnormally rapid respiration rate?

    a. apnea

    b. bradypnea

    c. dyspnea

    d. tachypnea

Matching

Match the common term in Column I with the anatomical term in Column II.

Column I	Column II
1. _____ windpipe	a. pharynx
2. _____ throat	b. thorax
3. _____ voice box	c. larynx
4. _____ chest	d. trachea
5. _____ nostril	e. naris
6. _____ cilia	f. abnormally large amounts of fluid in the intercellular tissue spaces
7. _____ edema	g. spitting up of blood from the lower respiratory tract
8. _____ atelectasis	h. incomplete expansion of the alveoli
9. _____ sequela	i. tiny hairs
10. _____ anoxia	j. occurring over a short course
11. _____ antitussive	k. occurring over a long course
12. _____ hemoptysis	l. condition following as a consequence of a disease
13. _____ acute	m. act of listening
14. _____ chronic	n. substance that controls or prevents coughing
15. _____ auscultation	o. absence of oxygen

Fill in the Blanks

1. Ox/i, ox/o, and ox/y refer to _____.

2. Nas/o and rhin/o mean _____.

3. Pulm/o and pulmon/o mean _____.

4. Pneum/o, pneumon/o, and pneu mean _____.

5. Ventilation and breathing mean _____.

Spelling

Circle the term that is spelled correctly.

1. abnormal condition of the lung that usually involves inflammation and congestion of the lung: pnuemonia pneumonia pnuemohnia

2. interruption of breathing resulting in lack of oxygen: asfyxiation asphixiation asphyxiation

3. slimelike substance that is composed of glandular secretion, salts, cells, and leukocytes: mucous mukus mucus

4. chronic allergic disorder: astmah asthma asmah

5. muscle that separates the thoracic and abdominal cavities: diafragm diaphram diaphragm

CROSSWORD PUZZLE

Disease Terms Puzzle

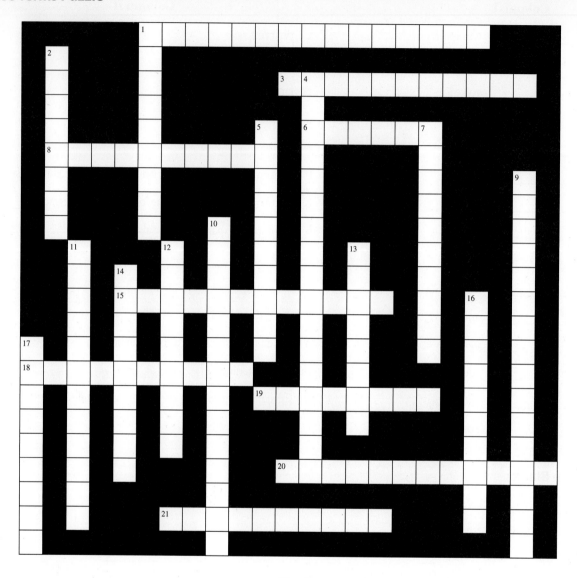

Across

1   abnormal condition of the pleura and the lung
3   incomplete expansion of the alveoli
6   absence of oxygen (almost complete lack of oxygen).
8   nasal discharge
15 abnormal accumulation of air or gas in the chest cavity
18 accumulation of blood in the chest cavity
19 inflammation of the nasal mucous membranes
20 interruption of breathing resulting in lack of oxygen
21 spitting of blood from the lower respiratory tract

Down

1   accumulation of pus in the chest cavity
2   inflammation of the pleura
4   inflammation of the trachea and bronchi
5   inflammation of the voice box
7   inhalation of a foreign substance into the upper respiratory tract.
9   abnormal condition of the bronchi and lung
10 accumulation of fluid in the lung tissue
11 localized protective response to destroy, dilute, or wall off injury
12 abnormal condition of the lung that usually involves inflammation and congestion of the lung
13 abnormal condition of blue discoloration
14 nosebleed
16 inflammation of the bronchi
17 act of producing sound

LABEL THE DIAGRAMS

Label the diagram in Figure 9–18.

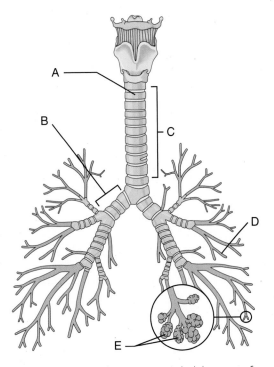

Figure 9–18   Respiratory tract. Label the parts of the lower respiratory tract as indicated by arrows. Provide the combining form for each part.

CASE STUDIES

Define the underlined terms in each case study.

An 8-mo-old F DSH was presented to the clinic for routine surgery. The cat was given an IV drug to sedate it so that an <u>endotracheal tube</u> could be placed. The endotracheal tube (ET tube) would serve as a delivery route for inhalant anesthesia. Upon attempts to <u>intubate</u> the cat, the larynx started to swell and spasm because of trauma. The cat had <u>inspiratory dyspnea</u>, a severe <u>cough</u>, <u>cyanotic</u> mucous membranes, and an elevated pulse rate. The <u>diagnosis</u> of <u>laryngeal spasm</u> was made. A drug (lidocaine) was placed in the laryngeal area to control the spasms, and endotracheal intubation was again attempted. After a few attempts, the ET tube was inserted, and the cat's breathing and mucous membrane color improved. If the ET tube could not have been placed, a <u>tracheotomy</u> would have been performed.

1. endotracheal tube _____

2. intubate _____

3. inspiratory dyspnea _____

4. cough _____

5. cyanotic _____

6. diagnosis _____

7. laryngeal spasm _____

8. tracheotomy _____

A group of beef cattle was moved into a feedlot about 2 wk ago. Some of the cattle are experiencing <u>anorexia</u>, are <u>pyrexic</u>, and have <u>mucopurulent nasal discharge</u>. Some of the cattle have rapid shallow breathing and a cough. <u>Auscultation</u> of the lungs revealed moist <u>rales</u>. Because of the history and respiratory signs, the veterinarian suspected <u>bovine</u> <u>pneumonic</u> pasteurellosis, or shipping fever, a severe respiratory disease seen in younger animals after shipping or stress. The affected cattle were isolated and treated with antibiotics. Management practices such as immunization and stress reduction were discussed with the owner.

9. anorexia _____

10. pyrexic _____

11. mucopurulent nasal discharge _____

12. auscultation _____

13. rales _____

14. bovine _____

15. pneumonic _____

CHAPTER 10

Skin Deep

Objectives

Upon completion of this chapter, the reader should be able to

- Identify the structures of the integumentary system
- Describe the functions of the integumentary system
- Recognize, define, spell, and pronounce the terms used to describe the diagnosis, pathology, and treatment of the integumentary system

FUNCTIONS OF THE INTEGUMENTARY SYSTEM

The **integumentary** (ihn-tehg-yoo-mehn-tah-rē) **system** consists of skin and its appendages. (Appendages include glands, hair, fur, wool, feathers, scales, claws, beaks, horns, hooves, and nails.) One of the largest organ systems in the body, the integumentary system is involved in many processes.

Skin plays a role in protecting animals from infection, waterproofing the body, preventing fluid loss, providing species-specific coloration, and providing a site for vitamin D synthesis. Exocrine glands, both sebaceous and sweat, are located in the integumentary system. Sebaceous glands lubricate the skin and discourage bacterial growth on the skin. Sweat glands regulate body temperature and excrete wastes through sweat. Hair and nails are other components of the integumentary system. Hair helps control body heat loss and is a sense receptor. Nails protect the dorsal surface of the distal phalanx.

STRUCTURES OF THE INTEGUMENTARY SYSTEM

The integumentary system is on the outside of the body and is resilient and versatile because it is subjected to a variety of insults such as trauma, toxic chemicals, and environmental conditions.

Skin

Skin covers the external surfaces of the body. The skin is composed of **epithelial** (ehp-ih-thē-lē-ahl) **tissue** and is sometimes called the **epithelium** (ehp-ih-thē-lē-uhm). The combining forms for skin are **cutane/o, derm/o,** and **dermat/o;** the suffix **-derma** means skin. **Dermatology** (dər-mah-tohl-ō-jē) is the study of skin.

Skin Stratification

The skin is made up of three layers: the epidermis, dermis, and subcutaneous layer (Figure 10–1). The outermost, or most superficial, layer of skin is the **epidermis** (ehp-ih-dər-mihs). The prefix **epi-** means above, and **dermis** means skin. The epidermis is several layers thick and does not contain blood vessels. The epidermis is sometimes called the avascular layer because it lacks blood vessels. The epidermis depends on the deeper layers for nourishment.

The thickness of the epidermis varies greatly from region to region in all animals. The thickest layers of the epidermis are found in the areas of greatest exposure, such as the foot pads and teats.

The epidermis is made up of squamous epithelium and the basal layer (Table 10–1). **Squamous epithelium** (skwā-muhs ehp-ih-thē-lē-uhm) is composed of flat, plate-like cells. Because these flat, platelike cells are arranged in many layers, this layer is called stratified squamous epithelium (Figure 10–2).

The basal layer is the deepest layer of the epidermis. Cells layer, multiply, and push upward into the basal layer. As the cells move superficially, they die and become filled with **keratin** (kehr-ah-tihn). Keratin is a protein that provides skin with its waterproofing properties. The combining form for keratin is **kerat/o.** Kerat/o also means horny or cornea. The basal layer also contains **melanocytes** (mehl-ah-nō-sītz), which produce and contain a black pigment. This black pigment is called **melanin** (mehl-ah-nihn). The combining form **melan/o** means black or dark. Melanin protects the skin from some of the harmful rays of the sun and is responsible for skin pigmentation. The absence of normal pigmentation is called **albinism** (ahl-bih-nihz-uhm); true albinism means that the hair, skin, and eyes have no pigmentation.

The **dermis** (dər-mihs) is the layer directly deep to the epidermis. The dermis also is called the **corium** (kō-rē-uhm). The dermis is composed of blood and lymph vessels, nerve fibers, and the accessory organs of the skin. The dermis also contains connective tissue, which is composed of the following cells:

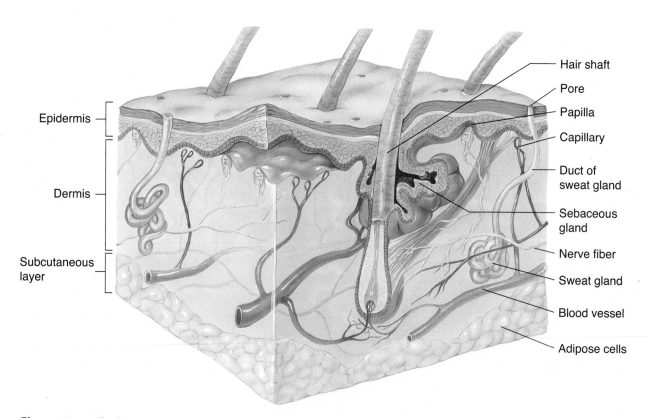

Epidermis

Dermis

Subcutaneous layer

Hair shaft
Pore
Papilla
Capillary
Duct of sweat gland
Sebaceous gland
Nerve fiber
Sweat gland
Blood vessel
Adipose cells

Figure 10–1 Skin layers.

Table 10–1 Layers of the Epidermis

The epidermis has five layers (from dermis to most superficial); *stratum* means layer or sheetlike mass.

stratum basale (strah-tuhm bā-sahl) or **stratum germinativum** (gər-mihn-ā-tihv-uhm) = deepest or basal layer that continually multiplies to replenish cells lost from the epidermal surface. **Cuboidal** (kyoo-boy-dahl) or cubelike cells are arranged in rows.

stratum spinosum (strah-tuhm spī-nō-suhm) = layer immediately superficial to the stratum basale, which is thickest in hairless regions and in areas of high wear and tear. **Keratinization** (kehr-ah-tihn-ah-zā-shuhn) and **desquamation** (deh-skwah-mā-shuhn) begin in this layer. Keratinization is the development of the hard, protein constituent of hair, nails, epidermis, horny structures, and tooth enamel. Desquamation is the process in which cell organelles gradually dissolve. The stratum spinosum also is called the **prickle** or **spinous layer.**

stratum granulosum (strah-tuhm grahn-yoo-lō-suhm) = layer immediately superficial to the stratum spinosum. Cells contain keratin granules in their cytoplasm.

stratum lucidum (strah-tuhm loo-sih-duhm) = layer immediately superficial to the stratum granulosum, which is clear because of the accumulation of keratin fibers in cell cytoplasm. This layer is not present in all species, but when present, it is found in areas of high wear and tear such as the foot pads.

stratum corneum (strah-tuhm kohr-nē-uhm) = most superficial layer of the epidermis, which consists of layers of dead, highly keratinized, and flattened cells; also called the **horny layer.**

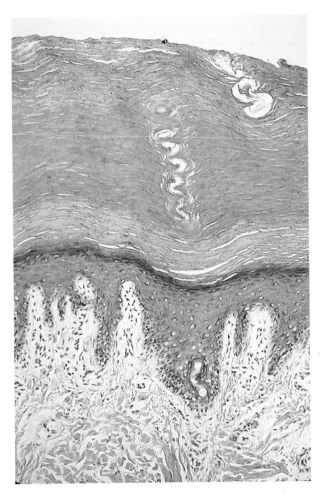

Figure 10–2 Photomicrograph of the specialized skin found in the foot pad of a cat. (Courtesy of William J. Bacha, PhD, and Linda M. Bacha, VMD.)

- **fibroblasts** (fī-brō-blahsts) = fiber-producing cells. Collagen (kohl-ah-jehn) is the major fiber in the dermis.
- **collagen** (kohl-ah-jehn) = tough, flexible, fibrous protein found in skin, bone, cartilage, tendons, and ligaments. *Kolla* in Greek means glue, and **-gen** means to produce.
- **histiocytes** (hihs-tē-ō-sīts) = phagocytic cells that engulf foreign substances; also called tissue **macrophages** (mahck-rō-fājs or mahck-rō-fahjs).
- **mast cells** = cells that respond to insult by producing and releasing histamine and heparin.
- **histamine** (hihs-tah-mēn) = chemical released in response to allergens that causes itching.
- **heparin** (hehp-ah-rihn) = anticoagulant chemical released in response to injury.
- **perception** (pər-sehp-shuhn) = ability to recognize sensory stimuli. Perception is received by nerve impulses that recognize temperature, touch, pain, and pressure. **Tactile** (tahck-tīl) perception is the ability to recognize touch sensation.

The **subcutaneous** (suhb-kyoo-tahn-ē-uhs) **layer,** or **hypodermis** (hī-pō-dər-mihs), is located deep to or under the dermis and is composed of connective tissue. The subcutaneous layer contains a large amount of **fat,** or **lipid** (lihp-ihd). Adipocytes (ahd-ih-pō-sīts) are fat cells that produce lipid. **Adip/o** is the combining form for fat (Figure 10–3).

Skin Association

Appendages or structures associated with the skin include glands, hair, fur, wool, feathers, scales, claws, beaks, horns, hooves, and nails.

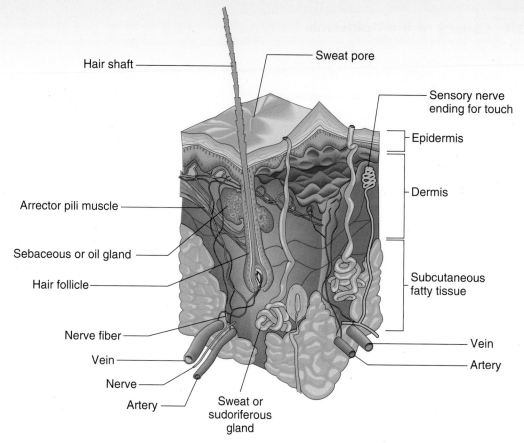

Figure 10–3 Skin structures.

Glands

There are two main categories of skin glands: sebaceous and sweat glands. **Sebaceous** (seh-bā-shuhs) **glands,** or **oil glands,** secrete an oily substance called **sebum** (sē-buhm). **Seb/o** is the combining form that means sebum or oily substance. Sebaceous glands are located in the dermis and are closely associated with hair follicles. Sebum is released from its gland through **ducts** (duhckts) that open into the hair follicles. Ducts are tubelike passages; tiny ducts are called **ductules.** Sebum moves from the hair follicle to the skin surface, where it lubricates the skin. Sebum is slightly acidic and retards bacterial growth on the skin. Sebaceous glands are considered **holocrine** (hō-lō-krihn) **glands** because the secreting cells and their secretions make up the discharge produced. Sebaceous glands are found in the anal sacs, glands that produce musk, and circumoral and supracaudal glands, which cats use to mark territory when they groom and rub their tail, respectively.

Sweat (sweht), or **sudoriferous** (soo-dohr-ihf-ohr-uhs), **glands** are aggregations of cells that are located in the dermis. Sweat glands are divided into **eccrine** (ē-krihn) **glands** and **apocrine** (ahp-ō-krihn) **glands.** Eccrine sweat glands produce and secrete water, salt, and waste (sweat) and are located in various regions of the body depending on the species. Eccrine sweat glands are tiny, coiled glands that have ducts that open directly onto the skin surface through pores. Apocrine glands produce and secrete a strong-smelling substance into the hair follicles. Apocrine glands are found throughout the body. Apocrine glands get their name from the fact that the free end or apical end of the cell is cast off along with the secretory products. Sweat glands help regulate body temperature against **hyperthermia** (hī-pər-thər-mē-ah), or high body temperature, and **hypothermia** (hī-pō-thər-mē-ah), or low body temperature.

Hidrosis (hī-drō-sihs) is the production and excretion of sweat. The combining form **hidr/o** means sweat. **Anhidrosis** (ahn-hī-drō-sihs) is the abnormal reduction of sweating; **hyperhidrosis** (hī-pər-hī-drō-sihs) is excessive sweating.

Ceruminous (seh-roo-mihn-uhs) glands are modified sweat glands that are located in the ear canal. The ceruminous glands secrete **cerumen** (seh-roo-mehn), a waxy substance of varying colors depending on the species; cerumen is commonly called **earwax.**

Hair

Hair is rodlike fibers made of dead protein cells filled with keratin. The combining forms for hair are **pil/i, pil/o,** and **trich/o.** The hair shaft is the portion of hair extending beyond the skin surface and is composed of the **cuticle** (kū-tih-kuhl), **cortex** (kōr-tehckz), and **medulla** (meh-doo-lah). The cuticle is one cell layer thick and appears scaly. The cortex is the main component of the

What is hair?

Animals have many different types of hair; therefore, many different terms describe the types of hair animals may have.

- **fur** = short, fine, soft hair.
- **pelt** = skin in addition to fur or hair.
- **guard hairs** = long, straight, stiff hairs that form the outer coat; also called **primary hairs** or **topcoat.** Guard hairs include tail and mane hair, bristly hair of swine, and most of the fur hair.
- **secondary hairs** = finer, softer, and wavy hair; also called the **undercoat.** Secondary hairs include wool and wavy hair located near the skin of rabbits.
- **tactile** (tahck-tīl) **hair** = long, brittle, extremely sensitive hairs usually located on the face; also called **vibrissae** (vī-brihs-ā), which are technically longer tactile hairs. An example of vibrissae are cat whiskers.
- **cilia** (sihl-ē-ah) = thin, short hairs. An example of cilia are the eyelashes.
- **simple pattern hair growth** = guard hairs that grow from separate follicular openings, as in cattle.
- **compound pattern hair growth** = multiple guard hairs that grow from single follicles, as in dogs.
- **shedding** = normal hair loss caused by temperature, hormones, photoperiod (light), nutrition, and other nondisease causes.

Things tend to be named for a reason. Sometimes bacteria, viruses, and fungi are named for the size, shape, or organism they affect. One example is the fungal genus *Trichophyton*. **Trich/o** means hair, and **phyt/o** means plant. **Fungi** are vegetative organisms (plants) that grow as branched, hairlike filaments (trich/o).

hair shaft, is several layers thick, and is responsible for coat color. The medulla is the innermost component of the hair shaft.

Hair **follicles** are sacs that hold the hair fibers. The **arrector pili** (ah-rehck-tər pī-lī) is a tiny muscle attached to the hair follicle that causes the hair to stand erect in response to cold temperatures or stress. When a dog contracts the arrector pili along the dorsal side of the neck and down the spine, it is called "raising the hackles." **Piloerection** (pī-lō-ē-rehck-shuhn) is the condition of the hair standing straight up.

Feathers and scales are discussed in Chapters 22 and 23, respectively.

Nails, Claws, and Hooves

The distal phalanx of animals is covered by nails, claws, or hooves. Nails, claws, and hooves all have a wall, sole, and pad, although they may be called different things. **Walls** usually are located dorsal and lateral to the distal phalanx. The **sole** is located ventral to the distal phalanx and usually is flaky. **Foot pads,** or **tori** (tohr-ē), provide cushioning and protection for the bones of the foot. Pads usually are thick and composed of keratinized epithelium. The pad has a subcutaneous layer that contains a large number of adipose cells and elastic connective tissue. Sweat glands are also found in most mammalian foot pads (Figure 10–4).

Dogs and cats have **digital pads** on the palmar and plantar surfaces of the phalanges. **Metacarpal** and **metatarsal pads** are singular pads located on the palmar and plantar surfaces of the metacarpal and metatarsal areas, respectively.

Carpal pads are located on the palmar surface of each carpus. Carpal pads do not bear weight when the animal is standing. Dogs and cats are called **digitigrade** (dihg-iht-ih-grād) animals because they walk on their toes, with only the digital and metacarpal and metatarsal pads making contact with the ground. **Plantigrade** (plahnt-ih-grād) animals have well-developed foot pads, such as those in primates. Plantigrade animals walk with phalanges, metacarpals and metatarsals, and carpal and tarsal bones making contact with the ground.

In dogs and cats, **nails** and **claws** are keratin plates covering the dorsal surface of the distal phalanx. The dorsal and lateral surface of the claw is covered by the wall, and the ventral surface is the sole. Beneath the wall and sole is the connective tissue dermis, which contains numerous blood

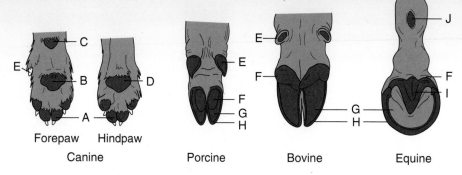

Figure 10–4 Comparison of animal feet. A = digital pad, B = metacarpal pad, C = carpal pad, D = metatarsal pad, E = dewclaw, F = bulb or heel, G = sole, H = wall, I = frog, and J = ergot.

The equine hoof is divided into various regions: the coronary band, periople, wall, bars, sole, bulb, and frog (Figure 10–5).

- **coronary** (kohr-ō-nār-ē) **band** = region where hoof meets the skin; analogous to the cuticle of the human nail. The coronary band is the site of hoof wall growth; also called the **coronet.**

- **periople** (pehr-ē-ō-puhl) = flaky tissue band located at the junction of the coronary band and the hoof wall and extends distally. The periople widens at the heel to cover the bulbs of the heels.

- **wall** = epidermal tissue that includes the toe (front), quarters (sides), and heels (back).

- **bars** (bahrz) = raised V-shaped structure on ventral surface of hoof. Bars are located on either side of the frog.

- **sole** (sōl) = softer hoof tissue located on the ventral surface of the hoof (bottom of the hoof).

- **frog** (frohg) = V-shaped pad of soft horn located in the central region of the ventral hoof surface of equine (located between the bars). When weight is put on the frog, blood is forced out of the foot to promote circulation of blood throughout the foot.

- **bulbs** (buhlbz) **of heel** = upward thickening of the frog above the heels of the wall.

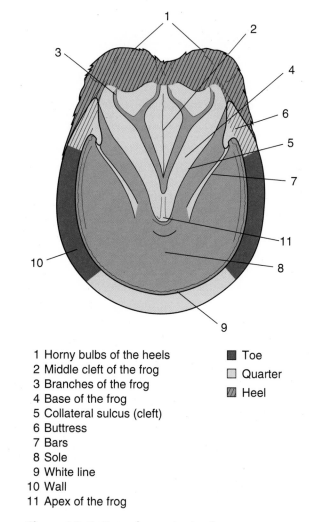

1 Horny bulbs of the heels
2 Middle cleft of the frog
3 Branches of the frog
4 Base of the frog
5 Collateral sulcus (cleft)
6 Buttress
7 Bars
8 Sole
9 White line
10 Wall
11 Apex of the frog

■ Toe
□ Quarter
▨ Heel

Figure 10–5 Parts of an equine hoof.

vessels and nerve endings. This sensitive tissue is known as the **quick** (Figure 10–6).

Quicking is the term used to describe trimming the nail or claw to the level of the dermis. Quicking results in bleeding and pain. The combining form for claw is **onych/o.**

Hooves are the horny covering of the distal phalanx in **ungulates** (uhng-yoo-lātz), or **hooved animals,** such as equine, ruminants, and swine. Some ungulates have a solid hoof, as in equine, and some have cloven or split hooves, as in

---

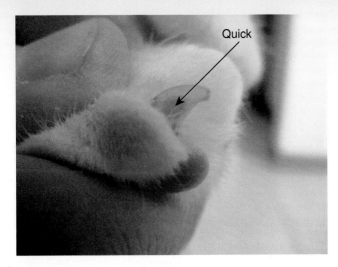

Figure 10–6   Quick of a cat nail.

ruminants and swine. The ventral surface of the hoof is the sole, which is large in equine and smaller in ruminants and swine. The combining form for hoof is **ungul/o.**

The pads of ungulates vary with the species. In ruminants and swine, the foot pad is called the bulb or heel. The pad of equine is called the frog and along with the bulb provides shock absorption. The **corium** (kōr-ē-uhm) is the dermis of the hoof and is located under the epidermal surface of the hoof wall, sole, and frog. The corium corresponds to the quick.

Vestigial Structures

Vestigial, or rudimentary, structures of the integumentary system include dewclaws, chestnuts, and ergots. **Dewclaws** (doo-klawz) are rudimentary bones. The dewclaw in dogs is the first digit, whereas in cloven-hoofed animals, the dewclaws are digits II and V. Dewclaws in dogs usually are found in the fore-paw (although they occasionally are seen in the hindpaw) and may be removed within a few days of birth to avoid trauma.

Chestnuts and ergots are vestigial pads in equine. **Chestnuts** are located on the medial surface of the leg; in the

front leg, they are located above the knee, and in the hind leg, they are located below the hock (Figures 10–7a and b). Chestnuts correspond to carpal pads in the dog. No two chestnuts are alike, and they do not change in size or shape throughout an

What is the difference between horns and antlers?

Horns and antlers are protective structures located in the head region of animals. **Horns** are permanent structures that grow continuously after birth. Horns grow from the frontal skull bones and originate from keratinized epithelium. **Cornification** (kohr-nih-fih-kā-shuhn) is the conversion of epithelium into keratin or horn. Horns may be located in different positions, as can be seen in the different ruminant species. Breeds that are naturally hornless are called **polled** (Figures 10–8a and b).

Antlers are not permanent structures and are shed and regrown annually. Antlers grow from the skull, as do horns. Antlers initially are covered with skin called **velvet,** which the animal rubs off after the skin dies. When the animal rubs off the velvet, the bone is exposed, the antlers lose their blood supply, and the antlers are eventually shed.

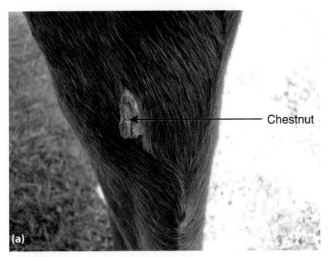

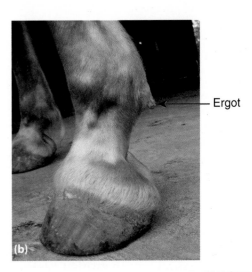

Figure 10–7   (a) Chestnut of a horse. (b) Ergot of a horse.

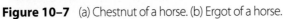

Figure 10–8   (a) Polled (b) Unpolled Hereford. (Courtesy of iStockphoto.)

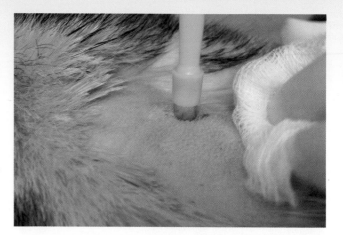

Figure 10–9   Punch biopsy of the skin. A punch biopsy is a type of incisional biopsy commonly used to obtain full-thickness skin samples. The punch has a circular opening that comes in different sizes. (Courtesy of Kimberly Kruse Sprecher, CVT.)

equine's life. **Ergots** are located in a tuft of hair on the fetlock joint. Ergots correspond to metacarpal and metatarsal pads in the dog. Digital pads of dogs are replaced by the bulbs of the heel (and the frog in equine) in ungulates.

TEST ME: INTEGUMENTARY SYSTEM

Diagnostic procedures performed on the integumentary system include the following:

- **biopsy** (bī-ohp-sē) = removal of living tissue for examination of life. The combining form **bi/o** means life; the suffix **-opsy** means view of. An **incisional biopsy** (ihn-sih-shuhn-ahl bī-ohp-sē) is the removal of a piece of a tumor or lesion for examination (Figure 10–9). An **incision** (ihn-sih-shuhn) is a cut into tissue. An **excisional biopsy** (ehcks-sih-shuhn-ahl bī-ohp-sē) is the removal of an entire tumor or lesion in addition to a margin of surrounding tissue for examination. An **excision** (ehcks-sih-shuhn) is a cut out of tissue. A needle biopsy is the insertion of

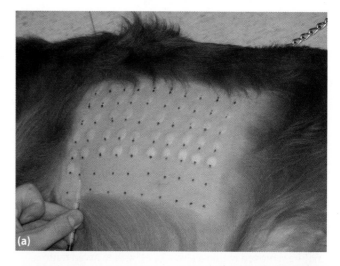

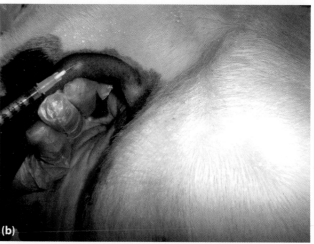

Figure 10–10   (a) Intradermal skin testing. Intradermal skin testing is used to assess atopy. (b) Tuberculosis caudal fold skin test in a bovine. [(a) Courtesy of Kimberly Kruse Sprecher, CVT. (b) Courtesy of James Meronek, DVM, MPH.]

a sharp instrument (needle) into a tissue for examination. Examination of biopsies involves the use of a microscope.

- **culture** = diagnostic or research procedure used to allow microbes to reproduce in predetermined media (nutrient source). Fungal and bacterial cultures are performed using media that contains specific nutrients necessary for optimal growth of these types of microbes. **Tissue culture** takes epithelial cells and grows them in a medium so that intracellular microbes such as viruses can replicate.
- **exfoliative cytology** (ehcks-fōl-ē-ah-tihv sī-tohl-ō-jē) = scraping of cells from tissue and examination under a microscope. **Exfoliative** means falling off.
- **intradermal** (ihn-trah-dər-mahl) **skin testing** = injection of test substances into the skin layer to observe a reaction (Figure 10–10a); used for diagnosis of **atopy** (ah-tō-pē) with the injection of multiple allergens or for tuberculosis testing by injecting tuberculin into the skin layer and observing the injection site for a 24-, 36-, and 72-hour postinjection reaction. (Tuberculosis testing is called **purified protein derivative,** or PPD, testing [Figure 10–10b].)
- **skin scrape** = microscopic examination of skin for the presence of mites; skin is sampled by scraping a scalpel blade across an area that is squeezed or raised so that the sample contains a deep skin sample (Figure 10–11).

PATHOLOGY: INTEGUMENTARY SYSTEM

Pathologic conditions of the integumentary system include the following:

- **abrasion** (ah-brā-shuhn) = injury in which superficial layers of skin are scraped.
- **abscess** (ahb-sehsz) = localized collection of pus (Figure 10–12).
- **acne** (ahck-nē) = skin inflammation caused by plugged sebaceous glands and comedone development from papules and pustules. **Chin acne** is a common condition in cats in which acne develops on the chin and lip area.
- **acute moist dermatitis** (ah-kūt moyst dər-mah-tī-tihs) = bacterial skin disease that is worsened by licking and scratching; also called **hot spot.**
- **alopecia** (ahl-ō-pē-shah) = hair loss resulting in hairless patches or complete lack of hair. **Shedding** is normal hair loss due to various causes (Figure 10–13).
- **atopy** (ah-tō-pē) = hypersensitivity reaction in animals involving pruritus with secondary dermatitis; commonly called **allergies** or **allergic dermatitis. Hypersensitization** is an increased response to an allergen. **Hyposensitization** is a decreased response to an allergen. Animals with atopy may undergo a series of hyposensitization injections to decrease their response to a specific allergen.

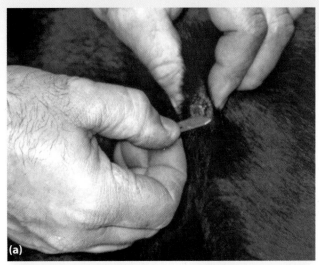

Figure 10–11 Skin scrape. Skin scrapings are used to detect mites. (Courtesy of Ron Fabrizius, DVM, Diplomate ACT.)

Figure 10–12 Draining an abscess on the side of the face of an anesthetized cat.

- **bullae** (buhl-ā) = multiple contained skin elevations filled with fluid that are greater than 0.5 cm in diameter. The singular form is **bulla** (buhl-ah).

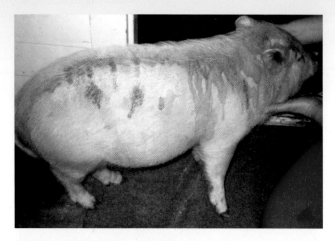

Figure 10–13 Alopecia in a pig with sarcoptic mange. (Courtesy of Ron Fabrizius, DVM, Diplomate ACT.)

(a)

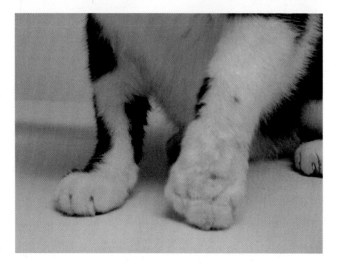

Figure 10–14 Cellulitis in cat paw (right side of image).

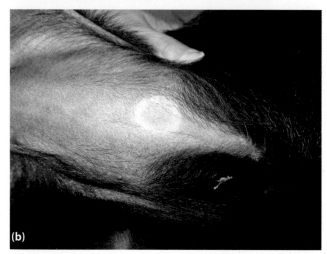

(b)

Figure 10–15 (a) Holstein heifer with ringworm. (b) Ringworm lesion on the abdomen of a dog.

- **burn** = tissue injury caused by heat, flame, electricity, chemicals, or radiation.
- **carbuncle** (kahr-buhng-kuhl) = cluster of furuncles.
- **carcinoma** (kahr-sih-nō-mah) = malignant neoplasm of epithelial tissue.
- **cellulitis** (sehl-yoo-lī-tihs) = inflammation of connective tissue (Figure 10–14). Inflammation may be **diffuse** (dih-fuhs), meaning widespread, or **localized** (lō-kahl-īzd), meaning within a well-defined area.
- **comedo** (kōm-eh-dō) = blackhead or buildup of sebum and keratin in a pore. Plural is **comedones** (kōm-eh-dō-nehz).
- **contusion** (kohn-too-shuhn) = injury that does not break the skin; characterized by pain, swelling, and discoloration.
- **crust** = collection of dried sebum and cell debris.
- **dermatitis** (dər-mah-tī-tihs) = inflammation of the skin. **Contact dermatitis** is inflammation of the skin caused by touching an irritant.

- **dermatocellulitis** (dər-mah-tō-sehl-yoo-lī-tihs) = inflammation of the skin and connective tissue.
- **dermatomycosis** (dər-mah-tō-mī-kō-sihs) = abnormal skin condition caused by superficial fungus; also called **dermatophytosis** (dər-mah-tō-fī-tō-sihs). **Dermatophytes** (dər-mah-tō-fītz) are superficial fungi that are found on the skin. An example of a dermatophyte is the fungus that causes ringworm (Figures 10–15a and b).
- **dermatosis** (dər-mah-tō-sihs) = abnormal skin condition. Plural is **dermatoses** (der-mah-tō-sēz).
- **discoid lupus erythematosus** (dihs-koyd loo-pihs eh-rih-thehm-ah-tō-sihs) = canine autoimmune disease in which the bridge of the nose (and sometimes the face and lips) exhibit depigmentation, erythema, scaling, and erosions; abbreviated DLE; may have been called collie nose or solar dermatitis in the past.
- **dyskeratosis** (dihs-kehr-ah-tō-sihs) = abnormal alteration in keratinization.

- **ecchymosis** (ehck-ih-mō-sihs) = purplish nonelevated patch of bleeding into the skin; also called a bruise; plural is **ecchymoses** (ehck-ih-mō-sēz).
- **ecthyma** (ehck-thih-mah) = skin infection with shallow eruptions caused by a pox virus (Figure 10–16); also known as soremouth.
- **eczema** (ehcks-zeh-mah) = general term for inflammatory skin disease characterized by erythema, papules, vesicles, crusts, and scabs either alone or in combination.
- **eosinophilic granuloma** (ē-ō-sihn-ō-fihl-ihck grahn-yoo-lō-mah) **complex** = collective name for autoimmune lesion of eosinophilic ulcer, eosinophilic plaque, and linear granuloma found in cats and rarely in dogs. This complex of diseases affects the skin, mucocutaneous junctions, and oral mucosa of cats, involving raised, ulcerated plaques (Figure 10–17). These lesions are named for their location: **eosinophilic ulcer,** or **rodent ulcer,** is located on the lip and oral mucosa

Figure 10–16 Contagious ecthyma (soremouth) in a sheep. (Courtesy of Ron Fabrizius, DVM, Diplomate ACT.)

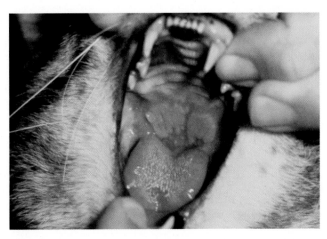

Figure 10–17 Eosinophilic ulcer in a cat.

of cats; **eosinophilic plaques** are raised pruritic lesions on the ventral abdomen of cats; and **linear granulomas** are located in a line usually on the caudal aspect of the hindlimb of cats.

- **erythema** (ehr-ih-thē-mah) = skin redness. **Erythematous** (ehr-ih-thehm-ah-tuhs) means pertaining to redness.
- **erythroderma** (eh-rihth-rō-dər-mah) = abnormal redness of skin occurring over a widespread area. Combining forms for red are **erythr/o, erythem/o,** and **erythemat/o.**
- **exanthema** (ehcks-ahn-thē-mah) = cutaneous rash caused by fever or disease. Singular is **exanthem** (ehcks-ahn-thuhm).
- **feline miliary dermatitis** (mihl-ē-ahr-ē dər-mah-tī-tihs) = skin disease of cats in which multiple crusts and bumps are present predominantly on the dorsum; the disease can be associated with many causes.
- **fissure** (fihs-sər) = cracklike sore (Figure 10–18).
- **fistula** (fihs-tyoo-lah) = abnormal passage from an internal organ to the body surface or between two internal organs. Plural is **fistulae** (fihs-tyoo-lā).
- **flea allergy dermatitis** (dər-mah-tī-tihs) = inflammation of the skin caused by an allergic reaction to flea saliva; abbreviated FAD. An **allergen** (ahl-ər-jehn) is a substance that produces an allergic response.
- **footrot** = bacterial (*Fusobacterium* sp.) hoof disease that spreads from the interdigital skin to the deeper foot structures.
- **frostbite** = tissue damage caused by extreme cold or contact with chemicals with extreme temperature (e.g., liquid nitrogen).
- **furuncle** (fyoo-ruhng-kuhl) = localized skin infection in a gland or hair follicle; also called a **boil. Furunculosis** (fyoo-ruhng-kuh-lō-sihs) is the abnormal condition of persistent boils over a period of time.
- **gangrene** (gahng-grēn) = necrosis associated with loss of circulation. **Necrosis** (neh-krō-sihs) is condition of dead tissue; **necrotic** (neh-krŏ-tihck) means pertaining to dead tissue. Decay that produces a foul smell is called **putrefaction** (pyoo-treh-fahck-shuhn).
- **granuloma** (grahn-yoo-lō-mah) = small area of healing tissue.
- **hemangioma** (hē-mahn-jē-ō-mah) = benign tumor composed of newly formed blood vessels.
- **hyperkeratosis** (hī-pər-kehr-ah-tō-sihs) = increased growth of the horny layer of the skin (Figure 10–19); also called **acanthokeratodermia** (ā-kahn-thō-kehr-ah-tō-dər-mah).
- **infestations** (ihn-fehs-tā-shuhns) = occupation and dwelling of a parasite on the external surface of tissue. **Ectoparasites** (ehck-tō-pahr-ah-sīts) live on the

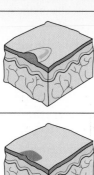 A **papule** is a small solid raised lesion that is less than 0.5 cm in diameter.

 A **plaque** is a solid raised lesion that is greater than 0.5 cm in diameter.

 A **macule** is a flat discolored lesion that is less than 1 cm in diameter.

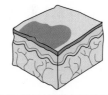 A **patch** is a flat discolored lesion that is greater than 1 cm in diameter.

 A **scale** is a flaking or dry patch made up of excess dead epidermal cells.

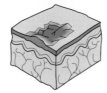 A **crust** is a collection of dried serum and cellular debris.

 A **wheal** is a smooth, slightly elevated swollen area that is redder or paler than the surrounding skin. It is usually accompanied by itching.

(a) Surface lesions

 A **cyst** is a closed sac or pouch containing fluid or semisolid material.

 A **pustule** is a small circumscribed elevation of the skin containing pus.

 A **vesicle** is a circumscribed elevation of skin containing fluid that is less than 0.5 cm in diameter.

 A **bulla** is a large vesicle that is more than 0.5 cm in diameter.

(b) Fluid-filled lesions

 An **ulcer** is an open sore or erosion of the skin or mucous membrane resulting in tissue loss.

 A **fissure** of the skin is a groove or crack-like sore.

(c) Erosive lesions

Figure 10–18   Skin lesions. Skin lesions may be (a) raised and discolored, (b) fluid-filled, or (c) erosive in nature.

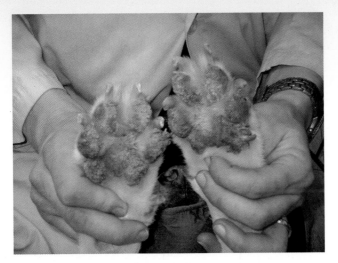

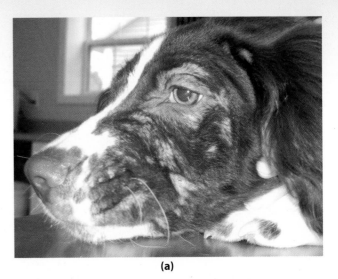

(a)

Figure 10–19 Hyperkeratosis of the foot pads of a dog. (Courtesy of Kimberly Kruse Sprecher, CVT.)

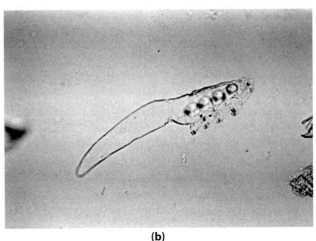

(b)

Figure 10–20 (a) Demodectic mange in a dog. (b) *Demodex canis.*

external surface; **ecto-** means outside. A **louse** (lows) is a wingless parasitic insect; plural is **lice** (līs). **Pediculosis** (pehd-ih-koo-lō-sihs) is lice infestation. A **mite** is an insect with a hard exoskeleton and paired, jointed legs. **Mange** (mānj) is a common term for skin disease caused by mites. There are different types of mange, such as **sarcoptic** (sahr-kohp-tihck) and **demodectic** (deh-mō-dehck-tihck), depending on the type of mite involved (Figures 10–20a and b). **Chiggers** (chihg-gərs) is infestation by mite larvae that results in severe pruritus. **Acariasis** (ahck-ah-rī-ah-sihs) is infestation with ticks or mites. **Maggots** (mah-gohts) are insect larvae found especially in dead or decaying tissue. **Myiasis** (mī-ī-ah-sihs) is infestation by fly larvae.

- **keratosis** (kehr-ah-tō-sihs) = abnormal condition of epidermal overgrowth and thickening. Plural is **keratoses** (kehr-ah-tō-sēs).
- **laceration** (lahs-ər-ā-shuhn) = accidental cut into the skin.
- **lesion** (lē-shuhn) = pathologic change of tissue; used to describe abnormalities in many locations.
- **lipoma** (lī-pō-mah) = benign growth of fat cells (Figure 10–21); also called **fatty tumor**; commonly seen in older dogs.
- **macule** (mahck-yool) = flat, discolored lesion less than 1 cm in diameter; also called **macula** (mahck-yoo-lah).
- **melanoma** (mehl-ah-nō-mah) = tumor or growth of pigmented skin cells (Figure 10–22). **Malignant melanoma** is the term used to describe cancer of the pigmented skin cells. One form of melanoma is **amelanotic melanoma,** which is an unpigmented malignant melanoma.
- **nodule** (nohd-yoo-uhl) = small knot protruding above the skin.
- **onychomycosis** (ohn-kē-ō-mī-kō-sihs) = superficial fungal infection of the claw.

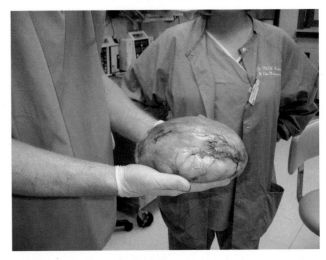

Figure 10–21 Lipoma that has been surgically removed from a dog. (Courtesy of Kimberly Kruse Sprecher, CVT.)

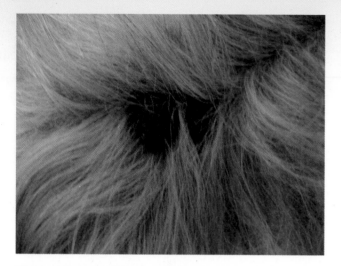

Figure 10–22 Melanoma in a dog.

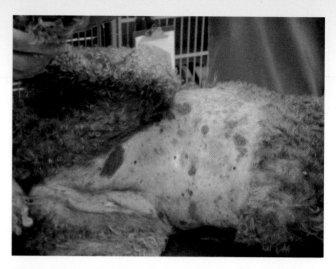

Figure 10–23 Petechiae in a dog. (Courtesy of Kimberly Kruse Sprecher, CVT.)

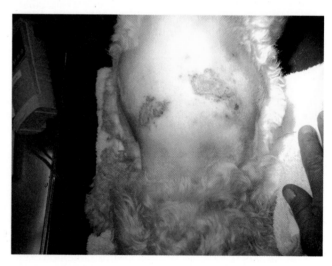

Figure 10–24 Pyoderma in a dog. (Courtesy of Kimberly Kruse Sprecher, CVT.)

- **pallor** (pahl-ohr) = skin paleness.
- **papilloma** (pahp-ih-lō-mah) = benign epithelial growth that is lobed.
- **papule** (pahp-yool) = small, raised skin lesion less than 0.5 cm in diameter.
- **parakeratosis** (pahr-ah-kehr-ah-tō-sihs) = lesion characterized by thick scales, cracking, and red raw surface caused by the persistence of keratinocyte nuclei in the horny layer of skin.
- **paronychia** (pahr-ohn-kē-ah) = bacterial or viral infection of the claw.
- **patch** = localized skin color change greater than 1 cm in diameter.
- **pemphigus** (pehm-fih-guhs) = group of immune mediated skin diseases characterized by vesicles, bullae, and ulcers. The most common form is **pemphigus vulgaris** (pehm-fih-guhs vuhl-gahr-ihs), which consists of shallow ulcerations frequently involving the oral mucosa and mucocutaneous junctions. *Pemphix* is Greek for blister.
- **petechiae** (peh-tē-kē-ā) = small, pinpoint hemorrhages (Figure 10–23). Singular is **petechia** (peh-tē-kē-ah).
- **plaque** (plahck) = solid raised lesion greater than 0.5 cm in diameter.
- **polyp** = growth from mucous membranes.
- **pruritus** (proo-rī-tuhs) = itching.
- **purpura** (pər-pə-rah) = condition characterized by hemorrhage into the skin that causes bruising. The two types of purpura are ecchymosis and petechia.
- **pustule** (puhs-tyool) = small, circumscribed, pus-filled skin elevation. **Circumscribed** (sehr-kuhm-skrībd) means contained in a limited area.
- **pyoderma** (pī-ō-dər-mah) = skin disease containing pus (Figure 10–24). **Pus** (puhs) is an inflammatory product made up of leukocytes, cell debris, and fluid. **Purulent** (pər-ū-lehnt) means containing or producing pus. **Puppy pyoderma** is a skin disease in puppies characterized by pus-containing lesions. **Juvenile**

pyoderma is a skin disease in puppies that progresses to a systemic disease characterized by fever, anorexia, and enlarged and abscessing lymph nodes; juvenile pyoderma also is called **puppy strangles.**
- **sarcoma** (sahr-kō-mah) = malignant neoplasm of soft tissue arising from connective tissue.
- **scale** = flake.
- **scar** (skahr) = mark left by a healing lesion where excess collagen was produced to replace injured tissue; also called **cicatrix** (sihck-ah-trihcks) or **cicatrices** (sihck-ah-trih-sēz), which are multiple scars.
- **sebaceous cyst** (seh-bā-shuhs sihst) = closed sac of yellow fatty material. A **cyst** (sihst) is a closed sac containing fluid or semisolid material.
- **seborrhea** (sehb-ō-rē-ah) = skin condition characterized by overproduction of sebum (oil).
- **skin tag** = small growth that hangs from the body by stalks.

- **ulcer** (uhl-sihr) = erosion of skin or mucous membrane. **Decubital ulcers** (dē-kyoo-bih-tahl uhl-sihrz) are erosions of skin or mucous membranes as a result of prolonged pressure; also called **bedsores.**
- **urticaria** (ər-tih-kā-rē-ah) = localized areas of swelling that itch; also called **hives.**
- **verrucae** (veh-roo-sē) = warts.
- **vesicle** (vehs-ih-kuhl) = contained skin elevation filled with fluid that is greater than 0.5 cm in diameter; also called a **blister, bulla** (buhl-ah), or **bleb.**
- **wheal** (whēl) = smooth, slightly raised swollen area that itches.

PROCEDURES: INTEGUMENTARY SYSTEM

Procedures performed on the integumentary system include the following:

- **cauterization** (kaw-tər-ī-zā-shuhn) = destruction of tissue using electric current, heat, or chemicals (Figure 10–25).
- **cryosurgery** (krī-ō-sihr-jər-ē) = destruction of tissue using extreme cold.
- **debridement** (dē-brīd-mehnt) = removal of tissue and foreign material to aid healing.

- **lance** (lahnz) to open or pierce with a lancet (scalpel blade) to allow drainage; abscesses are lanced to drain the pus present in an area.
- **laser** (lā-zər) = device that transfers light into an intense beam for various purposes; acronym for *light amplification by stimulated emission of radiation.*

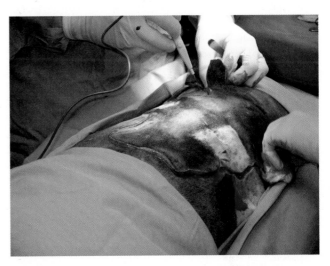

Figure 10–25 Cauterization destroys tissue through the use of electric current, heat, or chemicals. (Courtesy of Kimberly Kruse Sprecher, CVT.)

REVIEW EXERCISES

Multiple Choice
Choose the correct answer.

1. Pruritus is commonly called
    a. hair loss
    b. dry skin
    c. itching
    d. pus

2. Skin redness is called
    a. cellulitis
    b. erythema
    c. scleroderma
    d. scarring

3. Hypersensitivity reaction in animals involving pruritus with secondary dermatitis is called
    a. atrophy
    b. allergen
    c. antigen
    d. atopy

4. Hair loss resulting in hairless patches or complete lack of hair is called
    a. shedding
    b. lesion
    c. alopecia
    d. plaque

5. Occupation and dwelling of parasites on the external skin surfaces is called
    a. parasitism
    b. ectoparasites
    c. infestation
    d. myiasis

6. A skin disease containing pus is
    a. pyometra
    b. pyoderma
    c. pyoerythema
    d. pyosis

7. Producing or containing pus is called

    a. abscess

    b. purulent

    c. mucocutaneous

    d. polyp

8. Inflammation of connective tissue is

    a. connectitis

    b. dermatitis

    c. cutaneitis

    d. cellulitis

9. Putrefaction is

    a. foul-smelling decay

    b. homogenation

    c. granulomatous

    d. cellulitis

10. Large tactile hair is

    a. sensogenic

    b. plantigrade

    c. vibrissa

    d. cerumen

11. A skin condition characterized by overproduction of sebum (oil) is known as

    a. alopecia

    b. seborrhea

    c. hemangioma

    d. abscess

12. The term for a skin flake is

    a. patch

    b. crust

    c. scale

    d. wheal

13. Which term means skin paleness?

    a. petechiae

    b. pallor

    c. pustule

    d. purpura

14. The protein that provides skin with its waterproofing properties is known as

    a. dermis

    b. squamous

    c. sebum

    d. keratin

15. Removal of tissue and foreign material to aid healing is

    a. cauterization

    b. cryosurgery

    c. debridement

    d. laser treatment

16. A localized collection of pus is a(n)

    a. bleb

    b. abscess

    c. nodule

    d. vesicle

17. A benign growth of fat cells is known as a(n)

    a. sarcoma

    b. carcinoma

    c. lipoma

    d. adoma

18. Which term means pertaining to dead tissue?

    a. abrasion

    b. carbuncle

    c. purulent

    d. necrotic

19. Which term means erosion of skin or mucous membrane?

    a. fistula

    b. ulcer

    c. fissure

    d. urticaria

20. A tumor or growth of pigmented skin is a(n)

    a. melanoma

    b. ecthyma

    c. exanthema

    d. eczema

Matching

Match the common dermatologic term in Column I with the medical term in Column II.

Column I

1. _____ hive
2. _____ hot spot
3. _____ fatty tumor
4. _____ pale
5. _____ blister
6. _____ scar
7. _____ oil
8. _____ scrape
9. _____ boil
10. _____ redness
11. _____ itching
12. _____ flake
13. _____ warts
14. _____ allergic dermatitis
15. _____ hornless
16. _____ crack
17. _____ adipocytes
18. _____ dermis
19. _____ epidermis
20. _____ subcutaneous
21. _____ collagen
22. _____ pus
23. _____ urticaria
24. _____ carcinoma
25. _____ sudoriferous glands
26. _____ sebaceous glands

Column II

a. polled
b. cicatrix
c. furuncle
d. urticaria
e. abrasion
f. acute moist dermatitis
g. pruritus
h. atopy
i. lipoma
j. scale
k. erythema
l. pallor
m. sebum
n. verrucae
o. fissure
p. vesicle
q. malignant neoplasm of epithelial tissue
r. most superficial layer of skin
s. protein material found in skin, hair, and nails
t. sweat glands
u. fat cells
v. hives
w. middle layer of skin
x. deepest layer of the skin
y. oil glands
z. inflammatory product made up of leukocytes, cell debris, and fluid

Fill in the Blanks

1. Cutane/o, derm/o, dermat/o, and -derma mean _____ .

2. Pil/i, pil/o, and trich/o mean _____ .

3. The dermis and corium are names for _____ .

4. The subcutaneous and hypodermis are names for _____ .

5. A cyst, vesicle, pustule, and bulla are all _____ .

Spelling

Circle the term that is spelled correctly.

1. itching:	pruritis	pruritus	puritis
2. buildup of sebum and keratin in a pore:	komedo	comedo	comedoe
3. hair loss resulting in hairless areas:	alopecia	ahlopesia	alopecea
4. production and secretion of sweat:	hidrosis	hydrosis	hihdrosis
5. common term for skin disease caused by mites:	manje	mainge	mange

CROSSWORD PUZZLE

Skin Anatomy and Disease Puzzle

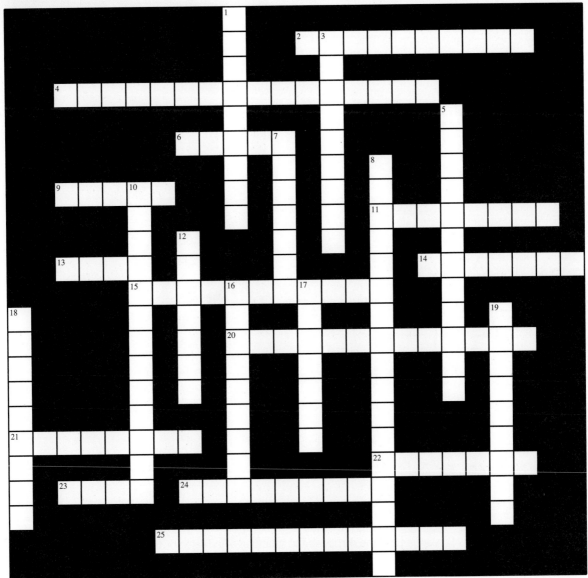

Across

2 accidental cut into the skin
4 removal of a piece of a tumor or lesion for examination
6 growth from mucous membrane
9 dermis under the nail or hoof that has a blood supply and is sensitive
11 small, pinpoint hemorrhage
13 closed sac containing fluid or semisolid material
14 earwax
15 pertaining to falling off
20 contained within a limited area
21 normal hair loss
22 localized collection of pus
23 permanent structure originating from the skull
24 secondary hairs that are soft, thin, and wavy
25 superficial fungal infection of the claw

Down

1 small area of healing tissue
3 fat cell
5 act of hair standing upright
7 condition characterized by hemorrhage into the skin that causes bruising
8 decreased response to an allergen
10 destruction of tissue using electric current, heat, or chemicals
12 widespread
16 within a well-defined area
17 pertaining to touch
18 injury that does not break the skin
19 abnormal condition of epidermal overgrowth and thickening

LABEL THE DIAGRAMS

Label the diagram in Figure 10–26.

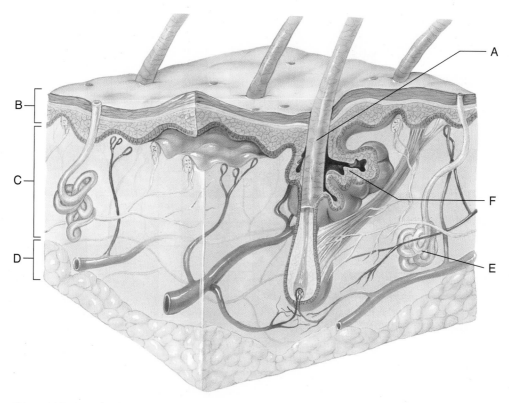

Figure 10–26 Skin layers. Label the skin layers and structures as indicated by the arrows.

CASE STUDIES

Define the underlined terms in each case study.

A 3-yr-old F/S golden retriever was presented with clinical signs of <u>pruritus</u>, abdominal <u>dermatitis</u>, and <u>otitis</u>. <u>Skin scrapes</u> were negative for external parasites. Ear cytology revealed a large number of yeast. The dog was referred to a <u>dermatologist</u>, who diagnosed <u>atopy</u> via <u>intradermal skin testing</u>. The dog was put on a <u>hypoallergenic</u> diet and was given <u>hyposensitization</u> injections. Medications were prescribed to control the pruritus and secondary <u>pyoderma</u>.

1. pruritus _____

2. dermatitis _____

3. otitis _____

4. skin scrapes _____

5. dermatologist _____

6. atopy _____

7. intradermal skin testing _____

8. hypoallergenic _____

9. hyposensitization _____

10. pyoderma _____

The veterinarian was on a farm performing a routine herd check when the farmer pointed out a cow to the veterinarian. The cow was a 2-yr-old Holstein cow that was treated 5 days ago with an <u>IM</u> injection of long-acting penicillin for mastitis (inflammation of the mammary gland). The cow's mastitis improved, and she was eating and <u>ruminating</u> normally. However, the cow now had an enlarged swelling on her <u>R</u> hind limb (Figure 10–27). On <u>PE</u>, the cow's <u>HR</u> and <u>RR</u> were normal and the temperature was elevated. The cow was not painful upon <u>palpation</u> of the swelling. The swelling had a soft center that appeared to be fluid-filled. The veterinarian diagnosed an <u>abscess</u> and treated the cow by cleaning the wound area, <u>lancing</u> the abscess, draining the pus from the swelling, and flushing the abscess with <u>antiseptic</u> solution. The farmer was advised to monitor the cow to make sure the swelling does not recur.

11. IM _____

12. ruminating _____

13. R _____

14. PE _____

15. HR _____

16. RR _____

17. palpation _____

18. abscess   _____

19. lancing   _____

20. antiseptic _____

Figure 10–27   Leg abscess in a Holstein.

CHAPTER 11

THE GREAT COMMUNICATOR

O b j e c t i v e s

Upon completion of this chapter, the reader should be able to

- Identify and describe the major structures and functions of the endocrine system
- Describe the role of the hypothalamus and pituitary gland in hormone secretion
- Recognize, define, spell, and pronounce terms related to the diagnosis, pathology, and treatment of the endocrine system

FUNCTIONS OF THE ENDOCRINE SYSTEM

The endocrine (ehn-dō-krihn) system is composed of ductless glands that secrete chemical messengers, called hormones, into the bloodstream. The prefix **endo-** means within, and the suffix **-crine** means to secrete or separate. Hormones enter the bloodstream and are carried throughout the body, affecting a variety of tissues and organs. Tissues and organs on which the hormones act are called **target organs. Hormone** (hōr-mōn) is a Greek term meaning impulse or to set in motion; hormones may excite or inhibit a motion or an action.

STRUCTURES OF THE ENDOCRINE SYSTEM

The glands of the normal endocrine system include the following (Figure 11–1):

- One pituitary gland (with two lobes)
- One thyroid gland (right and left lobes fused ventrally)
- Parathyroid glands (four in most species)
- Two adrenal glands
- One pancreas
- One thymus
- One pineal gland
- Two gonads (ovaries in females, testes in males)

The Pituitary Gland

The **pituitary** (pih-too-ih-tār-ē) **gland** is located at the base of the brain just below the hypothalamus. The pituitary gland also is called the **hypophysis** (hī-poh-fī-sihs) because it is a growth located beneath or ventral to the hypothalamus (part of the brain). (*Physis* is Greek for growth.) The pituitary gland is known as the master gland because it secretes many hormones that control or master other endocrine glands. The combining form for pituitary gland is **pituit/o.**

The pituitary gland acts in response to stimuli from the **hypothalamus** (hī-pō-thahl-ah-muhs). The hypothalamus is located below the thalamus in the brain and secretes releasing and inhibiting factors that affect the release of substances from the pituitary gland (Figure 11–2). The hypothalamus also produces antidiuretic hormone and oxytocin, which are released from the posterior pituitary gland. The hypothalamus is connected to the pituitary gland via a stalk called the **infundibulum** (ihn-fuhn-dihb-yoo-luhm). *Infundibulum* means funnel-shaped passage or opening.

The pituitary gland is made up of two lobes: the anterior (cranial) and posterior (caudal). The anterior lobe also is known as the **adenohypophysis** (ahd-ehn-ō-hī-pohf-ih-sihs) because it produces hormones (that is, has glandular function; **aden/o** = gland). Hormones released from the anterior pituitary gland sometimes are called **indirect-acting hormones** because they cause their target organ to produce a second hormone.

The posterior lobe also is known as the **neurohypophysis** (nū-rō-hī-pohf-ih-sihs) because it responds to a neurologic stimulus and does not produce hormones, but stores and secretes them. Hormones released from the posterior pituitary gland sometimes are called **direct-acting hormones** because they produce the desired effect directly in the target organ (Figure 11–3).

Secretions of the Anterior Pituitary Gland

- **thyroid** (thī-royd) **-stimulating hormone** = augments growth and secretions of the thyroid gland; abbreviated TSH.

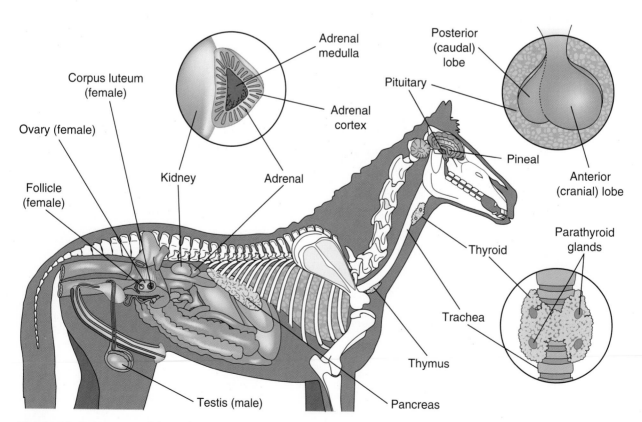

Figure 11–1 Locations of the endocrine glands. The relative locations of the endocrine glands are shown in this horse.

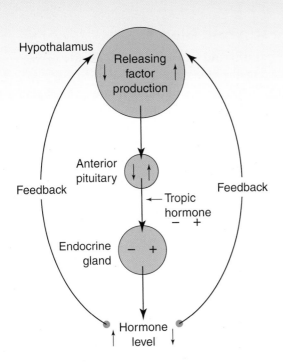

Figure 11–2 Feedback control mechanisms. Negative feedback mechanisms control the levels of most hormones in the blood by secreting, releasing, or inhibiting factors that affect hormone release.

- **adrenocorticotropic** (ahd-rēn-ō-kōr-tih-kō-trō-pihck) **hormone** = augments the growth and secretions of the adrenal cortex; abbreviated ACTH.
- **follicle** (fohl-lihck-kuhl) **-stimulating hormone** = augments the secretion of estrogen and growth of eggs in the ovaries (female) and the production of sperm in the testes (male); abbreviated FSH. FSH is a type of **gonadotropic** (gō-nahd-ō-trō-pihck) **hormone.** Gonadotropic can be divided into **gonad/o,** which means gamete-producing gland (ovary or testes), and **-tropic,** which means having an affinity for.
- **luteinizing** (loo-tehn-īz-ihng) **hormone** = augments ovulation and aids in the maintenance of pregnancy in females; abbreviated LH. **Lute/o** is the combining form for yellow. Luteinizing hormone transforms an ovarian follicle into a corpus luteum, or yellow body. LH is a type of gonadotropic hormone.
- **interstitial** (ihn-tər-stihsh-ahl) **cell-stimulating hormone** = stimulates testosterone secretion in males; abbreviated ICSH. ICSH is now considered to be LH.
- **prolactin** (prō-lahck-tihn) = augments milk secretion and influences maternal behavior; also known as **lactogenic hormone** or **luteotropin** (the

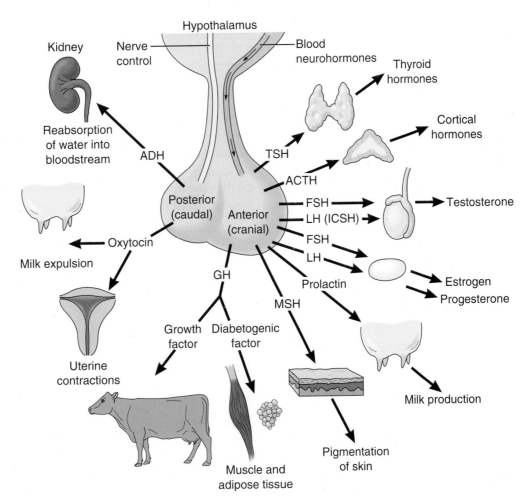

Figure 11–3 Secretions of the pituitary gland.

combining form **lact/o** means milk); sometimes abbreviated LTH.

- **growth hormone** = accelerates body growth; abbreviated GH; also known as **somatotropin** (the combining form **somat/o** means body); sometimes abbreviated STH.
- **melanocyte** (mehl-ah-nō-sīt) **-stimulating hormone** = augments skin pigmentation; abbreviated MSH.

Secretions of the Posterior Pituitary Gland

- **antidiuretic** (ahn-tih-dī-yoo-reht-ihck) **hormone** = maintains water balance in the body by augmenting water reabsorption in the kidneys (**anti-** means against; **diuretic** means pertaining to increased urine secretion); abbreviated ADH; also known as **vasopressin** (**vas/o** is the combining form for vessel, and **press/i** is the combining form for tension; vasopressin affects blood pressure).
- **oxytocin** (ohcks-ē-tō-sihn) = stimulates uterine contractions during parturition and milk letdown from the mammary ducts.

Thyroid Gland

The **thyroid** (thī-royd) **gland** is a butterfly-shaped gland, with the right and left lobes fused ventrally by an isthmus. (The isthmus may be rudimentary in horses and dogs.) The thyroid gland is located on either side of the larynx. The thyroid gland regulates metabolism, iodine uptake, and blood calcium levels. The combining forms for thyroid are **thyr/o** and **thyroid/o.**

Secretions of the Thyroid Gland

- **triiodothyronine** (trī-ī-ō-dō-thī-rō-nēn) = one of the thyroid hormones that regulates metabolism; abbreviated T$_3$.
- **thyroxine** (thī-rohcks-ihn) = one of the thyroid hormones that regulates metabolism; abbreviated T$_4$.
- **calcitonin** (kahl-sih-tō-nihn) = thyroid hormone that promotes the absorption of calcium from blood into bones.

Parathyroid Glands

The **parathyroid** (pahr-ah-thī-royd) **glands** are four glands located on the surface of the thyroid gland that secrete parathyroid hormone (or **parathormone** [pahr-ah-thōr-mōn] or PTH). The prefix **para-** means near or before. PTH helps regulate blood calcium and phosphorus levels. PTH increases blood calcium levels by reducing bone calcium levels. Calcium is regulated in the body by the antagonistic actions of PTH and calcitonin.

PTH also regulates phosphorus content of blood and bones. The combining form for the parathyroid glands is **parathyroid/o.**

Adrenal Glands

Two small glands, known as the **adrenal** (ahd-rē-nahl) or **suprarenal** (soop-rah-rē-nahl) glands, are located cranial to each kidney. The prefix **ad-** means toward, the combining form **ren/o** means kidney, and the prefix **supra-** means above. The adrenal gland regulates electrolytes, metabolism, sexual functions, and the body's response to injury. The combining forms **adren/o** and **adrenal/o** refer to the adrenal glands.

Each adrenal gland consists of two parts: the **adrenal cortex** (ahd-rē-nahl kōr-tehcks), or outer portion, and the **adrenal medulla** (ahd-rē-nahl meh-doo-lah), or inner portion. The combining form **cortic/o** means outer, and the combining form **medull/o** means inner or middle portion (Figure 11–4).

Secretions of the Adrenal Cortex

The adrenal cortex hormones are classified as steroids. A **steroid** (stehr-oyd) is a substance that has a specific chemical structure of carbon atoms in four interlocking rings. **Corticosteroids** (kōr-tih-kō-stehr-oydz) are produced by the adrenal cortex.

- **mineralocorticoids** (mihn-ər-ahl-ō-kōr-tih-koydz) = group of corticosteroids that regulates electrolyte and water balance by affecting ion transport in the kidney.

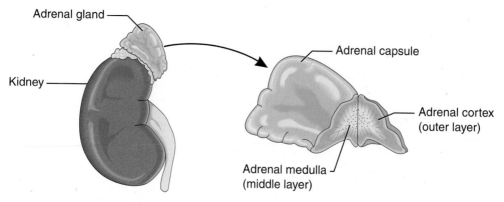

Figure 11–4 Portions of the adrenal gland.

The principal mineralocorticoid is **aldosterone** (ahl-dohs-tər-ōn), which is a hormone that regulates sodium and potassium.

- **glucocorticoids** (gloo-kō-kōr-tih-koydz) = group of corticosteroids that regulate carbohydrate, fat, and protein metabolism; resistance to stress; and immunologic functioning. An example of a glucocorticoid is **hydrocortisone** (hī-drō-kōr-tih-zōn), or **cortisol** (kōr-tih-zohl), which regulates carbohydrate, fat, and protein metabolism.
- **androgens** (ahn-drō-jehnz) = group of corticosteroids that aid in the development and maintenance of male sex characteristics. The combining form **andr/o** means male, and the suffix **-gen** means producing. **Anabolic** (ahn-ah-bohl-ihck) **steroids** are synthetic medications similar in structure to testosterone that are used to increase strength and muscle mass. The prefix **ana-** means up or excessive, and the combining form **bol/o** means throwing. Anabolic means pertaining to building up. The opposite of anabolism is catabolism. The prefix **cata-** means down; therefore, **catabolism** (kaht-ah-bōl-ihz-uhm) means breaking down.

Secretions of the Adrenal Medulla

- **epinephrine** (ehp-ih-nehf-rihn) = **catecholamine** (kaht-ih-kōl-ih-mēn) that stimulates the sympathetic nervous system (the fight-or-flight system) and increases blood pressure, heart rate, and blood glucose. (Catecholamines are nitrogen-containing compounds that act as hormones or neurotransmitters.) Epinephrine also is known as **adrenaline** (ahd-rehn-ah-lihn). Because epinephrine increases blood pressure, it is known as a vasopressor. A **vasopressor** (vahs-ō-prehs-ōr) is a substance that stimulates blood vessel contraction and increases blood pressure.
- **norepinephrine** (nōr-ehp-ih-nehf-rihn) = catecholamine that stimulates the sympathetic nervous system (the fight-or-flight system) and increases blood pressure, heart rate, and blood glucose. Norepinephrine also is known as **noradrenaline.**

Pancreas

The **pancreas** (pahn-krē-ahs) is an aggregation of cells located near the proximal duodenum that has both exocrine and endocrine functions. The exocrine function of the pancreas involves the secretion of digestive enzymes, which is covered in Chapter 6. The endocrine function of the pancreas involves the secretion of blood glucose–regulating hormones. (Blood glucose also is regulated in conjunction with other hormones such as thyroid hormones.) The combining form for pancreas is **pancreat/o.**

What is in a name?

It may seem as though medical terminology is a bunch of long words strung together with no relationship to each other. However, consider the names of the hormones released from the adrenal medulla:

- **epinephrine** = the prefix **epi-** means above. The combining form **nephr/o** means kidney. Its name indicates that epinephrine is made in the gland above the kidney, which is known as the adrenal gland.
- **adrenaline** = the prefix **ad-** means toward. The combining form **ren/o** means kidney. Its name indicates that adrenaline is made in the gland toward the kidney, which is known as the adrenal gland.
- **norepinephrine** = the prefix **nor-** means normal and denotes the parent compound in a pair of related substances. Its name indicates that norepinephrine is a substance very similar in chemical structure and function to epinephrine.

Specialized cells in the pancreas called the **islets of Langerhans** (ī-lehts ohf lahng-ər-hahnz) secrete the hormones that help regulate blood glucose (Figure 11–5).

Endocrine Secretions of the Pancreas

- **insulin** (ihn-suh-lihn) = hormone that decreases blood glucose levels by transporting blood glucose into body cells or into storage as glycogen (glī-kō-jehn). Glycogen is the main carbohydrate storage unit in animals.
- **glucagon** (gloo-kah-gohn) = hormone that increases blood glucose levels by breaking down glycogen.

Thymus

The **thymus** (thī-muhs) is a gland predominant in young animals located near midline in the cranioventral portion of the thoracic cavity. The thymus gland has an immunologic

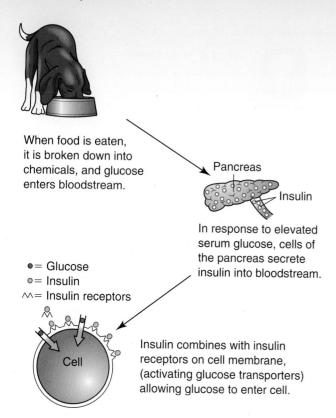

When food is eaten, it is broken down into chemicals, and glucose enters bloodstream.

Pancreas

Insulin

In response to elevated serum glucose, cells of the pancreas secrete insulin into bloodstream.

● = Glucose
◦ = Insulin
⋏ = Insulin receptors

Cell

Insulin combines with insulin receptors on cell membrane, (activating glucose transporters) allowing glucose to enter cell.

Figure 11–5 How insulin works.

function through its role in the maturation of T-lymphocytes. The combining form for the thymus is **thym/o.** Thym/o also refers to the soul; some people believed that the thymus gland was the seat of the soul because it was so close to the heart.

Secretion of the Thymus

- **thymosin** (thī-mō-sihn) = augments maturation of T-lymphocytes; therefore, it plays a role in the immune system, which is discussed in Chapter 15.

Pineal Gland

The **pineal** (pih-nē-ahl or pī-nē-ahl) **gland,** or **pineal body,** is an aggregation of cells located in the central portion of the brain. The pineal gland's functions are not fully understood, but one of its functions is secretion of hormones that affect circadian rhythm. The combining form **pineal/o** refers to the pineal gland.

Secretion of the Pineal Gland

- **melatonin** (mehl-ah-tōn-ihn) = controls circadian rhythm and reproductive timing. **Circadian** can be divided into **circa** (around) and **diem** (day). *Circadian* refers to the events occurring within a 24-hour period.

Gonads

The **gonads** (gō-nahdz) are gamete-producing glands. A **gamete** (gahm-ēt) is a sex cell. Gamete-producing glands

are the **ovaries** in females and **testes** in males. **Gonad/o** is the combining form for gonad or gamete-producing gland. **Gonadotropic hormones** stimulate the gonads.

Secretions of the Ovary

Secretions of the ovary are stimulated by **human chorionic** (kōr-ē-ohn-ihck) **gonadotropin,** or **hCG.** hCG is secreted by the embryo and by the placenta after implantation has occurred. Other gonadotropins also influence ovarian secretion.

- **estrogen** (ehs-trō-jehn) = hormone that aids in the development of secondary sex characteristics (an example is mammary gland development) and regulates ovulation in females.
- **progesterone** (prō-jēhs-tər-own) = hormone that aids in the maintenance of pregnancy. Progesterone also is secreted from the corpus luteum and placenta.

Secretion of the Testes

- **testosterone** (tehs-tohs-tər-own) = augments the development of secondary sex characteristics. Examples of secondary sex characteristics in animals include horns in rams (not horns in all species), boar tusks, and shoulder girth in cattle and horses. Testosterone also is thought to be secreted from the ovaries and adrenal cortex, although those secretions are in very small amounts.

TEST ME: ENDOCRINE SYSTEM

Diagnostic procedures performed on the endocrine system include the following:

- **ACTH stimulation test** = blood analysis for cortisol levels after administration of synthetic adrenocorticotropic hormone; used to differentiate pituitary-dependent hyperadrenocorticism from adrenal-dependent hyperadrenocorticism.
- **assays** (ahs-āz) = laboratory technique used to determine the amount of a particular substance in a sample. Assays are discussed in Chapter 16.
- **dexamethasone** (dehcks-ah-mehth-ah-zōn) **suppression test** = blood analysis for cortisol levels after administration of synthetic glucocorticoid (dexamethasone); used to differentiate pituitary-dependent hyperadrenocorticism from adrenal-dependent hyperadrenocorticism. Also called **dex suppression test.** Dex suppression tests may be high-dose or low-dose.
- **radioactive** (rā-dē-ō-ahck-tihv) **iodine uptake test** = analysis of thyroid function after induction of radioactive iodine has been given orally or intravenously. Absorption of the radioactive iodine is measured with a counter for a specific time period.
- **thyroid stimulation test** = blood analysis for thyroid hormone levels after administration of synthetic

thyroid-stimulating hormone; used to differentiate pituitary-dependent from thyroid-dependent dysfunction. **Synthetic** (sihn-theh-tihck) means pertaining to artificial production. TSH levels also can be measured before and after administration of thyrotropin-releasing hormone.

PATHOLOGY: ENDOCRINE SYSTEM

Pathologic conditions of the endocrine system include the following:

- **acromegaly** (ahck-rō-mehg-ah-lē) = enlargement of the extremities caused by excessive secretion of growth hormone after puberty. **Acr/o** means extremities.
- **Addison's** (ahd-ih-sohnz) **disease** = disorder caused by deficient adrenal cortex function; also called **hypoadrenocorticism** (hī-pō-ahd-rēn-ō-kōr-tih-kihz-uhm).
- **adrenopathy** (ahd-rēn-ohp-ah-thē) = disease of the adrenal glands.
- **aldosteronism** (ahl-doh-stər-ōn-ihzm) = disorder caused by excessive secretion of aldosterone by the adrenal cortex, resulting in electrolyte imbalance. An **electrolyte** (ē-lehck-trō-līt) is a charged substance found in blood.
- **Cushing's** (kuhsh-ihngz) **disease** = disorder caused by excessive adrenal cortex production of glucocorticoid, resulting in increased urination, drinking, and redistribution of body fat; also called **hyperadrenocorticism** (hī-pər-ahd-rēn-ō-kōr-tih-kihz-uhm) (Figure 11–6).
- **diabetes insipidus** (dī-ah-bē-tēz ihn-sihp-ih-duhs) = insufficient antidiuretic hormone production or the inability of the kidneys to respond to ADH stimuli. A **stimulus** (stihm-yoo-luhs) is an agent, act, or influence that produces a reaction. The plural of *stimulus* is **stimuli.**
- **diabetes mellitus** (dī-ah-bē-tēz mehl-ih-tuhs) = metabolic disorder of inadequate secretion of insulin or recognition of insulin by the body, resulting in increased urination, drinking, and weight loss. Severe insulin deficiency may result in **ketoacidosis** (kē-tō-ah-sih-dō-sihs), which is an abnormal condition of low pH accompanied by ketones (by-products of fat metabolism). **Acidosis** (ah-sih-dō-sihs) is an abnormal condition of low pH.
- **endocrinopathy** (ehn-dō-krih-nohp-ah-thē) = disease of the hormone-producing system.
- **gynecomastia** (gī-neh-kō-mahs-tē-ah) = condition of excessive mammary development in males.
- **hypercrinism** (hī-pər-krī-nihzm) = condition of excessive gland secretion.
- **hyperglycemia** (hī-pər-glī-sē-mē-ah) = abnormally elevated blood glucose.
- **hypergonadism** (hī-pər-gō-nahd-ihzm) = abnormal condition of excessive hormone secretion by the sex glands (ovaries in females; testes in males).
- **hyperinsulinism** (hī-pər-ihn-suh-lin-ihzm) = disorder of excessive hormone that transports blood glucose to body cells.
- **hyperparathyroidism** (hī-pər-pahr-ah-thī-royd-ihzm) = abnormal condition of excessive parathyroid hormone secretion resulting in hypercalcemia.
- **hyperpituitarism** (hī-pər-pih-too-ih-tahr-ihzm) = condition of excessive secretion of the pituitary gland.
- **hyperthyroidism** (hī-pər-thī-royd-ihzm) = condition of excessive thyroid hormone. Signs of hyperthyroidism include increased metabolic rate, weight loss, polyuria, and polydipsia (Figure 11–7).

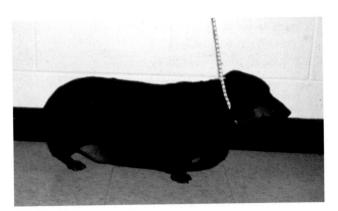

Figure 11–6 Dog with hyperadrenocorticism. Hyperadrenocorticism or Cushing's disease results in polyuria, polydipsia, and redistribution of body fat. (Courtesy of Mark Jackson, DVM, PhD, Glasgow University.)

Figure 11–7 Cat with diabetes mellitus. Increased metabolic rate, weight loss, polyuria, and polydipsia may indicate hyperthyroidism or diabetes mellitus in cats. (Courtesy of Mark Jackson, DVM, PhD, Glasgow University.)

- **hypocrinism** (hī-pō-krī-nihzm) = condition of deficient gland secretion.
- **hypoglycemia** (hī-pō-glī-sē-mē-ah) = abnormally low blood glucose.
- **hypogonadism** (hī-pō-gō-nahd-ihzm) = abnormal condition of deficient hormone secretion by the sex glands (ovaries in females; testes in males).
- **hypoparathyroidism** (hī-pō-pahr-ah-thī-royd-ihzm) = abnormal condition of deficient parathyroid hormone secretion resulting in hypocalcemia. **Hypocalcemia** (hī-pō-kahl-sē-mē-ah) is abnormally low levels of blood calcium. **Hypo-** is the prefix for deficient, **calc/i** is the combining form for calcium, and **-emia** is the suffix for blood condition. Hypercalcemia is abnormally high blood calcium levels because **hyper-** is the prefix meaning excessive.
- **hypophysitis** (hī-pō-fi-sī-tihs) = inflammation of the pituitary gland.
- **hypothyroidism** (hī-pō-thī-royd-ihzm) = condition of thyroid hormone deficiency. Signs of hypothyroidism include decreased metabolic rate, poor hair coat, lethargy, and increased sensitivity to cold. **Euthyroidism** (yoo-thī-royd-ihzm) is the condition of normal thyroid function. The prefix **eu-** means good, well, or easily.
- **insulinoma** (ihn-suh-lihn-ō-mah) = tumor of the islet of Langerhans of the pancreas.
- **pancreatitis** (pahn-krē-ah-tī-tihs) = inflammation of the pancreas (Figure 11–8).

Figure 11–8 Pancreas of a dog with diabetes mellitus. (Courtesy of Mark Jackson, DVM, PhD, Glasgow University.)

- **pheochromocytoma** (fē-ō-krō-mō-sī-tō-mah) = tumor of the adrenal medulla resulting in increased secretion of epinephrine and norepinephrine. **Phe/o** is the combining form for dusky; **chrom/o** is the combining form for color; **cyt/o** is the combining form for cell; **-oma** is the suffix meaning tumor. A pheochromocytoma is a tumor that takes on a dark (dusky) color because it is composed of chromaffin (or colored) cells.
- **pinealopathy** (pīn-ē-ah-lohp-ah-thē) = disorder of the pineal gland.
- **pituitarism** (pih-too-ih-tār-ihzm) = any disorder of the pituitary gland.
- **thymoma** (thī-mō-mah) = tumor of the thymus.
- **thyroiditis** (thī-roy-dī-tihs) = inflammation of the thyroid gland.
- **thyromegaly** (thī-rō-mehg-ah-lē) = enlargement of the thyroid gland.
- **thyrotoxicosis** (thī-rō-tohck-sih-kō-sihs) = abnormal life-threatening condition of excessive, poisonous quantities of thyroid hormones. The combining form **toxic/o** means poison.

PROCEDURES: ENDOCRINE SYSTEM

Procedures performed on the endocrine system include the following:

- **adrenalectomy** (ahd-rē-nahl-ehck-tō-mē) = surgical removal of one or both adrenal glands.
- **chemical thyroidectomy** (thī-royd-ehck-tō-mē) = administration of radioactive iodine to suppress thyroid function; also called radioactive iodine therapy.
- **hypophysectomy** (hī-pohf-ih-sehck-tō-mē) = surgical removal of the pituitary gland.
- **lobectomy** (lō-behck-tō-mē) = surgical removal of a lobe or a well-defined portion of an organ.
- **pancreatectomy** (pahn-krē-ah-tehck-tō-mē) = surgical removal of the pancreas.
- **pancreatotomy** (pahn-krē-ah-toht-ō-mē) = surgical incision into the pancreas.
- **parathyroidectomy** (pahr-ah-thī-royd-ehck-tō-mē) = surgical removal of one or more parathyroid glands.
- **pinealectomy** (pīn-ē-ahl-ehck-tō-mē) = surgical removal of the pineal gland.
- **thymectomy** (thī-mehck-tō-mē) = surgical removal of the thymus.
- **thyroidectomy** (thī-royd-ehck-tō-mē) = surgical removal of all or part of the thyroid gland.

REVIEW EXERCISES

Multiple Choice

Choose the correct answer.

1. The gland known as the master gland that helps maintain the appropriate levels of hormone in the body is the

   a. hypothalamus
   b. pituitary gland
   c. thyroid gland
   d. pancreas

2. The chemical substance secreted by the posterior pituitary gland that stimulates uterine contractions during parturition is

   a. prolactin
   b. luteinizing hormone
   c. ACTH
   d. oxytocin

3. The regulator of the endocrine system is the

   a. thyroid gland
   b. calcitonin
   c. hypothalamus
   d. parathyroid gland

4. Thyromegaly is

   a. enlargement of the thyroid gland
   b. augmentation of the thymus
   c. dissolution of the parathyroid glands
   d. radioactive iodine treatment of the thyroid gland

5. Surgical removal of a well-defined portion of an organ is a

   a. sacculectomy
   b. lumpectomy
   c. lobectomy
   d. cystectomy

6. An aggregation of cells specialized to secrete or excrete materials not related to their function is a

   a. hormone
   b. gland
   c. hypoadenum
   d. hyperadenum

7. Hypoadrenocorticism, a disorder caused by deficient adrenal cortex production of glucocorticoid, also is known as

   a. Cushing's disease
   b. Graves' disease
   c. Addison's disease
   d. Langerhans' disease

8. A tumor of the islets of Langerhans of the pancreas is called

   a. diabetes mellitus
   b. diabetes insipidus
   c. ketoacidosis
   d. insulinoma

9. The chemical substance that helps control circadian rhythm is

   a. circadianin
   b. pinealin
   c. melatonin
   d. thymin

10. Excessive mammary development in males is called

    a. feminum
    b. gynecomastia
    c. gyneconium
    d. feminomastia

11. Which of the following is a function of the thyroid gland?

    a. secretes thymosin
    b. secretes thyroid-stimulating hormone
    c. secretes corticosteroids
    d. secretes triiodothyronin, thyroxine, and calcitonin

12. What is another name for the anterior lobe of the pituitary gland?

    a. hypophysis
    b. epiphysis
    c. adenohypophysis
    d. neurohypophysis

13. Which hormone maintains water balance in the body by increasing water reabsorption in the kidney?

    a. insulin
    b. epinephrine
    c. antidiuretic hormone
    d. FSH

14. What is the name of a laboratory technique used to determine the amount of a particular substance in a sample?

    a. stimulation test
    b. synthetic test
    c. uptake test
    d. assay

15. An abnormally low blood glucose level is known as

    a. hyperglycemia
    b. hypoglycemia
    c. glucosemia
    d. diabetes mellitus

16. The condition of normal thyroid function is known as

    a. thyrotoxicosis
    b. thyromegaly
    c. euthyroidism
    d. thyroiditis

17. Which hormone transports blood glucose to the cells?

    a. glucagon
    b. insulin
    c. vasopressin
    d. melatonin

18. Glucocorticoids are secreted by the

    a. adrenal cortex
    b. adrenal medulla
    c. parathyroid gland
    d. pancreas

19. The pituitary gland acts in response to stimuli from the

    a. hypothalamus
    b. thymus
    c. parathyroid glands
    d. thyroid gland

20. The word part that means to secrete or separate is

    a. endo
    b. -crine
    c. hormon/o
    d. aden/o

Matching

Match the endocrine term in column I with the correct description in column II.

<table>
<tr><td>

Column I

1. _____ adrenals
2. _____ thyroid
3. _____ thymus
4. _____ pancreas
5. _____ pituitary
6. _____ parathyroid
7. _____ pineal
8. _____ gonads
9. _____ thyroidectomy
10. _____ hyperthyroidism
11. _____ diabetes mellitus
12. _____ diabetes insipidus
13. _____ adrenopathy
14. _____ acidosis
15. _____ glucagon
16. _____ hypocalcemia
17. _____ insulinoma
18. _____ hypercrinism

</td><td>

Column II

a. butterfly-shaped gland on either side of the larynx

b. gland located dorsal to the sternum

c. contains specialized cells that secrete hormones that affect sugar and starch metabolism

d. secretes melatonin

e. gamete-producing glands

f. small gland at the base of the brain

g. secretes hormone that reduces bone calcium levels and regulates phosphorus

h. two small glands located on top of each kidney

i. tumor of the islet of Langerhans cells of the pancreas

j. abnormally low blood calcium levels

k. insufficient antidiuretic hormone

l. insufficient secretion of insulin or recognition of insulin

m. surgical removal of the thyroid gland

n. condition of excessive gland secretion

o. condition of excessive thyroid hormone

p. disease of the adrenal gland

q. abnormal condition of low pH

r. hormone that increases blood glucose

</td></tr>
</table>

Fill in the Blanks

1. Thyr/o and thyroid/o mean _____.
2. Adren/o and adrenal/o mean _____.
3. -trophic means _____.
4. Lute/o means _____.
5. Somat/o means _____.

Spelling

Circle the term that is spelled correctly.

1. gland that has both exocrine and endocrine function: pancrease   pancreus   pancreas

2. substance that increases urine secretion: dyuretic   diyuretic   diuretic

3. gland located at the base of the brain: pitwoitary   pituitary   pitootary

4. part of the brain that controls secretions of the pituitary gland: hypothalamus   hypothalmus   hypothalmis

5. gamete-producing glands: gonids   gonads   goknads

Word Scramble

Use the definitions to unscramble the terms relating to the endocrine system.

1. condition of building up   mlsioaanb   _____

2. condition of breaking down   mlsioaactb   _____

3. abnormally elevated blood glucose   aieyeyhprglcm   _____

4. inflammation of the pancreas   iittseaacrnp   _____

5. gamete-producing   naodg   _____

6. surgical removal of the thyroid gland   dyorihtetcmoy   _____

7. study of the hormonal system   glnooyndeocri   _____

8. production of new glucose   ssgiluncoeogeen   _____

9. artificially produced   nttheicys   _____

10. abnormal condition of deficient thyroid hormone secretion   smidioryhtophy   _____

CROSSWORD PUZZLE

Endocrine Terms Puzzle

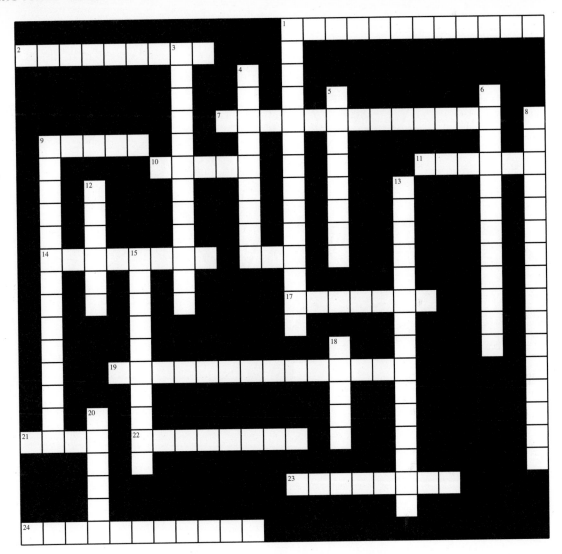

Across

1 condition of normal thyroid function
2 pertaining to artificial production
7 surgical incision into the pancreas
9 prefix meaning excessive, above normal, or elevated
10 combining form for gland
11 sex cell
14 abnormally low blood calcium levels
17 chemical substance produced by an organ which is transported by the bloodstream to regulate another organ
19 hormone secreted by the adrenal cortex used to suppress immunity
21 suffix meaning deficient, less than normal, or decreases
22 agent, act, or influence that produces a reaction
23 abnormal condition of low pH
24 substance that stimulates blood vessel contraction and increases blood pressure

Down

1 disease of hormone-producing system
3 funnel-shaped passage or opening
4 condition of building up
5 drowsiness or indifference
6 condition of excessive gland secretion
8 tumor of the adrenal medulla
9 surgical removal of the pituitary gland
12 suffix meaning affinity for
13 another term for hyperadrenocorticism
15 condition of breaking down
18 suffix meaning to secrete or separate
20 combining form for body

LABEL THE DIAGRAMS

Label the diagram in Figure 11–9.

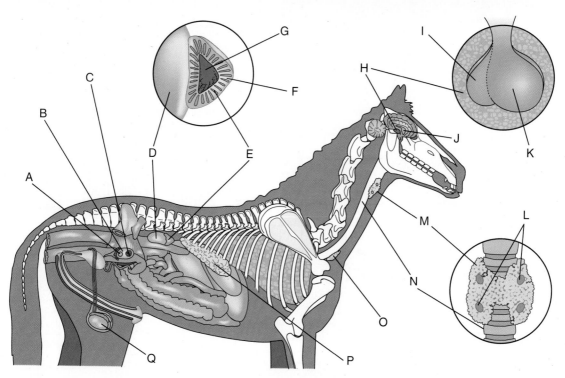

Figure 11–9   Endocrine organs. Label the endocrine organs and related structures. Provide the combining form for each endocrine gland.

CASE STUDIES

Define the underlined terms in each case study.

A 7-yr-old M/N golden retriever, Goldie, was presented to the clinic for bite wounds from another dog in the household. After the dog's vaccination status was checked, he was examined. The dog's vital signs included T = 103°F, HR = 120 BPM, RR = 42 breaths/min. Upon examination, it was noted that the dog weighed 85#, the haircoat had areas of alopecia, and the dog was lethargic. The wounds from the dog fight consisted of 2 lacerations; 1 laceration was 4 cm long located on the lateral side of the L forelimb proximal to the elbow, and the other laceration was 2 cm long located on the medial surface of the L hock. The owner was questioned as to whether these two dogs usually got along in the house, and the owner stated that they had before Goldie's recent history of lethargy and unwillingness to go out for walks during the past month. The owner thought that Goldie did not want to go out on walks because of the change of seasons and the cold, damp weather. Goldie was admitted to the clinic and anesthetized, and the lacerations were cleaned and sutured. Antibiotics were dispensed 2 T BID for 7 d. Goldie came back for suture removal in 10 days. The lacerations had healed; but hair had not begun to regrow in the areas of the lacerations, and more areas of alopecia were noted. A thyroid panel was recommended to assess Goldie's thyroid function. Blood was drawn, and a thyroid panel was run. The thyroid panel revealed that Goldie was hypothyroid, and thyroid supplementation was begun.

1. yr _____
2. M/N _____
3. T _____
4. HR _____
5. BPM _____
6. # _____
7. alopecia _____
8. lethargic _____
9. cm _____
10. lateral _____
11. proximal _____
12. elbow _____
13. medial _____
14. hock _____
15. anesthetized _____
16. BID _____
17. d _____
18. hypothyroid _____

A 15-yr-old F/S DSH was brought to the clinic for weight loss. Examination revealed that the cat was tachycardic, emaciated, and lethargic. Vital signs were T = 102°F, HR = 260 BPM, RR = 60 breaths/min. Abdominal palpation revealed a moderately full urinary bladder, slightly small kidneys, and gas-filled loops of bowel. Palpation of the neck revealed enlargement of thyroid glands. Blood was collected from the jugular vein for a CBC, chem screen, and T_4. Blood work revealed a slightly elevated BUN and creatinine, hypercholesterolemia, and elevated T_4 levels. It was recommended that the cat receive radioactive iodine treatments; however, the options of thyroidectomy and pharmaceutical management also were discussed. The cat was scheduled for radiation treatments at a referral center.

19. yr

_____

20. F/S

_____

21. DSH

_____

22. tachycardic

_____

23. emaciated

_____

24. lethargic

_____

25. T

_____

26. HR

_____

27. BPM

_____

28. RR

_____

29. palpation

_____

30. vein

_____

31. CBC

_____

32. chem screen

_____

33. T_4

_____

34. BUN

_____

35. creatinine

_____

36. hypercholesterolemia

_____

37. radioactive iodine treatments

_____

38. thyroidectomy

_____

CHAPTER 12

1 + 1 = 3
(OR MORE)

Objectives

Upon completion of this chapter, the reader should be able to

- Identify and describe the functions and structures of the male and female reproductive systems
- Use medical terminology to describe the estrous cycle in females
- Recognize, define, spell, and pronounce the terms related to the diagnosis, pathology, and treatment of the reproductive systems

THE REPRODUCTIVE SYSTEM

The **reproductive** (rē-prō-duhck-tihv) **system** is responsible for producing offspring. The reproductive system needs both male- and female-specific organs to complete offspring production. The term **theriogenology** (thēr-ē-ō-jehn-ohl-ō-jē) is used to describe animal reproduction or the study of producing beasts. **Theri/o** is the combining form for beast, **gen/o** is the combining form for producing, and **-logy** is the suffix meaning to study.

The reproductive organs, whether male or female, are called the **genitals** (jehn-ih-tahlz), or **genitalia** (jehn-ih-tā-lē-ah). The combining form **genit/o** refers to the organs of reproduction. The genitalia include external and internal organs.

The male reproductive system will be described first, followed by the female reproductive system.

FUNCTIONS OF THE MALE REPRODUCTIVE SYSTEM

The functions of the male reproductive system are to produce and deliver sperm to the egg to create life.

STRUCTURES OF THE MALE REPRODUCTIVE SYSTEM

The structures of the male reproductive system produce sperm, transport sperm out of the body, and produce hormones.

Scrotum

The **scrotum** (skrō-tuhm), or **scrotal sac,** is the external pouch that encloses and supports the testes. The scrotum encloses the testes outside the body so that the testes are at a temperature lower than body temperature. This lower temperature is needed for sperm development. *Scrotum* is the Latin term for bag. The combining form **scrot/o** means scrotum.

The area between the scrotum and the anus in males is the **perineum** (pehr-ih-nē-uhm). **Perine/o** is the combining form for the area between the scrotum (or vulva in females) and anus.

Testes

The **testes** (tehs-tēz), or **testicles** (tehs-tih-kuhlz), are the male sex glands that produce spermatozoa. Testes refers to glands, and **testis** (tehs-tihs) refers to a single gland.

The orientation of the testes in the scrotum varies between species. Some species have open inguinal rings that allow the testes to be withdrawn from the scrotum and into the abdomen. Sex glands are called **gonads** (gō-nahdz). The combining forms for testes are **orch/o, orchi/o, orchid/o, test/o,** and **testicul/o** (Figure 12–1).

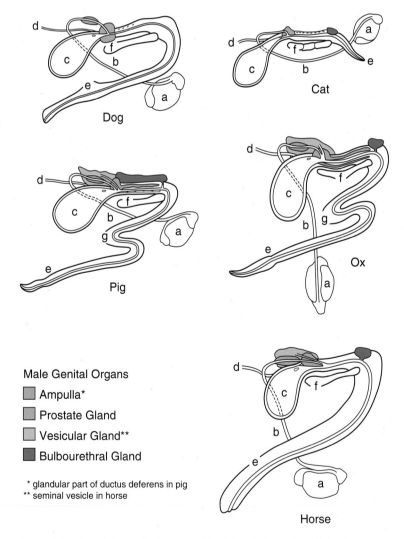

Male Genital Organs

- ▨ Ampulla*
- ▨ Prostate Gland
- ▨ Vesicular Gland**
- ▨ Bulbourethral Gland

* glandular part of ductus deferens in pig
** seminal vesicle in horse

Figure 12–1 Male genital organs. (a) Right testis and epididymis, (b) ductus deferens, (c) urinary bladder, (d) ureter, (e) penis and extrapelvic urethra, (f) pelvic symphysis, and (g) sigmoid flexure.

The testes develop in the fetal abdomen and descend into the scrotum before birth. The testes are suspended in the scrotum by the spermatic cord. The testicle is divided into compartments that contain coiled tubes called the **seminiferous tubules** (seh-mih-nihf-ər-uhs too-buhlz), and cells between the spaces are called **interstitial** (ihn-tər-stih-shahl) cells. The interstitial cells of the testes are called the **Leydig's** (lih-dihgz) **cells.** Leydig's cells have endocrine function. **Sertoli** (sihr-tō-lē) **cells** are specialized cells in the testes that support and nourish sperm growth. The seminiferous tubules are channels in the testes in which sperm are produced and through which the sperm leave the testes. **Sperm** (spərm), or **spermatozoa** (spər-mah-tō-zō-ah), are the male gametes, or sex cells. **Spermatozoon** (spər-mah-tō-zō-uhn) is one gamete. The combining forms for spermatozoa are **sperm/o** and **spermat/o.**

Ejaculated semen occasionally is evaluated microscopically to determine whether the spermatozoa have normal morphology and motility and whether they are present in adequate numbers (Figures 12–2a and b). A spermatozoon has a head, midpiece, and tail (Figure 12–3). The head contains the nucleus. At the top of the head is a structure called the **acrosome** (ahck-rō-zōm), which contains enzymes that allow the spermatozoon to penetrate the ovum. The midpiece contains mitochondria to provide energy to the sperm. The tail is actually a flagellum, providing movement for the spermatozoon to reach the ovum. **Spermatogenesis** (spər-mah-tō-jehn-eh-sihs) is production of male gametes.

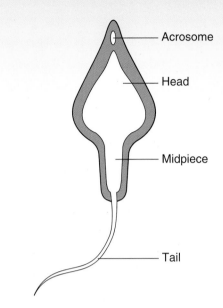

Figure 12–3 Parts of sperm.

Gamete comes from the Greek terms *gametes*, which means husband, and *gamete*, which means wife. *Gamein* is Greek for to marry. Johann Gregor Mendel was the first to apply the term *gamete* in biology to mean sex cells. Gametes are the sperm in males and ovum (egg) in females.

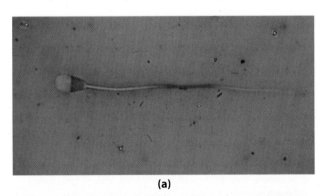

(a)

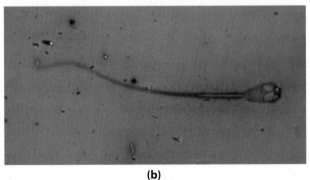

(b)

Figure 12–2 Sperm with abnormal shapes are not desirable. (a) Normal. (b) Defective. (Courtesy of Dr. Ben Bracket, Department of Physiology, School of Veterinary Sciences, The University of Georgia.)

Epididymis

The seminiferous tubules join together to form a cluster. Ducts emerge from this cluster and enter the epididymis. The **epididymis** (ehp-ih-dihd-ih-mihs) is the tube at the upper part of each testis that secretes part of the semen, stores semen before ejaculation, and provides a passageway for sperm. The epididymis is divided into head (or caput), body, and tail portions. The epididymis runs down the length of the testicle, turns upward, and becomes a narrower tube called the ductus deferens (Figure 12–4). Sperm are collected in the epididymis, where they become **motile** (mō-tihl), or capable of spontaneous motion. The combining form for epididymis is **epididym/o.**

Ductus Deferens

The **ductus deferens** (duhck-tuhs dehf-ər-ehnz) is a tube connected to the epididymis that carries sperm into the pelvic region toward the urethra. Each ductus deferens is encased by the spermatic cord. (The spermatic cord also encases nerves, blood and lymph vessels, and the cremaster muscle along with the ductus deferens.) The ductus deferens is the excretory duct

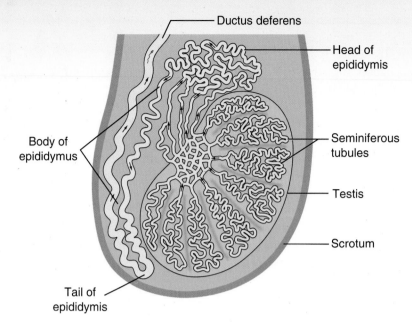

Figure 12–4 Cross section of scrotal structures.

of the testes. The ductus deferens in swine has a glandular portion called the **ampulla** (ahmp-yoo-lah). An ampulla is an enlarged part of a tube or canal.

Accessory Sex Glands

The male accessory sex glands include the seminal vesicles, prostate gland, and bulbourethral glands. Not all glands are present in all species (Table 12–1). The accessory sex glands add secretions to the sperm and flush urine from the urethra before sperm enter it.

The **seminal vesicles** (sehm-ih-nahl vehs-ih-kuhls), or **vesicular** (vehs-ih-koo-lahr) **glands,** are two glands that open into the ductus deferens where it joins the urethra. The seminal vesicles secrete a thick, yellow substance that nourishes sperm and adds volume to the ejaculated semen. **Semen** (sē-mehn) is the ejaculatory fluid that contains sperm and the secretions of the accessory sex glands. **Semin/i** is the combining form for semen.

The **ejaculatory** (ē-jahck-yoo-lā-tōr-ē) **duct** is formed by the union of the ductus deferens and the duct from the seminal vesicles. The ejaculatory duct passes through the prostate and enters the urethra.

The **prostate** (proh-stāt) **gland** is a single gland that surrounds or is near the urethra and may be well defined or diffuse

> *Seminal vesicles* is the term used in horses; *vesicular gland* is used in the other species when the gland is present.

Table 12–1 Male Accessory Glands of Different Species

	Prostate	Vesicular gland or seminal vesicle	Bulbourethral
dog	+	−	−
cat	+	−	+
pig	+	+	+
ruminants	+	+	+
horse	+	+	+

Note: In some species, the ampulla of the ductus deferens is glandular (as in horses).

depending on the species. The prostate gland secretes a thick fluid that aids in the motility of sperm. The combining form **prostat/o** means prostate gland.

The **bulbourethral** (buhl-bō-yoo-rē-thrahl) **glands** are two glands located on either side of the urethra. The bulbourethral glands secrete a thick mucus that acts as a lubricant for sperm. These glands are called **Cowper's** (cow-pərz) **glands** in humans.

Urethra

The **urethra** (yoo-rē-thrah) is a tube passing through the penis to the outside of the body; it serves both reproductive and urinary systems. The combining form for urethra is **urethr/o.**

Penis

The **penis** (pē-nihs) is the male sex organ that carries reproductive and urinary products out of the body. The glans penis is the distal part of the penis on which the urethra opens. The **prepuce** (prē-pyoos) is the retractable fold of skin covering the glans penis (Figure 12–5). The prepuce is sometimes called the **foreskin.** Dogs have an **os penis** (ohs pē-nihs), which is

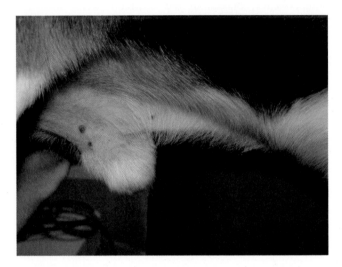

Figure 12–5 The prepuce is the retractable fold of skin covering the glans penis. (Courtesy of Terri Raffel, CVT.)

a bone encased in the penile tissue. All species except the cat have a cranioventrally directed penis. The combining forms **pen/i** and **priap/o** mean penis.

The penis is composed of erectile tissue that upon sexual stimulation fills with blood (under high pressure) and causes an erection. Some species, such as ruminants and swine, achieve an erection by straightening of the **sigmoid flexure** (sihg-moyd flehck-shər), an S-shaped bend in the penis. Other species, such as equine and canine, have a penis with almost all erectile tissue. Erection in these animals is caused by blood engorgement of the erectile tissue (Figure 12–6).

FUNCTIONS OF THE FEMALE REPRODUCTIVE SYSTEM

The functions of the female reproductive system are to create and support new life.

STRUCTURES OF THE FEMALE REPRODUCTIVE SYSTEM

The structures of the female reproductive system produce eggs, transport eggs for fertilization and implantation, house the embryo, deliver the fetus, and produce hormones.

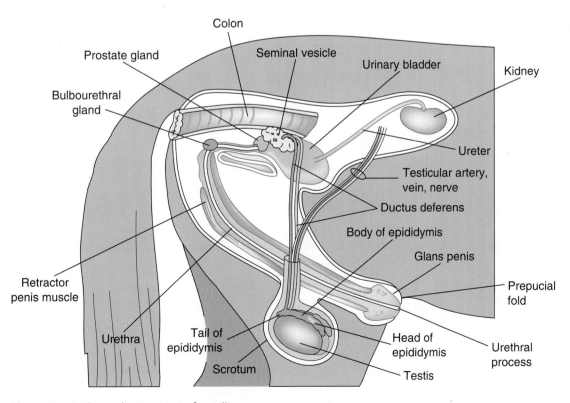

Figure 12–6 Reproductive tract of a stallion.

Ovaries

The **ovaries** (ō-vah-rēz) are a small pair of organs located in the caudal abdomen (Figure 12–7). An ovary is the female gonad that produces estrogen; progesterone; and **ova** (ō-vah), or **eggs.** The ovaries contain many small sacs called **graafian follicles** (grahf-ē-ahn fohl-lihck-kuhlz) (Figure 12–8). Each graafian follicle contains an ovum. The ova develop in the ovaries and are expelled (ovulated) when the egg matures. The combining forms **ovari/o** and **oophor/o** mean female gonad. The combining forms **oo/o, ov/i,** and **ov/o** mean egg. An egg cell is an **oocyte** (ō-ō-sīt).

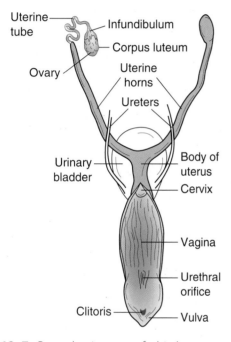

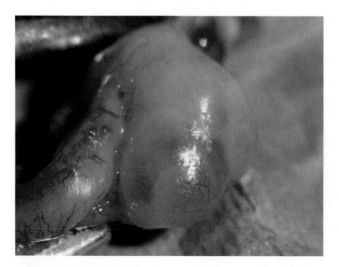

Figure 12–8 The follicle, which appears as a clear blister on the surface of the ovary, secretes a hormone called estrogen. (Courtesy of iStockphoto.)

Uterine Tubes

The **uterine** tubes are paired tubes that extend from the cranial portion of the uterus to the ovary (although they are not attached to the ovary). The uterine tubes also are called **oviducts** (ō-vih-duhckts) and **fallopian** (fah-lō-pē-ahn) **tubes.** The combining form for uterine tubes is **salping/o,** which means tube.

The distal end of each uterine tube is a funnel-shaped opening called the **infundibulum** (ihn-fuhn-dihb-yoo-luhm). The infundibulum contains fringed extensions called **fimbriae** (fihm-brē-ah) that catch ova when they leave the ovary. The fimbriae are not attached to the ovaries.

The proximal end of each uterine tube is connected to the uterine horns. The uterine tubes carry ova from the ovary to the uterus. The uterine tubes also transport sperm traveling up from the vagina and uterus. **Fertilization** (fər-tihl-ih-zā-shuhn), or egg and sperm union, usually occurs in the uterine tube.

Uterus

The **uterus** (yoo-tər-uhs) is a thick-walled, hollow organ with muscular walls and a mucous membrane lining that houses the developing embryo in pregnant females (Figure 12–9). The uterus is situated dorsal to the urinary bladder and ventral to the rectum. The combining forms **hyster/o, metr/o, metri/o,** and **uter/o** mean uterus. The uterus consists of three parts:

- **cornus** (kōr-nuhs) = horn. The cranial end of the uterus has two horns that travel toward the oviducts. **Cornu** means horn. Some animals are **bicornuate** (bī-kōrn-yoo-āt), which means having two large, well-defined uterine horns.
- **corpus** (kōr-puhs) = body. The middle portion of the uterus. **Corpu** means body.
- **cervix** (sihr-vihckz) = neck. The caudal portion of the uterus that extends into the vagina. **Cervic/o** means neck.

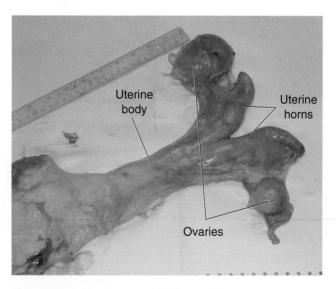

Figure 12–9 Bovine reproductive tract.

Figure 12–7 Reproductive tract of a bitch.

The uterus is composed of three major tissue types:

- **perimetrium** (pehr-ih-mē-trē-uhm) = membranous outer layer of the uterus. **Peri-** is the prefix for surrounding.
- **myometrium** (mī-ō-mē-trē-uhm) = muscular middle layer of the uterus. **My/o** is the combining form for muscle.
- **endometrium** (ehn-dō-mē-trē-uhm) = inner layer of the uterus. **Endo-** is the prefix meaning within.

Cervix

The **cervix** (sihr-vihckz) is the caudal continuation of the uterus and marks the cranial extent of the vagina. The cervix contains ringlike smooth muscle called **sphincters.** The main function of the cervix is to prevent foreign substances from entering the uterus. The cervix usually is closed tightly except during estrus, when it relaxes to allow entry of sperm. During pregnancy, the cervix is closed with a mucous plug. The mucous plug is released near parturition to allow fetal passage. The combining form **cervic/o** means neck or necklike structure.

Vagina

The **vagina** (vah-jī-nah) is the muscular tube lined with mucosa that extends from the cervix to the outside of the body. The vagina accepts the penis during copulation and serves as a passage for semen into the body and excretions and offspring out of the body. The combining forms **colp/o** and **vagin/o** mean vagina.

A membranous fold of tissue may partially or completely cover the external vaginal orifice. This fold is called the **hymen** (hī-mehn). An **orifice** (ōr-ih-fihs) is an entrance or outlet from a body cavity.

Vulva

The **vulva** (vuhl-vah), also known as the female external genitalia, or **pudendum** (pyoo-dehn-duhm), is the external opening to the urogenital tract and consists of the vaginal orifice, vestibular glands, clitoris, hymen, and urethral orifice. The combining forms for the vulva are **vulv/o** and **episi/o.** The **perineum** (pehr-ih-nē-uhm) is the region between the vaginal orifice and anus in females.

The **labia** (lā-bē-ah) are the fleshy borders or edges of the vulva and are occasionally called the **lips.** In animals, the vulva contains simple lips, whereas humans have major and minor labia. The **vaginal orifice** is the entrance from the vagina to the outside of the body. The **vestibular glands** (also known as **Bartholin's glands** in primates) are found in bovine, feline, and occasionally ovine species. The vestibular glands secrete mucus to lubricate the vagina. The **clitoris** (kliht-ə-rihs) is the sensitive erectile tissue of females located in the ventral portion of the vulva. The clitoris is the analog of the glans penis of the male. The **urethral orifice** is found where the vagina and vulva join and is sometimes associated with a vestigial hymen (Figure 12–10).

Mammary Glands

The **mammary** (mahm-mah-rē) **glands** are milk-producing glands in females. The number of mammary glands varies with the species: the mare, ewe, and doe (goat) have two; cows have four; sows have six or more pairs; and bitches and queens have four or more pairs. In litter-bearing species, the glandular structures usually are paired, located on the ventral surface, and called mammary glands, or **mammae** (mahm-ā). Each singular gland (mamma) is associated with one nipple. In large animals, the mammary gland is called an **udder** (uh-dər), is located in the inguinal area, and has two or four functional teats. The nipple area is called a **teat** (tēt) in large animals. In cows, the four mammae are called **quarters** (kwahr-tərz) (Figure 12–11).

Mammary glands are composed of connective and adipose tissue organized into lobes and lobules that contain milk-secreting sacs called **alveoli** (ahl-vē-ō-lī). Each lobe drains toward the teat or papilla via a **lactiferous** (lahck-tihf-ər-uhs) **duct. Lact/i** means milk. The lactiferous ducts come together to form the lactiferous, or teat, sinus. The **lactiferous sinus** is composed of the gland cistern (within the gland) and the teat cistern (within the teat). Milk travels from the gland cistern into the teat cistern. From the teat cistern, milk empties into the **papillary duct,** which is commonly called the **streak canal.**

The combining forms for mammary glands are **mamm/o** and **mast/o.**

THE ESTROUS CYCLE

The **estrous** (ehs-truhs) **cycle,** sometimes called the heat cycle, occurs at the onset of puberty and continues throughout an animal's life. The ability to sexually reproduce begins at puberty and varies among species. The estrous cycle prepares the uterus to accept a fertilized ovum.

The female reproductive system functions on cyclic intervals. Hormones secreted from the anterior pituitary gland and ovary control the estrous cycle. Although there is species variation during estrous cycle phases, the basic patterns are the same.

The estrous cycle starts when ova develop in ovarian follicles. One or more follicles continue to develop until they reach the ripened follicle size (called the graafian follicle). The graafian follicle ruptures, a process called **ovulation** (ohv-yoo-lā-shuhn). The ovum is expelled from the ovary into the oviduct. The ruptured follicle continues to grow and becomes filled with a yellow substance. The yellow ruptured follicle is called the **corpus luteum** (kōr-puhs loo-tē-uhm), or yellow body. The corpus luteum is abbreviated CL. The CL secretes progesterone. If the ovum is fertilized, the CL will continue to secrete progesterone to prevent future estrous cycling. If the ovum is not fertilized, the CL will shrink and reduce its progesterone secretion, and a new estrous cycle begins. The stage of the estrous cycle in which the graafian follicle is present is called the **follicular** (fohl-ihck-yoo-lahr) **phase.** Estrogen is the predominant hormone during the follicular phase.

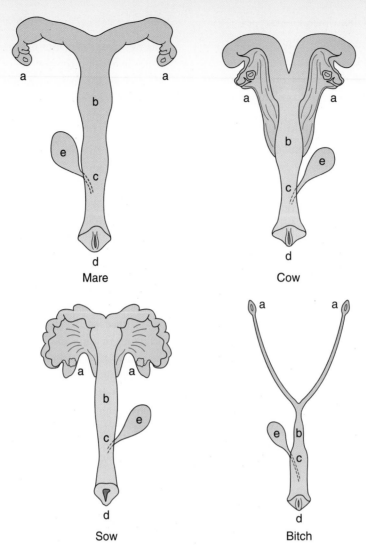

Mare

Cow

Sow

Bitch

Figure 12–10   Female genital organs and urinary bladder:
(a) ovaries, (b) uterus, (c) vagina, (d) vulva, and (e) urinary bladder.

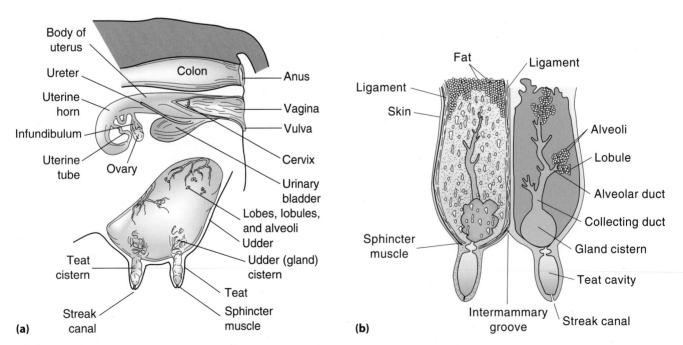

Figure 12–11   (a) Cross section of cow reproductive system lateral view. (b) Cross section of an udder cranial view.

The stage of the estrous cycle in which the corpus luteum is present is called the **luteal** (loo-tē-ahl) **phase.** Progesterone is the predominant hormone during the luteal phase.

The estrous cycle is also divided into phases. The phases of the estrous cycle are as follows:

- **proestrus** (prō-ehs-truhs) = period of the estrous cycle before sexual receptivity. The prefix **pro-** means before. Proestrus involves the secretion of follicle-stimulating hormone (FSH) by the anterior pituitary gland, which causes the follicles to develop in the ovary. FSH stimulates ovarian release of estrogen, which helps prepare the reproductive tract for pregnancy.
- **estrus** (ehs-truhs) = period of the estrous cycle in which the female is receptive to the male. During estrus, FSH levels decrease and luteinizing hormone (LH) levels increase, causing the graafian follicle to rupture and release its egg (ovulation). **Ovulation** (ohv-yoo-lā-shuhn) occurs, and the animal is said to be in **heat.** This also is called **standing heat.**
- **metestrus** (meht-ehs-truhs) = short period of the estrous cycle after sexual receptivity. The CL forms and produces progesterone during this phase. Progesterone ensures proper implantation and maintenance of pregnancy. If an animal is not pregnant, the CL will decrease in size and become a **corpus albicans** (kōr-puhs ahl-bih-kahnz), or white body. Metestrus may be followed by diestrus, estrus, pregnancy, or false pregnancy. Metestrus is more commonly used to describe the estrous cycle of cattle; the term is rarely used in dogs, cats, and horses.
- **diestrus** (dī-ehs-truhs) = period of the estrous cycle after metestrus. This short phase of inactivity and quietness is seen in polyestrous animals before the onset of proestrus.
- **anestrus** (ahn-ehs-truhs) = period of the estrous cycle when the animal is sexually quiet. This long phase of quietness is seen in seasonally polyestrous and seasonally monestrous animals.

MATING, PREGNANCY, AND BIRTH

Mating

For reproduction to occur, the male and female of the species must **copulate** (kohp-yoo-lāt) to allow the sperm from the male to be transferred into the female. Copulation and **coitus** (kō-ih-tuhs) are terms that mean sexual intercourse.

Intromission (ihn-trō-mihs-shuhn) is insertion of the penis into the vagina. During coitus, the male **ejaculates** (ē-jahck-yoo-lātz) into the female's vagina. Ejaculate is to

What do the following terms mean?

- **monestrous** (mohn-ehs-truhs) = having one estrous or heat cycle per year.
- **polyestrous** (pohl-ē-ehs-truhs) = having more than one estrous or heat cycle per year.
- **spontaneous ovulators** (spohn-tā-nē-uhs oh-vū-lā-tōrz) = animals in which ovum release occurs cyclically.
- **induced ovulators** (ihn-doosd oh-vū-lā-tōrz) = animals in which the ovum is released only after copulation; also called **reflex ovulators.** Examples include cats, rabbits, ferrets, llamas, and mink.
- **seasonally** = pertaining to a specific time of year. For example, queens are seasonally polyestrous. The queen will have multiple cycles from January through October (varies depending on photoperiod and geographic location).

release semen during copulation. **Ejaculat/o** means to throw or hurl out. Sperm travel through the vagina, into the uterus, and into the uterine tube. When a sperm penetrates the ovum that is descending down the uterine tube, fertilization occurs. **Fertilization** (fər-tihl-ih-zā-shuhn) is the union of ovum and sperm (Figure 12–12). If more than one ovum is passing down the oviduct when sperm are present, multiple fertilizations may take place.

Afterbirth

The **placenta** (plah-sehn-tah) is the female organ of mammals that develops during pregnancy and joins mother and offspring for exchange of nutrients, oxygen, and waste products. The placenta also is called the **afterbirth.** The **umbilical** (uhm-bihl-ih-kuhl) **cord** is the structure that forms where the fetus communicates with the placenta. The **umbilicus** (uhm-bihl-ih-kuhs) is the structure that forms on the abdominal wall where the umbilical cord was connected to the fetus.

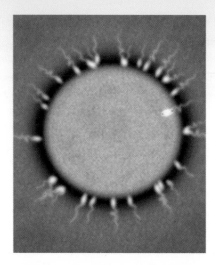

Figure 12–12 The outer membrane of the egg must be dissolved before the sperm can enter the egg. (Courtesy of Getty Images Inc./Chad Baker.)

 Twins

Twins are two offspring born from the same gestational period. Fraternal twins are two offspring born during the same labor resulting from fertilization of separate ova by separate sperm. Maternal twins are two offspring born during the same labor resulting from fertilization of a single ovum by a single sperm. (The fertilized egg separates into two parts.) The term *twin* is not used in litter-bearing species, even if only two offspring are produced.

Copulation and **coitus** are two words used to describe sexual intercourse. The following terms also are used in reference to mating:

- **mount** (mownt) = preparatory step to animal mating that involves one animal climbing on top of another animal or object; used as an indicator of heat (Figure 12–13).
- **tie** (tī) = period of copulation between a male and female canine during which the two animals are locked together by penile erectile tissue.
- **conception** (kohn-sehp-shuhn) = beginning of a new individual resulting from fertilization. The fertilized egg is called a **zygote** (zī-gōt). Division of sex cells in which the cell receives half the chromosomes from each parent is called **meiosis** (mī-ō-sihs).

Implantation (ihm-plahn-tā-shuhn) is the attachment and embedding of the zygote in the uterus. The developing zygote after implantation is called an **embryo** (ehm-brē-ō). An unborn animal is called a **fetus** (fē-tuhs); this term is used more toward the end of pregnancy. Pregnancy is the time period between conception and parturition.

The umbilicus also is called the **navel** (nā-vuhl). **Umbilic/o** is the combining form for navel.

The placenta and its associated structures are called the fetal membranes. The innermost membrane enveloping the embryo in the uterus is called the **amnion** (ahm-nē-ohn). The amnion forms the amniotic cavity and protects the fetus by engulfing it in amniotic fluid. The amnion may be called the **amniotic sac** or **bag of waters.** The **allantois** (ahl-ahn-tō-ihs) is the innermost layer of the placenta. It forms a sac between itself and the amnion, where fetal waste products accumulate. The **chorion** (kōr-ē-ohn) is the outermost layer of the placenta (Figure 12–14).

Figure 12–13 Cattle mounting. (Courtesy of Ron Fabrizius, DVM, Diplomate ACT.)

The embryo has distinct layers that give rise to various tissue types. Layers of the embryo include the **ectoderm** (ehck-tō-dərm), or outer layer of the embryo; the **mesoderm** (meh-sō-dərm), the middle layer of the embryo; and the **endoderm** (ehn-dō-dərm), the inner layer of the embryo.

The ruminant placenta has elevations on it that are located on the maternal or fetal surface. The **cotyledon** (koht-eh-lē-dohn) is the elevation of the ruminant placenta that is on the fetal surface and adheres to the maternal caruncle. Cotyledons also are called **buttons.** The **caruncle** (kahr-uhnck-uhl) is the fleshy mass on the maternal ruminant placenta that attaches to the fetal cotyledon. (Think c**a**runcle = m**a**ternal; cotyl**e**don = f**e**tus.)

Together the caruncle and cotyledon form the **placentome** (plahs-ehn-tohm) (Figures 12–15 and 12–16).

Pregnancy

Pregnancy (prehg-nahn-sē) is the condition of having a developing fetus in the uterus and is the time period between conception and parturition. The combining form for pregnancy is **pregn/o. Cyesis** (sī-eh-sihs) also means pregnancy. **Gestation** (jehs-tā-shuhn) is the period of development of the fetus in the uterus from conception to parturition and is the term more commonly used in reference to animals. The combining forms for gestation are **gest/o** and **gestat/o.** Gestation periods vary in length between animal species. A fetus is said to be **viable** (vī-ah-buhl) when it is capable of living outside the mother. Viability depends on the species and the age and weight of the fetus.

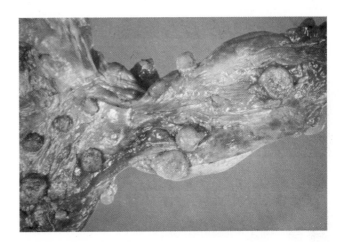

Figure 12–15   Bovine placenta. Caruncles are present on the maternal aspect of this bovine placenta. (Courtesy of Ron Fabrizius, DVM, Diplomate ACT.)

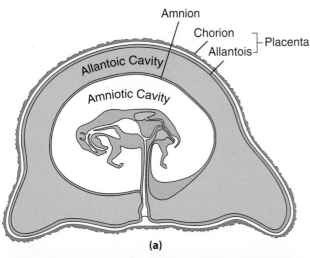

(a)

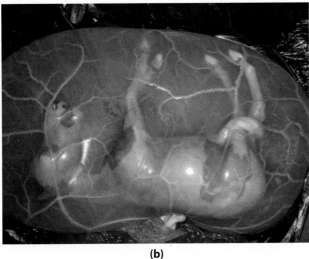

(b)

Figure 12–14   (a) Equine placenta and fetus. (b) Equine fetus within placenta. (Courtesy of Laura Lien, CVT, BS.)

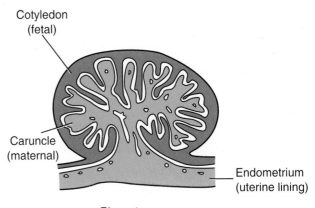

Placentome
1) Fetal attachment to uterus
2) Nutrients for fetus
3) Waste disposal for fetus

Figure 12–16   A placentome is the site of attachment between the fetal placenta and the maternal uterus in the cow. The blood supply of the two animals does not mix.

Terms Related to Gestational and Parturitional Status

- **gravid/o** (grahv-ih-dō) = combining form for pregnant.
- **-para** (pahr-ah) = suffix meaning bearing a live fetus.
- **nulligravida** (nuhl-ih-grahv-ih-dah) = one who has never been pregnant; **nulli-** means none; **nullipara** (nuhl-ih-pahr-ah) = female who has never borne a viable fetus.
- **primigravida** (preh-mih-grahv-ih-dah) = female during first pregnancy; **primi-** means first; **primipara** (prī-mihp-ah-rah) = female who has borne one offspring.
- **multigravida** (muhl-tih-grahv-ih-dah) = one who has had multiple pregnancies; **multi-** means many.
- **multiparous** (muhl-tihp-ah-ruhs) = female who has borne multiple offspring during different gestations. A **litter** (liht-tər) is a group of offspring born during the same labor.
- **viviparous** (vī-vihp-ahr-uhs) = bearing live young. **Vivi-** is the prefix for live.
- **oviparous** (ō-vihp-ahr-uhs) = bearing eggs. **Ovi** means egg.

Birth

The act of giving birth is called **parturition** (pahr-tyoo-rihsh-uhn) (Figure 12–17). Parturition also is called **labor. Part/o** is the combining form for giving birth. The period before the

Presentation

Presentation is the orientation of the fetus before delivery. Presentation varies with species. For example, in cattle and sheep, the fetus adopts a **cranial** (anterior) **presentation** in which the legs and head are directed toward the cervix. In swine, cranial and caudal presentations are considered normal. In a **caudal** (posterior) **presentation,** the pelvis and rear legs are directed toward the cervix. **Transverse presentation** involves the fetus lying across the cervix, and normal parturition is not achieved. In a **breech presentation,** the tail of the fetus is presented first and delivery may or may not be obstructed.

Figure 12–17 A foal birth normally has the front hooves presented first. (Courtesy of Bob Langrish.)

onset of labor is the **antepartum** (ahn-tē-pahr-tuhm) period; the period immediately after labor is the **postpartum** (pōst-pahr-tuhm) period.

Parturition is divided into stages. The first stage of labor involves **dilation** (dī-lā-shuhn) of the cervix. Dilation is the act of stretching. The second stage of labor involves uterine contractions of increasing frequency and strength and expulsion of the fetus. Expulsion of the fetus is called **delivery** (deh-lihv-ər-ē). The third stage of labor involves separation of the placenta from the uterus.

The postpartum period begins after delivery of the fetus. The newborn is called a **neonate** (nē-ō-nāt). The neonatal period varies from species to species but usually is less than 4 weeks. The first stool of a newborn that consists of material collected in the intestine of the fetus is called the **meconium** (meh-kō-nē-uhm).

After delivery of the fetus, the mother's uterus returns to its normal size. The process of the uterus returning to normal size is called **uterine involution** (yoo-tər-ihn ihn-vō-loo-shuhn). The mammary glands of the mother secrete **colostrum** (kuh-lohs-truhm), which is a thick fluid that contains nutrients and antibodies needed by the neonate. **Lactation** (lahck-tā-shuhn) is the process of forming and secreting milk.

TEST ME: REPRODUCTIVE SYSTEM

Diagnostic procedures performed on the reproductive system include the following:

- **amniocentesis** (ahm-nē-ō-sehn-tē-sihs) = surgical puncture with a needle through the abdominal and uterine walls to obtain amniotic fluid to evaluate the fetus.
- **radiography** (rā-dē-ohg-rah-fē) = procedure in which film is exposed as ionizing radiation passes through the patient and shows the internal body structures in profile.
- **ultrasound** (uhl-trah-sownd) = diagnostic test using high-frequency waves to evaluate internal structures. Ultrasonography works well in evaluating the uterus during pregnancy because the fluid present in the uterus helps define structures.

PATHOLOGY: REPRODUCTIVE SYSTEM

Pathologic conditions of the reproductive system include the following:

- **abortion** (ah-bōr-shuhn) = termination of pregnancy.
- **azoospermia** (ā-zō-ō-spǝr-mē-ah) = absence of sperm in the semen.
- **benign prostatic hypertrophy** (beh-nīn prohs-tah-tihck hī-pǝr-trō-fē) = abnormal noncancerous enlargement of the prostate; also called **prostatomegaly** or **enlarged prostate.**
- **cervicitis** (sihr-vih-sī-tihs) = inflammation of the neck of the uterus.
- **cryptorchidism** (krihp-tōr-kih-dihzm) = developmental defect in which one or both testes fail to descend into the scrotum; also called **undescended testicle(s).** Animals may be unilaterally or bilaterally cryptorchid. Unilaterally cryptorchid is sometimes called **monorchid** (mohn-ōr-kihd).
- **dystocia** (dihs-tōs-ah) = difficult birth; the female is having difficulty expelling the fetus.
- **ectopic pregnancy** (ehck-tohp-ihck prehg-nahn-sē) = fertilized ovum implanted outside the uterus.
- **epididymitis** (ehp-ih-dihd-ih-mī-tihs) = inflammation of the epididymis.
- **fetal defects** = abnormalities that occur in the development of the fetus (Figure 12–18). **Teratogens** (tǝr-ah-tō-jehnz) are substances that produce defects in the

Figure 12–18 A calf born with a fetal defect (a portion of two heads). (Courtesy of John Lynn.)

fetus. **Mutagens** (mū-tah-jehnz) are substances that produce change or that create genetic abnormalities.

- **fibroid** (fi-broyd) = benign tumor arising from the smooth muscle of the uterus; also called **leiomyoma** (lī-ō-mī-ō-mah).
- **hermaphroditism** (hǝr-mahf-rō-dih-tihzm) = condition of having both ovarian and testicular tissue. **Pseudohermaphroditism** (soo-dō-hǝr-mahf-rō-dih-tihzm) is the condition of having gonads of one sex but the physical characteristics of both sexes.
- **mastitis** (mahs-tī-tihs) = inflammation of the mammary gland(s) (Figures 12–19a and b).
- **metritis** (meh-trī-tihs) = inflammation of the uterus.
- **oligospermia** (ohl-ih-gō-spǝr-mē-ah) = deficient amount of sperm in semen; **oligo-** is the prefix for scant or few.
- **orchitis** (ōr-kī-tihs) = inflammation of the gonads of the male; also called **testitis.**
- **ovarian cyst** (ō-vahr-ē-ahn sihst) = collection of fluid or solid material in the female gonad.
- **paraphimosis** (pahr-ah-fih-mō-sihs) = retraction of the skin of the prepuce causing a painful swelling of the glans penis that prevents the penis from being retracted; the penis is extruded from the prepuce but cannot be returned to its normal position (Figure 12–20).
- **phimosis** (fih-mō-sihs) = narrowing of the skin of the prepuce so that it cannot be retracted to expose the glans penis; the penis cannot be extruded from the prepuce due to the small orifice.
- **pneumovagina** (nū-mō-vah-jī-nah) = conformational defect in the perineum of cows and mares that allows air to enter the vagina; also called **windsuckers.**

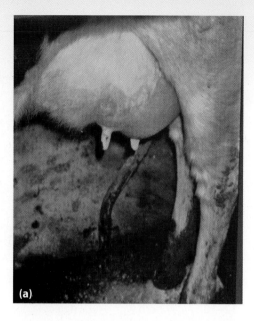

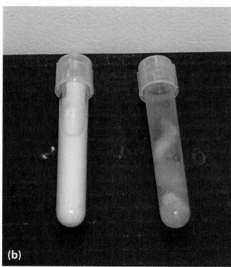

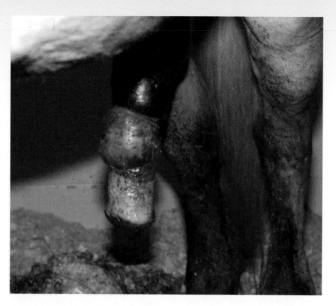

Figure 12–20 Paraphimosis in a horse.

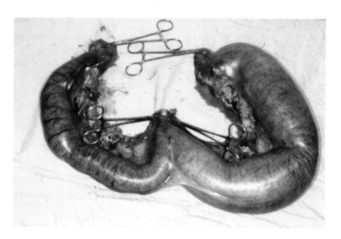

Figure 12–21 Pyometra in a bitch. (Courtesy of Ron Fabrizius, DVM, Diplomate ACT.)

Figure 12–19 (a) Mastitis in the rear quarter of a cow udder. (b) Normal white milk is shown in the left tube. Abnormal secretion from a cow with mastitis is shown in the right tube. [(a) Courtesy of Ron Fabrizius, DVM, Diplomate ACT.]

- **priapism** (prī-ahp-ihzm) = persistent penile erection not associated with sexual excitement.
- **prostatitis** (prohs-tah-tī-tihs) = inflammation of the prostate.
- **pseudocyesis** (soo-dō-sī-ē-sihs) = false pregnancy; also called **pseudopregnancy.** A behavioral and physical syndrome (most commonly seen in bitches 2–3 months after estrus) in which mammary glands develop, lactation occurs, and mothering behaviors occur. **Cyesis** means pregnancy; **pseudo-** is the prefix for false.
- **pyometra** (pī-ō-mē-trah) = pus in the uterus (Figure 12–21).

- **retained placenta** (rē-tānd plah-sehn-tah) = non-passage of the placenta after delivery of the fetus. A retained placenta can lead to metritis and infertility in the female.
- **scrotal hydrocele** (skrō-tahl hī-drō-sēl) = hernia of fluid in the testes or along the spermatic cord. **Hydro-** is the prefix for water; **-cele** is the suffix meaning hernia (Figure 12–22).
- **sterility** (stər-ihl-ih-tē) = inability to reproduce.
- **supernumerary** (soo-pər-nū-mahr-ē) = more than the normal number. Supernumerary teats is a condition in which an animal has more than the normal number of nipples (commonly seen in ruminants).
- **transmissible venereal tumor** (trahnz-mihs-ih-buhl vehn-ēr-ē-ahl too-mər) = naturally occurring, sexually

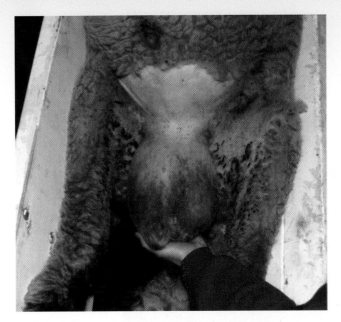

Figure 12–22 Scrotal hydrocele in a ram. (Courtesy of Ron Fabrizius, DVM, Diplomate ACT.)

transmitted tumor of dogs that affects the external genitalia and other mucous membranes; abbreviated TVT.

- **uterine prolapse** (yoo-tər-ihn prō-lahps) = protrusion of the uterus through the vaginal orifice (Figure 12–23). Commonly called cast her withers.

- **vaginal prolapse** (vah-jih-nahl prō-lahps) = protrusion of the vagina through the vaginal wall or vaginal orifice (Figure 12–24).
- **vaginitis** (vahj-ih-nī-tihs) = inflammation of the vagina.

PROCEDURES: REPRODUCTIVE SYSTEM

Procedures performed on the reproductive system include the following:

- **assisted delivery** = manual use of hands or equipment to aid in delivery of a fetus. In cattle, obstetric chains may be placed around the calf's legs and then force is applied to help extract the fetus (Figure 12–25).
- **cesarean section** (sē-sā-rē-ahn sehck-shuhn) = delivery of offspring through an incision in the maternal abdominal and uterine walls; also called a **C-section** (Figure 12–26).
- **electroejaculation** (ē-lehck-trō-ē-jahck-yoo-lā-shuhn) = method of collecting semen for artificial insemination or examination in which electrical stimulation is applied to the nerves to promote ejaculation. Electroejaculation is achieved by use of an **electroejaculator** (ē-lehck-trō-ē-jahck-yool-ā-tər), which is a probe and power

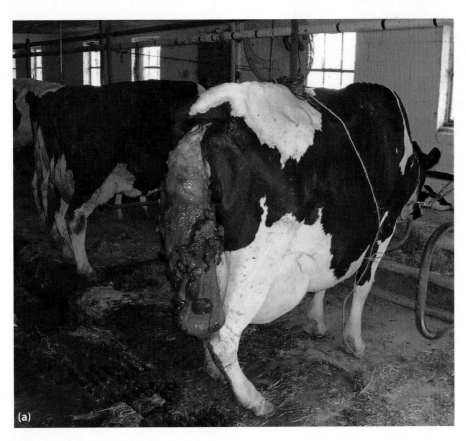

Figure 12–23 (a) Cow with prolapsed uterus.

Figure 12–23 (b) Close-up view of prolapsed uterus showing placentomes.

Figure 12–24 Vaginal prolapse in a ewe. (Courtesy of Ron Fabrizius, DVM, Diplomate ACT.)

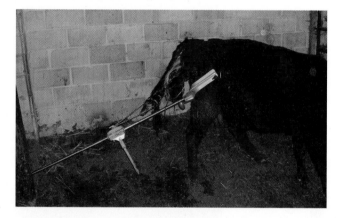

Figure 12–25 Assisted delivery can be accomplished in cows with a fetal extractor (calf jack), which applies pressure to the hind limbs of the cow as traction is applied to the calf.

source used to apply current to the nerves that promote ejaculation.

- **episiotomy** (eh-pihz-ē-oht-ō-mē) = surgical incision of the perineum and vagina to facilitate delivery of the fetus and to prevent damage to maternal structures.
- **fetotomy** (fē-toh-tō-mē) = cutting apart of a fetus to enable removal from the uterus; also called **embryotomy** (ehm-brē-ah-tō-mē).
- **hysterectomy** (hihs-tər-ehck-tō-mē) = surgical removal of the uterus.
- **mastectomy** (mahs-tehck-tō-mē) = surgical removal of the mammary gland or breast.
- **neuter** (nū-tər) = to sexually alter; usually used to describe the sexual altering of males. An animal that is not neutered is **intact** (ihn-tahckt), or has reproductive capability.

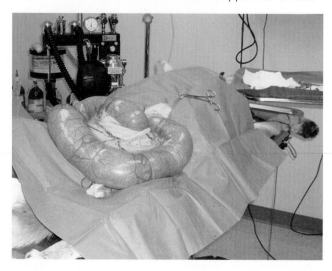

Figure 12–26 C-section in a bitch.

- **oophorectomy** (ō-ohf-ō-rehck-tō-mē) = surgical removal of the ovary (ovaries).
- **orchidectomy** (ōr-kih-dehck-tō-mē) = surgical removal of the testis (testes); also known as **orchectomy** (ōr-kehck-tō-mē), **orchiectomy** (ōr-kē-ehck-tō-mē), and **castration** (kahs-trā-shuhn) (Figure 12–27).
- **ovariohysterectomy** (ō-vahr-ē-ō-hihs-tər-ehck-tō-mē) = surgical removal of the ovaries, uterine tubes, and uterus; also called a **spay;** abbreviated OHE or OVH.
- **vasectomy** (vah-sehck-tō-mē) = sterilization of a male in which a portion of the ductus deferens is surgically removed, yet the animal may retain its libido. **Libido** (lih-bē-dō) is sexual desire.

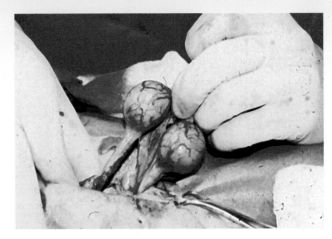

Figure 12–27 Canine castration. (Courtesy of Lodi Veterinary Hospital, SC.)

REVIEW EXERCISES

Multiple Choice

Choose the correct answer.

1. The inner layer of the uterus is called the
   a. endohysteria
   b. myometrium
   c. perimetrium
   d. endometrium

2. The area between the vaginal orifice or scrotum and the anus is called the
   a. clitoris
   b. perineum
   c. vulva
   d. inguinal area

3. Copulation also is called
   a. coitus
   b. impotence
   c. sterility
   d. zygote

4. The act of giving birth is
   a. freshening
   b. calving
   c. gestation
   d. parturition

5. A difficult birth is known as
   a. dystocia
   b. dyshernia
   c. dyspartia
   d. dyslaboratum

6. A false pregnancy also is called
   a. pseudo
   b. pseudopara
   c. pseudocyesis
   d. pseudogestia

7. A condition of an individual having both ovarian and testicular tissue is called
   a. hemisexual
   b. hermaphroditism
   c. supernumerary
   d. orchioovaris

8. Pyometra is
   a. pus in the uterus
   b. increased temperature of the uterus
   c. tumors in the uterus
   d. necrosis of the uterus

9. The innermost membrane enveloping the embryo in the uterus is the
   a. allantois
   b. umbilicus
   c. amnion
   d. chorion

10. Attachment and embedding of the zygote in the uterus is
    a. zygotion
    b. conception
    c. fertilization
    d. implantation

11. The ovum is the
    a. female gonad
    b. male gonad
    c. female gamete
    d. male gamete

12. The heat cycle in females is known as the
    a. estrus cycle
    b. estrous cycle
    c. lactogenic cycle
    d. follicular cycle

13. In large animals, the mammary gland is called the
    a. teat
    b. mammae
    c. lactiferous duct
    d. udder

14. The female organ of mammals that develops during pregnancy and joins the mother and offspring for exchange of nutrients, oxygen, and waste products is known as the
    a. umbilicus
    b. placenta
    c. mount
    d. navel

15. Substances that produce change or that create genetic abnormalities are known as
    a. teratogens
    b. priapisms
    c. dystociae
    d. mutagens

16. Reproductive organs, whether male or female, are called the
    a. theriogens
    b. genitals
    c. gametes
    d. perineum

17. An enlarged part of a tube or canal is called a(n)
    a. vesicle
    b. bulla
    c. ampulla
    d. motile

18. The term for surgical incision of the perineum and vagina to facilitate delivery of the fetus and to prevent damage to maternal structures is
    a. mastectomy
    b. cesarean section
    c. episiotomy
    d. oophorectomy

19. An animal that has not been neutered is referred to as
    a. sterile
    b. gravid
    c. intact
    d. cyesis

20. Another term for spay is
    a. orchidectomy
    b. ovariohysterectomy
    c. C-section
    d. hysterectomy

Matching

Match the term in Column I with the definition in Column II.

<table>
<tr><td colspan="2">Column I</td><td>Column II</td></tr>
<tr><td>1. _____ gestation</td><td></td><td>a. act of giving birth</td></tr>
<tr><td>2. _____ sterility</td><td></td><td>b. termination of pregnancy</td></tr>
<tr><td>3. _____ dystocia</td><td></td><td>c. thick fluid that contains nutrients and antibodies needed by neonates, which is secreted by the mother's mammary glands</td></tr>
<tr><td>4. _____ viviparous</td><td></td><td>d. period of pregnancy</td></tr>
<tr><td>5. _____ oviparous</td><td></td><td>e. outermost layer of the placenta</td></tr>
<tr><td>6. _____ allantois</td><td></td><td>f. innermost layer of the placenta</td></tr>
<tr><td>7. _____ chorion</td><td></td><td>g. bearing live young</td></tr>
<tr><td>8. _____ parturition</td><td></td><td>h. inability to reproduce</td></tr>
<tr><td>9. _____ colostrum</td><td></td><td>i. bearing eggs</td></tr>
<tr><td>10. _____ abortion</td><td></td><td>j. difficult birth</td></tr>
</table>

Fill in the Blanks

1. Colp/o and vagin/o mean _____ .

2. Hyster/o, metr/o, metri/o, and uter/o mean _____ .

3. Pen/i and priap/o mean _____ .

4. Orch/o, orchi/o, orchid/o, test/o, and testicul/o mean _____ .

5. Sperm/o and spermat/o mean _____ .

Spelling

Circle the term that is spelled correctly.

<table>
<tr><td>1. developmental defect in which one or both testes fail to descend into the scrotum:</td><td>cryptorchidism</td><td>criptorchidism</td><td>cryptorchydism</td></tr>
<tr><td>2. inflammation of the mammary glands:</td><td>mastis</td><td>mastitis</td><td>masitits</td></tr>
<tr><td>3. fertilized egg:</td><td>zigoat</td><td>zygoat</td><td>zygote</td></tr>
<tr><td>4. pregnancy:</td><td>syesis</td><td>cyesis</td><td>ciesis</td></tr>
<tr><td>5. to sexually alter:</td><td>neuter</td><td>nueter</td><td>newter</td></tr>
</table>

Word Building

Build the following terms using word parts.

1. pus in the uterus:

   word part for pus   _____

   word part for uterus   _____

   term for pus in the uterus   _____

2. near the ovary:

   word part for near   _____

   word part for ovary   _____

   term for near the ovary   _____

3. pertaining to the urinary and reproductive systems:

   word part for urinary   _____

   word part for reproductive system   _____

   word part for pertaining to   _____

   term for pertaining to the urinary and reproductive systems _____

4. capable of stimulating milk production:

   word part for milk   _____

   word part for producing   _____

   word part for pertaining to   _____

   term for capable of stimulating milk production   _____

5. surgical removal of the uterus:

   word part for uterus   _____

   word part for surgical removal   _____

   term for surgical removal of the uterus   _____

CROSSWORD PUZZLE

Reproductive System Puzzle

Across

1 eggs
6 muscular middle layer of the uterus
10 persistent penile erection not associated with sexual excitement
11 retractable fold of skin covering the glans penis
12 female external genitalia
13 fleshy borders or edges
14 inflammation of the gland that surrounds the urethra that secretes a thick fluid that aids in motility of sperm
18 to sexually alter (either sex)
19 capable of spontaneous motion
20 process of egg maturation and release
21 bearing eggs
22 common name for placenta

Down

2 bearing live young
3 structure that forms where the fetus communicates with the placenta
4 medical term for spay
5 condition of undescended testicles
7 ejaculatory fluid that contains sperm plus secretions from the secondary sex glands
8 preparatory step to animal mating that involves one animal jumping on top of another animal or object
9 union of ovum and sperm
10 retraction of the skin of the prepuce causing a painful swelling of the glans penis that prevents the penis from being retracted
15 entrance or outlet from a body cavity
16 funnel-shaped opening
17 male sex cells

CASE STUDIES

Define the underlined terms in each case study.

A 12-yr-old F miniature poodle was presented with lethargy, anorexia, and PU/PD. The bitch had been in proestrus 1 mo before presentation. On PE, the dog was pyrexic and tachypneic. Abdominal palpation yielded an enlarged uterus, and a purulent vaginal discharge was noted. Blood was collected for a CBC and chem panel. The CBC results included leukocytosis with a left shift. (A left shift is an increase in band neutrophils in peripheral blood due to many causes including infection.) The dx of pyometra was made, antibiotics were started, and the dog was scheduled for an emergency OHE.

1. yr _____

2. F _____

3. lethargy _____

4. anorexia _____

5. PU/PD _____

6. bitch _____

7 proestrus _____

8. PE _____

9. pyrexic _____

10. tachypneic _____

11. abdominal _____

12. palpation _____

13. uterus _____

14. purulent _____

15. vaginal _____

16. CBC _____

17. chem panel _____

18. leukocytosis _____

19. dx _____

20. pyometra _____

21. OHE _____

A 10-yr-old intact M German shepherd was presented with stranguria and hematuria. PE revealed pyrexia, anorexia, and a stiff gait. On rectal palpation, the veterinarian discovered that the prostate gland was bilaterally enlarged. Upon review of the record, it was noted that this dog has had a hx of recurrent UTIs. Radiographs were taken, and prostatomegaly was noted. After a dx of prostatitis was made, the dog was scheduled to be neutered the next day.

22. yr _____

23. intact _____

24. M _____

25. stranguria _____

26. hematuria _____

27. PE _____

28. pyrexia _____

29. anorexia _____

30. gait _____

31. rectal palpation _____

32. prostate gland _____

33. bilaterally _____

34. hx _____

35. recurrent _____

36. UTI _____

37. radiographs _____

38. prostatomegaly _____

39. dx _____

40. prostatitis _____

41. neutered _____

42. another term for neutering in male dogs _____

A 2-yr-old Holstein cow was examined because the farmer noticed that she was off feed. PE revealed that the cow had a slightly elevated rectal temperature. The farmer told the veterinarian that this cow had stepped on her teat previously, but it had appeared to be healing. The udder was palpated, and it was not warm to the touch or swollen. Milk was expressed from each quarter, and the milk appeared more watery than normal. A CMT paddle test was performed on the milk, and moderate precipitation was noted (Figure 12–28). The diagnosis of mastitis was made, and milk samples were taken for culture. Antibiotic treatment was started pending culture results, and milking hygiene was discussed with the farmer.

43. cow _____

44. off feed _____

45. medical term for off feed _____

46. teat _____

47. udder _____

48. quarter _____

49. CMT _____

50. diagnosis _____

51. mastitis _____

52. culture _____

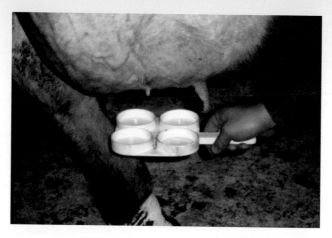

Figure 12–28   CMT paddle. The CMT (California mastitis test) paddle is a cow side test used to detect mastitis. (Courtesy of Ron Fabrizius, DVM, Diplomate ACT.)

A 5-yr-old quarter horse <u>mare</u> had <u>foaled</u> 10 hours earlier. The foal appeared normal, and the mare was letting the foal suckle. The mare had not moved since foaling, and the owner was concerned because the mare seemed quieter than normal. PE revealed a mildly elevated rectal temperature, <u>vaginal discharge</u> was noted, milk production seemed adequate, and the milk appeared normal. The owner was asked whether the mare had passed its <u>placenta</u>, and the owner did not know whether she had. The stall was examined for remnants of the placenta, and none were found. The veterinarian was concerned that the mare had a <u>retained placenta</u> and that an infection was starting. The equine placenta normally is passed within a few hours of foaling, and because it was at least 10 hours <u>postpartum</u>, an injection of oxytocin was given by slow <u>IV</u>. Blood was drawn for a <u>CBC</u> to assess <u>leukocyte</u> numbers. The owner was advised to watch for the passing of the placenta, and general hygiene was discussed with the client.

53. mare

54. foaled

55. vaginal discharge

56. placenta

57. retained placenta

58. postpartum

59. IV

60. CBC

61. leukocyte

CHAPTER 13

[NERVES OF STEEL]

Objectives

Upon completion of this chapter, the reader should be able to

- Identify and describe the structures and functions of the nervous system
- Identify the divisions of the nervous system and describe the structures and functions of each
- Recognize, define, spell, and pronounce terms related to the functions, disorders, and treatment of the nervous system

FUNCTIONS OF THE NERVOUS SYSTEM

The **nervous** (nər-vuhs) **system** coordinates and controls body activity. It detects and processes internal and external information and formulates appropriate responses.

STRUCTURES OF THE NERVOUS SYSTEM

The major structures of the nervous system are the brain, spinal cord, peripheral nerves, and sensory organs. The two major divisions of the nervous system are as follows:

- **central nervous system** (sehn-trahl nər-vuhs sihs-tehm) = portion of the nervous system that consists of the brain and spinal cord; abbreviated CNS.
- **peripheral nervous system** (pehr-ihf-ər-ahl nər-vuhs sihs-tehm) = portion of the nervous system that consists of the cranial and spinal nerves, autonomic nervous system, and ganglia; abbreviated PNS (Figure 13–1).

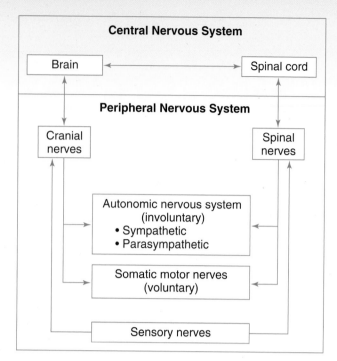

Figure 13–1 Divisions of the nervous system.

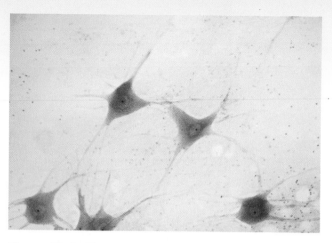

Figure 13–2 Photomicrograph of multiple motor neurons from the spinal cord of a bovine. (Courtesy of William J. Bacha, PhD, and Linda M. Bacha, MS, VMD.)

Back to Basics

The basic unit of the nervous system is the **neuron** (nū-rohn) (Figure 13–2). There are three types of neurons, based on their function.

- **sensory neurons** (sehn-sōr-ē nū-rohnz) = nerves that carry sensory impulses toward the CNS; also called **afferent** (ahf-fər-ahnt) or **ascending tracts** because they carry information toward the CNS. Sensory information such as sound and light is converted into electrical impulses so that the nerves can transport it.
- **associative neurons** (ahs-ō-shē-ah-tihv nū-rohnz) = nerves that carry impulses from one neuron to another; also called **connecting neurons.**
- **motor neurons** (mō-tər nū-rohnz) = nerves that carry impulses away from the CNS and toward the muscles and glands; also called **efferent** (ē-fər-ahnt) or **descending tracts** because they carry information away from the CNS.

The parts of a neuron are the cell body, one or more dendrites, one axon, and terminal end fibers. The cell body, or **soma** (sō-mah), has a nucleus and is responsible for maintaining the life of the neuron. The **dendrites** (dehn-drīts) are rootlike structures that receive impulses and conduct them toward the cell body. The combining form for dendrite is **dendr/o.** The **axon** (ahcks-ohn) is a single process that extends away from the cell body and conducts impulses away from the cell body. The combining form **ax/o** means axis or main stem. The terminal end fibers are the branching fibers that lead the impulse away from the axon and toward the synapse.

The dendrites and axons also are called **nerve fibers.** Bundles of nerve fibers bound together by specialized tissues are called **nerves** or **nerve trunks.** Nerve fibers are covered with a tubelike membrane called the **neurolemma** (nū-rō-lehm-ah) or neurilemma (nū-rih-lehm-ah). Neuron cell bodies grouped together within the CNS are called **nuclei** (nū-klē-ī), and those outside the CNS are called **ganglia** (gahng-glē-ah) (Figure 13–3).

The Gap

The junction between two neurons or between a neuron and a receptor is the **synapse** (sihn-ahps) (Figure 13–4). The combining forms for this space or point of contact are **synaps/o** and **synapt/o.** A synapse is the junction where neural impulses cause a release of a chemical substance called a **neurotransmitter** (nū-rō-trahnz-miht-ər) that allows the signal to move from one neuron to another. There are different neurotransmitters, each with specific functions.

Supporting Role

The **neuroglia** (nū-rohg-lē-ah), or **glial** (glē-ahl) **cells,** are the supportive cells of the nervous system. Glial cells consist of **astrocytes** (ahs-trō-sītz), **microglia** (mī-krō-glē-ah), **oligodendrocytes** (ohl-ih-gō-dehnd-rō-sītz), and **Schwann** (shwahn) **cells.** The combining form **gli/o** means glue, which helps explain the function of glial cells.

- **astr/o** means star. Astrocytes are star-shaped, cover the capillary surface of the brain, and help form the blood-brain barrier in the CNS.
- **micro-** means small. Microglia are small phagocytic cells that help fight infection in the CNS.
- **oligo-** means few, **dendr/o** means branching, and **-cyte** means cell. Oligodendrocytes are cells with few branches that hold the nerve fibers together and help form myelin in the CNS.
- Schwann cells (named after a German anatomist) help form myelin in the PNS.

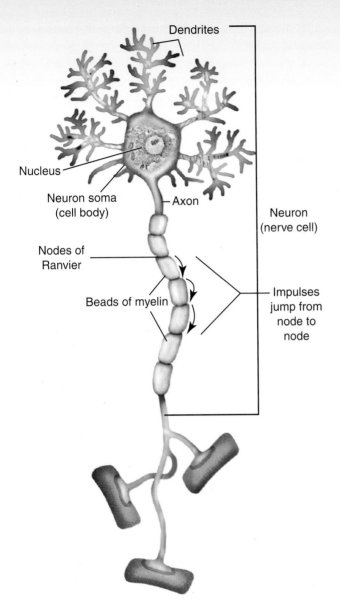

Figure 13–3 Structures of a neuron. Dendrites conduct impulses toward the neuron cell body, and axons conduct impulses away from the cell body.

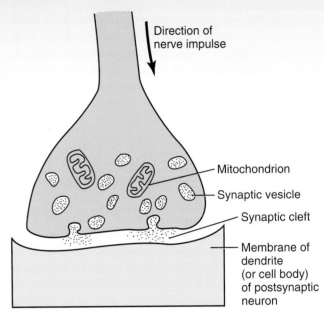

Figure 13–4 A synapse. Nerve impulse from the axon stimulates the release of a neurotransmitter.

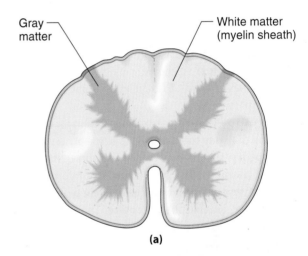

(a)

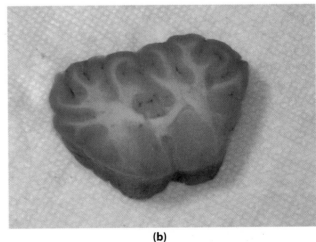

(b)

Figure 13–5 (a) Cross section of the spinal cord shows white matter and gray matter. The orientation of the white and gray matter in the spinal cord is opposite of that in the brain. (b) Cross section of the brain shows the white matter and gray matter.

Surrounding Tissue

Myelin (mī-eh-lihn) is a protective covering over some nerve cells, including parts of the spinal cord, white matter of the brain, and most peripheral nerves. Myelin also is called the **myelin sheath.** Myelin serves as an electrical insulator. Nerves that are described as **myelinated** (mī-lihn-āt-ehd) are surrounded by myelin and transport a signal much faster than nonmyelinated nerves. Myelin gives nerve fibers a white color, and myelinated nerves are called **white matter.** The **gray matter** does not contain myelinated fibers, so it is darker in color (Figures 13–5a and b). The gray matter is composed of cell bodies, branching dendrites, and neuroglia.

Myelin is interrupted at regular intervals along the length of a fiber by gaps called **nodes of Ranvier** (nōdz of rohn-vē-ā).

Ionic exchange takes place at the nodes of Ranvier. Some nerve fibers have a very thin layer of myelin and are referred to as **nonmyelinated.** The autonomic nervous system contains nonmyelinated fibers.

Carry an Impulse

A **nerve** (nərv) is one or more bundles of impulse-carrying fibers that connect the CNS to the other parts of the body. **Neur/i** and **neur/o** are combining forms for nerve or nerve tissue. Terms used in reference to nerves include the following:

- A **tract** (trahckt) is a group of nerve fibers located in the CNS. Ascending tracts (groups going *up,* or cranially) carry nerve impulses toward the brain, whereas descending tracts (groups going *down,* or caudally) carry nerve impulses away from the brain.
- A **ganglion** (gahng-glē-ohn) is a knotlike mass of neuron cell bodies located *outside* the CNS. More than one ganglion are called **ganglia,** or less often **ganglions.** The combining forms **gangli/o** and **ganglion/o** mean knotlike masses of nerve cell bodies located outside the CNS.
- A **plexus** (plehck-suhs) is a network of intersecting nerves or vessels. **Plexi** (plehck-sī) are networks of intersecting nerves or vessels. **Plex/o** is the combining form for a plexus, or network.
- **Innervation** (ihn-nər-vā-shuhn) is the supply or stimulation of a body part through the action of nerves.
- **Receptors** (rē-sehp-tərz) are sensory organs that receive external stimulation and transmit that information to the sensory neurons. There are different types of receptors. For example, **nociceptive** (nō-sih-sehp-tihv) receptors are pain receptors, whereas **proprioceptive** (prō-prē-ō-sehp-tihv) receptors are spatial orientation or perception of movement receptors.
- A **stimulus** (stihm-yoo-luhs) is something that excites or activates. Multiple excitations or activations are called **stimuli** (stihm-yoo-lī).
- An **impulse** (ihm-puhlz) is a wave of excitation transmitted through nervous tissue.
- A **reflex** (rē-flehcks) is an automatic, involuntary response to change. Reflex actions include the patellar and ulnar reflexes.

CENTRAL NERVOUS SYSTEM

The central nervous system (or cerebrospinal system or CNS) is made up of the brain and the spinal cord. The combining form for brain is **encephal/o,** and the combining form for spinal cord is **myel/o.** (Remember, myel/o also means bone marrow.) The CNS contains both white and gray matter; white matter contains myelinated fibers, and gray matter consists of nerve cell bodies.

Membrane

The brain and spinal cord are encased in connective tissue called the **meninx** (meh-nihncks). Because this connective tissue has three layers, it more commonly is referred to by its plural form of **meninges** (meh-nihn-jēz) (Figure 13–6).

Mening/o and **meningi/o** are combining forms for the layers of connective tissue enclosing the CNS.

The three layers of the meninges are as follows:

- **dura mater** (doo-rah mah-tər) = thick, tough, outermost layer of the meninges. **Dura** means tough, and this layer of the meninges is tough and strong. **Dur/o** is the combining form for dura mater. The dura mater also is called **pachymeninx** (pahck-ē-meh-nihcks); the prefix **pachy-** means thick.
- **arachnoid** (ah-rahck-noyd) **membrane** = second layer of the meninges. **Arachn/o** means spider, and the arachnoid membrane resembles a spider web. The arachnoid membrane is loosely attached to the other layers of the meninges to allow space between the layers.
- **pia mater** (pē-ah mah-tər) = third and deepest layer of the meninges. **Pia** means soft or tender, and this layer of the meninges is soft with a rich supply of blood vessels. The pia adheres to the CNS. The pia mater and arachnoid membranes collectively are called the **leptomeninges** (lehp-tō-meh-nihn-jēz) (Figure 13–7).

Certain terms are used to describe the location of structures in reference to the meninges. **Epidural** (ehp-ih-doo-rahl) means located above or superficial to the dura mater. (**Epi-** is the prefix meaning above or upon.) The **subdural** (suhb-doo-rahl) **space** is the area located below (deep to) the dura mater and above (superficial to)

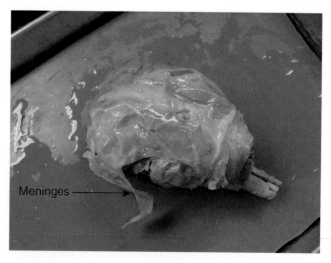

Meninges →

Figure 13–6 Meninges of the sheep.

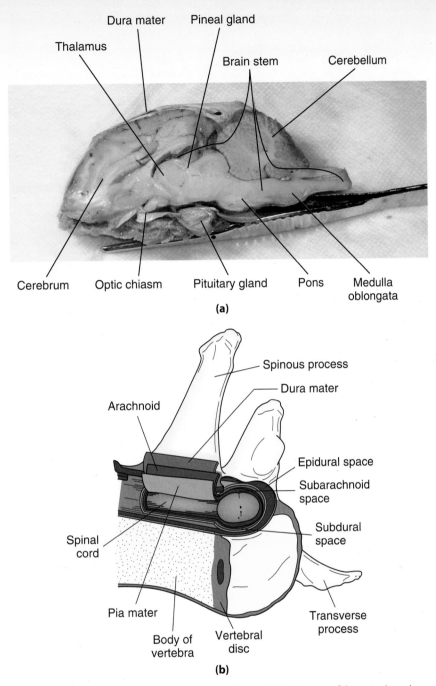

Figure 13–7 (a) Sagittal view of the sheep's brain. (b) Meninges of the spinal cord.

the arachnoid membrane. (The prefix **sub-** means below.) The **subarachnoid** (suhb-ah-rahck-noyd) **space** is the area located below (deep to) the arachnoid membrane and above (superficial to) the pia mater. The subarachnoid space contains cerebrospinal fluid.

Fluid

Cerebrospinal fluid (sər-ē-brō-spīn-ahl flū-ihd) is the clear, colorless ultrafiltrate that nourishes, cools, and cushions the CNS. Cerebrospinal fluid is abbreviated CSF. CSF is produced by special capillaries in the ventricles of the brain. The ventricles of the brain are cavities. Vascular folds of the pia mater in the

ventricles called the **choroid plexus** (kōr-oyd plehck-suhs) secrete CSF.

Brain

The brain is the enlarged and highly developed portion of the CNS that lies in the skull and is the main site of nervous control. The portion of the skull that encloses and protects the brain is called the **cranium** (krā-nē-uhm). **Crani/o** is the combining form for skull. **Intracranial** (ihn-trah-krā-nē-ahl) means within the cranium.

The brain is commonly divided into parts based on functional group or on location. **Encephal/o** is the combining

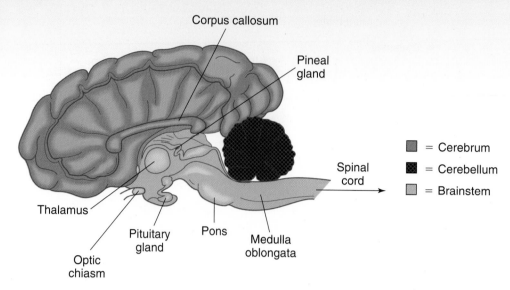

Corpus callosum

Pineal gland

Spinal cord

= Cerebrum
= Cerebellum
= Brainstem

Thalamus

Pituitary gland

Pons

Medulla oblongata

Optic chiasm

Figure 13–8 Sagittal section of the brain.

form for brain; however, other combining forms that refer to specific regions of the brain also are used.

The divisions of the brain (Figure 13–8) based on functional group include the following:

- **cerebrum** (sər-ē-bruhm). The combining form for cerebrum is **cerebr/o.** The cerebrum, the largest part of the brain, is responsible for receiving and processing stimuli, initiating voluntary movement, and storing information. The **cerebral cortex** (outer region) is made up of gray matter and is arranged in folds (Figure 13–9). The elevated portions of the cerebral cortex are known as **gyri** (jī-rī). The combining form **convolut/o** means coiled, and **gyr/o** means folding. The grooves of the cerebral cortex are called fissures or **sulci** (suhl-kī). The combining form **sulc/o** means groove. The medullary substance of the cerebrum is made up of white matter. The brain also has small cavities called **ventricles** (vehn-trih-kuhlz) (Figure 13–10). There are four ventricles of the brain: two lateral ventricles, a third ventricle, and a fourth ventricle. The ventricles of the brain (and central canal of the spinal cord) are lined with a membrane called the **ependyma** (eh-pehn-dih-mah).
- **cerebellum** (sehr-eh-behl-uhm). The combining form for cerebellum is **cerebell/o.** The cerebellum is the second largest part of the brain, and it coordinates muscle activity for smooth movement. The cerebellum has an inner portion, called the **vermis** (vər-mihs) because it is wormlike, and other portions divided into right and left cerebellar hemispheres.
- **brainstem** (brān-stehm). The brainstem is the stalklike portion of the brain that connects the cerebral hemispheres with the spinal cord. The brainstem consists of the pons, medulla oblongata, midbrain, and interbrain. The **interbrain** contains structures such as the pituitary

gland, hypothalamus, and thalamus. These structures are responsible for endocrine activity, regulation of thirst and water balance, and regulation of body temperature. The **midbrain** contains structures responsible for

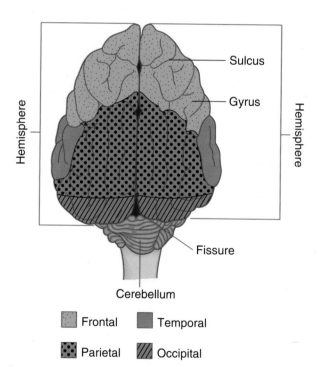

Sulcus

Gyrus

Hemisphere

Hemisphere

Fissure

Cerebellum

Frontal Temporal
Parietal Occipital

Figure 13–9 Dorsal view of the cerebral cortex. The cerebrum is divided into right and left hemispheres. Each hemisphere is further divided into lobes, and each lobe is named for the bone plate covering it: **frontal** (frohn-tahl) lobe = most cranial lobe that controls motor function; **parietal** (pahr-ī-ih-tahl) lobe = receives and interprets sensory nerve impulses; **occipital** (ohcks-ihp-ih-tahl) lobe = most caudal lobe that controls vision; **temporal** (tehmp-ruhl) lobe = laterally located lobe that controls hearing and smell.

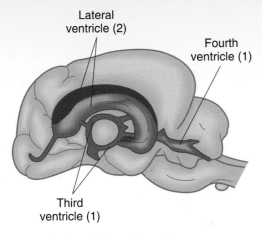

Figure 13–10 Lateral view of ventricles.

Some parts of the brain are named not based on location or division, but rather on how they look. The vermis of the cerebellum is one example. Another example is the hippocampus, a portion of the limbic system that involves memory. The **hippocampus** (hihp-ō-kahm-puhs) is shaped like a seahorse, which exists in mythology as a sea monster with the head of a horse and the tail of a fish and as an actual sea creature. The name *hippocampus* comes from the Greek *hippos,* meaning horse, and *kampos,* meaning sea monster. The arbor vitae (tree of life) of the cerebellum are the treelike outlines seen on sagittal views of the cerebellum.

visual and auditory reflexes, posture, and muscle control. The **pons** (pohnz) is the bridge at the base of the brain that allows nerves to cross over so that one side of the brain controls the opposite side of the body. The **medulla oblongata** (meh-duhl-ah ohb-lohng-gah-tah) is the cranial continuation of the spinal cord that controls basic life functions. Divisions of the brain also are based on location (Figure 13–11 and Table 13–1).

Spinal Cord

The **spinal cord** is the caudal continuation of the medulla oblongata. The spinal cord passes through an opening in the occipital bone called the **foramen magnum** (fər-ā-mehn mahg-nuhm). **Foramen** means passage or hole, and **magnum** means great. The spinal cord carries all of the tracts that influence the innervation of the limbs and lower part of the body. In addition, the spinal cord is the pathway for impulses going to and from the brain. **Myel/o** is the combining form for spinal cord. (Remember that myel/o also is the combining form for bone marrow.)

The spinal cord is part of the central nervous system, which means that it is surrounded by the meninges and is protected by cerebrospinal fluid. The gray matter of the spinal cord is located in the internal portion and is not protected by myelin.

The white matter of the spinal cord is located in the external portion and is myelinated.

The spinal cord contains two areas of swelling. The medical term for swelling (normal or abnormal) is **intumescence** (ihn-too-meh-sehns). The swelling is caused by an increase in white matter and cell bodies that are associated with the innervation of the limbs. One swelling occurs in the area of C6–T2, which is known as the **cervical intumescence.** The other swelling occurs in the area of the L4–caudal segment, which is known as the **lumbosacral intumescence.** The cranial parts of the spinal cord have tracts with fibers from the cranial and caudal portions, but the caudal parts of the spinal cord have only tracts with fibers from the caudal portions. Therefore, as the spinal cord proceeds caudally, its cross-sectional area decreases. At the level of the cranial lumbar vertebrae, the spinal cord becomes cone-shaped. This cone-shaped segment

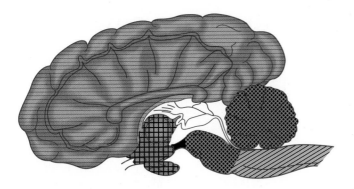
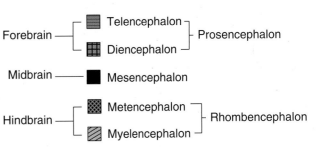

Figure 13–11 Brain divisions based on location.

Table 13–1 Brain Divisions Based on Location

Brain Part	Division	Some Components
forebrain; prosencephalon (prōs-ehn-sehf-ah-lohn)	**telencephalon** (tēl-ehn-sehf-ah-lohn)	cerebral cortex and olfactory brain, **limbic** (lihm-bihck) system (affects emotion and behavior); limbic means border
	diencephalon (dī-ehn-sehf-ah-lohn)	thalamus, epithalamus, and hypothalamus limbic system
		optic chiasm (crossing of the vision nerves); **chiasm** means crossing
midbrain; mesencephalon (mēz-ehn-sehf-ah-lohn)	**mesencephalon**	vision and hearing bodies, posture, and muscle control; limbic system
hindbrain; rhombencephalon (rohmb-ehn-sehf-ah-lohn)	**metencephalon** (meht-ehn-sehf-ah-lohn)	cerebellum and pons
	myelencephalon (mī-lehn-sehf-ah-lohn)	medulla oblongata

is called the **conus medullaris** (kō-nuhs mehd-yoo-lahr-ihs). **Conus** means cone. At the caudal end of the spinal cord (caudal lumbar, sacral, and coccygeal nerve segments), the tracts terminate and the spinal nerves fan outward and backward, giving the appearance of a horse's tail. This collection of

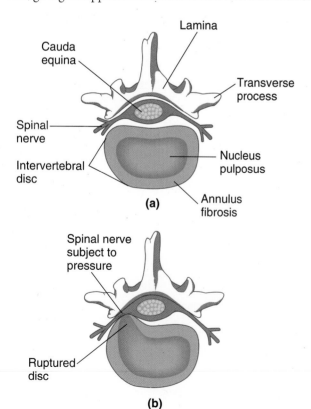

Figure 13–12 Intervertebral disc. (a) Normal intervertebral disc. (b) Ruptured disc.

spinal roots at the caudal part of the spinal cord is the **cauda equina** (kaw-dah ē-kwī-nah). The cauda equina includes the conus medullaris to the caudal vertebrae. The threadlike tapering section of the cauda equina is known as the **filum terminale** (fī-luhm tər-mih-nahl). **Filum** means threadlike structure. The filum terminale attaches the conus medullaris to the caudal vertebrae.

Cushions

The spinal cord is housed within the vertebrae to protect it from injury. The vertebrae are protected from each other by **intervertebral** (ihn-tər-vər-tə-brahl) discs that are located between vertebra (Figure 13–12). Intervertebral discs are layers of fibrocartilage that form pads separating and cushioning the vertebrae from each other. The center of the intervertebral disc is gelatinous (**nucleus pulposus**) (nū-klē-uhs puhl-pō-suhs), and the outer layer is fibrous (**annulus fibrosis**) (ahn-yoo-luhs fī-brō-sihs).

PERIPHERAL NERVOUS SYSTEM

The peripheral nervous system (PNS) consists of the cranial nerves, the autonomic nervous system, the spinal nerves, and ganglia.

Cranial Nerves

Twelve pairs of **cranial nerves** originate from the undersurface of the brain. The cranial nerves generally are named for the area or function they serve and are represented by Roman numerals (Figure 13–13 and Table 13–2).

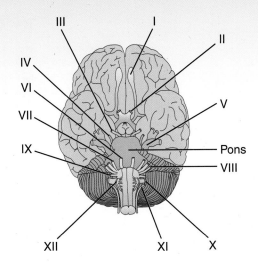

I. Olfactory
II. Optic Nerve
III. Oculomotor Nerve
IV. Trochlear Nerve
V. Trigeminal Nerve
VI. Abducent Nerve
VII. Facial Nerve
VIII. Vestibulocochlear Nerve
IX. Glossopharyngeal Nerve
X. Vagus Nerve
XI. Accessory Nerve
XII. Hypoglossal Nerve

Figure 13–13 Cranial nerves and ventral surface of the brain.

Table 13–2 Cranial Nerves

Cranial Nerve	Name	Function
I	olfactory	conducts sensory impulses from the nose to the brain (smell)
II	optic	conducts sensory impulses from the eyes to the brain (vision)
III	oculomotor	sends motor impulses to the external eye muscles (dorsal, medial, and ventral rectus; ventral oblique; and levator superioris) and to some internal eye muscles
IV	trochlear	sends motor impulses to one external eye muscle (dorsal oblique)
V	trigeminal	three branches: ophthalmic = sensory to cornea; maxillary = motor to upper jaw; mandibular = motor to lower jaw
VI	abducent	motor innervation to two muscles of the eye (retractor bulbi and lateral rectus)
VII	facial	motor to facial muscles, salivary glands, and lacrimal glands and taste sensation to anterior two-thirds of tongue
VIII	acoustic or vestibulocochlear	two branches: cochlear = sense of hearing; vestibular = sense of balance
IX	glossopharyngeal	motor to the parotid glands and pharyngeal muscles, taste sensation to caudal third of tongue, and sensory to the pharyngeal mucosa
X	vagus	sensory to part of the pharynx and larynx and parts of thoracic and abdominal viscera; motor for swallowing and voice production
XI	Accessory	accessory motor to shoulder muscles
XII	hypoglossal	motor to the muscles that control tongue movement

Spinal Nerves

The **spinal nerves** arise from the spinal cord. With the exception of some cervical and coccygeal nerves, the spinal nerves are paired and emerge caudal to the vertebra of the same number and name. The first cervical vertebra (C1, or the atlas) has a pair of spinal nerves emerging cranially and caudally to it; therefore, there are eight cervical spinal nerves. C1 exits the spinal canal in a foramen in the wing of the atlas, and C2 emerges at the C1–C2 intervertebral foramen. C8 emerges caudal to vertebra C7, and T1 emerges caudal to T1 vertebra. The coccygeal vertebrae usually have fewer pairs of nerves than the number of vertebrae (Figure 13–14).

Spinal nerves have dorsal and ventral roots. The **dorsal root** enters the dorsal portion of the spinal cord and carries afferent or sensory impulses from the periphery to the spinal cord. The **ventral root** emerges from the ventral portion of the spinal cord and carries efferent or motor impulses from the spinal cord to muscle fibers or glands (Figure 13–15).

Spinal nerves supply sensory and motor fibers to the body region associated with their emergence from the spinal cord. After spinal nerves exit the spinal cord, they branch to form the peripheral nerves of the trunk and limbs. Several spinal nerves may join together to form a single peripheral nerve. This network of intersecting nerves is called a **plexus.** Each appendage is innervated by a plexus. Each forelimb is supplied from nerves that arise from the **brachial** (brā-kē-ahl) **plexus** (C6–T2), and each hindlimb is supplied from nerves that arise from the **lumbosacral plexus** (L4–S3). Brachial means the arm, and lumbosacral means the loin and sacrum.

Spinal nerves are named for where they arise from the spinal cord. *C* represents cervical, *T* represents thoracic, *L* represents lumbar, *S* represents sacral, and *Co* and *Cy* represent coccygeal (or *Cd* represents caudal). These letter abbreviations are followed by numbers to represent the vertebral area from which the nerve exits the spinal cord. C4 represents cervical spinal nerve 4, T11 represents thoracic spinal nerve 11, and so on.

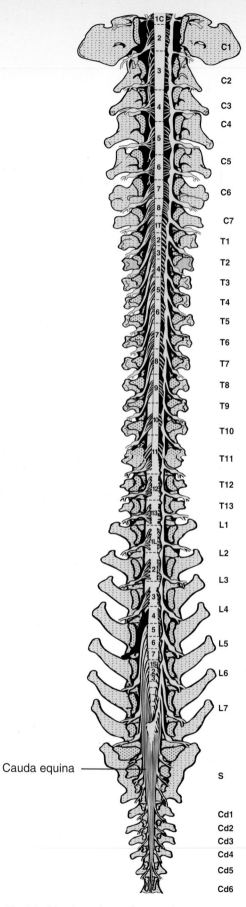

Cauda equina

Figure 13–14 Naming scheme for spinal nerves.

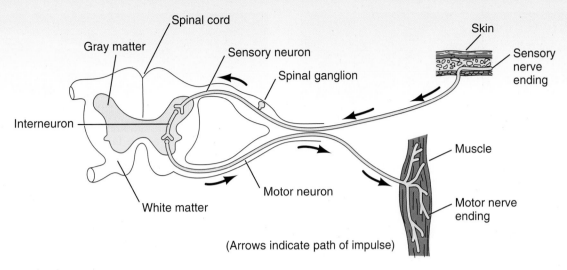

Figure 13–15 Spinal nerve.

Autonomic Nervous System

The **autonomic nervous system** (aw-tō-nah-mihck nər-vuhs sihs-tehm), or ANS, is the part of the peripheral nervous system that innervates smooth muscle, cardiac muscle, and glands. The two divisions of the autonomic nervous system are the sympathetic nervous system (sihm-pah-theh-tihck nər-vuhs sihs-tehm) and the **parasympathetic nervous system** (pahr-ah-sihm-pah-theh-tihck nər-vuhs sihs-tehm). The two divisions of the autonomic nervous system work together to maintain homeostasis within the body. **Homeostasis** (hō-mē-ō-stā-sihs) is the process of maintaining a stable internal body environment.

sympathetic	provides emergency and stress response; "fight or flight"	↑ heart rate, respiratory rate, and blood flow to muscles; ↓ gastrointestinal function; pupil dilation
parasympathetic	returns body to normal after stressful response; maintains normal body function	returns heart rate, respiratory rate, and blood flow to normal levels; returns normal gastrointestinal function; constricts pupil size to normal

TEST ME: NERVOUS SYSTEM

Diagnostic procedures performed on the nervous system include the following:

- **cerebrospinal fluid tap** (sər-ē-brō-spī-nahl flū-ihd tahp) = removal of cerebrospinal fluid; also called a CSF tap. CSF is obtained by inserting a needle or catheter into the **cisterna magna** (sihs-tər-nah mahg-nah) or lumbosacral area. The cisterna magna is the subarachnoid space located between the caudal surface of the cerebellum and the dorsal surface of the medulla oblongata. Intracranial pressures also may be measured before removal of CSF (Figure 13–16).

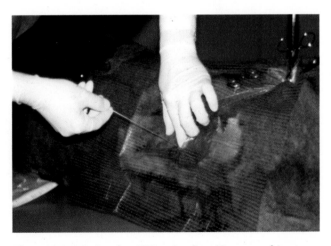

Figure 13–16 Lumbar CSF tap in a lion. (Courtesy of Anne E. Chauvet, DVM, Diplomate ACVIM—Neurology, University of Wisconsin School of Veterinary Medicine.)

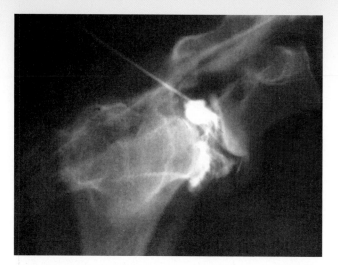

Figure 13–17 Discogram in a dog. (Courtesy of Anne E. Chauvet, DVM, Diplomate ACVIM—Neurology, University of Wisconsin School of Veterinary Medicine.)

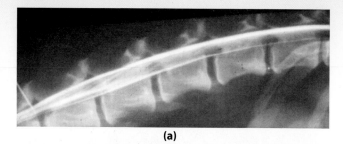

(a)

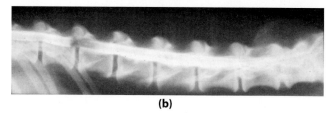

(b)

Figure 13–18 (a) Normal myelogram in a Labrador retriever. (b) Myelogram demonstrating disc disease. (Note that the dye does not flow continuously.) (Courtesy of Anne E. Chauvet, DVM, Diplomate ACVIM—Neurology, University of Wisconsin School of Veterinary Medicine.)

- **discography** (dihs-kō-grah-fē) = radiographic study of an intervertebral disc after injection of contrast material into the disc, also spelled *diskography*. The record of this procedure is called a **discogram** (Figure 13–17).
- **electroencephalography** (ē-lehck-trō-ehn-sehf-ah-lohg-rah-fē) = process of recording electrical activity of the brain; abbreviated EEG. An **electroencephalo-**graph (ē-lehck-trō-ehn-sehf-ah-lō-grahf) is the instrument used to record the electrical activity of the brain, and an **electroencephalogram** (ē-lehck-trō-ehn-sehf-ah-lō-grahm) is the record of the electrical activity of the brain.
- **magnetic resonance imaging** and **computed axial tomography** are covered in Chapter 16.
- **myelography** (mī-eh-lohg-rah-fē) = diagnostic study of the spinal cord after injection of contrast material. A **myelogram** (mī-eh-lō-grahm) is the record of the spinal cord after injection of contrast material (Figure 13–18).
- **pupillary light reflex** = response of pupil to a bright light source; abbreviated PLR. Light is shone in one eye, and that eye (direct) and the opposite eye (consensual) should constrict. It is used to assess neurologic damage.

levels of consciousness = descriptive terms used to describe mentation .

- **BAR** = bright, alert, and responsive.
- **coma** (kō-mah) = deep state of unconsciousness.
- **conscious** (kohn-shuhs) = awake, aware, and responsive; also known as **alert.**
- **disorientation** (dihs-ōr-ē-ehn-tā-shuhn) = condition in which the animal appears mentally confused.
- **lethargy** (lehth-ahr-jē) = drowsiness, indifference, and listlessness.
- **obtunded** (ohb-tuhn-dehd) = depressed.
- **stupor** (stoo-pər) = impaired consciousness with unresponsiveness to stimuli.

PATHOLOGY: NERVOUS SYSTEM

Pathologic conditions of the nervous system include the following:

- **amnesia** (ahm-nē-zē-ah) = memory loss.
- **astrocytoma** (ahs-trō-sī-tō-mah) = malignant intracranial tumor composed of astrocytes.
- **ataxia** (ā-tahck-sē-ah) = without coordination; "stumbling" (Figure 13–19).
- **catalepsy** (kaht-ah-lehp-sē) = waxing rigidity of muscles accompanied by a trancelike state.

Figure 13-20 CP deficit in a dog.

Figure 13-19 Ataxia in a cat. (Courtesty of Kimberly Kruse Sprecher, CVT.)

- **cataplexy** (kaht-ah-plehck-sē) = sudden attacks of muscular weakness and hypotonia triggered by an emotional response.
- **cerebellar hypoplasia** (sehr-eh-behl-ahr hī-pō-plā-zē-ah) = smaller-than-normal cerebellum; seen in cats secondary to feline panleukopenia virus, which leads to incoordination.
- **cervical vertebral malformation** = abnormal formation or instability of the caudal cervical vertebrae that causes ataxia and incoordination; seen more often in horses and dogs; also called **wobbler's syndrome.**
- **chorea** (kōr-ē-ah) = repetitive, rhythmic contraction of limb or facial muscles; also called **myoclonus** (mī-ō-klō-nuhs); usually the result of distemper viral infection in dogs.
- **choriomeningitis** (kōr-ē-ō-meh-nihn-jī-tihs) = inflammation of the choroid plexus and meninges.
- **concussion** (kohn-kuhsh-uhn) = shaking of the brain caused by injury. The combining form **concuss/o** means shaken together.
- **conscious proprioceptive deficit** (kohn-shuhs prō-prih-ō-sehp-tihv dehf-ih-siht) = neurologic defect in which the animal appears not to know where

its limbs are; abbreviated CP deficit, "knuckling" (Figure 13-20).
- **contusion** (kohn-too-zhuhn) = bruising. The combining form **contus/o** means bruise.
- **decerebration** (dē-sər-ē-brā-shuhn) = condition of loss of mental functions caused by damage to the midbrain.
- **demyelination** (dē-mī-eh-lih-nā-shuhn) = destruction or loss of myelin.
- **discospondylitis** (dihs-kō-spohn-dih-lī-tihs) = destructive inflammatory disorder that involves the intervertebral discs, vertebral end-plates, and vertebral bodies.
- **encephalitis** (ehn-sehf-ah-lī-tihs) = inflammation of the brain.
- **encephalocele** (ehn-sehf-ah-lō-sēl) = herniation of the brain through a gap in the skull.
- **encephalomalacia** (ehn-sehf-ah-lō-mah-lā-shē-ah) abnormal softening of the brain.
- **encephalomyelitis** (ehn-sehf-ah-lō-mī-ih-lī-tihs) inflammation of the brain and spinal cord.
- **encephalopathy** (ehn-sehf-ah-lohp-ah-thē) = any disease of the brain.
- **epilepsy** (ehp-ih-lehp-sē) = recurrent seizures of nonsystemic origin or of intracranial disease. Epilepsy may be described as **idiopathic** (ihd-ē-ō-pahth-ihck). *Idiopathic* means unknown cause or disease of an individual. **Idio-** is the prefix meaning individual.
- **hallucination** (hah-loo-sehn-ā-shuhn) = false sensory perception.
- **hematoma** (hē-mah-tō-mah) = mass or collection of blood. In the nervous system, a hematoma usually is

described by the area where it is found. An **epidural hematoma** (ehp-ih-doo-rahl hē-mah-tō-mah) is a collection of blood above or superficial to the dura mater. A **subdural hematoma** (suhb-doo-rahl hē-mah-to-mah) is a collection of blood below (deep to) the dura mater and above (superficial to) the arachnoid membrane.

- **hemiplegia** (hehm-ih-plē-jē-ah) = paralysis of one side of the body.
- **Horner's syndrome** (hōr-nərz sihn-drōm) = collection of signs relating to injury of the cervical sympathetic innervation to the eye; signs include sinking of the eyeball (enophthalmus), ptosis of the upper eyelid, pupil constriction, and prolapse of the third eyelid (Figure 13–21).
- **hydrocephalus** (hī-drō-sehf-ah-luhs) = abnormally elevated amount of cerebrospinal fluid in the ventricles of the brain; "water on the brain."
- **hyperesthesia** (hī-pər-ehs-thē-zē-ah) = excessive sensitivity.
- **hyperkinesis** (hī-pər-kihn-ē-sihs) = increased motor function or activity.
- **hypnosis** (hihp-nō-sihs) = condition of altered awareness; trancelike state.
- **intervertebral** (ihn-tər-vər-tə-brahl) **disc disease** = condition of pain and neurologic deficits resulting from the displacement of part or all of the material in the disc located between the vertebrae (see Figure 13–12).
- **leukoencephalomalacia** (loo-kō-ehn-sehf-ah-lō-mah-lā-shah) = abnormal softening of the white matter of the brain.
- **macrocephaly** (mahck-rō-sehf-ah-lē) = abnormally large skull.
- **meningioma** (meh-nihn-jē-ō-mah) = benign tumor of the meninges (Figure 13–22).

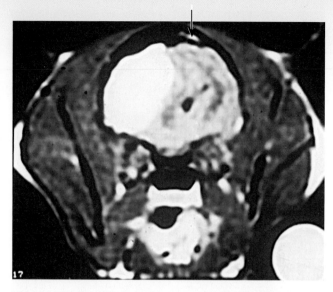

Figure 13–22 Computed tomogram of a cat with a meningioma. (Courtesy of Anne E. Chauvet, DVM, Diplomate ACVIM—Neurology, University of Wisconsin School of Veterinary Medicine.)

- **meningitis** (mehn-ihn-jī-tihs) = inflammation of the meninges.
- **meningocele** (meh-nihng-gō-sēl) = protrusion of the meninges through a defect in the skull or vertebrae.
- **meningoencephalitis** (meh-nihn-gō-ehn-sehf-ah-lī-tihs) = inflammation of the meninges and brain.
- **meningoencephalomyelitis** (meh-nihn-gō-ehn-sehf-ah-lō-mī-eh-lī-tihs) = inflammation of the meninges, brain, and spinal cord.
- **microcephaly** (mī-krō-sehf-ah-lē) = abnormally small skull.
- **monoplegia** (mohn-ō-plē-jē-ah) = paralysis of one limb.
- **myelitis** (mī-eh-lī-tihs) = inflammation of the spinal cord (or bone marrow).
- **myelopathy** (mī-eh-lah-pahth-ē) = disease of the spinal cord (or bone marrow).
- **myoparesis** (mī-ō-pahr-ē-sihs) = weakness of muscles. The suffix **-paresis** means weakness. As is the case with **-plegia,** the suffix -paresis is modified to describe the area of weakness. **Hemiparesis** (hehm-ih-pahr-ē-sihs) is weakness on one side of the body; **paraparesis** (pahr-ah-pahr-ē-sihs) is weakness of the lower body in bipeds or of hindlimbs in quadrupeds.
- **narcolepsy** (nahr-kō-lehp-sē) = syndrome of recurrent uncontrollable sleep episodes. The combining form **narc/o** means stupor, and the suffix **-lepsy** means seizure (episode).
- **neuralgia** (nū-rahl-jē-ah) = nerve pain.
- **neuritis** (nū-rī-tihs) = inflammation of the nerves.
- **opisthotonos** (ohp-ihs-thoht-ō-nohs) = tetanic spasm in which the head and tail are bent dorsally and the back is arched.

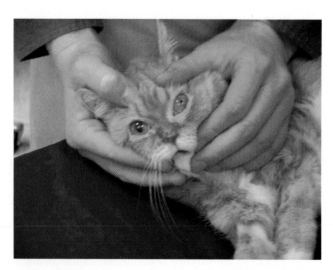

Figure 13–21 Horner's syndrome in a cat. (Courtesy of Kimberly Kruse Sprecher, CVT.)

- **paralysis** (pahr-ahl-ih-sihs) = loss of voluntary movement or immobility. The suffix **-plegia** means paralysis. Paralysis is further described by the areas it involves.
- **paraplegia** (pahr-ah-plē-jē-ah) = paralysis of the lower body in bipeds or of hindlimbs in quadrupeds (Figure 13–23). In reference to the nervous system, the prefix **para-** means hind or lower portion.
- **paresthesia** (pahr-ehs-thē-zē-ah) = abnormal sensation. The suffix **-esthesia** means sensation or feeling. Abnormal sensations may include tingling, numbness, or burning and may be difficult to assess in animals. The combining forms for burning are **caus/o** and **caust/o.**

Figure 13–23 Paraplegia in a rabbit.

Figure 13–24 Polioencephalomalacia in a sheep. Opisthotonos is one sign of PEM. (Courtesy of Ron Fabrizius, DVM, Diplomate ACT.)

- **polioencephalomalacia** (pō-lē-ō-ehn-sehf-ah-lō-mah-lā-shah) = abnormal softening of the gray matter of the brain; abbreviated PEM (Figure 13–24).
- **polioencephalomyelitis** (pō-lē-ō-ehn-sehf-ah-lō-mī-eh-lī-tihs) = inflammation of the gray matter of the brain and spinal cord.
- **poliomyelitis** (pō-lē-ō-mī-eh-lī-tihs) = inflammation of the gray matter of the spinal cord. The combining form **poli/o** means gray.
- **polyneuritis** (poh-lē-nū-rī-tihs) = inflammation of many nerves.
- **polyradiculoneuritis** (poh-lē-rah-dihck-yoo-lō-nū-rī-tihs) = inflammation of many peripheral nerves and spinal nerve roots that may lead to progressive paralysis; commonly called **coonhound paralysis** in coonhound dogs and **idiopathic polyradiculoneuropathy** in other dogs.
- **ptosis** (tō-sihs) = prolapse or drooping. The suffix **-ptosis** means prolapse, drooping, or falling downward; refers specifically to the upper eyelid.
- **radiculitis** (rah-dihck-yoo-lī-tihs) = inflammation of the root of a spinal nerve. **Radicul/o** is the combining form for root.
- **roaring** = noisy respiration caused by air passing through a narrowed larynx in horses; common term for **equine laryngeal hemiplegia** because of nerve fiber degeneration of the left recurrent laryngeal nerve (Figure 13–25).
- **seizure** (sē-zhər) = sudden, involuntary contraction of some muscles caused by a brain disturbance; also called

What side are you on?

Pathologic conditions of the nervous system may involve lesions that cause abnormal clinical signs on the same side or opposite side that the lesion occurs. In describing lesions of the nervous system, the terms *ipsilateral* and *contralateral* are used. **Ipsi-** is the prefix meaning the same, and **contra-** is the prefix meaning opposite.

Ipsilateral (ihp-sē-laht-ər-ahl) means on the same side, and **contralateral** (kohn-trah-laht-ər-ahl) means on the opposite side.

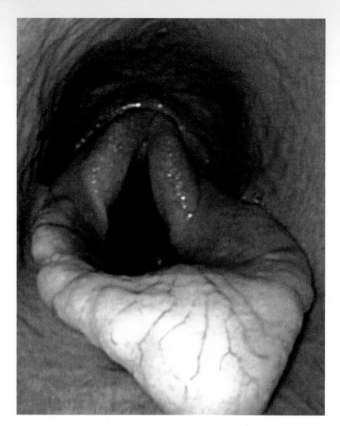

Figure 13–25 Laryngeal area of a horse with equine laryngeal hemiplegia (roaring). (Courtesy of Laura Lien, CVT, BS.)

Seizure stages

Seizures are divided into stages to aid in determining when they may start or end. These stages are as follows:

preictal (prē-ihck-tahl) = period before a seizure; also called the **aura** (aw-rah). An animal may pace, excessively lick, fly bite, or seem anxious during this stage.

ictus (ihck-tuhs) = attack or actual seizure. An animal may convulse, lose control of excretory functions, shake, and appear confused during this stage.

postictal (pōst-ihck-tahl) = period after a seizure. An animal may appear obtunded, tired, fearful, or anxious during this stage.

convulsions. The most common type of seizure in animals is **grand mal** (grahnd mahl), in which the animal experiences loss of consciousness and muscle contractions. Other types of seizures include partial, which have a seizure focus that does not spread, and petit mal, which is a mild generalized seizure in which loss of consciousness and generalized loss of muscle tone occur.

- **spasticity** (spahs-tihs-ih-tē) = state of increased muscular tone.
- **spina bifida** (spī-nah bihf-ih-dah) = congenital anomaly in which the spinal canal does not close over the spinal cord. The combining form **bifid/o** means split or cleft.
- **syncope** (sihn-kō-pē) = fainting; sudden fall in blood pressure or cardiac systole resulting in cerebral anemia and loss of consciousness.
- **tetraplegia** (teht-rah-plē-jē-ah) = paralysis of all four limbs; also called **quadriplegia** (kwohd-rih-plē-jē-ah).
- **tremor** (treh-mər) = involuntary trembling.
- **vestibular disease** (vehs-tihb-yoo-lahr dih-zēz) = neurologic disorder characterized by head tilt, nystagmus, rolling, falling, and circling. **Nystagmus** (nī-stahg-muhs) is involuntary, rhythmic movement of the eye and is discussed in Chapter 14.

PROCEDURES: NERVOUS SYSTEM

Procedures performed on the nervous system include the following:

- **analgesia** (ahn-ahl-jēz-ē-ah) = without pain. *Analgesia* is used to describe pain relief, which is different from *anesthesia* (absence of sensation). **Endorphins** (ehn-dōr-fihnz) are natural, opioid-like chemicals that are produced in the brain and that raise the pain threshold.
- **anesthesia** (ahn-ehs-thē-zē-ah) = absence of sensation. An **anesthetic** (ahn-ehs-theht-ihck) is a substance used to induce anesthesia. There are different types of anesthesia. Some types include **topical anesthesia** = absence of sensation after a substance has been applied to the skin or external surface; **local anesthesia** = absence of sensation after chemical injection to an adjacent area; **epidural anesthesia** = absence of sensation to a region after injection of a chemical into the epidural space; **general anesthesia** = absence of sensation and consciousness.
- **disc fenestration** (dihsk fehn-ih-strā-shuhn) = removal of intervertebral disc material by perforating and scraping out its contents (Figure 13–26).

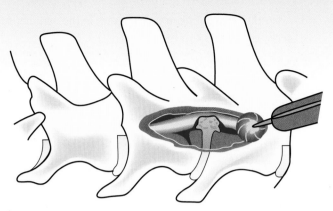

Figure 13–26 A disc fenestration involves removal of intervertebral disc material.

- **dysesthesia** (dihs-eh-stēsh-ah) = impaired sensation.
- **laminectomy** (lahm-ihn-ehck-tō-mē) = surgical removal of the lamina of the vertebra to relieve pressure on the spinal cord.
- **neurectomy** (nū-rehck-tō-mē) = surgical removal of a nerve.
- **neuroanastomosis** (nū-rō-ahn-ahs-tō-mō-sihs) = connecting nerves together.
- **neuroplasty** (nū-rō-plahs-tē) = surgical repair of a nerve.
- **neurorrhaphy** (nū-rōr-ah-fē) = suturing the ends of a severed nerve.
- **neurotomy** (nū-roht-ō-mē) = surgical incision or dissection of a nerve.

REVIEW EXERCISES

Multiple Choice

Choose the correct answer.

1. The space between two neurons or between a neuron and a receptor is a(n)

   a. synapse
   b. ganglion
   c. axon
   d. dendrite

2. Maintaining a constant internal environment is called

   a. osmosis
   b. homeosmosis
   c. homeostasis
   d. ipsistasis

3. Inflammation of the root of a spinal nerve is

   a. myelitis
   b. radiculitis
   c. polyneuritis
   d. poliomyelitis

4. The three-layered membrane lining the CNS is called the

   a. hippocampus
   b. pons
   c. myelin
   d. meninges

5. The protective sheath that covers some nerve cells of the spinal cord, white matter of the brain, and most peripheral nerves is called

   a. pia
   b. dura
   c. glia
   d. myelin

6. The division of the autonomic nervous system that is concerned with body functions under emergency or stress is the

   a. peripheral
   b. central
   c. sympathetic
   d. parasympathetic

7. A network of intersecting nerves is a

   a. bundle
   b. trunk
   c. tract
   d. plexus

8. What type of neuron carries impulses away from the CNS and toward the muscles?

   a. afferent (sensory)
   b. efferent (motor)
   c. sympathetic
   d. parasympathetic

9. An automatic, involuntary response to change is called a(n)

    a. impulse
    b. stimulus
    c. reflex
    d. receptor

10. The largest portion of the brain that is involved with thought and memory is the

    a. cerebrum
    b. cerebellum
    c. brainstem
    d. spinal cord

11. Elevated portions of the cerebral cortex are

    a. sulci
    b. plexuses
    c. gyri
    d. hemispheres

12. The term meaning without pain is

    a. analgesia
    b. endorphin
    c. idiopathic
    d. ptosis

13. Inflammation of the gray matter of the spinal cord is known as

    a. myelitis
    b. poliomyelitis
    c. leukomyelitis
    d. neuritis

14. The term for without coordination or "stumbling" is

    a. ataxia
    b. spasticity
    c. seizure
    d. epilepsy

15. A knotlike mass of neuron cell bodies in the peripheral nervous system is known as a

    a. microglia
    b. synapse
    c. receptor
    d. ganglion

16. A "depressed" animal is referred to as

    a. conscious
    b. disoriented
    c. obtunded
    d. depressed

17. Inflammation of the membranes surrounding the brain and spinal cord is known as

    a. meningitis
    b. encephalopathy
    c. encephalitis
    d. chorea

18. Conscious proprioceptive deficit is commonly called

    a. nystagmus
    b. roaring
    c. knuckling
    d. wobbler's syndrome

19. The medical term for crossing is

    a. intumescence
    b. plexus
    c. chiasm
    d. filum

20. Small cavities in the brain are known as

    a. gyri
    b. sulci
    c. ventricles
    d. ependyma

Matching

Match the term in Column I with the definition in Column II.

Column I

Column II

1. _____ conscious

2. _____ BAR

3. _____ coma

4. _____ lethargy

5. _____ obtunded

6. _____ disorientation

7. _____ stupor

a. depressed

b. impaired consciousness with unresponsiveness to stimuli

c. bright, alert, and responsive

d. deep state of unconsciousness

e. awake, aware, and responsive; also known as alert

f. condition in which the animal appears mentally confused

g. drowsiness, indifference, and listlessness

Match the term in Column I with the definition in Column II.

Column I

Column II

8. _____ homeostasis

9. _____ ganglia

10. _____ stimulus

11. _____ reflex

12. _____ synapse

13. _____ soma

14. _____ impulse

15. _____ neuroglia

16. _____ myelin

17. _____ myoclonus

a. supportive cells of the nervous system

b. protective covering over some nerve cells that serves as an electrical insulator

c. wave of excitation transmitted through nervous tissue

d. neuron cell bodies grouped together outside the CNS

e. repetitive, rhythmic contraction of limb or facial muscles

f. something that excites or activates

g. automatic, involuntary response to change

h. space between two neurons or between a neuron and a receptor

i. cell body

j. process of maintaining a constant internal body environment

Fill in the Blanks

1. Neur/i and neur/o mean _____.

2. Mening/o and meningi/o mean _____.

3. Pachy- means _____.

4. Gangli/o and ganglion/o mean _____.

5. Gli/o means _____.

Spelling

Circle the term that is spelled correctly.

1. fainting: sincope   syncopy   syncope

2. substance that produces absence of sensation: anestetic   anesthetic   anesthetick

3. loss of voluntary movement: paralysis   paralisis   parlysis

4. convulsions: seezure   siezure   seizure

5. prolapse or drooping: tosis   ptosis   ptoesis

Word Scramble

Use the definitions to unscramble the terms relating to the nervous system.

1. incision into a nerve   tronumoye   _____

2. period before a seizure   uaar   _____

3. disease of the spinal cord (or bone marrow)   pthayyelom   _____

4. passage or hole   fmnraoe   _____

5. paralysis of the lower limbs in bipeds or of hindlimbs in quadrupeds   ppaaaliger   _____

6. opposite   rcoatn   _____

7. recurrent seizures of nonsystemic origin   yspelipe   _____

CROSSWORD PUZZLE

Nervous System Terms Puzzle

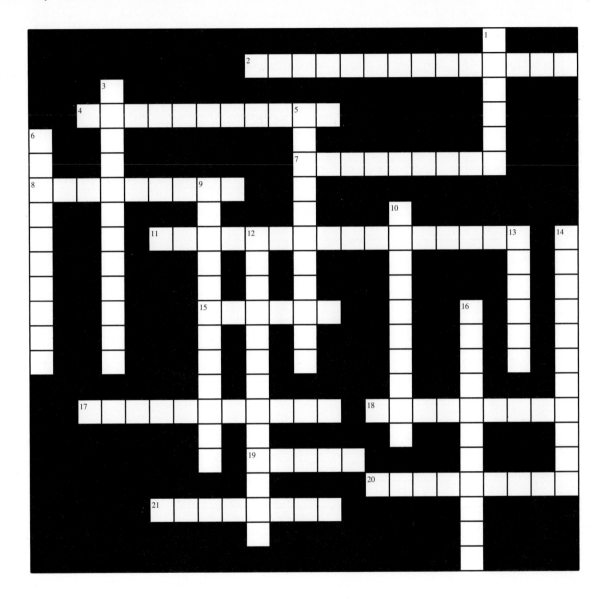

Across

2 spatial orientation or perception of movement
4 diagnostic radiographic study of the spinal cord after injection of contrast material into the subarachnoid space
7 repetitive, rhythmic contraction of skeletal muscle
8 surgical incision or dissection of a nerve
11 abnormal softening of the brain
15 crossing
17 the same side
18 star-shaped cell
19 period of an actual seizure
20 nerve pain
21 inflammation of the spinal cord (or bone marrow)

Down

1 prolapse or drooping
3 increased motor function or activity
5 weakness of one side of the body
6 shaking of brain caused by injury
9 abnormally small skull
10 syndrome of recurrent uncontrollable sleep episodes
12 inflammation of the gray matter of the spinal cord
13 without coordination or stumbling
14 abnormal sensation
16 surgical repair of a nerve

CASE STUDIES

Define the underlined terms in each case study.

A 3-yr-old F mixed breed dog was presented to the clinic for <u>convulsions</u> (Figure 13–27). A thorough history was taken, including a description of the convulsion episodes, possible toxin exposure, and eating history. PE revealed that the animal was <u>obtunded</u> but otherwise in normal health. The <u>neurologic</u> examination was normal. Blood was drawn for CBC and chem panel. The CBC was normal, and the chem panel did not show evidence of <u>renal</u> disease, <u>hepatic</u> disease, or <u>hypoglycemia</u>. Urine was collected for <u>UA</u> via <u>cystocentesis</u>. The results of the UA were normal. The dog's history was examined, and it was noted that the dog's vaccinations were current. The <u>signalment</u>, history, and <u>clinical</u> signs were used to <u>diagnose</u> <u>idiopathic</u> <u>epilepsy</u>. Diagnostic tests, including <u>CSF tap</u> and <u>EEG</u>, were recommended to the owners but were declined at this time. The owners were advised to have the dog <u>spayed</u> (hormone levels may contribute to <u>seizure</u> activity) and to monitor the animal's activity for recurrence of seizures. If the seizures recur and persist, <u>anticonvulsant</u> medication will be used in an attempt to control them.

1. convulsions _____

2. obtunded _____

3. neurologic _____

4. renal _____

5. hepatic _____

6. hypoglycemia _____

7. UA _____

8. cystocentesis _____

9. signalment _____

10. clinical _____

11. diagnose _____

12. idiopathic _____

13. epilepsy _____

14. CSF tap _____

15. EEG _____

16. spayed _____

17. seizure _____

18. anticonvulsant _____

Figure 13–27 A seizuring dog. During a grand mal seizure, dogs typically are unresponsive, lying on their sides, and paddling their legs.

A 5-yr-old <u>M/N</u> dachshund was presented to the clinic with a history of not being able to walk up stairs or jump on the bed. The owner stated that the dog had decreased its eating and was more <u>lethargic</u> than usual. PE revealed an <u>obese</u> dog that had normal <u>vital signs</u>. The neurologic examination revealed normal <u>cranial nerves</u> and <u>CP deficit</u> present on both hindlimbs. <u>Patellar reflexes</u> were <u>hyporeflexive</u>, with the right side worse than the left. <u>Anal</u> tone was adequate. <u>Radiographs</u> were recommended to assess whether the dog had calcified or herniated <u>intervertebral discs</u>. Abnormal discs were noted in the <u>lumbar</u> region. The dog was referred for <u>myelography</u>. The <u>myelogram</u> confirmed that the dog had <u>herniated</u> discs in the lumbar region, and surgery was recommended. <u>Disc fenestration</u> surgery was performed the following day. The dog went home with orders for strict cage rest and rechecks by the referring veterinarian.

19. M/N _____

20. lethargic _____

21. obese _____

22. vital signs _____

23. cranial nerves _____

24. CP deficit _____

25. patellar reflexes _____

26. hyporeflexive _____

27. anal _____

28. radiographs _____

29. intervertebral discs _____

30. lumbar _____

31. myelography _____

32. myelogram _____

33. herniated _____

34. disc fenestration _____

CHAPTER 14

SEEING AND HEARING

Objectives

Upon completion of this chapter, the reader should be able to

- Identify and describe the structures and functions of the eyes and ears
- Recognize, define, spell, and pronounce terms related to the diagnosis, pathology, and treatment of eye and ear disorders

FUNCTIONS OF THE EYE

The **ocular** (ohck-yoo-lahr) **system** is responsible for vision. The **eyes** are the receptor organs for sight. Combining forms for the eye or sight are **opt/i, opt/o, optic/o, ocul/o,** and **ophthalm/o. Extraocular** (ehcks-trah-ohck-yoo-lahr) means outside the eyeball, and **intraocular** (ihn-trah-ohck-yoo-lahr) means within the eyeball. **Periocular** (pehr-ē-ohck-yoo-lahr) means around the eyeball.

STRUCTURES OF THE EYE

The structures of the eye include the accessory structures and the eyeball.

Accessories

The accessory structures of an organ are called **adnexa** (ahd-nehck-sah). **Stroma** (strō-mah) is another term used to describe the supporting tissue of an organ. The adnexa of the eye include the orbit, eye muscles, eyelids, eyelashes, conjunctiva, and lacrimal apparatus (Figure 14–1).

- **orbit** (ohr-biht) = bony cavity of the skull that contains the eyeball. The term **periorbita** (pehr-ih-ōr-bih-tah) means eye socket.
- **eye muscles** = seven major muscles attached to each eye that make a range of movement possible (two oblique muscles, four rectus muscles, and the retractor bulbi)

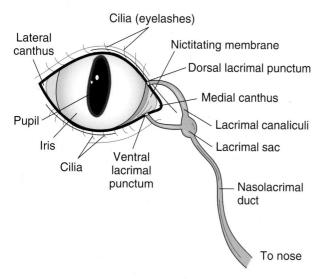

Figure 14–1 Adnexa of the eye.

(Figure 14–2). The muscles of both eyes work together in coordinated movements to make normal binocular vision possible. **Binocular** (bī-nohck-yoo-lahr) means both eyes. The **extrinsic muscles** (ehcks-trihn-sihck muhs-uhlz) are six muscles that attach the outside of the eyeball to the bones of the orbit. The **levator palpebrae muscles** (lē-vā-tər pahl-pē-brā muhs-uhlz) are muscles that raise the upper eyelid.

- **eyelids** = each eye has an upper and lower eyelid to protect the eye from injury, foreign material, and excessive light. The combining form **blephar/o** (blehf-ah-rō) means eyelid. **Palpebra** (pahl-pē-brah) is another term used for the eyelid; the plural form is **palpebrae** (pal-pē-brā). **Palpebral** (pahl-pē-brahl) means pertaining to the eyelid.
- The angle where the upper and lower eyelids meet is called the **canthus** (kahn-thuhs). The combining form **canth/o** means corner of the eye. The medial canthus is the corner of the eye nearer to the nose (also called the inner canthus). The lateral canthus is the corner of the eye farther away from the nose (also called the outer canthus).
- The **tarsal** (tahr-sahl) **plate** (or tarsus) is the platelike framework within the upper and lower eyelids that provides stiffness and shape. The combining form **tars/o** means edge of the eyelid or "ankle" joint.
- **Meibomian** (mī-bō-mē-ahn) **glands** are the sebaceous glands on the margins of each eyelid; also called tarsal glands.
- **eyelashes** = the edge of each eyelid has hairlike structures called **cilia** (sihl-ē-ah). The cilia, or eyelashes, protect the eye from foreign material.
- **conjunctiva** (kohn-juhnck-tī-vah) = mucous membrane that lines the underside of each eyelid. The conjunctiva forms a protective covering of the exposed surface of the eyeball when the eyelids are closed.

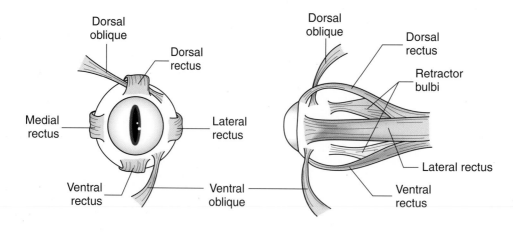

Front View   **Lateral View**

Figure 14–2 Extrinsic eye muscle of a dog.

Figure 14–3   Third eyelid, or nictitans, in a cat.

The combining form **conjunctiv/o** means conjunctiva. The **nictitating membrane** (nihck-tih-tāt-ihng mehm-brān) is the conjunctival fold attached at the medial canthus that moves across the cornea when the eyelids close; it also is called the **third eyelid** or **nictitans** (nihck-tih-tahnz) or **haws** (hawz) (Figure 14–3).

▪ **lacrimal apparatus** (lahck-rih-mahl ahp-ah-raht-tuhs) = structures that produce, store, and remove tears. **Lacrimation** (lahck-rih-mā-shuhn) is the condition of normal tear secretion. The combining forms **lacrim/o** and **dacry/o** mean teardrop, tear duct, or lacrimal duct. The lacrimal glands are glands that secrete tears. The **lacrimal canaliculi** (kahn-ah-lihck-yoo-lī) are the ducts at the medial canthus that collect tears and drain them into the lacrimal sac. The **lacrimal sac,** or **dacryocyst** (dahck-rē-ō-sihst), is the enlargement that collects tears at the upper portion of the tear duct.

The **nasolacrimal** (nā-sō-lahck-rih-mahl) **duct** is the passageway that drains tears into the nose. The **dorsal punctum** (dōr-sahl puhnck-tuhm) is the small spot near the upper medial canthus where the nasolacrimal duct begins; the **ventral punctum** (vehn-trahl puhnck-tuhm) is the small spot near the lower medial canthus where the nasolacrimal duct begins. A **punctum** (puhnck-tuhm) is a point or small spot.

Eyeball

The **eyeball,** or **globe,** is a sphere with multilayered walls. These walls are the sclera, choroid, and retina. Another term for the eyeball is **orb** (ōrb) (Figure 14–4).

Sclera

The **sclera** (sklehr-ah) is the fibrous outer layer of the eye that maintains the shape of the eye. It is sometimes called the **white of the eye.** The combining form **scler/o** means sclera, or hard.

The anterior portion of the sclera is transparent and is called the **cornea** (kōr-nē-ah). The cornea provides most of the focusing power of the eye. The combining forms **corne/o** and **kerat/o** mean cornea. **Descemet's membrane** (dehs-eh-māz mehm-brān) is the innermost or deepest layer of the cornea.

Choroid

The **choroid** (kō-royd) is the opaque middle layer of the eyeball that contains blood vessels and supplies blood for the entire eye. **Opaque** (ō-pāk) means that light cannot pass through. The **tapetum lucidum** (tah-pē-duhm loo-sehd-uhm) is the brightly colored iridescent reflecting tissue layer of the choroid of most species. The tapetum lucidum also is called **choroid tapetum.** The **tapetum nigrum** (tah-pē-duhm nī-gruhm) is the black pigmented tissue layer of the choroid in some species. *Tapetum* is the medical term for a layer of cells. The combining

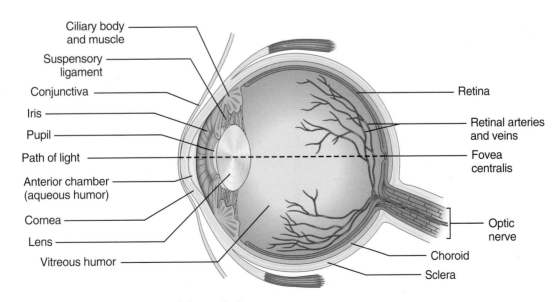

Figure 14–4   Cross section of the eyeball.

form **choroid/o** means choroid. Associated with the choroid are the iris, pupil, lens, and ciliary muscles. The **iris** (ī-rihs) is the pigmented muscular diaphragm of the choroid that surrounds the pupil. The iris is composed of muscle fiber rings that contract or relax to change the size of the pupil and thus regulate the amount of light entering the lens. The **corpora nigra** (kōr-pōr-ah nī-grah) is the black pigmentation at the edge of the iris in equine and ruminants. The combining forms **ir/i, ir/o, irid/o,** and **irit/o** refer to the iris of the eye.

The **pupil** (pū-pihl) is the circular opening in the center of the iris. The combining forms **pupill/o** and **core/o** mean pupil. Muscles in the iris control the amount of light entering the pupil. To decrease the amount of light entering the eye, the iris muscles contract and make the opening smaller. Making the opening smaller is called **constriction;** when used in reference to pupillary constriction, it is called **miosis** (mī-ō-sihs). To increase the amount of light entering the eye, the iris muscles relax and make the opening larger. Making the opening larger is called **dilation;** when used in reference to pupillary dilation, it is called **mydriasis** (mih-drī-ah-sihs).

The **lens** is the clear, flexible, curved capsule located behind the iris and pupil. The shape of the lens is altered by the ciliary muscles. Adjusting the shape of the lens to improve near or far vision is known as accommodation. The varying shape of the lens affects the angle at which light rays enter the retina. The combining form **phac/o** means lens of the eye.

The **ciliary** (sihl-ē-ər-ē) **body** is the thickened extension of the choroid that assists in accommodation or adjustment of the lens. The **ciliary muscles,** located in the ciliary body, are muscles that adjust the shape and thickness of the lens. These adjustments make it possible for the lens to refine the focus of light rays on the retina.

Working together

Parts of the sclera and choroid sometimes are referred to together. Examples of these terms are as follows:

- **iridocorneal** (ihr-ihd-ō-kōr-nē-ahl) = pertaining to the iris and cornea.
- **uvea** (yoo-vē-ah) = term used to describe the iris, ciliary body, and choroid.
- **limbus** (lihm-buhs) = term used for the corneoscleral junction.

Retina

The **retina** (reht-ih-nah) is the nervous tissue layer of the eye that receives images. The retina is located in the posterior chamber of the eye. The combining form **retin/o** means retina.

The retina contains specialized cells called rods and cones that convert visual images to nerve impulses that travel from the eye to the brain via the optic nerve. **Rods** are specialized cells of the retina that react to light, and **cones** are specialized cells of the retina that react to color and fine detail.

The **optic** (ohp-tihck) **disk** is the region of the eye where nerve endings of the retina gather to form the optic nerve. It also is called the **blind spot** because it does not contain any rods or cones.

The **macula lutea** (mahck-yoo-lah lū-tē-ah) is a centrally depressed, clearly defined yellow area in the center of the retina. The macula lutea surrounds a small depression called the fovea centralis. The **fovea centralis** (fō-vē-ah sehn-trah-lihs) contains the greatest concentration of cones in the retina. The combining form **macul/o** means spot; **lute/o** is the combining form for yellow; the combining form for pit is **fove/o.** The term *macula* is used in other parts of the body, such as the ear and kidney.

Eye Chambers

The eye is divided into parts to make identification and location of structures easier. The **anterior segment** (also known as the aqueous chamber) is the cranial one-third of the eyeball and is divided into anterior and posterior chambers. The **anterior chamber** is the eye cavity located between the caudal surface of the cornea and the cranial surface of the iris. The **posterior chamber** is the eye cavity located between the caudal surface of the iris and the cranial surface of the lens.

The anterior and posterior chambers of the eye are filled with a watery fluid called **aqueous humor** (ah-kwē-uhs hū-mər).

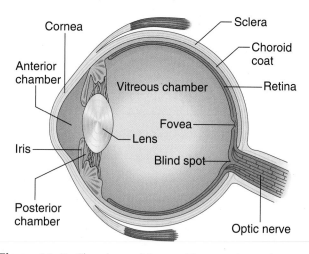

Figure 14–5 Chambers of the eye. The anterior and posterior chambers of the eye contain aqueous humor. The vitreous chamber of the eye contains vitreous humor.

Aqueous humor is the anterior segment fluid that nourishes the intraocular structures. The combining form **aque/o** means water. Humor is any clear body fluid.

The caudal two-thirds of the eyeball is called the **vitreous chamber. Vitreous humor** (viht-rē-uhs hū-mər), or vitreous, is the soft, clear, jellylike mass that fills the vitreous chamber. The combining form **vitre/o** means glassy (Figure 14–5).

VISION

Vision occurs when light rays enter the eye through the cornea, pass through the lens, and travel to the retina. The image is focused on the retina and is then transmitted to the optic nerve. Stimulations are transmitted from the optic nerve, to the optic chiasm, to the midbrain, and to the visual cortex of the occipital lobe of the cerebrum.

Accommodation (ah-kohm-ō-dā-shuhn) is the process of eye adjustments for seeing objects at various distances. These adjustments are accomplished through changes in lens shape.

Refraction (rē-frahck-shuhn) is the process of the lens bending the light rays to help them focus on the retina. Refraction also is called **focusing.**

Convergence (kohn-vər-jehns) is simultaneous inward movement of both eyes. Convergence usually occurs in an effort to maintain single binocular vision as an object approaches.

Acuity (ah-kū-ih-tē) means sharpness or acuteness, usually used in reference to vision.

TEST ME: EYES

Diagnostic tests performed on the eyes include the following:

- **conjunctival** (kohn-juhnck-tī-vahl) **scrape** = diagnostic test using an instrument to peel cells from the conjunctiva so that they can be viewed microscopically.
- **electroretinography** (ē-lehck-trō-reh-tihn-ohg-rah-fē) = procedure of recording the electrical activity of the retina. An **electroretinogram** (ē-lehck-trō-reh-tihn-ō-grahm) is the record of electrical activity of the retina; abbreviated ERG.
- **fluorescein** (fluhr-ō-sēn) **dye stain** = diagnostic test to detect corneal injury by placing dye on the surface of the cornea (Figure 14–6).
- **goniometry** (gō-nē-ah-meh-trē) = procedure to measure the drainage angle of the eye. **Gon/i** is the combining form for angle or seed.
- **menace response** (mehn-ahs rē-spohns) = diagnostic test to detect vision in which movement is made toward the animal to test whether it will see movement and try to close its eyelids.

Figure 14–6   Fluorscein dye stain is used to detect whether corneal defects are present in this dog. (Courtesy of Terri Raffel, CVT.)

Figure 14–7   An ophthalmoscope is used to examine the internal structures of the eye.

- **ophthalmoscope** (ohp-thahl-mō-skōp) = instrument used for ophthalmoscopy (Figure 14–7).
- **ophthalmoscopy** (ohp-thahl-mohs-kō-pē) = procedure used to examine the interior eye structures; may be direct or indirect (Figure 14–8).
- **palpebral** (pahl-pē-brahl) **reflex** = diagnostic test in which the eye should blink in response to touch to the medial canthus of the eye. This test is used to make neurologic assessment of cranial nerves V and VII and to assess depth of anesthesia.
- **pupillary light reflex** (pū-puhl-ār-ē līt rē-flehcks) = response of pupil to light; abbreviated PLR. When light is shown in the pupil, constriction should take place.

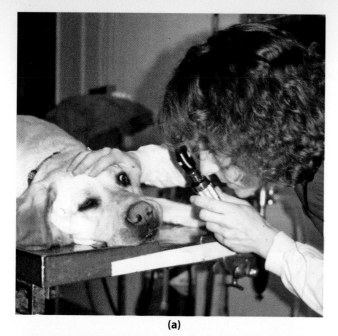

(a)

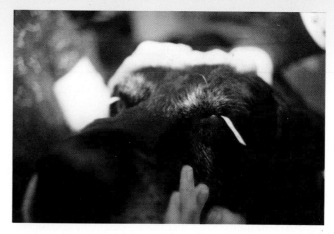

Figure 14–9 The Schirmer tear test is used to measure tear production in this dog. (Courtesy of Lodi Veterinary Hospital, SC.)

(b)

Figure 14–8 Ophthalmoscopy. Two methods of ophthalmoscopy are (a) direct and (b) indirect. (Courtesy of Lodi Veterinary Hospital, SC.)

In a description of the eyes, the abbreviations *OD*, *OS*, and *OU* are used. OD means right eye (oculus dexter), OS means left eye (oculus sinister), and OU means both eyes (oculus uterque). Similarly AD, AS, and AU are used to describe the right, left, and both ears, respectively. *A* stands for auris.

Examination of the eyes may entail treatment with chemicals. Anesthetics may be used so that tonometers can be placed on the cornea and intraocular pressure can be measured. Retinal examination may be aided with the use of cycloplegics or mydriatics. **Cycloplegics** (sī-klō-plē-jihcks) cause paralysis of the ciliary muscle that may aid in dilation of the pupil and ease the pain of ciliary muscle spasms. **Mydriatics** (mihd-rē-ah-tihcks) are agents that dilate the pupil.

- **Schirmer** (shər-mər) **tear test** = diagnostic test using a graded paper strip to measure tear production (Figure 14–9).
- **slit lamp examination** = visual testing of the cornea, lens, fluids, and membranes of the interior of the eye using a narrow beam of light.
- **tonometry** (tō-nohm-eh-trē) = procedure using an instrument to measure intraocular pressure indirectly.

Intraocular pressure is determined by the resistance of the eyeball to indentation by an applied force. Tonometry can be applanation, in which the instrument and weights are placed on the cornea to measure resistance, or pneumatic, in which a puff of air is blown against the cornea to flatten it slightly to measure resistance. A **Schiotz** (shē-ohtz) **tonometer** is an example of an applanation tonometer (Figure 14–10).

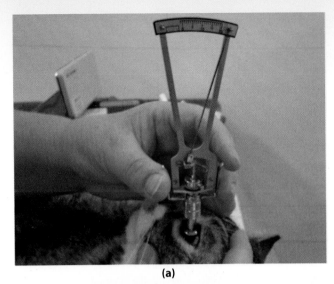

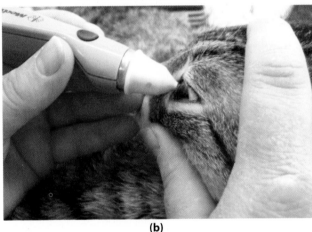

Figure 14–10 Detection of intraocular pressure by tonometry. (a) A Schiotz tonometer, an example of an applanation tonometer, measures corneal resistance to an applied force. (b) A pneumatic tonometer detects intraocular pressure by measuring corneal resistance to a puff of air. (Courtesy of Kelly Gilligan, DVM.)

Pathology: Eyes

Pathologic conditions of the eyes include the following:

- **amblyopia** (ahm-blē-ō-pē-ah) = dimness or loss of sight without detectable eye disease. The combining form **ambly/o** means dim.
- **anisocoria** (ahn-ih-sō-kō-rē-ah) = condition of unequal pupil size. **Anis/o** is the combining form meaning unequal (**an-** is not; **iso-** is equal) (Figure 14–11).
- **anophthalmos** (ahn-ohp-thahl-mōs) = without development of one or both eyes.
- **aphakia** (ah-fahk-ē-ah) = absence of the lens.
- **blepharitis** (blehf-ah-rī-tihs) = inflammation of the eyelid.
- **blepharoptosis** (blehf-ah-rō-tō-sihs) = drooping of the upper eyelid.

- **blepharospasm** (blehf-rō-spahzm) = rapid, involuntary contractions of the eyelid.
- **blindness** (blīnd-nehs) = inability to see.
- **buphthalmos** (boof-thahl-muhs) = abnormal enlargement of the eye.
- **cataract** (kaht-ah-rahckt) = cloudiness or opacity of the lens (Figure 14–12).
- **chalazion** (kah-lā-zē-ohn) = localized swelling of the eyelid resulting from the obstruction of a sebaceous gland of the eyelid.
- **conjunctivitis** (kohn-juhnck-tih-vī-tihs) = inflammation of the conjunctiva.
- **corneal ulceration** (kōr-nē-ahl uhl-sihr-ah-shuhn) = surface depression on the cornea (Figure 14–13).
- **cyclopia** (sī-klō-pē-ah) = congenital anomaly characterized by a single orbit.
- **dacryoadenitis** (dahck-rē-ō-ahd-ehn-ī-tihs) = inflammation of the lacrimal gland.
- **dacryocystitis** (dahck-rē-ō-sihs-tī-tihs) = inflammation of the lacrimal sac and abnormal tear drainage.
- **diplopia** (dih-plō-pē-ah) = double vision. **Dipl/o** is the combining form for double; **-opia** is the suffix meaning vision.
- **distichiasis** (dihs-tē-kē-ī-ah-sihs) = abnormal condition of a double row of eyelashes that usually result in conjunctival injury. **Distichia** (dihs-tēck-ē-ah) is a double row of eyelashes.

Figure 14–11 Anisocoria.

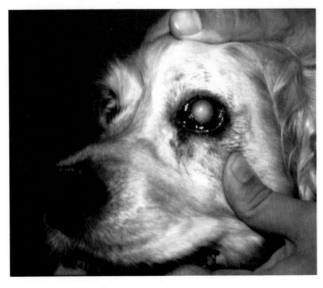

Figure 14–12 Cataract in a dog. (Courtesy of Mark Jackson, DVM, PhD.)

Figure 14–13 Corneal ulceration in a cat.

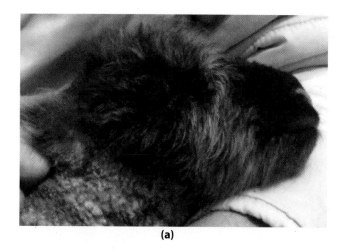

(a)

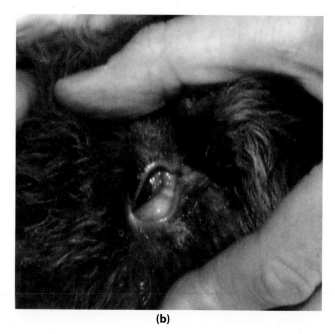

(b)

Figure 14–14 (a) Entropion in a lamb. (b) Note the severe conjunctivitis associated with entropion. (Courtesy of Ron Fabrizius, DVM, Diplomate ACT.)

- **ectropion** (ehck-trō-pē-ohn) = eversion, or turning outward, of the eyelid.
- **entropion** (ehn-trō-pē-ohn) = inversion, or turning inward, of the eyelid (Figure 14–14).
- **epiphora** (ē-pihf-ōr-ah) = excessive tear production.
- **episcleritis** (ehp-ih-sklehr-ī-tihs) = inflammation of the tissue of the cornea.
- **exophthalmos** (ehcks-ohp-thahl-mōs) = abnormal protrusion of the eyeball.
- **floaters** (flō-tərz) = particles that cast shadows on the retina suspended in the vitreous fluid; also called **vitreous floaters.**
- **glaucoma** (glaw-kō-mah) = group of disorders resulting from elevated intraocular pressure (Figure 14–15).
- **hordeolum** (hōr-dē-ō-luhm) = infection of one or more glands of the eyelid; also called a **stye** (stī).
- **hypertropia** (hī-pər-trō-pē-ah) = deviation of one eye upward.
- **hypopyon** (hī-pō-pē-ohn) = pus in the anterior chamber of the eye.
- **hypotropia** (hī-pō-trō-pē-ah) = deviation of one eye downward.
- **iritis** (ī-rī-tihs) = inflammation of the iris.
- **keratitis** (kehr-ah-tī-tihs) = inflammation of the cornea (Figure 14–16).
- **keratoconjunctivitis** (kehr-ah-tō-kohn-juhnck-tih-vī-tihs) = inflammation of the cornea and conjunctiva.
- **macular degeneration** (mahck-yoo-lahr dē-jehn-ər-ā-shuhn) = condition of central vision loss.
- **microphthalmia** (mī-krohf-thahl-mē-ah) = abnormally small eyes; also called **microphthalmos** (mī-krohp-thahl-mōs).
- **monochromatism** (mohn-ō-krō-mah-tihzm) = lack of ability to distinguish colors; also called **color blindness.**

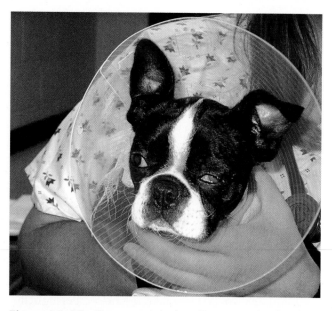

Figure 14–15 Glaucoma in a dog. (Courtesy of Kimberly Kruse Sprecher, CVT.)

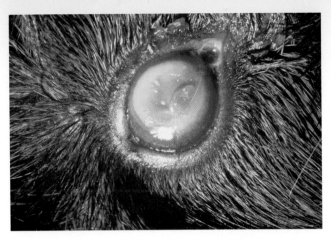

Figure 14–16 Ulcerative keratitis in a Guinea Pig.

Mono- is the prefix meaning one; **chrom/o** is the combining form for color.

- **nuclear sclerosis** (nū-klē-ahr sklehr-ō-sihs) = drying out of the lens with age.
- **nyctalopia** (nihck-tah-lō-pē-ah) = condition of inability or difficulty seeing at night; also called **night blindness.**
- **nystagmus** (nī-stahg-muhs) = involuntary, constant, rhythmic movement of the eye.
- **ophthalmoplegia** (ohp-thahl-mō-plē-jē-ah) = paralysis of eye muscles.
- **panophthalmitis** (pahn-ohp-thahl-mih-tī-tihs) = inflammation of all eye structures.
- **papilledema** (pahp-ehl-eh-dē-mah) = swelling of the optic disk.
- **photophobia** (fō-tō-fō-bē-ah) = fear or intolerance of light.
- **proptosis** (prohp-tō-sihs) = displacement of the eye from the orbit.
- **retinal detachment** (reht-ih-nahl dē-tahch-mehnt) = separation of the nervous layer of the eye from the choroid; also called **detached retina.**
- **retinopathy** (reht-ih-nohp-ah-thē) = any disorder of the retina.
- **scleral injection** (skleh-rahl ihn-jehck-shuhn) = dilation of blood vessels into the sclera.

(a) Divergent strabismus (exotropia)

(b) Convergent strabismus (esotropia)

Figure 14–17 Strabismus. (a) Divergent strabismus (exotropia). (b) Convergent strabismus (esotropia).

- **scleritis** (skleh-rī-tihs) = inflammation of the sclera.
- **strabismus** (strah-bihz-muhs) = disorder in which the eyes are not directed in a parallel manner; deviation of one or both eyes (Figure 14–17). **Convergent strabismus** (kohn-vər-jehnt strah-bihz-muhs) is deviation of the eyes toward each other; also known as **crossed eyes** and **esotropia** (ehs-ō-trō-pē-ah). The suffix **-tropia** means turning; the prefix **eso-** means inward. **Divergent strabismus** (dī-vər-jehnt strah-bihz-muhs) is deviation of the eyes away from each other; also called **exotropia** (ehck-sō-trō-pē-ah). The prefix **exo-** means outward.
- **synechia** (sī-nēk-ē-ah) = adhesion that binds the iris to an adjacent structure; plural is **synechiae** (sī-nēk-ē-ā).
- **uveitis** (yoo-vē-ī-tihs) = inflammation of the uvea.

PROCEDURES: EYES

Procedures performed on the eyes include the following:

- **blepharectomy** (blehf-ār-ehck-tō-mē) = surgical removal of all or part of the eyelid.
- **blepharoplasty** (blehf-ār-rō-plahs-tē) = surgical repair of the eyelid.
- **blepharorrhaphy** (blehf-ār-ōr-rah-fē) = suturing together of the eyelids; also called **tarsorrhaphy** (tahr-soh-rah-fē).
- **blepharotomy** (blehf-ār-ah-tō-mē) = incision of the eyelid; also called **tarsotomy** (tahr-soh-tō-mē).
- **canthectomy** (kahn-thehck-tō-mē) = surgical removal of the corner of the eyelid.
- **canthoplasty** (kahn-thō-plahs-tē) = surgical repair of the palpebral fissure.
- **canthotomy** (kahn-thoh-tō-mē) = incision into the corner of the eyelid.
- **conjunctivoplasty** (kohn-juhnck-tī-vō-plahs-tē) = surgical repair of the conjunctiva.
- **dacryocystectomy** (dahck-rē-ō-sihs-tehck-tō-mē) = surgical removal of the lacrimal sac.
- **dacryocystotomy** (dahck-rē-ō-sihs-tah-tō-mē) = incision into the lacrimal sac.
- **enucleation** (ē-nū-klē-ā-shuhn) = removal of the eyeball (Figure 14–18).
- **extracapsular extraction** (ehcks-trah-kahp-soo-lahr ehcks-trahck-shuhn) = removal of a cataract that leaves the posterior lens capsule intact.
- **goniotomy** (gō-nē-ah-tō-mē) = incision into the anterior chamber angle for treatment of glaucoma.
- **intracapsular extraction** (ihn-trah-kahp-soo-lahr ehcks-trahck-shuhn) = cataract removal that includes the surrounding capsule.
- **iridectomy** (ihr-ih-dehck-tō-mē) = surgical removal of a portion of the iris.
- **keratectomy** (kehr-ah-tehck-tō-mē) = surgical removal of part of the cornea.

Figure 14–18 Enucleation. (Courtesy of Kimberly Kruse Sprecher, CVT.)

- **keratocentesis** (kehr-ah-tō-sehn-tē-sihs) = puncture of the cornea to allow aspiration of aqueous humor.
- **keratoplasty** (kehr-ah-tō-plahs-tē) = surgical repair of the cornea (may include corneal transplant).
- **keratotomy** (kehr-ah-toh-tō-mē) = incision into the cornea.
- **lacromotomy** (lahck-rō-moh-tō-mē) = incision into the lacrimal gland or duct.
- **lensectomy** (lehn-sehck-tō-mē) = surgical removal of a lens (usually performed on cataracts).
- **tarsectomy** (tahr-sehck-tō-mē) = surgical removal of all or part of the tarsal plate of the third eyelid.
- **tarsorrhaphy** (tahr-sōr-ah-fē) = suturing together of the eyelids.

FUNCTIONS OF THE EAR

The **ear** is the sensory organ that enables hearing and helps to maintain balance. The combining forms **audit/o, aud/i,** and **ot/o** mean ear. **Acoust/o** and **acous/o** are combining forms for sound or hearing. Therefore, the term **auditory** pertains to the ear, and **acoustic** pertains to sound.

STRUCTURES OF THE EAR

The ear is divided into the outer, middle, and inner portions. Each part has its own unique structures (Figure 14–19).

Outer or External Ear

The **pinna** (pihn-ah) is the external portion of the ear that catches sound waves and transmits them to the external auditory canal. The pinna also is known as the **auricle** (awr-ihck-kuhl). The combining form **pinn/i** means external ear. The combining forms **aur/i** and **aur/o** mean external ear but are commonly used simply to mean ear.

The **external auditory canal** is the tube that transmits sound from the pinna to the tympanic membrane. The external auditory canal also is known as the **external auditory meatus.** Glands that line the external auditory canal secrete **cerumen** (seh-roo-mehn), which is commonly known as **earwax.**

Middle Ear

The middle ear begins with the **eardrum.** The medical term for eardrum is **tympanic membrane** (tihm-pahn-ihck mehm-brān). The tympanic membrane is the tissue that separates the external ear from the middle ear. When sound waves reach the tympanic membrane, it transmits sounds to the ossicles. The combining forms **tympan/o** and **myring/o** mean eardrum.

The **auditory ossicles** (aw-dih-tōr-ē ohs-ih-kulz) are three little bones of the middle ear that transmit sound vibrations. The three bones are as follows:

- **malleus** (mahl-ē-uhs) = auditory ossicle known as the hammer.
- **incus** (ihng-kuhs) = auditory ossicle known as the anvil.
- **stapes** (stā-pēz) = auditory ossicle known as the stirrup.

The **eustachian tube** (yoo-stā-shuhn toob) follows the auditory ossicles. The eustachian tube, or **auditory tube,** is the narrow duct that leads from the middle ear to the nasopharynx. It helps equalize air pressure in the middle ear with that of the atmosphere.

The **oval window** (ō-vahl wihn-dō), located at the base of the stapes, is the membrane that separates the middle and inner ear.

The **round window** (rownd wihn-dō) is the membrane that receives sound waves through fluid after they have passed through the cochlea.

The **tympanic bulla** (tihm-pahn-ihck buhl-ah) is the osseous chamber at the base of the skull. **Bulla** (buhl-ah) is the medical term for large vesicle.

Inner Ear

The inner ear contains sensory receptors for hearing and balance (Figure 14–20). The inner ear consists of three spaces in the temporal bone assembled in the **bony labyrinth** (lahb-ih-rihnth). The combining form **labyrinth/o** means maze, labyrinth, and inner ear. The bony labyrinth is filled with a waterlike fluid called **perilymph** (pehr-eh-lihmf). A membranous sac is suspended in the perilymph and follows the shape of the bony labyrinth. This membranous labyrinth is filled with a thicker fluid called **endolymph** (ehn-dō-lihmf).

The bony labyrinth is divided into three parts:

- **vestibule** = located adjacent to the oval window and between the semicircular canals and cochlea. The vestibule (and semicircular canals) contain specialized mechanoreceptors for balance and position.

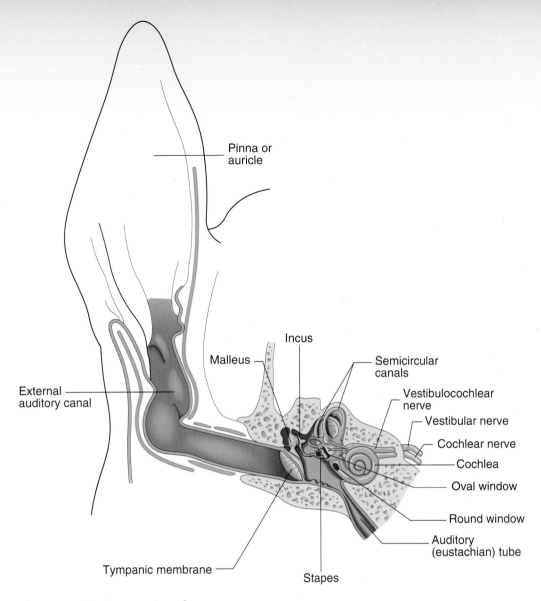

Figure 14–19 Cross section of ear structures.

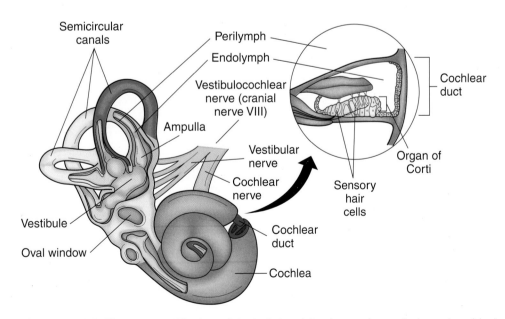

Figure 14–20 The inner ear. The bony labyrinth (semicircular canals, vestibule, and cochlea) is the hard outer wall of the entire inner ear. The membranous labyrinth is located in the bony labyrinth. The membranous labyrinth is surrounded by perilymph and is filled with endolymph.

- **semicircular canals** = located adjacent to the vestibule. The semicircular canals are oriented at right angles to each other. The three canals are the **vestibular, tympanic,** and **cochlear.** Each canal has a dilated area called the **ampulla** that contains sensory cells with hairlike extensions. These sensory cells are suspended in endolymph, and when the head moves, the hairlike extensions bend. This bending generates nerve impulses for the regulation of position.
- **cochlea** (kōck-lē-ah) = spiral-shaped passage that leads from the oval window to the inner ear. The combining form **cochle/o** means snail or spiral. The following tube and organ are located in the cochlea:
  - **cochlear duct** (kōck-lē-ahr duhckt) = membranous tube in the bony cochlea that is filled with endolymph. Endolymph vibrates when sound waves strike it.
  - **organ of Corti** (ōr-gahn ohf kōr-tē) = spiral organ of hearing located in the cochlea that receives and relays vibrations. Specialized cells in the organ of Corti generate nerve impulses when they are bent by the movement

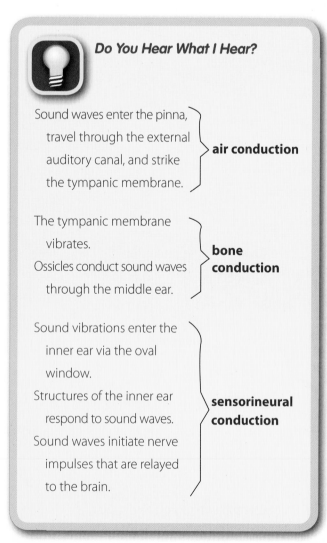

Do You Hear What I Hear?

Sound waves enter the pinna, travel through the external auditory canal, and strike the tympanic membrane. — **air conduction**

The tympanic membrane vibrates.
Ossicles conduct sound waves through the middle ear. — **bone conduction**

Sound vibrations enter the inner ear via the oval window.
Structures of the inner ear respond to sound waves.
Sound waves initiate nerve impulses that are relayed to the brain. — **sensorineural conduction**

of endolymph. Nerve impulses are relayed to the auditory nerve fibers that transmit them to the cerebral cortex.

MECHANISM OF HEARING

Sound waves enter the ear through the pinna, travel through the external auditory canal, and strike the tympanic membrane. This is called **air conduction.** As the tympanic membrane vibrates, it moves the ossicles. The ossicles conduct the sound waves through the middle ear. This is called **bone conduction.** Sound vibrations reach the inner ear via the round window. The structures of the inner ear respond to sound waves that displace fluid in the inner ear. Stimulation of hair cells in the organ of Corti initiates a nerve impulse that is transmitted to the vestibulocochlear nerve. This nerve impulse is then relayed to the brain. This is called **sensorineural conduction** (Figure 14–21).

MECHANISM OF EQUILIBRIUM

In addition to the structures listed previously, the ear also has structures that maintain equilibrium. **Equilibrium** (ē-kwihl-ih-brē-uhm) is the state of balance. The sense of equilibrium is a combination of static equilibrium (maintaining the position of the head relative to gravity) and dynamic equilibrium (maintaining balance in response to rotational or angular movement).

Static equilibrium is controlled by organs in the vestibule of the inner ear. The membranous labyrinth inside the vestibule has a **saccule** (sahck-yoo-uhl) and **utricle** (yoo-trih-kuhl), which are small, hairlike sacs. The saccule and utricle contain a **macula** (mahck-yoo-lah), an organ consisting of hair cells covered by a gelatinous mass containing **otoliths** (ō-tō-lihthz), which are small stones. When the head is in an upright position, the hairs are straight. When the head tilts or bends, the otoliths and gelatinous mass move in response to gravity (Figure 14–22). As the gelatinous mass moves, it bends some of the hairs on the receptor cells which in turn initiates an impulse that stimulates the vestibular branch of cranial nerve VIII. The impulse travels from the vestibular branch of cranial nerve VIII to the brain, which interprets the information and sends motor impulses to the appropriate muscles to maintain balance.

Dynamic equilibrium is controlled by the semicircular canals. Three semicircular canals in the inner ear lie in planes at right angles to each other. At the base of each canal is a swelling called the ampulla, which contains sensory organs called **cristae** (krihs-tā). Each crista contains sensory hair cells and a gelatinous mass. When the head turns rapidly, the semicircular canals move with the head but the endolymph within the membrane of the semicircular canals remains stationary. The fluid in the semicircular canals pushes against the gelatinous

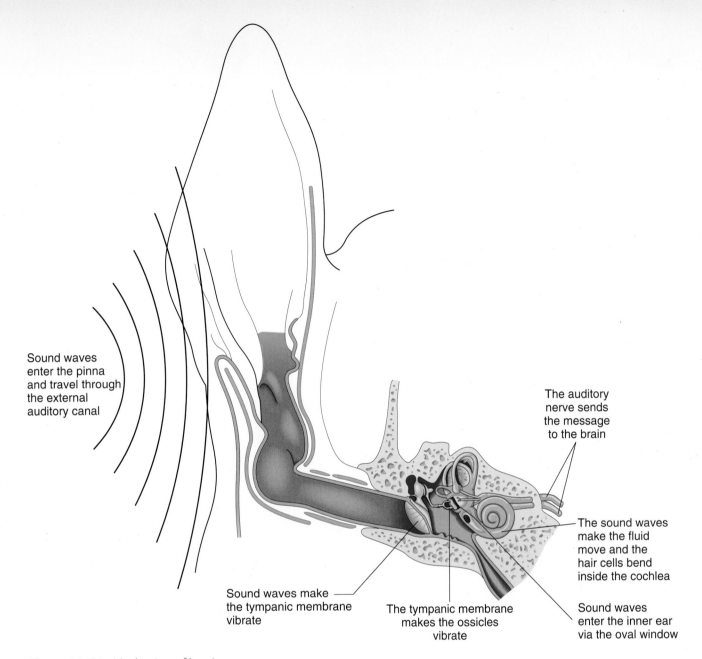

Sound waves enter the pinna and travel through the external auditory canal

The auditory nerve sends the message to the brain

The sound waves make the fluid move and the hair cells bend inside the cochlea

Sound waves make the tympanic membrane vibrate

The tympanic membrane makes the ossicles vibrate

Sound waves enter the inner ear via the oval window

Figure 14–21   Mechanism of hearing.

mass which bends some of the hairs on the hair cells. This triggers a sensory impulse to the vestibular branch of cranial nerve VIII. The impulse travels from the vestibular branch of cranial nerve VIII to the brain, which interprets the information and sends appropriate responses to maintain balance.

TEST ME: EARS

Diagnostic procedures performed on the ear include the following:

- **otoscopy** (ō-tohs-kō-pē) = procedure used to examine the ear for parasites, irritation to the ear lining, discharge, and the integrity of the tympanic membrane.

Otoscope (ō-tō-skōp) is the instrument used for otoscopy (Figure 14–23).

PATHOLOGY: EARS

Pathologic conditions of the ears include the following:

- **aural hematoma** (awr-ahl hē-mah-tō-mah) = collection or mass of blood on the outer ear. **Aural** means pertaining to the ear (external ear) (Figure 14–24).
- **deafness** (dehf-nehs) = complete or partial hearing loss.
- **myringitis** (mihr-ihn-jī-tihs) = inflammation of the eardrum.
- **otalgia** (ō-tahl-jē-ah) = ear pain.

Gravity

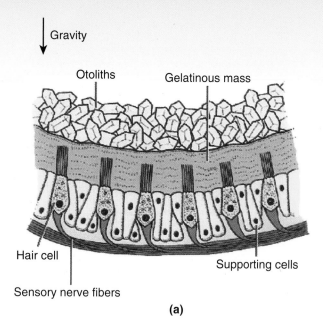

Otoliths   Gelatinous mass

Hair cell

Sensory nerve fibers

Supporting cells

(a)

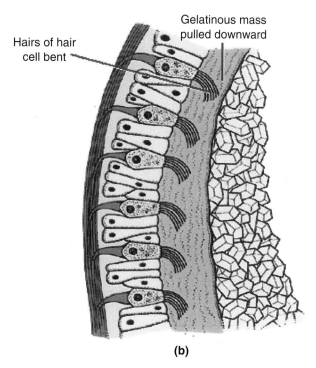

Hairs of hair
cell bent

Gelatinous mass
pulled downward

(b)

Figure 14–22 The role of the macula in maintaining static equilibrium. (a) When an animal's head is upright, gravity causes otoliths to press on hair cells, which initiate impulses to the brain. (b) When an animal's head is bent the gelatinous mass in the macula moves and bends the hair cells. This sends an impulse to the vestibular branch of cranial nerve VIII which in turn signals the brain to send motor impulses to the appropriate muscles to maintain balance.

- **otitis** (ō-tī-tihs) = inflammation of the ear; usually has a second term that describes the location: **otitis externa** (ō-tī-tihs ehcks-tər-nah) = inflammation of the outer ear; **otitis media** (ō-tī-tihs mē-dē-ah) = inflammation of

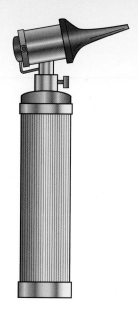

Figure 14–23 Otoscope.

Figure 14–24 Aural hematoma in a cat.

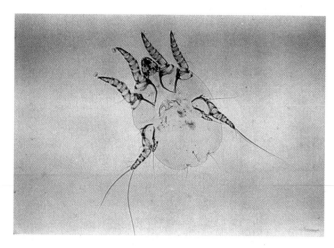

Figure 14–25 Otodectes mite.

- **panotitis** (pahn-ō-tī-tihs) = inflammation of all ear parts.
- **vertigo** (vər-tih-gō) = sense of dizziness.

Procedures: Ears

Procedures performed on the ears include the following:
- **ablation** (ah-blā-shuhn) = removal of a part (Figure 14–26).
- **myringectomy** (mihr-ihn-jehck-tō-mē) = surgical removal of all or part of the eardrum; also called tympanectomy (tihm-pahn-ehck-tō-mē).
- **otoplasty** (ō-tō-plahs-tē) = surgical repair of the ear.

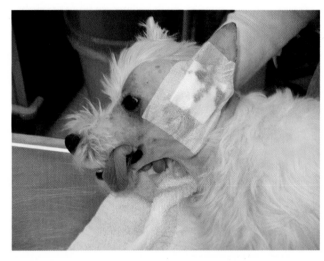

Figure 14–26 Ear ablation surgery in a dog. (Courtesy of Kimberly Kruse Sprecher, CVT.)

What is in a name?

The names of parasites may seem long and cumbersome at first glance. However, many times the name of something can tell a great deal about what it does or where it lives. The scientific name of the common ear mite of dogs and cats is *Otodectes cynotis*. From the combining form **ot/o** in the name, it can be concluded that *Otodectes* are parasites that infiltrate the ear. **Dectes** means biter; therefore, *Otodectes* are biting parasites found in the ear (Figure 14–25).

the middle ear; **otitis interna** (ō-tī-tihs ihn-tər-nah) = inflammation of the inner ear.
- **otomycosis** (ō-tō-mī-kō-sihs) = fungal infection of the ear.
- **otopathy** (ō-tohp-ah-thē) = disease of the ear.
- **otopyorrhea** (ō-tō-pī-ō-rē-ah) = pus discharge from the ear.
- **otorrhea** (ō-tō-rē-ah) = ear discharge.

Review Exercises

Multiple Choice
Choose the correct answer.

1. The state of balance is
   a. vertigo
   b. hemostasis
   c. vestibular
   d. equilibrium

2. The outer or external ear is separated from the middle ear by the
   a. oval window
   b. round window
   c. tympanic membrane
   d. pinna

3. Another term for earwax is
   a. pinna
   b. auricle
   c. cerumen
   d. corti

4. The fibrous tissue that maintains the shape of the eye is the
   a. choroid
   b. sclera
   c. white of the eye
   d. b and c

5. The term for corner of the eye is

    a. canthus

    b. cilia

    c. cerumen

    d. cornus

6. The colored muscular layer of the eye that surrounds the pupil is the

    a. cornea

    b. choroid

    c. lens

    d. iris

7. Involuntary, constant, rhythmic movement of the eyeball is called

    a. ectropion

    b. nystagmus

    c. strabismus

    d. entropion

8. A group of eye disorders resulting from increased intra-ocular pressure is

    a. ophthalmopathy

    b. glaucoma

    c. floaters

    d. hypertension

9. Opacity of the lens is called

    a. opaque

    b. turgid

    c. cataract

    d. diplopia

10. The process of the lens bending the light ray to help focus the rays on the retina is called

    a. convergence

    b. refraction

    c. humo

    d. fovea

11. The fibrous layer of clear tissue that extends over the anterior portion of the eye and is continuous with the white of the eye is the

    a. ciliary body

    b. pupil

    c. cornea

    d. iris

12. The region of the eye where nerve endings of the retina gather to form the optic nerve is called the

    a. optic disk

    b. posterior chamber

    c. sclera

    d. choroid

13. What eye structure is transparent and focuses light on the retina?

    a. conjunctiva

    b. lens

    c. vitreous humor

    d. aqueous humor

14. The meaning of palpebr/o is

    a. eyelid

    b. cornea

    c. conjunctiva

    d. eyelashes

15. Glaucoma usually is diagnosed by

    a. ophthalmoscopy

    b. slit lamp examination

    c. Schirmer tear test

    d. tonometry

16. The term for removal of a part is

    a. otalgia

    b. vertigo

    c. ablation

    d. acoustic

17. The term for pupillary dilation is

    a. miosis

    b. meiosis

    c. macula

    d. mydriasis

18. The condition of normal tear production is

    a. lacrimation

    b. nictitans

    c. vitreous

    d. acuity

19. The instrument used to visually examine the ear (the ear lining, presence or absence of discharge, and integrity of the tympanic membrane) is known as a(n)

    a. myringoscope
    b. mycoscope
    c. otoscope
    d. dectescope

20. What is the name of the spiral-shaped passage that leads from the oval window to the inner ear?

    a. organ of Corti
    b. semicircular canals
    c. ampulla
    d. cochlea

Matching

Match the ocular term in Column I with the definition in Column II.

Column I	Column II
1. _____ palpebra	a. iris, ciliary body, and choroid
2. _____ orbit	b. platelike frame within the upper and lower eyelids
3. _____ cilia	c. eyelid
4. _____ cornea	d. eyelashes
5. _____ conjunctiva	e. bony cavity of the skull that contains the eyeball
6. _____ tarsus	f. transparent anterior portion of the sclera
7. _____ uvea	g. mucous membrane that lines the underside of each eyelid

Match the auditory term in Column I with the definition in Column II.

Column I	Column II
8. _____ pinna	a. narrow duct that leads from the middle ear to the nasopharynx
9. _____ cerumen	b. auditory ossicle known as the hammer
10. _____ ampulla	c. auditory ossicle known as the anvil
11. _____ otoliths	d. auditory ossicle known as the stirrup
12. _____ tympanic bulla	e. earwax
13. _____ incus	f. external portion of the ear
14. _____ stapes	g. osseous chamber at the base of the skull
15. _____ malleus	h. tube that transmits sound from the pinna to the eardrum
16. _____ eustachian tube	i. dilated area in the semicircular canals
17. _____ external auditory canal	j. small stones in the saccule and utricle

Fill in the Blanks

1. Lacrim/o and dacry/o mean _____ .

2. Tympan/o and myring/o mean _____ .

3. Pinn/i, aur/i, and aur/o mean _____ .

4. Opt/i, opt/o, optic/o, ocul/o, and ophthalm/o mean _____ .

5. Acoust/o and acous/o mean _____ .

Spelling

Circle the term that is spelled correctly.

1. double row of eyelashes: dystichiasys   distechiasis   distichiasis

2. groups of disorders resulting from elevated intraocular pressure: glaukoma   glawcoma   glaucoma

3. inflammation of the iris, ciliary body, and choroid: uveitis   uveititis   uveytis

4. blocking the passage of light: opague   opaque   opaiue

5. narrow duct that leads from the middle ear to the nasopharynx: eustayshian tube   eustaichian tube   eustachian tube

CROSSWORD PUZZLES

Eye Terms Puzzle

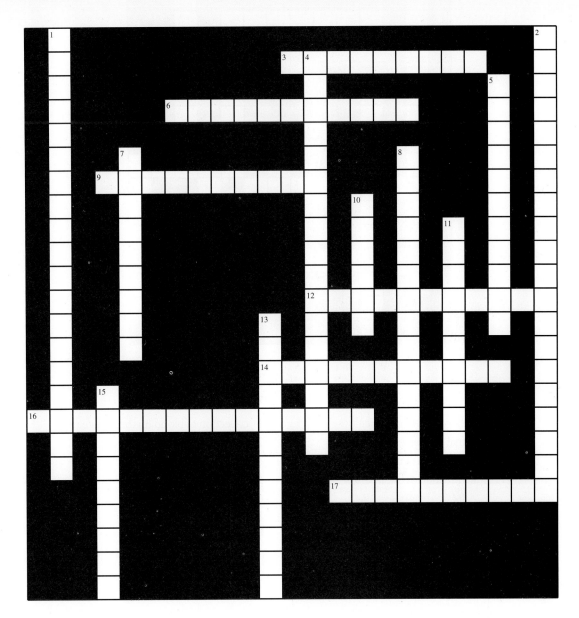

Across

3 eversion of the eyelid
6 inflammation of the eyelid
9 condition of unequal pupil size
12 removal of the eyeball
14 any disorder of the retina
16 incision into the lacrimal sac
17 disorder in which the eyes are not directed in a parallel manner

Down

1 another term for third eyelid
2 inflammation of the cornea and mucous membranes that line the eyelid
4 surface depression on the cornea
5 intolerance of light
7 inversion of the eyelid
8 paralysis of the eye muscles
10 corneoscleral junction
11 incision into the cornea
13 suturing together of the eyelids
15 displacement of the eye from the orbit

Ear Terms Puzzle

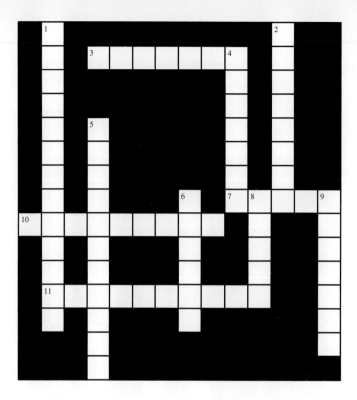

Across

3 another term for hammer
7 external portion of the ear
10 another term for inner ear
11 inflammation of the eardrum

Down

1 collection of blood on the outer ear
2 removal of a part by cutting
4 another term for stapes
5 pus discharge from the ear
6 inflammation of the ear
8 another term for anvil
9 another term for pinna

CASE STUDIES

Define the underlined terms in each case study.

A 6-yr-old DSH M/N cat was presented for inappetence and <u>blepharospasm</u>. PE revealed T = 103.4°F, HR = 200 bpm, RR = 40 breaths/min, mm = pink and dry, CRT = 1.5 sec. The <u>conjunctiva</u> was reddened, and the <u>sclera</u> was infected. Heart and lungs auscultated normally. Abdominal palpation was normal. Oral examination was normal. <u>Ocular</u> exam: <u>Anterior chamber</u> was cloudy. <u>Iris</u> appeared normal. <u>Tonometer</u> readings were <u>OS</u> 7 <u>mm Hg</u>, <u>OD</u> 10 mm Hg. Dx was <u>uveitis</u>—anterior chamber. Possible causes include infections (FIP, FeLV, toxoplasmosis), immune mediated, systemic disease, and trauma. Further diagnostic tests are being pursued by the veterinarian.

1. blepharospasm   _____

2. conjunctiva   _____

3. sclera   _____

4. ocular   _____

5. anterior chamber   _____

6. iris   _____

7. tonometer   _____

8. OS   _____

9. mm Hg   _____

10. OD   _____

11. uveitis   _____

An 8-mo-old M black Labrador retriever was presented to the clinic for <u>ocular</u> discharge and rubbing at the eyes. Upon the animal's presentation to the clinic, the vital signs were normal, attitude was normal for a puppy, and <u>bilateral mucopurulent ocular discharge</u> was noted. <u>Ophthalmic</u> examination revealed normal <u>Schirmer tear test</u> values, a normal-looking <u>retina</u>, and no stain retention via <u>fluorescence staining</u>. Upon examination of the eyelids, it was noted that the dog had <u>blepharospasm</u> and <u>entropion</u> was noted. The entropion probably was the cause of the eye infection because the eyelashes were brushing against the <u>cornea</u>. <u>Blepharoplasty</u> was recommended to this owner. <u>Topical</u> antibiotics were dispensed pending surgery.

12. ocular   _____

13. bilateral mucopurulent ocular discharge   _____

14. ophthalmic   _____

15. Schirmer tear test   _____

16. retina   _____

17. fluorescence staining   _____

18. blepharospasm   _____

19. entropion   _____

20. cornea   _____

21. blepharoplasty   _____

22. topical   _____

Buddy, a 3-yr-old M/N springer spaniel, was presented to the clinic with <u>recurrent</u> ear problems. Previously diagnosed <u>bilateral</u> ear problems responded well to treatment. Buddy had been seen 2 wk earlier for ear problems, and <u>otitis externa</u> was diagnosed. Ear <u>cytology</u> revealed that Buddy had a severe yeast infection of both ears, and an antifungal drug was prescribed. On today's examination, <u>mucopurulent</u> discharge was noted <u>AU</u>. <u>Otoscopic</u> examination revealed that <u>AD</u> was <u>hyperemic</u> and <u>hyperkeratotic</u>, and the ear was so swollen that the <u>tympanic membrane</u> could not be visualized. <u>AS</u> was hyperemic and hyperkeratotic, and the tympanic membrane was intact. Ear cytology revealed yeast, and the dog was getting worse. Oral antifungal drugs and anti-inflammatory drugs were prescribed. The owner was advised that if the problem did not resolve, <u>ear ablation</u> surgery may be warranted.

23. recurrent _____

24. bilateral _____

25. otitis externa _____

26. cytology _____

27. mucopurulent _____

28. AU _____

29. otoscopic _____

30. AD _____

31. hyperemic _____

32. hyperkeratotic _____

33. tympanic membrane _____

34. AS _____

35. ear ablation _____

CHAPTER 15

FEED AND PROTECT ME

Objectives

Upon completion of this chapter, the reader should be able to

- Identify and describe the major structures and functions of the hematologic, lymphatic, and immune systems
- Recognize, define, spell, and pronounce the major terms related to diagnosis, pathology, and treatment of the hematologic, immune, and lymphatic systems
- Recognize, define, spell, and pronounce terms related to oncology

HEMATOLOGIC SYSTEM

The hematologic system encompasses the production of blood and the transport of blood throughout the body.

Functions of Blood

Blood supplies body tissues with oxygen, nutrients, and various chemicals. Blood transports waste products to various organs for removal from the body. Blood cells also play important roles in the immune and endocrine systems.

Structures of Blood

Blood is composed of 55 percent liquid plasma and 45 percent formed elements. Formed elements include red blood cells, white blood cells, and clotting cells. The combining forms for blood are **hem/o** and **hemat/o.**

Blood is formed in the bone marrow. **Hematopoiesis** (hē-mah-tō-poy-ē-sihs) is the medical term for formation of blood. The suffix **-poiesis** means formation. The components of blood can be separated clinically and examined microscopically. A blood sample is collected with a needle and syringe. **Drawing blood** is a common term for collecting a blood sample. Blood can be collected in a tube that has an **anticoagulant** (ahn-tih-kō-ahg-yoo-lahnt). An anticoagulant is a substance that prevents clotting of blood. **EDTA** (ethylenediaminetetraacetic acid) and **heparin** (hehp-ahr-ihn) are types of anticoagulants found in blood tubes and

are used clinically as drugs. **Coagulation** (kō-ahg-yoo-lā-shuhn) is the process of clotting. Sometimes coagulation of blood is desired after blood is placed in a collection tube. When blood coagulates, a layer of leukocytes and thrombocytes forms, which appears at the interface of the erythrocytes and plasma after blood is centrifuged. This layer that appears at the interface of the erythrocytes and plasma is called the **buffy coat** (buhf-ē kōt) (Figure 15–1).

What is normal?

Blood is about 55 percent liquid and 45 percent cellular material, but what is the normal blood volume of an animal? Approximately 9 percent of an animal's body weight is blood.

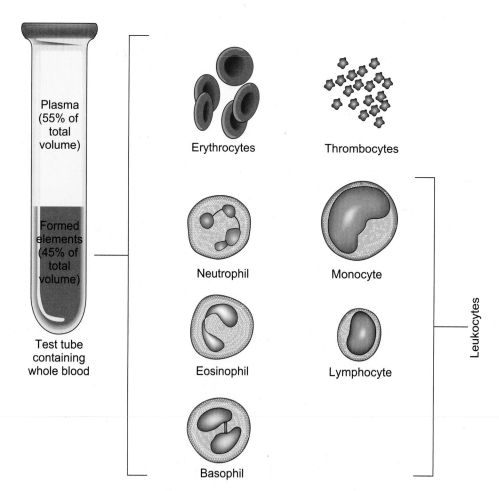

Figure 15–1 Blood components.

Liquid

The liquid portion of blood consists of the following:

- **Serum** (sē-ruhm) is the liquid portion of blood with clotting proteins removed (Figure 15–2).
- **Plasma** (plahz-mah) is the straw-colored fluid portion of blood that transports nutrients, hormones, and waste products. Plasma contains clotting proteins.
- Clotting proteins found in plasma include **fibrinogen** (fih-brihn-ō-jehn) and **prothrombin** (prō-throhm-bihn). The combining form **fibrin/o** means fibrin or threads of a clot, the prefix **pro-** means before, and the combining form **thromb/o** means clot. **Albumin** (ahl-byoo-mihn) is another example of a plasma protein.
- Fats also circulate in plasma. **Cholesterol** (kō-lehs-tər-ohl) and **triglyceride** (trī-glihs-ər-īd) are types of lipids that circulate in blood.

Formed Elements

ERYTHROCYTES

An **erythrocyte** (eh-rihth-rō-sīt) is a mature red blood cell (oxygen-carrying cell) and is abbreviated RBC. The combining form **erythr/o** means red, and the suffix **-cyte** means cell. Erythrocytes have a biconcave disc shape and contain

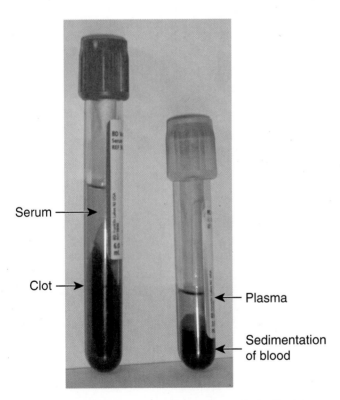

Serum

Clot

Plasma

Sedimentation of blood

Figure 15–2 Serum versus plasma. Serum is the liquid portion of blood seen in the sample collected in the red-topped tube without anticoagulant (left), and plasma is the liquid portion of blood seen in the sample collected in the purple-topped tube containing anticoagulant (right). Although serum and plasma look similar, plasma contains clotting proteins while serum does not.

hemoglobin (hē-mō-glō-bihn), a blood protein that transports oxygen. **Heme** (hēm) is the nonprotein, iron-containing portion of hemoglobin. The central pale area of an RBC that represents the thinnest part of the biconcave disc is the area of **central pallor** (sehn-trahl pahl-ohr). RBCs that have more central pallor than normal have decreased hemoglobin concentration in their cells. These RBCs are described as hypochromic and appear pale. It would seem logical then that RBCs that have less central pallor than normal would have increased hemoglobin concentration in their cells and would be described as hyperchromic; however, that term is misleading. RBCs can hold only so much hemoglobin before their cell membrane is altered. This alteration of the RBC cell membrane makes the membrane shrink so that the amount of hemoglobin is contained in a smaller space, making the cells appear darker (bluer) in color. These cells are more commonly called polychromatic. Polychromatic RBCs also may be young erythrocytes that have been released early from the bone marrow. They are typically larger and bluer in color than mature erythrocytes due to the presence of organelles such as ribosomes and mitochondria that are still present in immature cells.

Erythrocytes are produced in the bone marrow. The combining form for bone marrow (and spinal cord) is **myel/o.** Erythrocytes vary in appearance from species to species.

A **reticulocyte** (reh-tihck-yoo-lō-sīt) is an immature, non-nucleated erythrocyte characterized by polychromasia (Wright's stain) or a meshlike pattern of threads (new methylene blue stain).

When RBCs are no longer useful, they are destroyed by macrophages. A **macrophage** (mahck-rō-fahj or mahck-rō-fāj) is a large cell that destroys by eating (engulfing). The combining form **macr/o** means large, and the suffix **-phage** means eating. A **phagocyte** (fā-gō-sīt) is "a cell that eats." The formal definition of a phagocyte is a leukocyte that ingests foreign material.

Hematology (hē-mah-tah-lō-jē) is the study of blood. When blood cells are studied, it is important to note the cell's morphology. **Morphology** (mōr-fah-lō-jē) is the study of form (Table 15–1 and Figure 15–3).

LEUKOCYTES

A **leukocyte** (loo-kō-sīt) is a white blood cell and is abbreviated WBC. The combining form **leuk/o** means white. Leukocytes are produced in the bone marrow (and other places) and function primarily in fighting disease in the body. **Leukocytopoiesis** (loo-kō-sīt-ō-poy-ē-sihs) is the production of white blood cells. The production of leukocytes is also called **leucopoiesis** (loo-kō-poy-ē-sihs).

A **granulocyte** (grahn-yoo-lō-sīt) is a cell that contains prominent grainlike structures in its cytoplasm; an **agranulocyte** (ā-grahn-yoo-lō-sīt) is a cell that does not contain prominent grainlike structures in its cytoplasm. Agranulocytes often are referred to as other leukocytes due to the fact that they have less prominent cytoplasmic granules and are not truly agranulocytic. Types of WBCs are summarized in Table 15–2 and Figure 15–3.

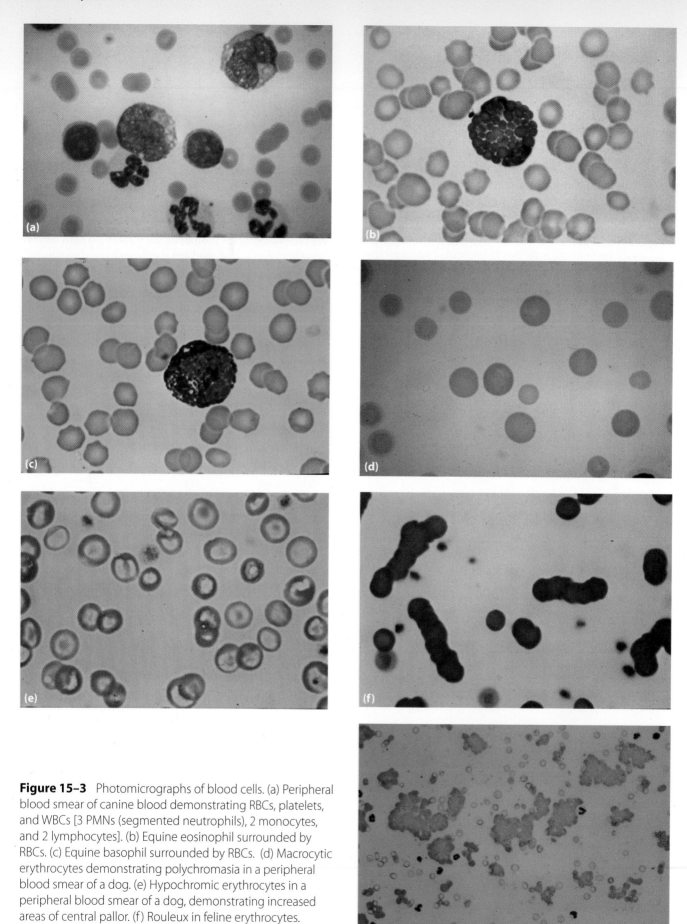

Figure 15–3 Photomicrographs of blood cells. (a) Peripheral blood smear of canine blood demonstrating RBCs, platelets, and WBCs [3 PMNs (segmented neutrophils), 2 monocytes, and 2 lymphocytes]. (b) Equine eosinophil surrounded by RBCs. (c) Equine basophil surrounded by RBCs. (d) Macrocytic erythrocytes demonstrating polychromasia in a peripheral blood smear of a dog. (e) Hypochromic erythrocytes in a peripheral blood smear of a dog, demonstrating increased areas of central pallor. (f) Rouleux in feline erythrocytes. (g) Agglutination in canine erythrocytes.

Table 15–1 Terms Used to Describe Erythrocytes

Term	Pronunciation	Description
normocytic	(nohr-mō-siht-ihck)	normal cell size
macrocytic	(mahck-rō-siht-ihck)	larger-than-normal cell size
microcytic	(mī-krō-siht-ihck)	smaller-than-normal cell size
poikilocytosis	(poy-kē-loh-sī-tō-sihs)	condition of irregular cells; clinically means varied shapes of erythrocytes
anisocytosis	(ahn-eh-sō-sī-tō-sihs)	condition of unequal cell size; excessive variation in RBC size
normochromic	(nohr-mō-krō-mihck)	normal RBC color (has area of central pallor of normal size)
hypochromic	(hī-pō-krō-mihck)	less-than-normal color (has enlarged area of central pallor)
hyperchromic	(hī-pər-krō-mihck)	more-than-normal color (term is not commonly used)
polychromasia	(poh-lē-krō-mah-zē-ah)	"condition of many colors" that appears as an overall blue tint of Wright-stained RBCs due to shrinking of the cell membrane; indicates slight immaturity of the erythrocyte
rouleaux	(roo-low)	RBCs that are arranged like stacks of coins on the peripheral blood smear; may be artifact or may be normal in species such as cats and horses
agglutination	(ah-gloo-tih-nā-shuhn)	clumping of RBCs due to the presence of an antibody directed against RBC surface antigens that forms a latticework that links them together

Table 15–2 Major Groups of Leukocytes

Type of Leukocyte	Pronunciation	Description
lymphocyte	(lihm-fō-sīt)	class of "agranulocytic" leukocyte that contains a diverse set of cells including those that can directly attack specific pathogens or produce antibodies
monocyte	(mohn-ō-sīt)	class of "agranulocytic" leukocyte that has a phagocytic function and participates in the inflammatory response
neutrophil	(nū-trō-fihl)	class of granulocytic leukocyte that has a phagocytic function (mainly against bacteria); also called **polymorphonuclear** (poh-lē-mōr-fō-nū-klē-ahr) **leukocyte,** or PMN; *polymorphonuclear* means multishaped nucleus
eosinophil	(ē-ō-sihn-ō-fihl)	class of granulocytic leukocyte that detoxifies allergens and controls parasitic infections by damaging parasite membranes
basophil	(bā-sō-fihl)	class of granulocytic leukocyte that promotes the inflammatory response and contains histamine in its cytoplasmic granules

A **band cell** (bahnd sehl) or banded neutrophil is an immature polymorphonuclear leukocyte. A mature neutrophil is a segmented neutrophil.

Some cells are described as basophilic or eosinophilic. **Basophilic** (bā-sō-fihl-ihck) means stained readily with basic, or blue, dyes in many commonly used stains such as hematoxylin and eosin (H&E), Giemsa, and Wright's. **Eosinophilic** (ē-ō-sihn-ō-fihl-ihck) means stained readily with acidic, or pink, dyes in many commonly used stains such as H&E, Giemsa, and Wright's.

Clotting Cells

Clotting cells also are produced in the bone marrow and play a part in the clotting of blood. A **thrombocyte** (throhm-bō-sīt) is a nucleated clotting cell, and a **platelet** (plāt-leht) is an anucleated clotting cell. Occasionally, those terms are used interchangeably; however, they have different meanings.

A **megakaryocyte** (mehg-ah-kahr-ē-ō-sīt) is a large, nucleated cell found in the bone marrow from which platelets are formed.

TEST ME: HEMATOLOGIC SYSTEM

Diagnostic tests performed on the hematologic system include the following:

- **blood smear** (smēr) = blood specimen for microscopic examination in which blood is spread thinly across a microscope slide and typically stained (Figure 15–4).
- **bone marrow biopsy** (bōn mahr-ō bī-ohp-sē) = sample of bone marrow obtained by needle aspiration for examination of cells. Bone marrow samples are taken from long bones, ribs, or the sternum.
- **laboratory tests,** which are covered in Chapter 16.

PATHOLOGY: HEMATOLOGIC SYSTEM

Pathologic conditions of the hematologic system include the following:

- **anemia** (ah-nē-mē-ah) = blood condition of less-than-normal levels of red blood cells and/or hemoglobin.
- **basopenia** (bā-sō-pē-nē-ah) = deficiency in the number of basophils in the blood; **-penia** is a suffix that indicates a decrease in a particular type of cell.
- **basophilia** (bā-sō-fihl-ē-ah) elevation in the number of basophils in the blood; **-philia** is a suffix that indicates an increase in a particular type of cell.
- **dyscrasia** (dihs-krā-zē-ah) = any abnormal condition of the blood.
- **edema** (eh-dē-mah) = accumulation of fluid in the intercellular space (Figure 15–5). **Edemateous** (eh-deh-mah-tuhs) is the adjective form of edema.
- **eosinopenia** (ē-ō-sihn-ō-pē-nē-ah) = deficiency in the number of eosinophils in the blood.
- **eosinophilia** (ē-ō-sihn-ō-fihl-ē-ah) = elevation in the number of eosinophils in the blood.
- **erythrocytosis** (eh-rihth-rō-sī-tō-sihs) = abnormal increase in red blood cells. The suffix **-cytosis** means condition of cell but implies elevated cell numbers.
- **exudate** (ehcks-yoo-dāt) = material that has escaped from blood vessels and is high in protein, cells, or solid materials derived from cells.
- **hemolytic** (hē-mō-liht-ihck) = removing and destroying red blood cells. Hemolytic anemia is excessive RBC destruction, resulting in lower-than-normal levels of RBCs. Hemolytic serum contains red blood cell components that are released when erythrocytes are damaged due to a variety of causes (such as improper specimen processing, collection, or transport) and appears pink or blood-tinged (Figure 15–6). **Hemolysis** (hē-mohl-eh-sihs) is the breaking down of blood cells. **Lysis** (lī-sihs) is the medical term for destruction or breakdown.

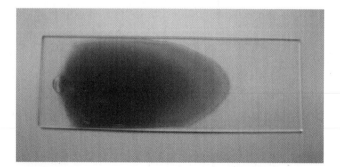

Figure 15–4   A blood smear is used to examine blood cells under a microscope.

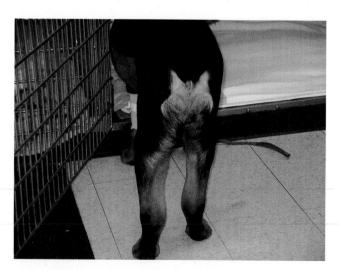

Figure 15–5   Edema in the hind legs of a dog. (Courtesy of Kimberly Kruse Sprecher, CVT.)

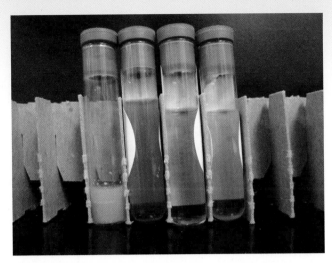

Figure 15–6 Lipemic serum contains an excessive amount of fat and appears white (tube on the left), while hemolytic serum contains red blood cell components that are released when erythrocytes are damaged due to a variety of causes (such as improper specimen processing, collection, or transport) and appears pink or blood-tinged (second tube from the left). Icteric serum is yellow-tinged due to increased bilirubin levels (third tube from the left), while normal serum is clear (tube on the right).

- **hemophilia** (hē-mō-fihl-ē-ah) = hereditary condition of deficient blood coagulation.
- **hemorrhage** (hehm-ōr-ihdj) = loss of blood (usually in a short period of time). **Hemostasis** (hē-mō-stā-sihs) is the act of controlling blood or bleeding. A **hemostat** (hē-mō-staht) is an instrument used to control bleeding.
- **hyperalbuminemia** (hī-par-ahl-byoo-mih-nē-mē-ah) = blood condition of abnormally high albumin levels.
- **hyperemia** (hī-par-ē-mē-ah) = excess blood in a part; engorgement.
- **hyperlipidemia** (hī-par-lihp-ih-dē-mē-ah) = blood condition of abnormally high fat levels; more accurately means abnormally high fat levels caused by fat metabolism. **Lip/o** is the combining form for fat.
- **icteric** (ihck-tər-ihck) **serum** = serum that has yellow pigmentation that is suggestive of hyperbilirubinemia (Figure 15–6). Hyperbilirubinemia may be caused by increased erythrocyte damage, liver disease, and disruption of bile flow.
- **left shift** (lehft shihft) = common term for an alteration in the distribution of leukocytes in which there are increases in band forms, usually in response to bacterial infection.
- **leukemia** (loo-kē-mē-ah) = elevation in the number of malignant white blood cells. Leukemias may be classified according to the concentration of neoplastic cells that are circulating in blood (such as granulocytic leukemia and lymphocytic leukemia). Leukemias are further classified as either acute or chronic based on the majority or degree of neoplastic cell differentiation and clinical course.

- **leukocytosis** (loo-kō-sī-tō-sihs) = elevation in the number of white blood cells.
- **leukopenia** (loo-kō-pē-nē-ah) = deficiency of white blood cells; sometimes called leukocytopenia (loo-kō-sīt-ō-pē-nē-ah).
- **lipemia** (lī-pē-mē-ah) = excessive amount of fats in the blood.
- **lipemic serum** (lī-pē-mihck sē-ruhm) = fats from blood that have settled in the serum. Clinically the serum appears cloudy and white (Figure 15–6).
- **lymphocytosis** (lihm-fō-sī-tō-sihs) = elevated numbers of lymphocytes in the blood.
- **lymphopenia** (lihm-fō-pē-nē-ah) = deficiency of lymphocytes in the blood.
- **monocytopenia** (moh-nō-sīt-ō-pē-nē-ah) = deficiency of monocytes in the blood.
- **monocytosis** (moh-nō-sī-tō-sihs) = elevated numbers of monocytes in the blood.
- **myelodysplasia** (mī-ehl-ō-dihs-plā-zē-ah) = hematologic disorder characterized clinically and morphologically by ineffective hematopoiesis that results in some form of cytopenia such as anemia, neutropenia, and/or thrombocytopenia.
- **neutropenia** (nū-trō-pē-nē-ah) = deficiency in the number of neutrophils in the blood.
- **neutrophilia** (nū-trō-fihl-ē-ah) = elevation in the number of neutrophils in the blood.
- **pancytopenia** (pahn-sīt-ō-pē-nē-ah) = deficiency of all types of blood cells.
- **phagocytosis** (fahg-ō-sī-tō-sihs) = condition of engulfing or eating cells.
- **polycythemia** (poh-lē-sī-thē-mē-ah) = condition of many cells; clinically means excessive erythrocytes.
- **septicemia** (sehp-tih-sē-mē-ah) = blood condition in which pathogenic microorganisms (bacteria) and their toxins are present. The suffix **-emia** means blood condition. **Pathogenic** (pahth-ō-jehn-ihck) means producing disease. **Bacteremia** (bahck-tər-ē-mē-ah) is the blood condition in which bacteria are present.
- **thrombocytopenia** (throhm-bō-sīt-ō-pē-nē-ah) = abnormal decrease in the number of clotting cells. The suffix **-penia** means less than normal or deficiency.
- **thrombocytosis** (throhm-bō-sī-tō-sihs) = elevation in the number of clotting cells.
- **transudate** (trahnz-yoo-dāt) = material that has passed through a membrane and is high in fluidity and low in protein, cells, or solid materials derived from cells.

LYMPHATIC SYSTEM

Functions of the Lymphatic System

The lymphatic system functions as part of the immune system, returns excess lymph to the blood, and absorbs fats and fat-soluble vitamins from the digestive system and transports them to cells. The combining form **lymph/o** means lymph fluid, lymph vessels, and lymph nodes. **Lymphoid** (lihm-foyd) pertains to lymph or tissue of the lymphatic system.

Structures of the Lymphatic System

The major structures of the lymphatic system include lymph vessels, lymph nodes, lymph fluid, tonsils, spleen, thymus, and lymphocytes.

Lymph Fluid

Interstitial fluid (ihn-tər-stihsh-ahl flū-ihd) is the clear, colorless tissue fluid that leaves the capillaries and flows in the spaces between the cells of a tissue or an organ. Interstitial fluid functions to bathe and nourish the cells. *Interstitial* pertains to the spaces in a tissue or an organ. **Lymph** (lihmf) is formed when interstitial fluid moves into the capillaries of the lymphatic system. Lymph brings nutrients and hormones to cells and carries waste products from tissue back to the bloodstream (Figure 15–7).

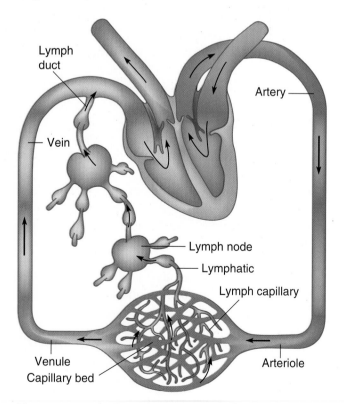

Figure 15–7 Fluids that leave circulation through the capillaries are returned to venous circulation by the lymphatic system.

Lymph Vessels

Lymph is carried from the tissue space via thin-walled tubes called **lymph capillaries.** Lymph capillaries take the lymph to the lymphatic vessels. **Lymphatic vessels** (lihm-fah-tihck vehs-uhlz) are similar to veins in that they have valves to prevent the backflow of lymph. Lymph always travels toward the thoracic cavity. In the thoracic cavity, the right lymphatic duct and thoracic duct empty lymph into veins. Lymph ducts release lymph (and whatever is in lymph) into venous blood, where it is quickly passed to the lungs and then throughout the body. This mechanism can spread infection, cancers, and other diseases. The **cisterna chyli** (sihs-tər-nah kī-lē) is the origin of the thoracic duct and saclike structure for the lymph collection.

Lacteals (lahck-tē-ahls), located in the small intestine, are specialized lymph vessels that transport fats and fat-soluble vitamins (Figure 15–8).

Lymph Nodes

Lymph nodes (lihmf nōdz) are small bean-shaped structures that filter lymph and store B and T lymphocytes (Figure 15–9a). The primary function of lymph nodes is to filter lymph to remove harmful substances such as bacteria and viruses. Because cells are destroyed in lymph nodes, swollen lymph nodes often are an indication of disease. Lymph nodes are described according to their location: Mandibular lymph nodes are located near the mandible, parotid lymph nodes are located near the ear (**para-** means near, and **otos** is Greek for ear), mesenteric lymph nodes are located in the mesentery (Figure 15–9b), etc.

Tonsils

The **tonsils** (tohn-sahlz) are masses of lymphatic tissue that protect the nose and cranial (upper) throat. Tonsils are described according to their location: Lingual tonsils are located near the tongue, palatine tonsils are located near the palate or roof of the mouth, and pharyngeal tonsils are located near the throat. **Tonsill/o** is the combining form for tonsil.

Spleen

The **spleen** (splēn) is an organ located in the cranial abdomen that filters foreign material from the blood, stores red blood cells, and maintains an appropriate balance of cells and plasma in the blood (Figure 15–10). The spleen also is a secondary lymphoid tissue (as opposed to the primary lymphoid tissues which are the thymus and bone marrow) where mature, differentiated B and T lymphocytes reside and wait for antigenic stimulation. Upon antigenic stimulation, these lymphocytes proliferate. In addition to lymphocytes, macrophages also are found in the spleen. Macrophages line the sinusoids of the spleen (called sinusoidal lining cells) where

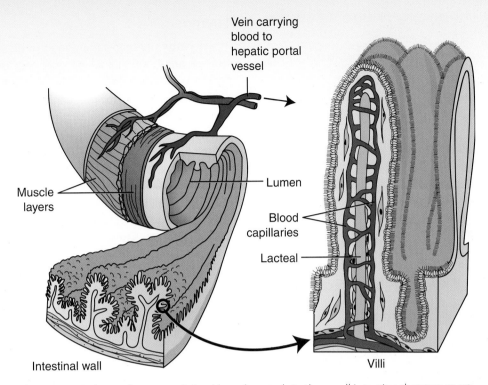

Figure 15–8 Lacteals are specialized lymph vessels in the small intestine that transport fats and fat-soluble vitamins.

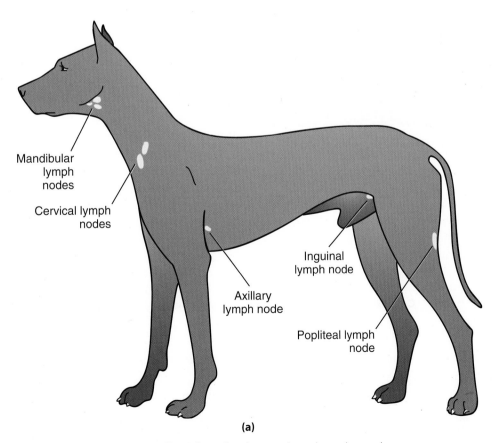

(a)

Figure 15–9 (a) Location of lymph nodes that can be palpated on a dog.

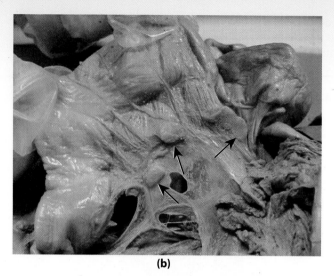

(b)

Figure 15–9 (b) Mesenteric lymph nodes are located in the mesentery.

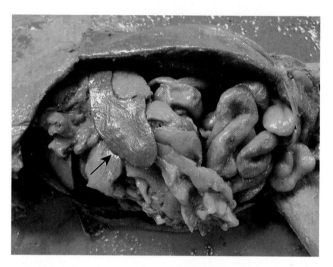

Figure 15–10 The spleen is a mass of lymphatic tissue located in the cranial abdomen.

they will phagocytize foreign material, break it down, and present antigenic parts on their surface for the helper T lymphocytes to recognize and initiate an immune response. The combining form **splen/o** means spleen.

Thymus

The **thymus** (thī-muhs) is a gland that has an immunologic function and is found predominantly in young animals. The thymus is located near midline in the cranioventral portion of the thoracic cavity. The immunologic role of the thymus is development of T cells.

Some of the lymphocytes formed in the bone marrow migrate to the thymus, where they multiply and mature into T cells. T cells play an important role in the immune response. The thymus also has an endocrine function. The combining form for thymus is **thym/o.**

IMMUNE SYSTEM

The immune system functions to protect the body from harmful substances. The term **immunity** comes from the Latin term *immunitas,* which means exemption. Immunity was used to imply that an animal was exempt from or protected against foreign substances. The combining form **immun/o** means protected. **Immunology** (ihm-yoo-nohl-ō-jē) is the study of the immune system.

The immune system is not contained in one set of organs or in one area. Many structures from different body systems aid in protecting the body. The lymphatic system, respiratory tract, gastrointestinal tract, integumentary system, and others work together to prevent the body from being harmed from foreign invaders.

Specialized Cells

Some cells are specialized for immune reactions. The **lymphocyte** is a type of white blood cell that is involved in the immune response and works against specific antigens. Lymphocytes are formed in the bone marrow and mature in lymphatic tissue throughout the body, such as the spleen, thymus, and bone marrow. There are two subpopulations of lymphocytes—the T lymphocytes, which are responsible for cell-mediated immunity, and the B lymphocytes, which are responsible for humoral immunity (Figure 15–11). **B lymphocytes** are produced and mature in the bone marrow and are responsible for antibody-mediated or humoral immunity.

Cell-Mediated and Humoral (Antibody-Mediated) Immunity	
Cell-Mediated Immunity	**Humoral (Antibody-Mediated) Immunity**
T cells are responsible for cell-mediated immunity.	**B cells** are responsible for humoral immunity.
T cells directly attack the invading antigen.	Differentiated B cells produce antibodies that react with the antigen or substances produced by the antigen.
Cell-mediated immunity is most effective against viruses that infect body cells, cancer cells, and foreign tissue cells.	Humoral immunity is most effective against bacteria, viruses that are outside body cells, and toxins. It is also involved in allergic reactions.

Figure 15–11 Comparison of cell-mediated and antibody-mediated immunity.

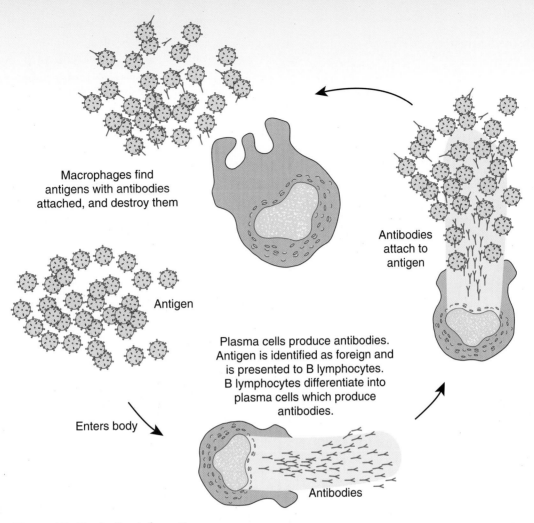

Macrophages find antigens with antibodies attached, and destroy them

Antibodies attach to antigen

Antigen

Plasma cells produce antibodies. Antigen is identified as foreign and is presented to B lymphocytes. B lymphocytes differentiate into plasma cells which produce antibodies.

Enters body

Antibodies

Figure 15–12 Antibody formation.

In the presence of a specific antigen, B lymphocytes differentiate into plasma cells and memory B cells. Memory B cells remember a specific antigen and stimulate a faster and more intense response when that same antigen is presented to the body. A **plasma cell** (plahz-mah sehl) is an immune cell that produces and secretes a specific antibody for a specific antigen. Plasma cells also are called **plasmocytes** (plahz-mō-sītz). The antibodies made by plasma cells are called **immunoglobulins** (ihm-yoo-nō-glohb-yoo-lihnz). Immunoglobulin is abbreviated Ig. There are five distinct immunoglobulins (Table 15–3 and Figure 15–12).

T lymphocytes are small, circulating lymphocytes produced in the bone marrow. These cells mature in the thymus, so they are called **T cells.** The primary function of T lymphocytes is to coordinate immune defenses and kill organisms on contact. T cells are involved in cell-mediated immunity. Cell-mediated immunity involves T cell activation and cellular secretions. Although cell-mediated immunity does not involve antibody production, the antibodies produced during humoral immunity may play a role in some cell-mediated responses.

There are different types of T lymphocytes. **Helper T cells** secrete substances such as lymphokines that stimulate the

Table 15–3	Types of Immunoglobulin and Their Functions
IgA	found in the mucous membrane lining of intestines, bronchi, saliva, and tears; protects those areas
IgD	found in large amounts on the surface of B cells; unknown function
IgE	found in lungs, skin, and cells of mucous membranes; provides defense against the environment and is involved with allergies
IgG	synthesized in response to invading germs such as bacteria, fungi, and viruses; most abundant antibody and only class that can cross the placenta
IgM	found in circulating fluids; first immunoglobulin produced when antigens invade and provides protection in the earliest stages of infection

Anti- **Means Against:** The terms *antigen* and *antibody* must be distinguished. **Antigen** (ahn-tih-jehn) is a substance that the body regards as foreign (such as a virus, bacterium, or toxin). **Antibody** (ahn-tih-boh-dē) is a disease-fighting protein produced by the body in response to the presence of a specific antigen.

production of B lymphocytes and cytotoxic T cells. **Cytotoxic T cells** destroy intracellular pathogens. **Suppressor T cells** stop B and T lymphocyte activity when this activity is no longer needed. **Memory T cells** remember a specific antigen and stimulate a faster and more intense response when that same antigen is presented to the body.

Monocytes are another type of leukocyte formed in the bone marrow and transported to other parts of the body. Monocytes migrate to tissues such as the spleen to become macrophages. A **macrophage** is a phagocytic cell that protects the body by eating invading cells and by interacting with other cells of the immune system. **Histiocytes** (hihs-tē-ō-sītz) are large macrophages found in loose connective tissue.

How Does the Immune System Work?

The body has many ways to protect itself from invading organisms and substances. The initial defenses against pathogen invasion are nonspecific. The first line of defense limits access to internal tissues and organs of the body. Anatomical barriers such as intact skin help protect the animal's body from infection. **Intact** (ihn-tahckt) means having no cuts, scrapes, openings, or alterations. Intact skin makes it more difficult for invading organisms and substances to obtain access to an animal. Oil secreted by sebaceous glands discourages bacterial growth on the skin.

The respiratory system has its own line of defense. Foreign material breathed in is trapped in the cilia of the nares and the moist mucous membrane that lines the respiratory tract. Mucus continually flushes away trapped debris. Coughing and sneezing also remove foreign material.

The digestive system destroys invading organisms that are swallowed by the acidic nature of the stomach.

An animal's health, age, and heredity also play roles in protecting the body. Some animals may be immunodeficient or hypersensitive.

Infectious agents that penetrate the first line of defense are usually destroyed by other nonspecific responses such as fever,

inflammation, chemicals, and complement. The complement system is a nonspecific defense mechanism, and its activation can result in initiation of inflammation, activation of leukocytes, lysis of pathogens, and increased phagocytosis. **Complement** is a series of enzymatic proteins that occur in normal plasma. Complement is thus named because it is complementary to the immune system and it aids phagocytes in destroying antigens and causes cell lysis.

The immune system is activated when the previously listed defenses fail. Activation of the immune system is outlined here:

Organisms invade the body. Macrophages ingest the invading organisms and present antigen fragments to its surface. Helper T cells are activated.

↓

Helper T cells multiply and produce lymphokines that stimulate cytotoxic T cell production. Complement goes to the affected area. B cells sensitive to the organism multiply. B cells transform into plasma cells that produce antibody.

↓

Complement proteins lyse affected cells. Antibodies bind to organisms.

↓

If infection is contained, suppressor T cells stop the immune response. Memory B and T cells remain ready if this particular antigen enters the body again.

Immunity (ihm-yoo-nih-tē) is the state of being resistant to a specific disease. Different forms of immunity are obtained during life. The types of immunity are as follows:

- **naturally acquired passive immunity** = resistance to a specific disease by the passing of protection from mother to offspring before birth or through colostrum.
- **naturally acquired active immunity** = resistance to a specific disease after the development of antibodies during the actual disease.
- **artificially acquired passive immunity** = resistance to a specific disease by receiving antiserum-containing antibodies from another host.
- **artificially acquired active immunity** = resistance to a specific disease through vaccination.

TEST ME: IMMUNE AND LYMPHATIC SYSTEMS

Diagnostic tests performed on the immune and lymphatic systems are described in Chapter 16.

Words used when discussing the immune system

resistant (rē-zihs-tahnt) = not susceptible.

heredity (hər-eh-dih-tē) = genetic transmission of characteristics from parent to offspring.

vaccination (vahck-sihn-ā-shuhn) = administration of antigen (vaccine) to stimulate a protective immune response against a specific infectious agent; also called **immunization** (ihm-yoo-nih-zā-shuhn).

vaccine (vahck-sēn) = preparation of pathogen (live, weakened, or killed) or a portion of pathogen that is administered to stimulate a protective immune response against the pathogen.

multiplication (muhl-tih-plih-kā-shuhn) = reproduction.

inhibit (ihn-hihb-iht) = to slow or stop.

opportunistic (ohp-ər-too-nihs-tihck) = able to cause disease due to debilitation when disease normally would not be produced.

debilitated (dē-bihl-ih-tāt-ehd) = weakened.

PATHOLOGY: IMMUNE AND LYMPHATIC SYSTEMS

Pathologic conditions of the immune and lymphatic systems include the following:

- **allergy** (ahl-ər-jē) = overreaction by the body to a particular antigen; also called **hypersensitivity** (hī-pər-sehn-sih-tihv-ih-tē). An **allergen** (ahl-ər-jehn) is a substance capable of inducing an allergic reaction.
- **anaphylaxis** (ahn-ah-fih-lahck-sihs) = severe response to a foreign substance. Signs develop acutely and may include swelling, blockage of airways, tachycardia, and ptylism.

- **autoimmune disease** (aw-tō-ihm-yoon dih-zēz) = disorder in which the body makes antibodies directed against itself.
- **immunosuppression** (ihm-yoo-nō-suhp-prehsh-uhn) = reduction or decrease in the state of resistance to disease. An **immunosuppressant** (ihm-yoo-nō-suh-prehs-ahnt) is a chemical that prevents or reduces the body's normal reaction to disease.
- **lymphadenitis** (lihm-fahd-eh-nī-tihs) = inflammation of the lymph nodes; also called **swollen glands.**
- **lymphadenopathy** (lihm-fahd-eh-nohp-ah-thē) = disease of the lymph nodes; an example of incorrect use of *aden/o* because nodes are not glands.
- **lymphangioma** (lihm-fahn-jē-ō-mah) = abnormal collection of lymphatic vessels forming a mass (usually benign).
- **splenomegaly** (splehn-ō-mehg-ah-lē) = enlargement of the spleen.
- **tonsillitis** (tohn-sih-lī-tihs) = inflammation of the tonsils.

PROCEDURES: IMMUNE AND LYMPHATIC SYSTEMS

Procedures performed on the immune and lymphatic systems include the following:

- **splenectomy** (splehn-ehck-tō-mē) = surgical removal of the spleen.
- **thymectomy** (thī-mehck-tō-mē) = surgical removal of the thymus.
- **tonsillectomy** (tohn-sih-lehck-tō-mē) = surgical removal of the tonsils.

ONCOLOGY

Oncology (ohng-kohl-ō-jē) is the study of tumors. The combining form **onc/o** means tumor. The term **tumor** does not mean cancerous. When a tumor is cancerous, the term **malignant** is used to describe it. Nonmalignant tumors are called **benign.** The term *neoplasm* is another word used to describe abnormal growths. A **neoplasm** (nē-ō-plahz-uhm) is any abnormal new growth of tissue in which the multiplication of cells is uncontrolled, more rapid than normal, and progressive.

Tumors often are described by their appearance. **Pedunculated** (peh-duhnck-yoo-lā-tehd) means having a peduncle, or stalk. **Well-circumscribed** means that the mass has well-defined borders. **Invasive** means that the mass does not have well-defined borders and is spreading.

Malignant growths tend to spread to distant body sites. **Metastasis** (meh-tahs-tah-sihs) is a pathogenic growth distant from the primary disease site. Metastasis literally means beyond control. The plural form of metastasis is **metastases** (meh-tahs-tah-sēz). The term **metastasize** (meh-tahs-tah-sīz) is used to describe invasion by the pathogenic growth to a point distant from the primary disease site.

Malignant growths also may be described by their tissue of origin. A **carcinoma** (kahr-sih-nō-mah) is a malignant growth of epithelial cells, whereas a **sarcoma** (sahr-kō-mah) is a malignant neoplasm arising from any type of connective tissue. The combining form **carcin/o** means cancer, and the combining form **sarc/o** means flesh. Connective tissue sometimes is referred to as fleshy. The combining form *carcin/o* is used to form the term *carcinogen*. A **carcinogen** is a substance that produces cancer. The suffix **-gen** means producing.

TEST ME: ONCOLOGY

Diagnostic tests performed on tumors include the following:

- **biopsies** = described in Chapter 10.
- **radiographs** = records of ionizing radiation used to visualize internal body structures. Radiographs are taken in oncology patients to assess the extent of some tumors or to check for metastases (Figure 15–13).
- **touch preps** = collections of cells on a glass slide pressed against a part of the mass. The slide is then examined under a microscope.

PATHOLOGY: ONCOLOGY

Tumors are named for the tissues involved. Some examples include the following:

- **adenocarcinoma** (ahd-eh-nō-kahr-sih-nō-mah) = malignant growth of epithelial glandular tissue.
- **blastoma** (blahs-tō-mah) = neoplasm composed of immature undifferentiated cells.
- **hemangioma** (hē-mahn-jē-ō-mah) = benign neoplasm composed of newly formed blood vessels.
- **hemangiosarcoma** (hē-mahn-jē-ō-sahr-kō-mah) = malignant tumor of vascular tissue (Figure 15–14).
- **lymphoma** (lihm-fō-mah) = general term for neoplasm composed of lymphoid tissue (usually malignant); also called **lymphosarcoma;** abbreviated LSA.
- **mast cell tumor** = malignant growth of tissue mast cells (cells that release histamine); abbreviated MCT. Mast cell tumors are associated with vomiting, anorexia, and various signs depending on the tissue involved (Figure 15–15).
- **melanoma** (mehl-ah-nō-mah) = neoplasm composed of melanin-pigmented cells (see Figure 10–22).
- **myeloma** (mī-eh-lō-mah) = malignant neoplasm composed of bone marrow.

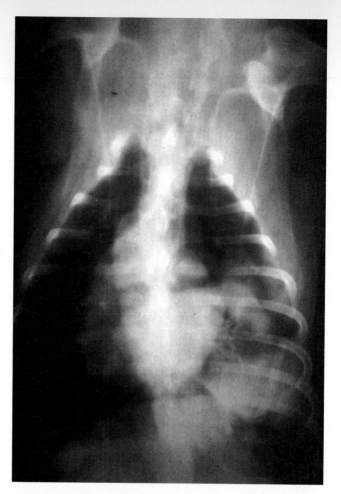

Figure 15–13 Ventrodorsal radiograph of pulmonary metastases. (Courtesy of Anne E. Chauvet, DVM, Diplomate ACVIM—Neurology Department, University of Wisconsin School of Veterinary Medicine.)

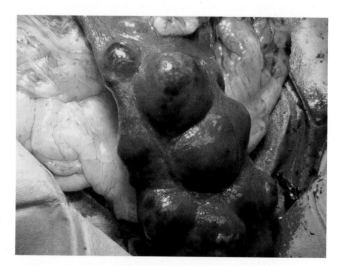

Figure 15–14 Intraoperative photograph of a spleen with hemangiosarcoma.

- **myosarcoma** (mī-ō-sahr-kō-mah) = malignant neoplasm composed of muscle.
- **myxoma** (mihcks-ō-mah) = tumor of connective tissue.
- **neuroblastoma** (nū-rō-blahs-tō-mah) = malignant neoplasm of nervous tissue origin.

- **osteosarcoma** (ohs-tē-ō-sahr-kō-mah) = malignant neoplasm composed of bone (Figure 15–16).
- **squamous cell carcinoma** (skwā-mohs sehl kahr-sih-nō-mah) = malignant tumor developed from squamous epithelial tissue; abbreviated SCC (Figure 15–17).

PROCEDURES: ONCOLOGY

Procedures performed on tumors include the following:

- **chemotherapy** (kē-mō-thehr-ah-pē) = treatment of neoplasm through the use of chemicals.
- **lymphadenectomy** (lihm-fahd-ehn-ehck-tō-mē) = surgical removal of a lymph node. The name for removal of any tissue that may have a mass or tumor can be derived by adding the combining form for the area being removed.
- **radiation therapy** (rā-dē-ah-shuhn thehr-ah-pē) = treatment of neoplasm through the use of X-rays.
- **surgical excision** = removal of the entire mass in addition to some normal tissue to ensure that the entire mass is removed.

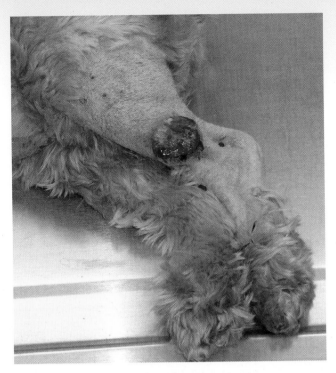

Figure 15–15 A mast cell tumor on a dog's leg.

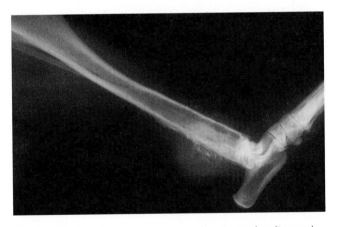

Figure 15–16 Osteosarcoma in a dog. Lateral radiograph of a dog with osteosarcoma of the distal tibia.

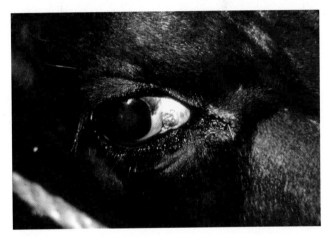

Figure 15–17 SCC of a bovine eye. (Courtesy of Ron Fabrizius, DVM, Diplomate ACT.)

REVIEW EXERCISES

Multiple Choice

Choose the correct answer.

1. Red blood cells are called

    a. erythrocytes
    b. leukocytes
    c. thrombocytes
    d. platelets

2. White blood cells are called

    a. erythrocytes
    b. leukocytes
    c. thrombocytes
    d. platelets

3. An elevation in white blood cells is called

    a. leukopenia
    b. leukocytosis
    c. leukemia
    d. leukosis

4. An immature, non-nucleated erythrocyte that is characterized by polychromasia (Wright's stain) or a meshlike pattern of threads (new methylene blue stain) is called a

    a. macrophage
    b. prothrombocyte
    c. reticulocyte
    d. phagocyte

5. An immature polymorphonuclear leukocyte is called a

    a. baby PMN
    b. band cell
    c. left shift
    d. micro PMN

6. Lacteals are located in the

    a. groin
    b. loin
    c. small intestine
    d. stomach

7. Lymph also is known as

    a. serum
    b. plasma
    c. interstitial fluid
    d. a and b

8. T cells are responsible for

    a. allergies
    b. humoral immunity
    c. cell-mediated immunity
    d. thymus production

9. B cells are responsible for

    a. allergies
    b. humoral immunity
    c. cell-mediated immunity
    d. thymus production

10. The spleen is

    a. hemolytic
    b. an organ where lymphocytes are differentiated
    c. a storage area for RBCs
    d. all of the above

11. A malignant tumor that developed from epithelial tissue is known as a(n)

    a. squamous cell carcinoma
    b. myxoma
    c. osteosarcoma
    d. blastoma

12. What is the term for a malignant tumor of vascular tissue?

    a. hemangioma
    b. hemangiosarcoma
    c. neuroblastoma
    d. lymphoma

13. Malignant neoplasms arising from connective tissue are known as

    a. carcinomas
    b. sarcomas
    c. melomas
    d. myxomas

14. When a growth does not have well-defined borders, it is described as being

    a. circumscribed
    b. pedunculated
    c. invasive
    d. metastasized

15. A deficiency in white blood cells is called
   a. leukopenia
   b. leukocytosis
   c. erythrocytosis
   d. anemia

16. Hemolysis is
   a. increasing the RBC count
   b. decreasing the RBC count
   c. breaking down blood cells
   d. excess amounts of fat in the blood

17. A substance that prevents clotting is known as a(n)
   a. hematopoietic agent
   b. anticoagulant
   c. anisocytosis
   d. poikiloctosis

18. Larger-than-normal RBCs are described as being
   a. hyperchromic
   b. hypochromic
   c. microcytic
   d. macrocytic

19. What term is used to describe the condition in which microorganisms and their toxins are present in the blood?
   a. septicemia
   b. bacteremia
   c. anemia
   d. edema

20. The name for a disease-fighting protein produced by the body in response to a particular foreign substance is
   a. antigen
   b. antibody
   c. antipathogen
   d. dysrasia

Matching

Match the immunologic term in Column I with the definition in Column II.

Column I

1. _____ resistant
2. _____ heredity
3. _____ vaccination
4. _____ immunization
5. _____ vaccine
6. _____ multiplication
7. _____ inhibit
8. _____ opportunistic
9. _____ debilitated

Column II

a. weakened

b. able to cause disease due to debilitation when disease would not be produced normally

c. administration of antigen to stimulate a protective immune response against a specific infectious agent

d. reproduction

e. not susceptible

f. genetic transmission of characteristics from parent to offspring

g. to slow or stop

h. preparation of pathogen (live, weakened, or killed) or a portion of pathogen that is administered to stimulate a protective immune response against the pathogen

Match the term in Column I with the definition in Column II.

	Column I		Column II
10.	_____ benign	a.	abnormal new growth of tissue
11.	_____ malignant	b.	cancerous
12.	_____ neoplasm	c.	enlarged spleen
13.	_____ metastasis	d.	condition of eating cells
14.	_____ splenomegaly	e.	breakdown
15.	_____ edema	f.	noncancerous
16.	_____ phagocytosis	g.	accumulation of fluid in the intercellular space
17.	_____ lysis	h.	pathogenic growth distant from its primary disease site
18.	_____ leukocytosis	i.	abnormal increase in WBCs
19.	_____ leukopenia	j.	abnormal decrease in WBCs

Fill in the Blanks

1. _____ means no cuts, scrapes, openings, or alterations.

2. The state of being resistant to a specific disease is _____.

3. A substance that the body regards as foreign is a(n) _____.

4. A deficiency of red blood cells or hemoglobin is called _____.

5. Excessive blood in a part is called _____.

6. _____ is a severe response to a foreign substance.

7. A substance capable of inducing an allergic reaction is called a(n) _____.

8. The study of tumors is _____.

9. A general term for a malignant neoplastic disorder of lymphoid tissue is _____.

10. A blood condition in which pathogenic microorganisms or their toxins are present is called _____.

11. -poiesis means _____.

12. Hem/o and hemat/o mean _____.

13. Immun/o means _____.

14. -emia means _____.

15. Lip/o means _____.

Spelling

Circle the term that is spelled correctly.

1. mature red blood cell: erithrocyte   erythrocyte   erythrosite

2. study of form: morfology   morphology   morphlogy

3. condition of unequal cell size: anistocytosis   anisocytosis   anystocytosis

4. tumor of connective tissue: mixoma   myxtoma   myxoma

5. malignant neoplasm composed of bone: ostesarcoma   osteosarkoma   osteosarcoma

CROSSWORD PUZZLE

Hematologic, Lymphatic, and Immune System Terms Puzzle

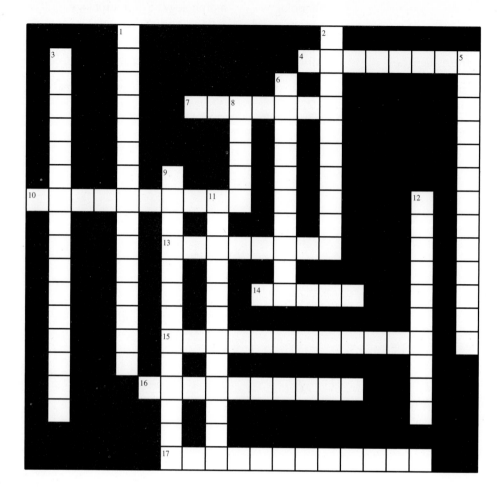

Across

4 abnormal increase in the number of malignant WBCs
7 malignant neoplasm of bone marrow
10 abnormal decrease in the number of WBCs
13 disease-fighting protein produced by the body in response to the presence of a foreign substance
14 destruction or break down
15 malignant neoplasm of bone
16 another name for polymorphonuclear leukocyte
17 enlargement of the spleen

Down

1 malignant neoplasm of vascular tissue
2 study of blood
3 another term for allergy
5 substance that prevents clotting
6 study of form
8 accumulation of fluid in the intercellular space
9 formation of blood
11 pertaining to the spaces within a tissue or organ
12 pathogenic growth distant from the primary disease site

CHAPTER 16

[TESTING TESTING]

Objectives

Upon completion of this chapter, the reader should be able to

- Describe terms and equipment for the basic physical examination
- Recognize, define, spell, and pronounce terms associated with physical examinations
- Recognize, define, spell, and pronounce terms associated with laboratory analysis
- Describe positioning for radiographic and imaging procedures
- Recognize, define, spell, and pronounce terms associated with radiographic and imaging procedures

BASIC PHYSICAL EXAMINATION

Physical examinations are performed to assess a patient's condition. **Assessment** is the term used to describe the evaluation of a condition. After a patient is assessed, the information is written in a medical record. The animal's signalment should always be included. A **signalment** is a description of the animal with information about the animal, including the species, breed, age, and sexual status (intact or neutered).

Vital Signs

Vital signs are parameters taken from the animal to assess its health.

Temperature is the vital sign that tells about the degree of heat or cold. An animal's temperature is recorded in degrees Fahrenheit or Celsius. Different species have different normal temperature ranges. An elevated body temperature is called a **fever. Febrile** (fē-brīl or feh-brīl) is the medical term for fever; **afebrile** (ā-fē-brīl or ā-feh-brīl) means without a fever. **Pyrexia** (pī-rehck-sē-ah) is another medical term for fever. **Pyr/o** means fire. A decrease in body temperature is known as **hypothermia.**

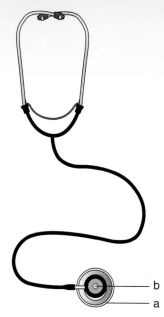

Figure 16–1 Parts of a stethoscope. (a) Diaphragm. The diaphragm is the flat, circular portion of the chestpiece covered with a thin membrane. The diaphragm transmits high-pitched sounds, such as those produced by the bowel, lungs, and heart. (b) Bell. The bell is not covered by a membrane. The bell facilitates auscultation of lower-frequency sounds, such as the third and fourth sounds of the heart.

Pulse, another vital sign, tells the number of times the heart beats per minute. The pulse also is called the **pulse rate.** Pulse is taken by palpating an artery. Heart rate also may be considered a vital sign and is taken by auscultating the heart with a stethoscope (Figures 16–1 and 16–2). **Heart rate** is the number of times the heart contracts and relaxes per minute. Heart rate is abbreviated HR.

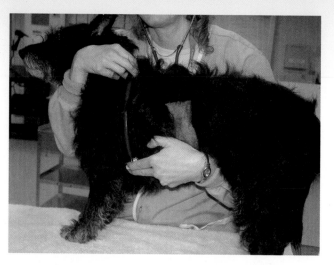

Figure 16–2 A stethoscope is used to listen to body sounds. (Courtesy of Teri Raffel, CVT.)

Respiration is the number of respirations per minute. Respiration is one total inhale and one total exhale. Respiration also is called the **respiration rate** and is abbreviated RR (Figure 16–3).

Blood pressure is another vital sign that may be taken on veterinary patients. A **sphygmomanometer** (sfihg-mō-mah-nohm-eh-tər) is an instrument used to measure blood pressure (Figure 16–4). A Doppler is used to listen to blood sounds during the measurement of blood pressure in animals.

Listening

Auscultation (aws-kuhl-tā-shuhn) is the act of listening, which usually involves the use of a stethoscope to listen to body sounds. Auscultation can be used to assess the condition of the heart, lungs, pleura, and abdomen.

	Heart rate (beats/min)	Respiratory rate (breaths/min)	Rectal temperature (°C)	Rectal temperature (°F)
Dogs	70–160	8–20	38–39	100.5–102.5
Cats	150–210	8–30	38–39	100.5–102.5
Hamsters	250–500	35–135	37–38	99–100.5
Guinea pigs	230–280	42–104	37–39.5	99–103
Rabbits	130–325	30–60	38.5–40	101.5–104
Horses	28–50	8–16	37.5–38.5	99.5–101.5
Cattle	40–80	12–36	38–39	100.5–102.5
Sheep	60–120	12–50	39–40	102.5–104
Goats	70–135	12–50	38.5–40.5	101.5–105
Pigs	58–100	8–18	39–40	102.5–104
Llamas	60–90	10–30	37–39	99–102.5
Ferrets	230–250	33–36	38–40	100.5–104

Figure 16–3 Normal vital sign ranges.

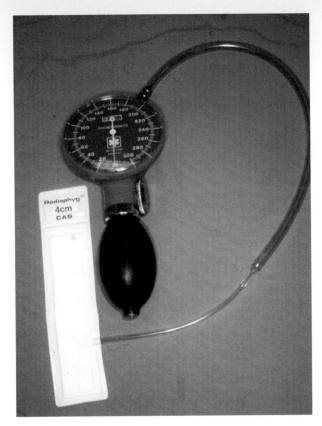

Figure 16–4   Sphygmomanometer. (Courtesy of Teri Raffel, CVT.)

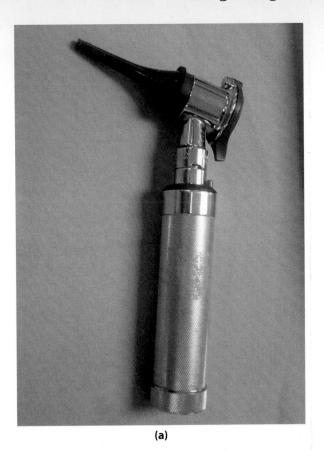

(a)

(b)

Figure 16–5   Specula. (a) A speculum is attached to an otoscope to allow better visualization of a canal or cavity. (b) A mouth speculum is used to enhance visualization of the oral cavity. [(a) Courtesy of Teri Raffel, CVT; (b) Courtesy of Ron Fabrizius, DVM, Diplomate ACT.]

Touching

Palpation (pahl-pā-shuhn) is examination by feeling. During palpation, one can feel the texture, size, consistency, and location of body parts or masses.

Percussion (pər-kuhsh-uhn) is examination by tapping the surface to determine density of a body area. Sound may be produced when the surface is tapped with a finger or an instrument. The sound produced by percussion varies depending on the amount of fluid, solid, or air present in the area being examined.

Looking

Various instruments can be used during the physical examination to obtain a better view of a body system. Examples include an ophthalmoscope and otoscope. A **speculum** (spehck-yoo-luhm) is an instrument used to enlarge the opening of a canal or cavity. A speculum is attached to an otoscope (or another scope) to provide a better view of a canal or cavity. A mouth speculum is used to better visualize the oral cavity (Figures 16–5a and b).

LABORATORY TERMINOLOGY

Specialized terminology has been developed to describe tests and results of laboratory tests. Blood for laboratory tests usually is collected via venipuncture. **Venipuncture** (vehn-ih-puhnk-tər) is withdrawing blood from a vein (usually with a needle and syringe).

Terminology of laboratory tests is given in Table 16–1.

Records are kept under many different methods. One method is the SOAP method. **SOAP** is an acronym for subjective, objective, assessment, and plan analysis (Figure 16–6). An acronym is a word formed by the initial letters of the major parts of the name. Some diseases, structures, and procedures are derived from a person's name, and the name is known as an **eponym.**

Table 16–1   Blood Test Terminology

Term	Pronunciation	Definition
agglutination	(ah-gloo-tih-nā-shuhn)	clumping together of cells or particles
assay	(ahs-ā)	assessment or test to determine the number of organisms, cells, or amount of a chemical substance found in a sample
complete blood count		diagnostic evaluation of blood to determine the number of erythrocytes, leukocytes, and thrombocytes per cubic millimeter of blood; abbreviated CBC
differential	(dihf-ər-ehn-shahl)	diagnostic evaluation of the number of white blood cell types per cubic millimeter of blood
diluent	(dihl-yoo-ehnt)	liquid used to make a dilution
hematocrit	(hē-maht-ō-kriht)	percentage of erythrocytes in blood; "to separate blood"; also called **crit, PCV,** or **packed cell volume** (Figure 16–7)
hemogram	(hē-mō-grahm)	record of the findings in examination of blood especially with reference to the numbers, proportions, and morphology of the blood cells
immunofluorescence	(ihm-yoo-nō-floo-rehs-ehns)	method of tagging antibodies with a luminating dye to detect antigen–antibody complexes
leukogram	(loo-kō-grahm)	numeric and descriptive data in the WBC distribution used to identify a pathologic process
profile	(prō-fīl)	group of laboratory tests performed on serum; also called **screen** or **panel;** includes tests that measure levels of glucose, liver enzymes, and kidney enzymes
prothrombin time	(prō-throhm-bihn)	diagnostic evaluation of the number of seconds needed for thromboplastin to coagulate plasma
radioimmunoassay	(rā-dē-ō-ihm-yoo-nō-ahs-ā)	laboratory technique in which a radioactively labeled substance is mixed with a blood specimen to determine the amount of a particular substance in the mixture; also called **radioassay**
red cell count		number of erythrocytes per cubic millimeter of blood
serology	(sē-rohl-ō-jē)	laboratory study of serum and the reactions of antigens and antibodies
white cell count		number of leukocytes per cubic millimeter of blood

Animal Medical Hospital

Ⓢ Dog presented BAR c̄ hx of persistent
  cough, anorexia, and wt loss.
Ⓞ T = 103.4 °F HR = 120 bpm   RR = 30 breaths/min
  CRT = 1 sec   mm = pink. Lungs auscultated
  harshly, Heart sounds appear Ⓝ.
  Rest of PE - WNL.
Ⓐ Respiratory disease — R/o Kennel cough, pneumonia.
Ⓟ Radiograph chest
  CBC
  Medication pending results.

Figure 16–6 SOAP method of record keeping.

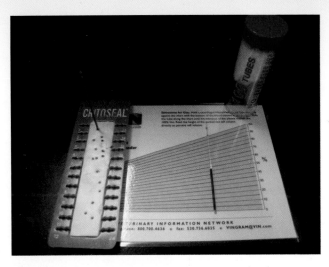

Figure 16–7 Microhematocrit tubes are used to determine a patient's hematocrit.

Special equipment is needed to prepare the samples for many laboratory tests. A **refractometer** (rē-frahck-tah-mē-tər) is an instrument used to determine the deviation of light through objects. Refractometers are used to measure solute (particle) concentration of serum, urine, and other body fluids (Figure 16–8).

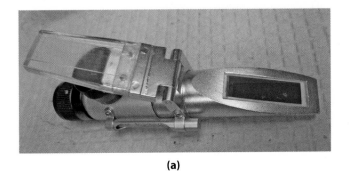

(a)

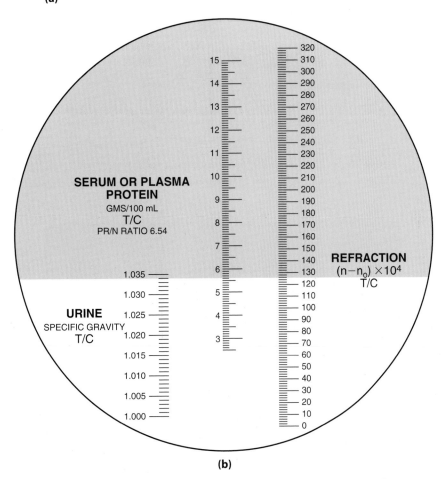

(b)

Figure 16–8 Refractometer. (a) A refractometer measures specific gravity optically. (b) This refractometer scale shows a urine specific gravity of 1.034 (lower left scale) and serum or plasma protein of 5.6 (middle scale).

(a)

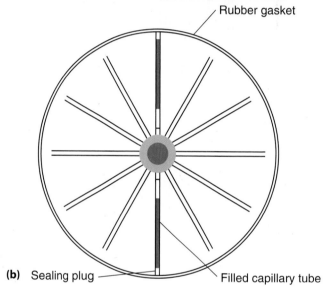

Rubber gasket

(b) Sealing plug   Filled capillary tube

Figure 16–9 Centrifuge. (a) Microhematocrit centrifuge. (b) Proper placement of sealed capillary tubes in a microhematocrit centrifuge.

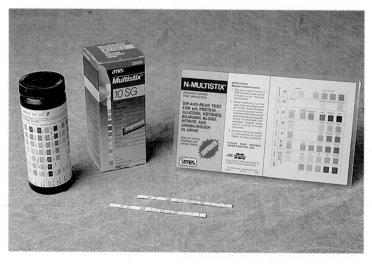

Figure 16–10 Reagent strips (dipsticks) for chemical testing of urine.

Table 16–2   Basic Medical Terms

Diagnosis (dī-ahg-nō-sihs) is the determination of the cause of disease; to "know completely"; plural is **diagnoses** (dī-ahg-nō-sēz). To **diagnose** (dī-ahg-nōs) is to determine the cause of disease.	**Differential diagnosis** (dihf-ər-ehn-shahl dī-ahg-nō-sihs) is the determination of possible causes of diseases; a list of possible causes of disease.	**Prognosis** (prohg-nō-sihs) is the prediction of the outcome of disease; to "know before."
Sign (sīn) is a characteristic of disease that can be observed by others.	**Symptom** (sihmp-tuhm) is a characteristic of disease that can be sensed only by the patient; incorrect term in veterinary medicine.	A **syndrome** (sihn-drōm) is a set of signs that occur together.
Acute (ah-kūt) means having a short course with a sudden onset; implies severe. **Peracute** (pər-ah-kyūt) means having an excessively acute onset.	**Chronic** (krohn-ihck) means having a long course with a progressive onset; persisting for a long time.	**Remission** (rē-mih-shuhn) is partial or complete disappearance of disease signs.
Endemic (ehn-dehm-ihck) is the ongoing presence of disease in a group; also called **enzootic** (ehn-zō-oh-tihck) if the disease is always present in an animal community.	**Epidemic** (ehp-ih-dehm-ihck) is the sudden and widespread outbreak of disease in a group; also called **epizootic** (ehp-ih-zō-oh-tihck) if the outbreak attacks many animals in a group.	**Pandemic** (pahn-dehm-ihck) is disease outbreak occurring over a large geographic area; also called **panzootic** (pahn-zō-oh-tihck) if the widespread outbreak affects many animals.

A **centrifuge** (sehn-trih-fūj) is a machine that spins samples very rapidly to separate elements based on weight. A centrifuge is used to separate the formed elements of blood from the liquid portion of blood. A centrifuge also separates the liquid portion of urine from the heavier solids (Figures 16–9a and b).

The pH of a sample also may provide information about a patient's status. The terms *acid* and *alkaline* are used to describe a patient's pH status. **Acid** (ah-sihd) is the property of low pH, or high number of hydrogen ions. **Alkaline** (ahl-kah-lihn) is the property of high pH, or low number of hydrogen ions. Alkaline also is called **basic** (Figure 16–10).

Urinalysis terms are discussed in Chapter 7.

BASIC MEDICAL TERMS

Basic medical terms are listed in Table 16–2.

PATHOGENIC ORGANISMS

A **pathogen** (pahth-ō-jehn) is a microorganism that produces disease. A **microorganism** (mī-krō-ōr-gahn-ihzm) is a living organism of microscopic dimensions (Table 16–3). Not all microorganisms are pathogens. The term **virulence** (vihr-yoo-lehns) is used to describe the ability of an organism to cause disease.

TYPES OF DISEASES

Disease is deviation from normal. There are different types of diseases, depending on how they are spread and what causes them. Types of diseases include the following:

- **contagious** (kohn-tā-juhs) **disease** = disease that can be spread from one animal to another by direct or indirect contact. Direct contact is spread from animal to animal, whereas indirect contact is spread through contact with contaminated objects. Contagious disease also may be referred to as **communicable** (kuh-mū-nih-kuh-buhl).
- **iatrogenic** (ī-aht-rō-jehn-ihck) **disease** = disorder caused by physicians or veterinarians (and the treatment ordered).
- **idiopathic** (ihd-ē-ō-pahth-ihck) **disease** = disorder of unknown cause. Idiopathic disease is a disease peculiar to an individual and not likely to be seen in others.
- **infectious disease** (ihn-fehck-shuhs dih-zēz) = disorder caused by pathogenic organisms.
- **noncontagious** (nohn-kohn-tā-juhs) = disease that cannot be spread to another animal by contact or contact with a contaminated object.
- **noninfectious disease** (nohn-ihn-fehck-shuhs dih-zēz) = disorder not caused by organisms (examples include genetic, traumatic, and iatrogenic).
- **nosocomial infection** (nōs-ō-kō-mē-ahl ihn-fehck-shuhn) = disorder caused by pathogenic organisms contracted in a facility or clinic.

Table 16–3   Types of Organisms

Organism	Examples
bacterium (bahck-tē-rē-uhm) is a microscopic, microscopic, prokaryotic unicellular organism; plural is **bacteria** (bahck-tē-rē-ah). A prokaryote (prō-kahr-ē-ōt) is an organism without a membrane-bound nucleus.	**staphylococci** (stahf-ih-lō-kohck-sī) are grapelike clusters of spherical bacteria; **coccus** (kohck-uhs) means sphere. (cocci is plural) Staphylococcus
	streptococci (strehp-tō-kohck-sī) are spherical bacteria that form chains. Streptococcus
	bacilli (bah-sihl-ī) are rod-shaped bacteria. (bacillus is singular) Bacillus
	spirochetes (spī-rō-kētz) are spiral-shaped bacteria that are tightly coiled. Spirochete
	endospore (ehn-dō-spōr) is a resistant, oval body formed in some bacteria. Endospore in bacterium Free endospore
	rickettsia (rih-keht-sē-ah) is a small rod-shaped bacterium transmitted by lice, fleas, ticks, or mites.
fungus (fuhng-guhs) is a eukaryotic organism without chlorophyll; plural is **fungi** (fuh-jī). A eukaryote (yoo-kahr-ē-ōt) is an organism with a membrane-bound nucleus.	**yeast** (yēst) is a budding form of fungus. **mold** (mohld) is a filamentous form of fungi. Yeast Mold
parasite (pahr-ah-sīt) is an organism that lives on or in another living organism.	Parasite
virus (vī-ruhs) is a small organism that is not visualized via microscopy; viruses live only by invading cells.	Virus

DISEASE TERMINOLOGY

- **asymptomatic** (ā-sihmp-tō-mah-tihck) = without signs of disease.
- **atraumatic** (ā-traw-mah-tihck) = pertaining to, resulting from, or caused by a noninjurious route.
- **carrier** (kahr-ē-ər) = animal that harbors an infectious agent without displaying clinical signs and who may transmit the infectious agent to others.
- **clinical** (klihn-ih-kahl) = visible, readily observed, pertaining to treatment.
- **contract** (kohn-trahckt) = to catch a disease.
- **disease** (dih-zēz) = deviation from normal health.
- **epidemiology** (ehp-ih-dē-mē-ohl-ō-jē) = study of relationships determining frequency and distribution of diseases.
- **etiology** (ē-tē-ohl-ō-jē) = study of disease causes.
- **excessive** (ehck-sehs-ihv) = more than normal.
- **focus** (fō-kuhs) = localized region.
- **germ** (jərm) = common term for any pathogenic microorganism, but especially bacterial and viral organisms.
- **incidence** (ihn-sih-dehns) = number of new cases of disease occurring during a given time.
- **labile** (lā-bīl) = unstable.
- **lethal** (lē-thahl) = causing death.
- **morbid** (mōr-bihd) = afflicted with disease.
- **morbidity** (mōr-bihd-ih-tē) = ratio of diseased animals to well animals in a population.
- **moribund** (mōr-ih-buhnd) = near death.
- **mortality** (mōr-tahl-ih-tē) = ratio of diseased animals that die to diseased animals.
- **palliative** (pah-lē-ah-tihv) = able to relieve but not cure a condition.
- **phobia** (fō-bē-ah) = extreme fear.
- **prevalence** (preh-vah-lehns) = number of cases of disease in a population at a certain time.
- **prophylaxis** (prō-fihl-ahcks-sihs) = prevention.
- **sequela** (sē-kwehl-ah) = condition occurring as a consequence of another condition.
- **subclinical** (suhb-klihn-ih-kahl) = without showing signs of disease.
- **susceptible** (sah-sehp-tih-buhl) = lacking resistance.
- **swollen** (swohl-ehn) = enlarged by fluid retention.
- **transmissible** (trahnz-mihs-ih-buhl) = ability to transfer from one animal to the next.
- **transmission** (trahnz-mih-shuhn) = transfer from one animal to the next. There are different types of disease transmission, including **bloodborne transmission** (spread of disease via blood or body fluids), **sexual transmission** (spread of disease via contact with reproductive areas or through copulation), **airborne transmission** (spread of disease via respiratory droplets), and **fecal–oral transmission** (spread of disease via eating, drinking, or licking contaminated food, water, or objects).
- **traumatic** (traw-mah-tihck) = pertaining to, resulting from, or causing injury.
- **zoonosis** (zō-ō-nō-sihs) = disease that can be transmitted between animals and humans.

ENDOSCOPY

Endoscopy is the visual examination of the interior of any cavity of the body by means of an endoscope. The procedures and instruments are named for the body parts involved.

Specific endoscopic procedures are covered in the chapter of the body system on which they are used.

Endoscopic surgery is a procedure using an endoscope to aid in surgical procedures so that only very small incisions are made. Some instruments used in endoscopic surgery include a trocar and cannula. A **trocar** (trō-kahr) is a sharp, needlelike instrument that has a cannula (tube) that is used to puncture the wall of a body cavity to withdraw fluid or gas (Figure 16–11). A **cannula** (kahn-yoo-lah) is a hollow tube.

CENTESIS

Centesis is the surgical puncture to remove fluid or gas for diagnostic purposes or for treatment. Specific centesis procedures are covered in the chapter of the body system in which they are used.

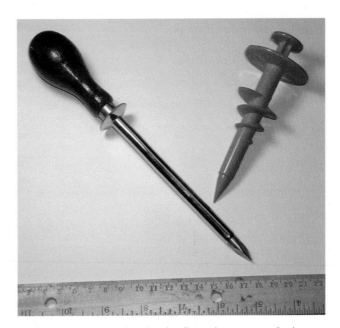

Figure 16–11 Metal and indwelling plastic trocar. Both trocars have a solid pointed spike in the center that is removed once the trocar is in place. With the spike removed, the tube portion of the trocar allows gas to escape.

IMAGING TECHNIQUES

Imaging techniques are used to visualize and examine internal structures of the body.

Radiology

The first imaging technique developed involves the use of ionizing radiation, or X-rays, to produce an image. **Radiography** (rā-dē-ohg-rah-fē) is the procedure in which film is exposed as ionizing radiation passes through the patient and shows the internal body structures in profile. A **radiograph** (rā-dē-ō-grahf), or X-ray, is the record of ionizing radiation used to visualize internal body structures. Note that **graph** (as opposed to gram) is used to mean *record* in this case.

Radiographs are composed of shades of gray. Hard tissues such as bone are called radiopaque. **Radiopaque** (rā-dē-ō-pāk) means appearing white or light gray on a radiograph. Air and soft tissues are called radiolucent. **Radiolucent** (rā-dē-ō-loo-sehnt) means appearing black or dark gray on a radiograph.

Radiology (rā-dē-ohl-ō-jē) is the study of internal body structures after exposure to ionizing radiation. A radiologist

X-rays were discovered in 1895 by Wilhelm Konrad Roentgen, a German physicist.

Radiology comes from the Latin word **radius,** meaning a rod, and the suffix **-logy,** meaning the science of or the study of. Radiology also is known as roentgenology as a tribute to its discoverer. A **roentgen** (rehnt-gehn) is the international unit of radiation. Another term, **rad** (an acronym for *radiation absorbed dose*), is a unit by which absorption of ionizing radiation is measured.

Two abbreviations commonly used in radiography are kVp and MAS. **kVp** stands for kilovoltage peak and represents the strength of the X-ray beam. **MAS** stands for milliamperes per second and represents the number of X-ray beams (because it is based on time).

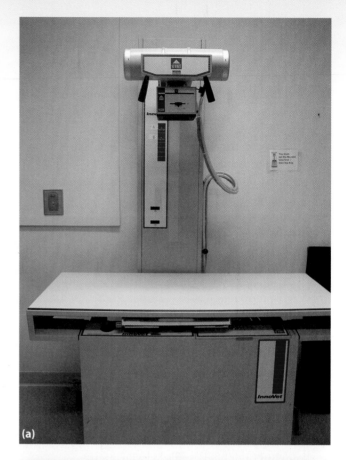
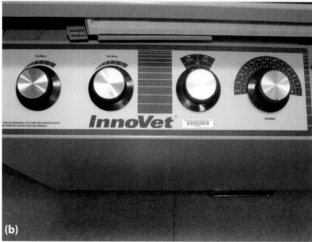

Figure 16–12 Radiographic machine. (a) Radiographic table and X-ray tube. (b) Radiographic control panel. (Courtesy of Teri Raffel, CVT.)

(rā-dē-ohl-ō-jihst) is a specialist who studies internal body structures after their exposure to ionizing radiation (Figure 16–12).

There are different types of radiographs depending on how they are taken and whether any additional diagnostic tool is used. A **scout film** is a plain radiograph made without the use of a contrast medium. A scout radiograph indicates whether abnormalities exist in the patient or whether there is need for further testing or for more specialized radiographic studies.

Some body structures, such as the intestinal lumen, are difficult to assess using X-rays alone. In those cases, a radiographic contrast medium may be used. **Radiographic contrast** (rā-dē-ō-grahf-ihck kohn-trahst) **medium** or material is a substance used to show structures on X-ray that are otherwise difficult to see. **Barium sulfate** (bār-ē-uhm suhl-fāt), or **barium,** is one example of contrast material.

The type of contrast radiograph taken depends on which structures are to be visualized (Figure 6–32). The route of administration of a contrast medium also varies depending on which structures are to be visualized.

A **lower GI** is a type of contrast radiograph used to visualize the structures of the lower gastrointestinal tract. In a lower GI, an enema is used to introduce contrast material into the colon; therefore, it also is called a **barium enema** (bār-ē-uhm ehn-ah-mah). An **upper GI** is a type of contrast radiograph used to visualize the structures of the upper gastrointestinal tract. In an upper GI, contrast material is swallowed; therefore, it also is called a **barium swallow** (bār-ē-uhm swahl-ō) (see Figure 6–19).

Contrast material also can be injected intravenously. An intravenous contrast medium is injected into the vein to make visible the flow of blood through the blood vessels and organs.

Radiographic techniques also are named for the vessels or organs involved. An example is a **lymphangiography** (limh-fahn-jē-ohg-rah-fē), which is a radiographic examination of the lymphatic vessels after injection of contrast material.

Projection and Positioning

Two terms frequently used in radiography are *projection* and *positioning*. **Projection** (prō-jehck-shuhn) is the path of the X-ray beam (Table 16–4). **Positioning** (pō-sih-shuhn-ihng) is the specified body position and the part of the body closest to the film (Figures 16–13a and b). **Recumbency** is used in reference to positioning. (Refer to Chapter 2 for positional terms.) **Anatomical position** (ahn-ah-tohm-ihck-ahl pō-sih-shuhn) refers to the animal in its normal standing position.

Computed Tomography

Computed tomography (kohm-puh-tehd tō-moh-grah-fē) is the procedure in which ionizing radiation with computer assistance passes through the patient and shows the internal body structures in cross-sectional views. It also

Table 16–4 Types of Projection

Projection	Pronunciation	Definition
craniocaudal projection	(krā-nē-ō-kaw-dahl prō-jehck-shuhn)	X-ray beam passes from cranial to caudal; formerly called **anteroposterior projection** (ahn-tēr-ō-pohs-tēr-ē-ər prō-jehck-shuhn), or A/P projection.
caudocranial projection	(kaw-dō-krā-nē-ahl prō-jehck-shuhn)	X-ray beam passes from caudal to cranial; formerly called **posteroanterior projection** (pohs-tēr-ō-ahn-tēr-ih-ər prō-jehck-shuhn), or P/A projection.
dorsoventral projection	(dōr-sō-vehn-trahl prō-jehck-shuhn)	X-ray beam passes dorsally to ventrally; abbreviated D/V.
ventrodorsal projection	(vehn-trō-dōr-sahl prō-jehck-shuhn)	X-ray beam passes ventrally to dorsally; abbreviated V/D.
lateral projection	(laht-ər-ahl prō-jehck-shuhn)	X-ray beam passes from side to side, with the patient at right angles to the film; when used for the skull, cervical spine, thorax, and abdomen, they are described as right or left lateral projections; when used for extremities, they are described as lateromedial or mediolateral projections.
oblique projection	(ō-blēk prō-jehck-shuhn)	X-ray beam passes through the body on an angle. May be further described by the direction of the beam and the degree of the angle (i.e., 60 degrees DP oblique of the hoof) or simply the direction (i.e., dorsomedial-palmarolateral oblique [DMPaLO] of the carpus).

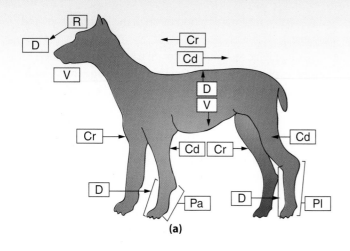

(a)

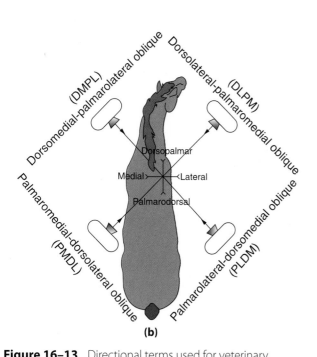

(b)

Figure 16–13 Directional terms used for veterinary radiography with their standard abbreviations (based on American College of Veterinary Radiologists) are shown in these figures. Cr = cranial, D = dorsal, V = ventral, R = rostral, Pa = palmer, Pl = plantar, L = lateral, and M = medial. (a) When radiographic views of given structures are written, medial and lateral should be subservient when used in combination with other terms (e.g., dorsomedial); (b) on the limbs, dorsal, palmar, plantar, cranial, and caudal should take precedence when used in combination with other terms (e.g., dorsoproximal).

is called **CT scan** or **CAT** (computed axial tomography) **scan. Tomography** (tō-moh-grah-fē) is a recording of the internal body structures at predetermined planes. Information obtained by radiation detectors is downloaded to a computer, analyzed, and converted to grayscale images corresponding to anatomical body slices. Those images are viewed on a monitor or as a printed hard copy (Figure 16–14).

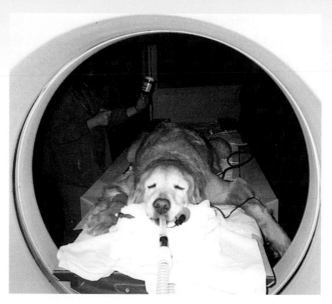

Figure 16–14 CT scan. (Courtesy of Kimberly Kruse Sprecher, CVT.)

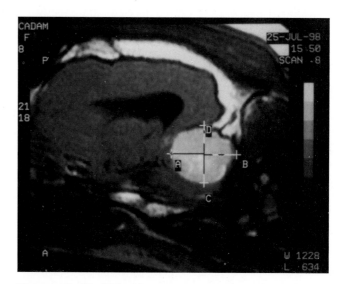

Figure 16–15 MRI with contrast material in a dog with a cerebellar tumor. (Courtesy of Anne E. Chauvet, DVM, Diplomate ACVIM—Neurology, University of Wisconsin School of Veterinary Medicine.)

Magnetic Resonance Imaging

Magnetic resonance imaging (mahg-neh-tihck reh-sohn-ahns ih-mah-jihng) is the procedure in which radio waves and a strong magnetic field pass through the patient and show the internal body structures in three-dimensional views. Magnetic resonance imaging is abbreviated MRI. MRI is used for imaging the brain, spine, and joints (Figure 16–15).

Fluoroscopy

Fluoroscopy (floor-ohs-kō-pē) is the procedure used to visually examine internal body structures in motion using radiation

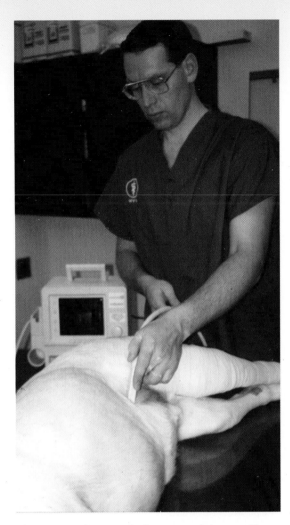

Figure 16–16 Ultrasound of a canine abdomen. (Courtesy of Lodi Veterinary Hospital, SC.)

The following terms are related to ultrasound techniques:

- **amplitude** (ahm-plih-tood) = intensity of an ultrasound wave.
- **anechoic** (ahn-eh-kō-ihck) = ultrasonic term for when waves are transmitted to deeper tissue and none are reflected back.
- **attenuation** (ah-tehn-yoo-ā-shuhn) = loss of intensity of the ultrasound beam as it travels through tissue.
- **echoic** (eh-kō-ihck) = ultrasound property of producing adequate levels of reflections (echoes) when sound waves are returned to the transducer and displayed.
- **frequency** (frē-kwehn-sē) = number of cycles per unit of time.
- **hyperechoic** (hī-pər-eh-kō-ihck) = tissue that reflects more sound back to the transducer than the surrounding tissues; appears bright (Figure 16–17).
- **hypoechoic** (hī-pō-eh-kō-ihck) = tissue that reflects less sound back to the transducer than the surrounding tissues; appears dark.
- **isoechoic** (ī-sō-eh-kō-ihck) = tissue that has the same ultrasonic appearance as that of the surrounding tissue.
- **resolution** (rehs-ō-loo-shuhn) = ability to separately identify different structures on radiograph or ultrasound.
- **velocity** (vehl-oh-sih-tē) = speed at which something travels through an object.
- **wavelength** (wāv-lehngth) = length that a wave must travel in one cycle.

to project images on a fluorescent screen. The combining form **fluor/o** means luminous. **Luminous** (loo-mih-nuhs) means giving off a soft light.

Ultrasound

Ultrasound (uhl-trah-sownd), or **ultrasonography** (uhl-trah-soh-noh-grah-fē), is the imaging of internal body structures by recording echoes of high-frequency waves. Ultrasound is most effective for viewing solid organs or soft tissues not blocked by bone or air. Ultrasound also is effective for viewing body parts through fluid, as in an ultrasound of a gravid uterus. A **sonogram** (soh-nō-grahm) shows the internal body structures by recording echoes of pulses of sound waves above the range of human hearing (Figure 16–16).

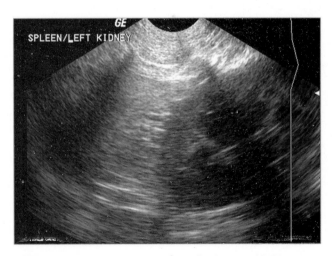

Figure 16–17 Ultrasound image of spleen and kidney showing hyperechoic and hypoechoic areas. The kidney is hypoechoic to the spleen. (Courtesy of Fern Delany, RDMS, VMTH University of Wisconsin-Madison.)

REVIEW EXERCISES

Multiple Choice

Choose the correct answer.

1. Examination by tapping the surface to determine density of a body area is called

    a. palpation
    b. auscultation
    c. percussion
    d. tapping

2. The percentage of RBCs in blood is called a

    a. hematocrit
    b. PCV
    c. crit
    d. all of the above

3. A machine that spins samples very rapidly to separate elements based on weight is a

    a. counter
    b. centrifuge
    c. refractometer
    d. cannula

4. The quality of appearing white or light gray on a radiograph is called

    a. radiopaque
    b. radiolucent
    c. radiodense
    d. radiopale

5. The determination of the cause of disease is the

    a. prognosis
    b. diagnosis
    c. symptom
    d. sign

6. A set of signs that occur together is called a(n)

    a. sign
    b. symptom
    c. endemic
    d. syndrome

7. The medical term for sphere is

    a. bacilli
    b. spirochete
    c. coccus
    d. strept/o

8. The term for relief of condition but not a cure is

    a. prognosis
    b. diagnosis
    c. prevalence
    d. palliative

9. Pertaining to fever is

    a. febrile
    b. friable
    c. lethal
    d. morbid

10. A disease that can be transmitted between animals and humans is said to be

    a. animalistic
    b. zoonotic
    c. humanistic
    d. sequela

11. What is the term for the ultrasound property of producing adequate levels of reflections (echoes) when sound waves are returned to the transducer and displayed?

    a. echoic
    b. anechoic
    c. hyperechoic
    d. hypoechoic

12. The acronym used to describe a unit by which absorption of ionizing radiation is measured is

    a. MAS
    b. kVp
    c. CT
    d. rad

13. Surgical puncture to remove fluid or gas for diagnostic purposes or for treatment is known as

    a. cannula
    b. centesis
    c. contagious
    d. communicable

14. Examination by feeling is

    a. percussion
    b. auscultation
    c. assessment
    d. palpation

15. Clumping of cells is known as

    a. an assay
    b. agglutination
    c. serology
    d. a profile

16. The medical term for prevention is

    a. prevalence
    b. palliative
    c. prophylaxis
    d. phobia

17. The description of an animal with information about the animal, including the species, breed, age, and sexual status, is called the animal's

    a. symptoms
    b. signalment
    c. assessment
    d. auscultation

18. A decrease in body temperature is known as

    a. afebrile
    b. febrile
    c. hypothermia
    d. prognosis

19. The term for having a short course is

    a. remission
    b. acute
    c. chronic
    d. endemia

20. The study of disease causes is

    a. etiology
    b. toxicology
    c. biology
    d. pathology

Matching

Match the testing term in Column I with the definition in Column II.

Column I

1. _____ infectious disease
2. _____ contagious disease
3. _____ noncontagious disease
4. _____ noninfectious disease
5. _____ communicable disease
6. _____ iatrogenic disease
7. _____ nosocomial infection
8. _____ zoonotic disease
9. _____ diagnosis
10. _____ susceptible
11. _____ trocar
12. _____ projection
13. _____ roentgen
14. _____ bacillus
15. _____ coccus
16. _____ chronic

Column II

a. disorder caused by physicians or veterinarians and the treatment ordered

b. disorder caused by pathogenic organisms contracted in a facility or clinic

c. disease spread from one animal to another by direct or indirect contact

d. disorder caused by pathogenic organisms

e. disease that can be transmitted between animals and humans

f. disorder transmitted from animal to animal or through contact with contaminated objects

g. disorder not caused by organisms (examples include genetic, traumatic, and iatrogenic)

h. disease that cannot be spread to another animal by contact or contact with an infected object

i. international unit of radiation

j. path of the X-ray beam

k. spherical bacterium

l. determination of the cause of disease

m. having a long course

n. sharp, needlelike instrument that has a tube that is used to puncture the wall of a body cavity to withdraw fluid or gas

o. rod-shaped bacterium

p. lacking resistance

Fill in the Blanks

1. Pyr/o means _____.
2. Fluor/o means _____.
3. a- means _____.
4. HR is the abbreviation for _____.
5. RR is the abbreviation for _____.
6. V/D is the abbreviation for _____.
7. MLO is the abbreviation for _____.
8. DLPM is the abbreviation for _____.
9. DMPL is the abbreviation for _____.
10. DP is the abbreviation for _____.

Spelling

Circle the term that is spelled correctly.

1. ability of an organism to cause disease: virilence   virulence   virulense

2. able to be spread from one animal to another
   by direct or indirect contact: contagious   kontagious   contagous

3. laboratory study of serum: sirology   sarology   serology

4. fever: pyrexia   pyrexa   pirexa

5. instrument used to enlarge the opening of a canal or cavity: spekulum   speculum   speckulum

CROSSWORD PUZZLE

Testing Terminology Puzzle

Across

3   tissue that reflects less sound back to the transducer than the surrounding tissues; appears dark

8   number of leukocytes per cubic milliliter of blood

10   ultrasound property of producing adequate levels of reflections (echoes) when sound waves are returned to the transducer and displayed

11   number of cycles per unit of time

17   high pH property

18   spiral-shaped bacteria

20   concentration of hydrogen ions

22   diagnostic evaluation of blood to determine the number of rbc, wbc, and thrombocytes per cubic milliliter of blood

23   rod-shaped bacteria

25   combining form for chain

26   loss of intensity of the ultrasound beam as it travels through tissue

27   tissue that reflects more sound back to the transducer than the surrounding tissues; appears bright

Down

1   ability to separately identify different structures on radiograph or ultrasound

2   international unit of radiation

5   path of the X-ray beam

6   ultrasonic term for when waves are transmitted to deeper tissue and none are reflected back

7   microscopic living organism

9   plant-like eukaryotic organism without chlorophyll

12   budding form of fungus

13   combining form for grapelike clusters

14   speed at which something travels through an object

15   tissue that has the same ultrasonic appearance as that of the surrounding tissue

16   low pH property

19   specified body position and the part of the body closest to the film

21   length that a wave must travel in one cycle

24   intensity of an ultrasound wave

CHAPTER 17

[DRUGS, DISEASES, AND DISSECTION]

Objectives

Upon completion of this chapter, the reader should be able to

- Recognize, define, spell, and pronounce the terms associated with pharmacology and drugs used in various treatments
- Recognize, define, spell, and pronounce the terms associated with pathological procedures and processes
- Recognize, define, spell, and pronounce the terms associated with different types of surgery and the instruments used in surgery

PHARMACOLOGIC TERMS

Pharmacology (fahrm-ah-kohl-lō-jē) is the study of the nature, uses, and effects of drugs. Some drugs are dispensed only by a licensed professional, and other drugs are not. A **prescription** (prē-skrihp-shuhn) **drug** is a medication that may be purchased by prescription or from a licensed professional. An **over-the-counter** (ō-vər theh kount-ər) **drug** is a medication that may be purchased without a prescription. A **generic** (jehn-ār-ihck) **drug** is a medication not protected by a brand name or trademark. (It also is called a **nonproprietary drug.**)

Pharmacology

- **agonist** (ā-gohn-ihst) = substance that produces effect by binding to an appropriate receptor.
- **antagonist** (ahn-tā-gohn-ihst) = substance that inhibits a specific action by binding with a particular receptor instead of allowing the agonist to bind to the receptor.
- **antiserum** (ahn-tih-sēr-uhm) = serum containing specific antibodies extracted from a hyperimmunized animal, or an animal that has been infected with the microorganisms containing antigen.
- **antitoxin** (ahn-tih-tohcks-sihn) = specific antiserum aimed at a poison that contains a concentration of antibodies extracted from the serum or plasma of a healthy animal.
- **bacterin** (bahck-tər-ihn) = bacterial vaccine.
- **chelated** (kē-lā-tehd) = bound to and precipitated out of solution.
- **contraindication** (kohn-trah-ihn-dih-kā-shuhn) = recommendation not to use.
- **diffusion** (dih-fū-shuhn) = movement of solutes from an area of high concentration of particles to one of low concentration of particles (Figure 17–1).
- **dosage** (dō-sahj) = amount of medication based on units per weight of animal (such as 10 mg/lb and 2 mg/kg).
- **dosage interval** (dō-sahj ihn-tər-vahl) = time between administrations of a drug (such as bid or q12h).
- **dose** (dōs) = amount of medication measured (such as milligrams, milliliters, units, and grams).
- **drug** (druhg) = agent used to diagnose, prevent, or treat a disease.
- **efficacy** (ehf-ih-kah-sē) = extent to which a drug causes the intended effects; effectiveness.
- **endogenous** (ehn-dah-jehn-uhs) = originating within the body.
- **exogenous** (ehcks-ah-jehn-uhs) = originating outside the body.
- **hydrophilic** (hī-drō-fihl-ihck) = water-loving; ionized form.
- **hyperkalemia** (hī-pər-kā-lē-mē-ah) = excessive level of blood potassium.
- **hypernatremia** (hī-pər-nā-trē-mē-ah) = excessive level of blood sodium.
- **hypertonic** (hī-pər-tohn-ihck) **solution** = solution that has more particles than the solution or cell to which it is being compared. The tonicity of solutions usually is compared to blood cells (Figure 17–2).
- **hypokalemia** (hī-pō-kā-lē-mē-ah) = deficiency of blood potassium.
- **hyponatremia** (hī-pō-nā-trē-mē-ah) = deficiency of blood sodium.
- **hypotonic** (hī-pō-tohn-ihck) **solution** = solution that has fewer particles than the solution or cell to which it is being compared (Figure 17–2).

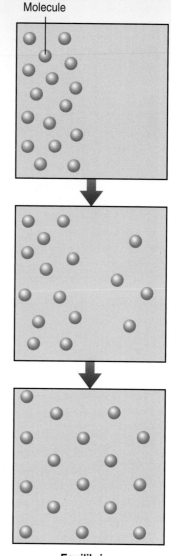

Molecule

Equilibrium

Figure 17–1 Diffusion is the random movement of molecules from an area of high concentration of particles to an area of low concentration of particles. Diffusion can occur in the presence or absence of a membrane.

- **hypovolemia** (hī-pō-vō-lē-mē-ah) = low circulating blood volume.
- **ionized** (ī-ohn-īzd) = electrically charged.
- **isotonic** (ī-sō-tohn-ihck) **solution** = solution that has equal particles to the solution or cell to which it is being compared (Figure 17–2).
- **lipophilic** (lihp-ō-fihl-ihck) = fat-loving; nonionized form.
- **monovalent** (mohn-ō-vā-lehnt) = vaccine, antiserum, or antitoxin developed specifically for a single antigen or organism.
- **nonionized** (nohn-ī-ohn-īzd) = not charged electrically.
- **osmosis** (ohz-mō-sihs) = movement of water across a selectively permeable membrane along its concentration gradient (Figure 17–3).

Hypertonic solution
(more particles
outside cell)

Hypotonic solution
(fewer particles
outside cell)

Isotonic solution
(equal particles inside
and outside cell)

Figure 17–2 Solution tonicity.

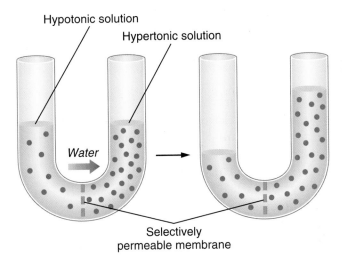

Hypotonic solution

Hypertonic solution

Water

Selectively
permeable membrane

Figure 17–3 Osmosis is the movement of water across a selectively permeable membrane along its concentration gradient. In this example, water moves across the membrane to equalize the concentration of particles.

- **pharmacodynamics** (fahrm-ah-kō-dī-nahm-ihcks) = physiological effects of drugs and their mechanisms of action.
- **pharmacokinetics** (fahrm-ah-kō-kihn-eht-ihcks) = movement of drugs or chemicals; consists of absorption, distribution, biotransformation, and elimination.
- **placebo** (plah-sē-bō) = inactive substance that is given for its suggestive effects or substance used as a control in experimental setting.
- **polyvalent** (poh-lē-vā-lehnt) = vaccine, antiserum, or antitoxin that is active against multiple antigens or organisms; mixed vaccine.
- **prevention** (prē-vehn-shuhn) = avoidance; also called **prophylaxis** (prō-fih-lahck-sihs).
- **regimen** (reh-jeh-mehn) = course of treatment.
- **turgor** (tər-gər) = degree of fullness or rigidity caused by fluid content.

Routes of Administration

- **inhalation** (ihn-hah-lā-shuhn) = vapors and gases taken in through the nose and mouth and absorbed into the bloodstream through the lungs (Figure 17–4).
- **intra-arterial** (ihn-trah-ahr-tehr-ē-ahl) = within the artery; abbreviated IA.
- **intradermal** (ihn-trah-dər-mahl) = within the skin; abbreviated ID.

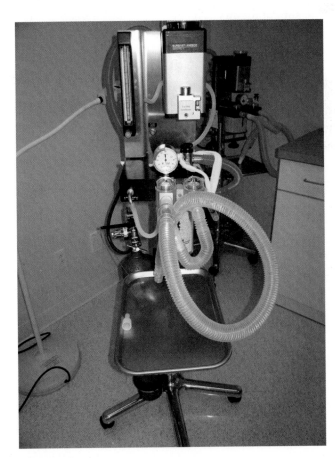

Figure 17–4 Inhalation anesthesia is given to the patient via an anesthetic machine.

- **intramuscular** (ihn-trah-muhs-kyū-lahr) = within the muscle; abbreviated IM.
- **intraocular** (ihn-trah-ohck-yoo-lahr) = within the eye.
- **intraosseous** (ihn-trah-ohs-ē-uhs) = within the bone (medullary cavity of a long bone).
- **intraperitoneal** (ihn-trah-pehr-ih-tohn-ē-ahl) = within the peritoneal cavity; abbreviated IP.
- **intrathecal** (ihn-trah-thē-kahl) = within a sheath; injection of a substance through the spinal cord and into the subarachnoid space.
- **intratracheal** (ihn-trah-trā-kē-ahl) = within the trachea, or windpipe.
- **intravenous** (ihn-trah-vehn-uhs) = within the vein; abbreviated IV.
- **nebulization** (nehb-yoo-lih-zā-shuhn) = process of making a fine mist; a method of drug administration.
- **nonparenteral** (nohn-pah-rehn-tər-ahl) = administration via the gastrointestinal tract.
- **oral** (ōr-ahl) = by mouth; abbreviated PO or p.o. *Nothing orally* is abbreviated NPO or n.p.o.
- **parenteral** (pah-rehn-tər-ahl) = through routes other than the gastrointestinal tract (Figure 17–5).
- **percutaneous** (pehr-kyoo-tā-nē-uhs) = through the skin.
- **rectal** (rehck-tahl) = by rectum.
- **subcutaneous** (suhb-kyoo-tā-nē-uhs) = under the skin, or dermal layer; abbreviated SQ, SC, or subq.
- **sublingual** (suhb-lihng-wahl) = under the tongue.
- **transdermal** (trahnz-dər-mahl) = across the skin. Medication is stored in a patch placed on the skin, and the medication is absorbed through the skin.

Drug Categories

- **analgesic** (ahn-ahl-jē-zihck) = substance that relieves pain without affecting consciousness.
- **anesthetic** (ahn-ehs-theht-ihck) = substance that produces a lack of sensation (Figure 17–6).

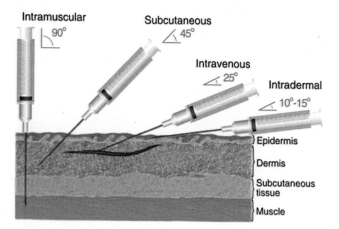

Figure 17–5 Examples of parenteral routes of drug administration.

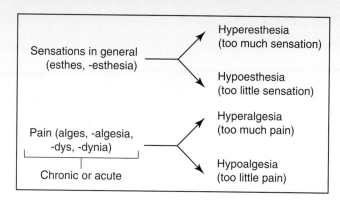

Figure 17–6 Word parts for sensation and pain.

- **anthelmintic** (ahn-thehl-mihn-tihck) = substance that works against intestinal worms.
- **antibiotic** (ahn-tih-bī-ah-tihck) = substance that inhibits the growth of or kills bacteria. Antibiotics can be **bacteriostatic** (bahck-tē-rē-ō-stah-tihck), which means controlling bacterial growth (inhibiting bacterial replication), or **bactericidal** (bahck-tē-rē-sī-dahl), which means killing bacteria.
- **anticoagulant** (ahn-tih-kō-ahg-yoo-lahnt) = substance that inhibits clot formation.
- **anticonvulsant** (ahn-tih-kohn-vuhl-sahnt) = substance that prevents seizures.
- **antidiarrheal** (ahn-tih-dī-ər-rē-ahl) = substance that prevents watery, frequent bowel movements.
- **antiemetic** (ahn-tih-ē-meh-tihck) = substance that prevents vomiting.
- **antineoplastic agent** (ahn-tih-nē-ō-plah-stihck ā-jehnt) = substance that treats neoplasms; usually used against malignancies.
- **antipruritic agent** (ahn-tih-prər-ih-tihck ā-jehnt) = substance that controls itching.
- **antipyretic** (ahn-tih-pī-reh-tihck) = substance that reduces fever.
- **antiseptic** (ahn-tih-sehp-tihck) = chemical agent that kills or prevents the growth of microorganisms on living tissue.
- **antitussive** (ahn-tih-tuhs-ihv) = substance that reduces coughing.
- **asepsis** (ā-sehp-sihs) = state without infection.
- **chronotrope** (krohn-ō-trōp) = substance that changes heart rate. Positive chronotropes increase heart rate, while negative chronotropes decrease heart rate.
- **cytotoxic agent** (sī-tō-tohcks-ihck ā-jehnt) = substance that kills or damages cells.
- **disinfectant** (dihs-ehn-fehck-tahnt) = chemical agent that kills or prevents the growth of microorganisms on inanimate objects.
- **emetic** (ē-meh-tihck) = substance that induces vomiting.
- **endectocide** (ehnd-ehck-tō-sīd) = agent that kills both internal and external parasites.

- **immunosuppressant** (ihm-yoo-nō-suhp-prehsahnt) = substance that prevents or decreases the body's reaction to invasion by disease or foreign material.
- **inotrope** (ihn-ō-trōp) = substance affecting muscle contraction. Positive inotropes increase myocardial contractility, while negative inotropes decrease myocardial contractility.
- **miotic agent** (mī-ah-tihck ā-jehnt) = substance used to constrict the pupils.
- **mucolytic** (mū-kō-lih-tihck) = substance that breaks up mucus and reduces its viscosity.
- **mydriatic agent** (mihd-rē-ah-tihck ā-jehnt) = substance used to dilate the pupils.
- **sterilize** (stehr-ih-līz) = to destroy all organisms including bacterial endospores.

Weights and Measures

- **centimeter** (sehn-tih-mē-tər) = metric unit of length equal to one one-hundredth of a meter; abbreviated cm.
- **dram** (drahm) = apothecary unit of measure used for prescription vials. One dram equals 1.8 ounce (by weight), and 1 fluid dram equals 4 mL.
- **gram** (grahm) = metric base unit of weight equal to 0.035 ounce; abbreviated g.
- **kilogram** (kihl-ō-grahm) = metric unit of weight that is 1000 grams; 1 kilogram is approximately 2.2 pounds; abbreviated kg.
- **liter** (lē-tər) = metric base unit of volume equal to 0.2642 gallons ; abbreviated L.
- **meter** (mē-tər) = metric base unit of length equal to 1.09 yards; abbreviated m.
- **milligram** (mihl-ih-grahm) = metric unit of weight equal to one one-thousandth of a gram; abbreviated mg.
- **milliliter** (mihl-ih-lē-tər) = metric unit of volume equal to 0.034 of an ounce or one one-thousandth of a liter; abbreviated mL; equivalent to 1 cubic centimeter (cc) (Table 17–1).
- **millimeter** (mihl-ih-mē-tər) = metric unit of length equal to one one-thousandth of a meter; abbreviated mm.
- **percent** (pər-sehnt) = part per 100 parts; represented by %.

SURGICAL TERMS

Surgery is the branch of science that treats diseases, injuries, and deformities by manual or operative methods. Surgical terms were developed to describe concisely many surgical procedures. Some surgical terms include the following:

- **appositional** (ahp-ō-sih-shuhn-ahl) = placed side to side.
- **aseptic technique** (ā-sehp-tihck tehck-nēk) = precautions taken to prevent contamination of a surgical wound.
- **avulsion** (ə-vuhl-shuhn) = tearing away of a part.

Table 17–1 Frequently Used Drug Abbreviations

Abbreviation	Definition
bid	twice daily (bis in die)
c̄	with
cc	cubic centimeter (same as mL)
gt	drop (gutta); drops is gtt (guttae)
mL	milliliter
NPO/n.p.o	nothing orally (non per os)
p̄	after
PO/p.o.	orally (per os)
prn	as needed
q	every
q4h	every 4 hours
q6h	every 6 hours
q8h	every 8 hours
q12h	every 12 hours
q24h	every 24 hours
qd	every day (same as sid)
qh	every hour
qid	four times daily
qn	every night
qod or eod	every other day
s̄	without
sid	once daily (qd or q24h is the preferred abbreviation)
T	tablespoon or tablet
tab	tablet (also abbreviated T)
tid	three times daily (ter in die)

- **coaptation** (kō-ahp-tā-shuhn) = act of approximating.
- **curettage** (kyoo-reh-tahj) = removal of material or growths from the surface of a cavity.

- **debridement** (deh-brēd-mehnt) = removal of foreign material and devitalized or contaminated tissue.
- **dehiscence** (dē-hihs-ehns) = disruption or opening of the surgical wound (Figure 17–7).
- **dissect** (dī-sehckt) = separation or cutting apart; **dissecare** is Latin for "to cut up."
- **enucleation** (ē-nə-klē-ā-shuhn) = removal of an organ in whole; usually used for removal of the eyeball (Figure 14–18).
- **epithelialization** (ehp-ih-thē-lē-ahl-ih-zā-shuhn) = healing by growth of epithelium over an incomplete surface.
- **eversion** (ē-vər-shuhn) = turning outward (Figure 17–8).
- **eviscerate** (ē-vihs-ər-āt) = removal or exposure of internal organs.
- **excise** (ehck-sīz) = to surgically remove.
- **exteriorize** (ehcks-tēr-ē-ōr-īz) = to move an internal organ to the outside of the body (Figure 12–26 which shows exteriorization of the female reproductive tract).

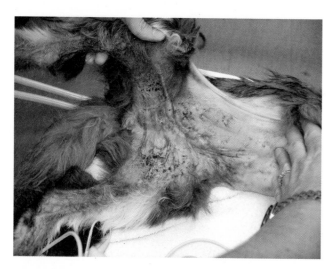

Figure 17–7   Wound dehiscence in a cat. (Courtesy of Kimberly Kruse Sprecher, CVT.)

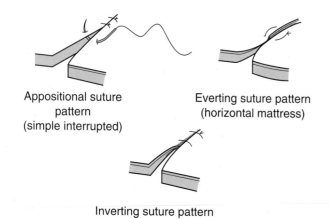

Appositional suture pattern (simple interrupted)

Everting suture pattern (horizontal mattress)

Inverting suture pattern (vertical mattress)

Figure 17–8   Appositional versus everting versus inverting suture patterns.

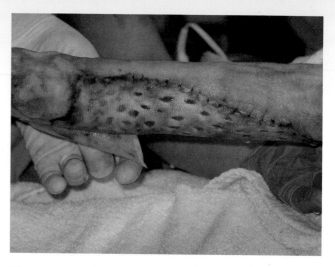

Figure 17–9   Skin graft in a dog. (Courtesy of Kimberly Kruse Sprecher, CVT.)

- **fenestration** (fehn-ih-strā-shuhn) = perforation.
- **flap** (flahp) = mass of tissue for grafting in which part of the tissue still adheres to the body; used to repair defects adjacent to the mass site.
- **fracture** (frahck-shər) = breaking of a part, especially a bone.
- **friable** (frī-ah-buhl) = easily crumbled.
- **fulguration** (fuhl-gər-ā-shuhn) = destruction of living tissue by electric sparks generated by a high-frequency current.
- **graft** (grahft) = tissue or organ for transplantation or implantation (Figure 17–9). There are different types of grafts. An **allograft** (ah-lō-grahft) is a graft from another individual of the same species. An **autograft** (awt-ō-grahft) is a graft from the same individual. An **isograft** (ī-sō-grahft) is a graft from genetically identical animals, such as twins or inbred strains.
- **imbrication** (ihm-brih-kā-shuhn) = overlapping of apposing surfaces to realign organs and provide extra support.
- **implant** (ihm-plahnt) = material inserted or grafted into the body.
- **incise** (ihn-sīz) = to surgically cut into.
- **intraop** (ihn-trah-ohp) = common term for *during surgery*; intraoperatively.
- **inversion** (ihn-vər-shuhn) = turning inward.
- **involucrum** (ihn-voh-loo-kruhm) = covering or sheath that contains a sequestrum of bone.
- **laceration** (lah-sihr-ā-shuhn) = act of tearing.
- **lavage** (lah-vahj) = irrigation of tissue with fluid.
- **ligate** (lī-gāt) = to tie or strangulate. A **ligature** (lihg-ah-chūr) is any substance used to tie or strangulate. Ligatures usually are made of suture material.
- **lumpectomy** (luhmp-ehck-tō-mē) = general term for surgical removal of a mass.
- **pinning** (pihn-ihng) = insertion of a metal rod into the medullary cavity of a long bone.

- **postop** (pōst-ohp) = common term for after surgery; postoperatively.
- **preop** (prē-ohp) = common term for before surgery; preoperatively.
- **resect** (rē-sehckt) = to remove an organ or tissue. Resect is used in reference to holding tissue or an organ out of the surgical field.
- **rupture** (ruhp-chuhr) = forcible tearing.
- **sacculectomy** (sahk-yoo-lehck-tō-mē) = surgical removal of a saclike part; usually refers to surgical removal of the anal sacs.
- **seroma** (sehr-ō-mah) = accumulation of serum beneath the surgical incision.
- **suction** (suhck-shuhn) = aspiration of gas or fluid by mechanical means (Figures 17–10a and b).
- **suture** (soo-chuhr) = to stitch or close an area; also refers to the material used in closing a surgical or traumatic

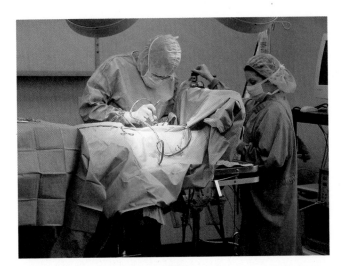

(a)

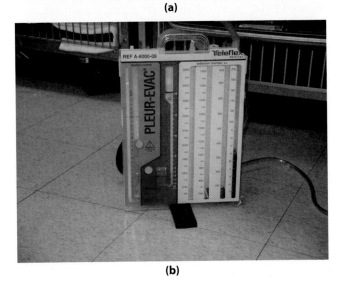

(b)

Figure 17–10 (a) Suction is used to remove fluid (or gas) during surgery. (b) Suction unit showing the collection chamber. [(a) Courtesy of Dr. David Sweet and Ann Zackim; (b) Courtesy of Kimberly Kruse Sprecher, CVT.]

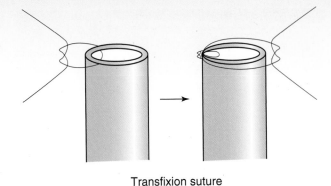

Transfixion suture

Figure 17–11 Transfixion suture. Transfixion sutures are used for large, isolated vessels and organs to prevent slippage of the ligature.

wound with stitches. Suture material may be absorbable or nonabsorbable. (Suture also is a type of joint.)

- **transect** (trahn-sehckt) = to cut across; a cross section or section made across a long axis. **Sect** means to cut.
- **transfix** (trahnz-fihcks) = pierce through and through. A transfixion suture pierces through an organ before ligation (Figure 17–11).
- **transplant** (trahnz-plahnt) = to transfer tissue from one part to another part.
- **wicking** (wihck-ihng) = applying material to move liquid from one area to another.

Surgical Equipment

- **autoclave** (aw-tō-klāv) = apparatus for sterilizing by steam under pressure (Figure 17–12).
- **bandage** (bahn-dahj) = to cover by wrapping or the material used to cover by wrapping (Figure 17–13).
- **belly band** (behl-ē bahnd) = common term for abdominal wrap; circumferentially wrapping the abdomen with bandages to apply pressure to the area.
- **bone plate** = flat metal bar with screw holes that is used in bone fracture repair (Figures 17–14a and b).
- **bone screw** = screw that holds bone fragments together to repair bone fractures.
- **boxlock** (bohcks-lohck) = movable joint of any ringed instrument (Figure 17–15).
- **cast** (kahst) = stiff dressing used to immobilize various body parts.
- **cautery** (caw-tər-ē) = application of a burning substance, a hot instrument, an electric current, or another agent to destroy tissue.
- **cerclage** (sihr-klahj) **wire** = band of metal that completely (cerclage) or partially (hemicerclage) goes around the circumference of bone that is used in conjunction with other stabilization techniques to repair bone fractures (Figure 17–16).

(a)

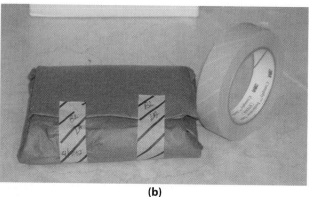

(b)

Figure 17–12 (a) An autoclave provides high-pressure steam heat to sterilize surgical instruments. (b) In a surgery pack (a collection of instruments used for one procedure) that has been properly autoclaved, the indicator tape will change from light-colored lines (roll on right) to dark-colored lines (tape on pack).

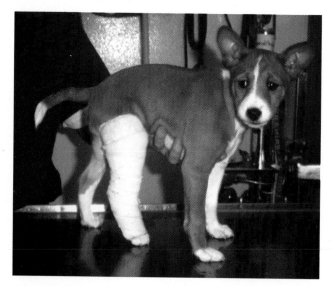

Figure 17–13 Puppy with a Robert Jones bandage. (Courtesy of Lodi Veterinary Hospital, SC.)

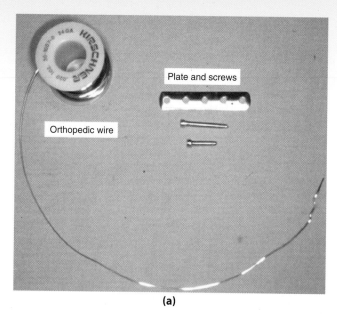

Plate and screws

Orthopedic wire

(a)

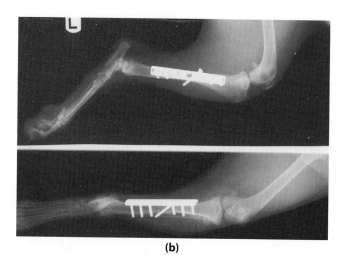

(b)

Figure 17–14 (a) Examples of orthopedic equipment (b) Radiograph showing the repair of a tibial fracture using a bone plate and screws. [(a)Courtesy of Teri Raffel, CVT.]

- **chuck** (chuhck) = clamping device for holding a drill bit.
- **clamp** (klahmp) = instrument used to secure or occlude things.
- **curette** (kyoor-reht) = instrument with cupped head to scrape material from cavity walls.
- **drain** (drān) = device by which a channel may be established for the exit of fluids from a wound (Figures 17–17a and b).
- **drape** (drāp) = cloth arranged over a patient's body to provide a sterile field around the area to be examined, treated, or incised.
- **dressing** (drehs-sihng) = various materials used to cover and protect a wound.
- **elastrator** (ē-lahs-trā-tər) = bloodless castration device using small elastic bands.

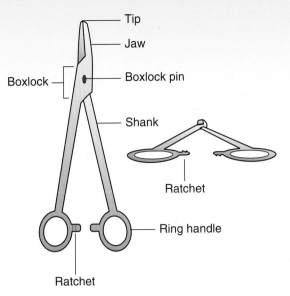

Figure 17–15 Parts of surgical instruments.

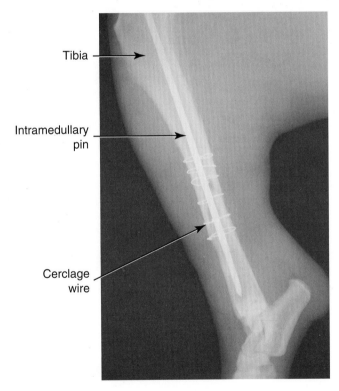

Figure 17–16 Radiograph showing the repair of a fractured tibia. The bone was repaired with an intramedullary pin and six cerclage wires.

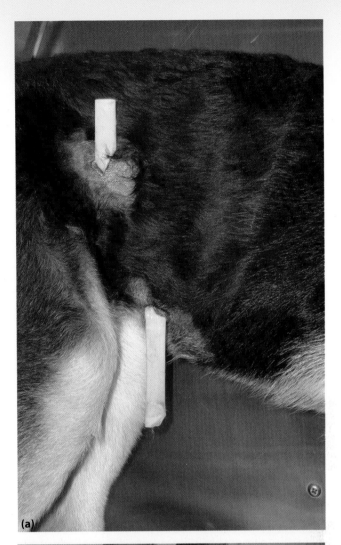

(a)

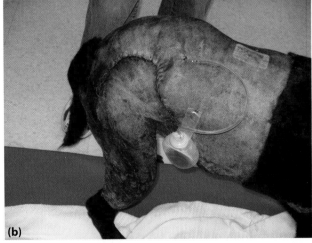

(b)

Figure 17–17 Surgical drain. (a) A Penrose drain was placed in the flank area of this dog to maintain an opening in the skin that allows accumulating fluid to drain to the exterior. The fluid does not drain through the center of the tubing, but is allowed to leak from the skin openings. (b) Dog with an active drain and reservoir unit which allows measurement of drainage volume and description of the fluid appearance. (Courtesy of Kimberly Kruse Sprecher, CVT.)

- **elevator** (ehl-eh-vā-tər) = instrument used to reflect tissue from bone.
- **emasculatome** (ē-mahs-kyoo-lah-tōm) = instrument used to crush and sever the spermatic cord through intact skin.
- **emasculator** (ē-mahs-kyoo-lā-tər) = instrument used in closed castrations to crush and sever the spermatic cord.

- **hemostatic forcep** (hē-mō-stah-tihck fōr-sehp) = locking instrument used to grasp and ligate vessels and tissues to control bleeding; also called **hemostat** (hē-mō-staht).
- **intramedullary pins** (ihn-trah-mehd-yoo-lahr-ē pihnz) = metal rods that are inserted into the medullary cavity of long bones to repair stable fractures (Figure 17–16).
- **prosthesis** (prohs-thē-sihs) = artificial substitute for a diseased or missing part of the body.
- **ratchet** (rah-cheht) = graded locking portion of an instrument located near the finger rings.
- **retractor** (rē-trahck-tər) = instrument used to hold back tissue (Figure 17–18).
- **rongeurs** (rohn-jūrz) = forceps with cupped jaws used to break large bone pieces into smaller ones.
- **scalpel** (skahl-puhl) = small, straight knife with a thin, sharp blade used for surgery and dissection.
- **serration** (sihr-ā-shuhn) = sawlike edge or border.
- **sling** (slihng) = bandage for supporting part of the body.
- **splint** (splihnt) = rigid or flexible appliance for fixation of movable or displaced parts (Figure 17–19).
- **tissue forceps** (tihs-yoo fōr-sehps) = tweezerlike, nonlocking instruments used to grasp tissue.

Surgical Approaches

In surgery, the specific procedure by which an organ or a part is exposed is called the **approach** (ah-prōch). Different approaches allow the best exposure to different parts of the body. Following are examples of different surgical approaches (Figure 17–20):

- **flank incision** (flahnk ihn-sihz-shuhn) = surgical cut perpendicular to the long axis of the body, caudal to the last rib.
- **paracostal incision** (pahr-ah-kah-stahl ihn-sihz-shuhn) = surgical cut oriented parallel to the last rib.
- **paramedian incision** (pahr-ah-mē-dē-ahn ihn-sihz-shuhn) = surgical cut lateral and parallel to the ventral midline but not on the midline.
- **ventral midline incision** (vehn-trahl mihd-līn ihn-sihz-shuhn) = surgical cut along the midsagittal plane of the abdomen along the linea alba.

Biopsies

The term **biopsy** (bī-ohp-sē) means removing living tissue to examine. Biopsy also is used for the specimen removed during the procedure. The first definition is more correct; however, the term *biopsy* is commonly used both ways.

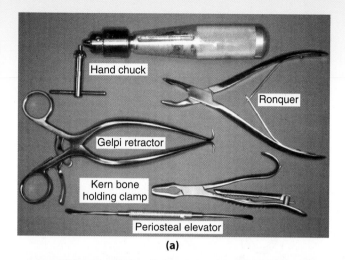

(a)

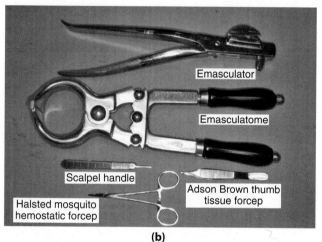

(b)

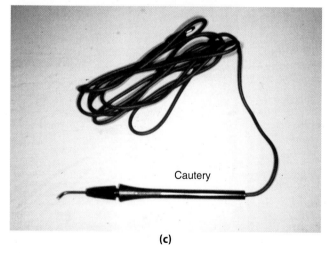

(c)

Figure 17–18 (a) and (b) Examples of surgical instruments. (c) Cautery unit for hemostasis. (Courtesy of Teri Raffel, CVT.)

Types of biopsies include the following:

- **excisional biopsy** (ehcks-sih-shuhn-ahl bī-ohp-sē) = removing entire mass, tissue, or organ to examine.
- **incisional biopsy** (ihn-sih-shuhn-ahl bī-ohp-sē) = cutting into and removing part of a mass, a tissue, or an organ to examine.

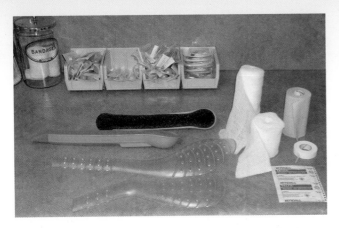

Figure 17–19 Selection of bandage materials and splints.

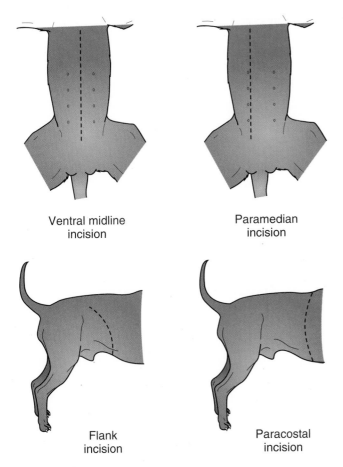

Ventral midline incision

Paramedian incision

Flank incision

Paracostal incision

Figure 17–20 Incision types.

- **needle biopsy** (nē-dahl bī-ohp-sē) = insertion of a sharp instrument (needle) into a tissue for extraction of tissue to be examined.

A **necropsy** (nē-krohp-sē) is a postmortem examination that consists of a thorough examination of a dead animal to determine the cause and manner of death and to evaluate any disease or injury that may be present.

Needles and Sutures

Suture material and needles are used by surgeons to close wounds or to tie things (Figure 17–21). Terms used in reference to suture material and needles include the following:

- **blunt** (bluhnt) = dull, not sharp; used to describe needles or instrument ends.
- **cutting** (kuht-ihng) **needle** = needle that has two or three opposing cutting edges.
- **ligation** (lī-gā-shuhn) = act of tying.
- **ligature** (lihg-ah-chūr) = substance used to tie a vessel or strangulate a part.
- **monofilament** (mohn-ō-fihl-ah-mehnt) = single strand of material; used to describe suture.
- **multifilament** (muhl-tī-fihl-ah-mehnt) = several strands that are twisted together; used to describe suture.
- **stapling** (stā-plihng) = method of suturing that involves the use of stainless steel staples to close a wound.
- **surgical clip** (sihr-jih-kahl klihp) = metal staplelike device used for vessel ligation.
- **swaged** (swehgd) **needle** = needle joined with suture material in a continuous unit; eyeless needle.
- **taper** (tā-pər) **needle** = needle with a rounded tip that is sharp to allow piercing but not cutting of tissue.

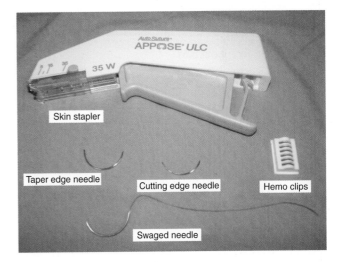

Skin stapler

Taper edge needle   Cutting edge needle   Hemo clips

Swaged needle

Figure 17–21 Needles and staples. (Courtesy of Teri Raffel, CVT.)

REVIEW EXERCISES

Multiple Choice
Choose the correct answer.

1. A monovalent vaccine, antiserum, or antitoxin is one that is developed for

    a. many organisms
    b. one organism
    c. monocyte activation
    d. subcutaneous injection

2. Hydrophilic substances are

    a. solid substances
    b. liquid substances
    c. water-loving substances
    d. fat-loving substances

3. Movement of water across a selectively permeable membrane along its concentration gradient is

    a. diffusion
    b. efficacy
    c. chelation
    d. osmosis

4. Substances that control itching are called

    a. antipruritic
    b. antimiotic
    c. antitussive
    d. antiseptic

5. To surgically cut out is to

    a. incise
    b. excise
    c. ligate
    d. evert

6. The abbreviation for nothing orally is

    a. PO
    b. qid
    c. prn
    d. NPO

7. The surgical term for side-by-side placement is

    a. inversion
    b. eversion
    c. fulguration
    d. apposition

8. A device by which a channel may be established for the exit of fluids from a wound is a

    a. sling
    b. cast
    c. drain
    d. dressing

9. A sawlike edge or border is a

    a. serration
    b. cautery
    c. chuck
    d. cerclage

10. Another name for an eyeless needle is

    a. monofilament
    b. multifilament
    c. swaged
    d. taper

11. The term for an excessive level of sodium in the blood is

    a. hyperkalemia
    b. hypercalcemia
    c. hypernatremia
    d. hypersodemia

12. A solution that is less concentrated than what it is being compared with is known as

    a. hypotonic
    b. isotonic
    c. hypertonic
    d. equivalent

13. The route of drug administration via the gastrointestinal tract is

    a. parenteral
    b. nonparenteral
    c. rectal
    d. transdermal

14. An emetic is a substance that

    a. induces vomiting
    b. controls vomiting
    c. eases constipation
    d. controls diarrhea

15. The term meaning to tie or strangulate is

    a. lavage
    b. gavage
    c. ligate
    d. excise

16. The graded locking portion of an instrument located near the finger rings is the

    a. boxlock
    b. ratchet
    c. chuck
    d. clamp

17. A needle that has a rounded tip is called a

    a. swaged needle
    b. cutting needle
    c. stapling needle
    d. taper needle

18. An instrument that controls bleeding is a

    a. cerclage
    b. sling
    c. hemostat
    d. prosthesis

19. Irrigation of tissue with fluid is called

    a. lavage
    b. gavage
    c. ligate
    d. excise

20. The abbreviation *bid* stands for

    a. once daily
    b. twice daily
    c. three times daily
    d. four times daily

Matching

Match the abbreviation in Column I with its meaning in Column II.

Column I	Column II
1. _____ tid	a. drop
2. _____ gt	b. as needed
3. _____ c̄	c. nothing orally
4. _____ IV	d. subcutaneously
5. _____ IM	e. three times daily
6. _____ SQ	f. every
7. _____ prn	g. intravenously
8. _____ NPO	h. cubic centimeter (same as milliliter)
9. _____ cc	i. intramuscularly
10. _____ q	j. with

Fill in the Blanks

1. A term meaning to remove an organ or tissue is _____ .

2. An accumulation of serum beneath the surgical incision is _____ .

3. The removal or exposure of internal organs is _____ .

4. Another word for perforation is _____ .

5. The degree of fullness or rigidity caused by fluid content is _____ .

Spelling

Circle the term that is spelled correctly.

1. metric unit of volume equivalent to 1 cc: mililiter   milliliter   mililitir

2. excessive levels of potassium in the blood: hypercalemia   hyperkallemia   hyperkalemia

3. administration of drug through routes other than
   the gastrointestinal tract: parental   parentral   parenteral

4. amount of medication based on units per weight of animal: doseage   doesage   dosage

5. water-loving: hydrofillic   hydrophillic   hydrophilic

Word Scramble

Use the definitions to unscramble the terms.

1. act of tying   ntoiaigl   _____

2. blunt point   rtpae   _____

3. cut into   eiincs   _____

4. stiff dressing   satc   _____

5. bandage for supporting a body part   nligs   _____

6. graded locking portion of an instrument tthecar   _____

7. movable joint of any ringed instrument   ooxkclb   _____

8. forcible tearing   uutprer   _____

9. removal of an organ in whole   eeucnlaiont   _____

10. act of approximating   aattoinpco   _____

11. exposure of internal organs   eevtiarsce   _____

12. pierce through and through   xtriansf   _____

CROSSWORD PUZZLE

Pharmacology Terms Puzzle

Across

3 without infection
9 substance that induces vomiting
11 as needed
12 non-GI route of administration
13 amount of medication based on units/weight
14 course of treatment
16 drop
17 substance that reduces coughing
18 substance that dilates pupils

Down

1 to destroy all organisms
2 substance that relieves pain
4 to avoid
5 electrically charged
6 bound to and precipitated out of solution
7 within the skin
8 excessive blood potassium levels
10 agent that kills internal and external parasites
15 killed bacterial vaccine

CHAPTER 18

DOGS AND CATS

Objectives

Upon completion of this chapter, the reader should be able to

- Recognize, define, spell, and pronounce terms related to dogs and cats
- Analyze case studies and apply medical terminology in a practical setting

DOGS AND CATS

For many years, people have used dogs and cats for different purposes. Originally, dogs and cats were domesticated for work such as herding and controlling rodents. Although dogs and cats still may be used for work, they are more commonly kept as pets.

Many of the anatomy and physiology concepts and medical terms related to dogs and cats have been covered in previous chapters. The lists in this chapter apply more specifically to the care and treatment of dogs and cats.

ANATOMY AND PHYSIOLOGY TERMS

- **anal** (ā-nahl) **glands** = secretory tissues composed of apocrine and sebaceous glands located in the anal sac; secretion of the anal glands is stored in the anal sacs and may play a role in territorial marking, as a defense mechanism, or as a pheromone for sexual behavior.
- **anal sacs** (ā-nahl sahcks) = pair of pouches that store an oily, foul-smelling fluid secreted by the anal glands located in the skin between the internal and external anal sphincters (located at the five o'clock and seven o'clock positions); each sac has a duct that opens to the skin at the anal orifice, and fluid is expressed during defecation, excitement, and social interaction (Figures 18–1 and 18–2).
- **carnassial** (kahr-nā-zē-ahl) **tooth** = large, shearing cheek tooth; the upper fourth premolar and lower first molar in dogs (Figure 18–3a) and the upper third premolar and lower first molar in cats; carnassial teeth may develop abscesses secondary to trauma or disease (Figure 18–3b).
- **constitution** (kohn-stih-too-shuhn) = physical makeup of an animal.
- **coprophagy** (kohp-rohf-ah-jē) = ingestion of feces.
- **debarking** (dē-bahrk-ihng) = surgical procedure that cuts vocal folds to soften a dog's bark; also called **devocalization** (dē-vō-kahl-ih-zā-shuhn).
- **Elizabethan** (ē-lihz-ah-bēth-ahn) **collar** = device placed around the neck and head of dogs or cats to prevent them from traumatizing an area; commonly called an E-collar (Figure 18–4).
- **hepatic lipidosis** (heh-pah-tihck lihp-ih-dō-sihs) = syndrome characterized by excess fat accumulation in the liver of cats that typically occurs after a period of anorexia (Figure 18–5).
- **induced ovulator** (ihn-doosd ohv-yoo-lā-tər) = species that ovulates only as a result of sexual activity (cats, rabbits, ferrets, llamas, camels, and mink).

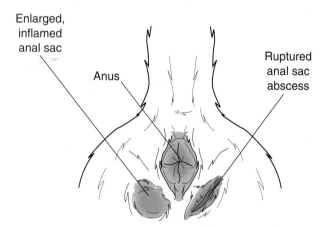

Enlarged, inflamed anal sac

Anus

Ruptured anal sac abscess

Figure 18–1 Line drawing of anal sac location.

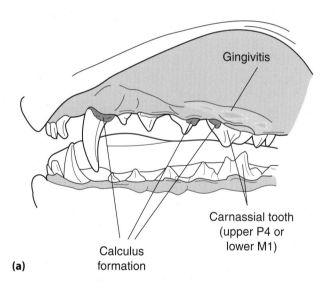

Gingivitis

Carnassial tooth (upper P4 or lower M1)

Calculus formation

(a)

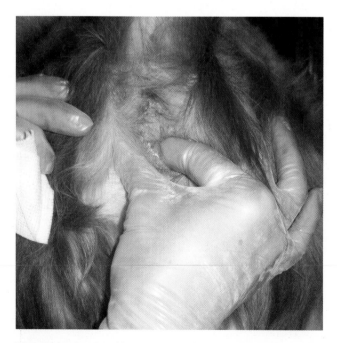

Figure 18–2 Digital expression of fluid from anal sac.

(b)

Figure 18–3 (a) Carnassial teeth of dogs. (b) Carnassial tooth abscess in a dog.

Figure 18–4 Elizabethan collar on a dog. (Courtesy of Kimberly Kruse Sprecher, CVT.)

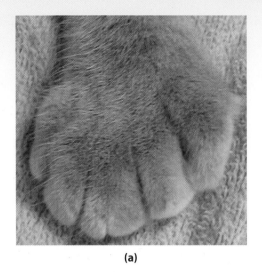

(a)

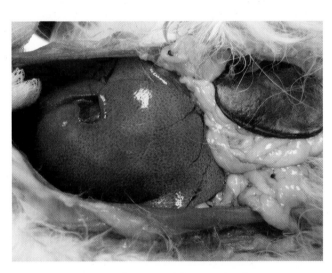

Figure 18–5 Liver of a cat with hepatic lipidosis.

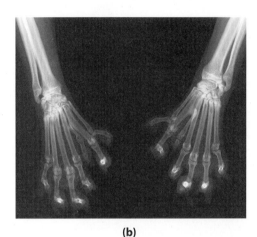

(b)

Figure 18–6 Polydactyly in a cat. (a) Paw of a polydactyl cat. (b) Radiograph of a polydactyl cat. [(a) Courtesy of Linda L. Kratochwill, DVM; (b) Courtesy of University of Wisconsin Veterinary Hospital–Radiology).]

- **polydactyly** (poh-lē-dahck-tih-lē) = more than the normal number of digits (Figures 18–6a and b).
- **spraying** (sprā-ihng) = urination on objects to mark territory.
- **steatitis** (stē-aht-ī-tihs) = inflammation of fat, usually caused by feeding cats too much oily fish; also called yellow fat disease.

BREED-RELATED TERMS

- **angora** (ahn-gōr-ah) = type of long fur on cats (and other species).
- **calico** (kahl-ih-kō) = cat with three colors of fur (black, orange, and white); usually female; if male, they have a genotype of XXY (Figure 18–7a).

- **domestic** (dō-mehs-tihck) **longhair** = cat breed that has long guard hairs; abbreviated DLH.
- **domestic** (dō-mehs-tihck) **shorthair** = cat breed that has short guard hairs; abbreviated DSH.
- **mackerel tabby** (mahck-ər-ehl tah-bē) = two-toned feline fur with stripes.
- **mongrel** (mohn-grehl) = mixed breed of any animal.
- **purebred** (pər-brehd) = member of a recognized breed.
- **ruddy** (ruhd-dē) = orange-brown color with ticking of dark brown or black.
- **self** (sehlf) = one-color fur.
- **tabby** (tah-bē) = feline fur with two colors that may be in stripes or spots (Figure 18–7b).
- **ticked coat** (tihckd kōt) = fur in which darker colors are found on the tips of each guard hair.

(a)

(b)

(c)

Figure 18–7 (a) Calico cat, (b) Tabby cat, (c) Tortoiseshell cat. (Courtesy of iStock Photo.)

- **ticking** (tihck-ihng) = fur coat that has guard hairs with darker tips mixed in.
- **tortoiseshell** (tōr-tihs-shehl) = feline fur with two colors (orange and black) producing a spotted or blotched pattern (Figure 18–7c).

DESCRIPTIVE TERMS

- **docile** (doh-sīl) = tame and easygoing.
- **dull** (duhl) = lacking shine to haircoat; also used to describe behavior more lethargic than normal.
- **euthanasia** (yoo-thehn-ā-zhah) = inducing death of an animal quickly and painlessly; "putting an animal to sleep."
- **feral** (fehr-ahl) = wild; not domesticated.
- **gait** (gāt) = way an animal moves.
- **gloves** (gluhvz) = white paws.
- **luster** (luhs-tər) = shine.
- **obesity** (ō-bē-siht-ē) = excessive fat accumulation in the body.
- **points** (poyntz) = color of nose (mask), ears, tail, and feet of an animal (Figure 18–8).
- **quarantine** (kwahr-ehn-tēn) = isolation of animals to determine whether they have or carry a disease.
- **retractile** (rē-trahck-tīl) = ability to draw back; feline claws can be drawn back.
- **sheen** (shēn) = shininess or luster.
- **staunch** (stawnch) = strong and steady while on point.
- **stud** (stuhd) = male animal used for breeding purposes.
- **temperament** (tehm-pər-ah-mehnt) = emotional and mental qualities of an individual.
- **thorough** (thər-ō) = working every bit of ground and cover.
- **throwback** (thrō-bahck) = offspring that shows an ancestor's characteristic that has not appeared in previous generations.

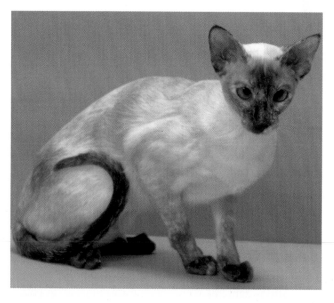

Figure 18–8 Points in a colorpoint shorthair cat. (Photo by Isabelle Francais.)

- **timid** (tihm-ihd) = showing lack of confidence or shy.
- **underfur** (uhn-dər-fər) = very dense, soft, short hair that is found beneath the longer, coarser guard hairs.
- **unthrifty** (uhn-thrihf-tē) = not thriving.
- **vigor** (vihg-ər) = healthy growth; also means high energy.

VACCINATIONS OF DOGS AND CATS

- **canine adenovirus** (ahd-nō-vī-ruhs) **2** = adenovirus infection in canines that causes signs of upper respiratory disease; abbreviated CAV-2.
- **canine distemper** (dihs-tehm-pər) **virus** = highly contagious paramyxovirus infection in canines that is associated with respiratory, digestive, muscular, and neurologic signs; abbreviated CDV.
- **canine hepatitis** (hehp-ah-tī-tihs) **virus** = highly contagious adenovirus 1 infection in canines that is associated with ocular ("blue eye"), abdominal, and liver signs; abbreviated ICH or CAV-1.
- **canine herpesvirus** (hər-pēz-vī-ruhs) = herpesvirus infection in canines that primarily affects newborn puppies and is associated with listlessness, nasal discharge, rash, neurologic signs, and death; abbreviated CHV.
- **canine parainfluenza** (pahr-ah-ihn-flū-ehn-zah) **virus** = paramyxovirus infection of canines that contributes to upper respiratory infections and causes subclinical bronchitis.
- **canine parvovirus** (pahr-vō-vī-ruhs) = highly contagious parvovirus infection in canines that is associated with severe diarrhea, vomiting, dehydration, and leukopenia.
- **coronavirus** (kō-rō-nah-vī-ruhs) = coronavirus that causes gastrointestinal disease in dogs and gastrointestinal and systemic disease in cats that is usually spread through contaminated feces; known as feline infectious peritonitis (FIP) in cats; abbreviated CCV in dogs.
- *Crotalus atrox* **toxoid** (krō-tah-luhs ah-trohcks tohcks-oyd) = inactivated toxin from the Western diamondback rattlesnake used in dogs to reduce morbidity and mortality due to envenomation by this snake.
- **feline calicivirus** (kah-lē-sē-vī-ruhs) = picornavirus infection in felines that is associated with upper respiratory and ocular infections.
- **feline chlamydia** (klah-mihd-dē-ah) = bacterial infection in felines caused by the bacterium *Chlamydophila psittaci* (formerly known as *Chlamydia psittaci*) that is associated with upper respiratory and ocular infections.
- **feline immunodeficiency virus** = lentivirus infection in felines that initially presents with fever and lymphadenopathy that over a long period of time progresses to a wide range of clinical signs such as anemia, lethargy, weight loss, and secondary infections; abbreviated FIV.
- **feline infectious peritonitis** = coronavirus infection in felines that is characterized by an insidious onset, fever, and weight loss. The wet form has peritoneal or pleural effusions (or both), whereas the dry form has pyogranulomas in any location; abbreviated FIP.
- **feline leukemia** (loo-kē-mē-ah) **virus** = feline retrovirus that may produce elevated numbers of abnormal leukocytes, immune suppression, cancer, and illness associated with immune suppression; abbreviated FeLV.
- **feline panleukopenia** (pahn-loo-kō-pē-nē-ah) **virus** = parvovirus infection of felines that is associated with fever, vomiting, diarrhea, and a decrease in all types of white blood cells; abbreviated FPV; commonly called feline distemper.
- **feline rhinotracheitis** (rī-nō-trā-kē-ī-tihs) **virus** = herpesvirus infection in felines that is associated with upper respiratory and ocular infections.
- *Giardia lamblia* (gē-ahr-dē-ah lahmb-lē-ah) = protozoan that may cause asymptomatic disease or cause diarrhea in dogs and cats.
- **infectious tracheobronchitis** (ihn-fehck-shuhs trā-kē-ō-brohng-kī-tihs) = upper respiratory infection caused by the bacterium *Bordetella bronchiseptica* (bōr-dih-tehl-ah brohnk-ō-sehp-tih-kā) that produces a severe hacking cough; also called kennel cough.
- **leptospirosis** (lehp-tō-spī-rō-sihs) = bacterial disease caused by various serotypes of *Leptospira*; signs include renal failure, jaundice, fever, and abortion.
- **Lyme** (līm) **disease** = bacterial disease caused by the bacterium *Borrelia burgdorferi* transported by a tick vector; associated with fever, anorexia, joint disorders, and occasionally neurologic signs; also called Lyme borreliosis.
- **rabies** (rā-bēz) **virus** = fatal zoonotic rhabdovirus infection of all warm-blooded animals that causes neurologic signs; transmitted by a bite or infected body fluid; abbreviated RV. RV is required by public health agencies for licensure of dogs; RV is recommended as a public health measure for cats.

REVIEW EXERCISES

Multiple Choice
Choose the correct answer.

1. The term meaning ingestion of feces is
    a. stoolophagia
    b. fecophagia
    c. coprophagy
    d. bowelophagy

2. The term for more than the normal number of digits is
    a. timid
    b. throwback
    c. polydactyly
    d. mackerel

3. Cats that have white paws are said to have
    a. points
    b. luster
    c. staunch
    d. gloves

4. A cat with three colors of fur is called a(n)
    a. angora
    b. calico
    c. tortoiseshell
    d. ticked coat

5. A male animal used for breeding purposes is known as a
    a. sheen
    b. vigor
    c. stud
    d. feral

6. Urinating on objects to mark territory is called
    a. urination
    b. voiding
    c. micturition
    d. spraying

7. Inflammation of fat is
    a. liposis
    b. lipoma
    c. steatitis
    d. adiposis

8. Devocalization in dogs is commonly called
    a. vocalectomy
    b. cordectomy
    c. silencing
    d. debarking

9. The pouches that store an oily, foul-smelling fluid in dogs and cats are called
    a. anal glands
    b. anal sacs
    c. anal bullae
    d. anal goblets

10. The large, shearing cheek tooth in dogs is commonly called a
    a. premolar tooth
    b. carnassial tooth
    c. polyrooted tooth
    d. deciduous tooth

11. What is the device placed around the neck and head of dogs and cats to prevent them from traumatizing an area?
    a. throwback collar
    b. Elizabethan collar
    c. point collar
    d. gaited collar

12. Inducing death of an animal quickly and painlessly is
    a. quarantine
    b. staunch
    c. euthanasia
    d. unthrifty

13. Feline claws can be drawn back because they are
    a. distal
    b. covered with underfur
    c. unthrifty
    d. retractile

14. A fur coat that has guard hairs with darker tips is a
    a. tortoiseshell
    b. mackerel tabby
    c. ticked coat
    d. calico

15. A mixed breed of any animal is called a

    a. mongrel
    b. purebred
    c. self
    d. domestic

16. A wild animal is called

    a. docile
    b. unthrifty
    c. throwback
    d. feral

17. Cats ovulate as a result of sexual activity and are called

    a. seasonal
    b. polyestrus
    c. induced ovulators
    d. polydactyl

18. Vigor is to timid as

    a. stud is to staunch
    b. luster is to dull
    c. thorough is to bred
    d. tabby is to ruddy

19. The way an animal moves is its

    a. ticking
    b. self
    c. gait
    d. staunch

20. Isolation of animals to determine whether they have or carry a disease is

    a. quarantine
    b. euthanasia
    c. domestic
    d. temperament

CROSSWORD PUZZLE

Dog and Cat Terms Puzzle

Across

4 ingestion of feces
6 giving birth to cats
7 fur coat that has guard hairs with darker tips mixed in
8 the way an animal moves
9 tame
10 one color fur
14 not thriving
16 giving birth to dogs
17 urination on objects to mark territory
18 more than the normal number of digits

Down

1 mixed breed of any animal
2 cat having three colors of fur
3 male animal used for breeding
5 color of nose, ears, tail, and feet
11 inducing death of an animal quickly and painlessly
12 wild
13 white paws
15 intact male cat

CASE STUDIES

Define the underlined terms in each case study.

A 3-yr-old F/S black Labrador retriever was presented to the clinic for removal of a round bone from the mandible. Hx: The dog had been chewing on the bone during the day and had gotten the bone stuck on its mandible. The dog had been pawing at the bone for the past 45 min. On PE, the dog was tachycardic, anxious, and tachypnic. The rostral end of the mandible was swollen. The dog was sedated with an IV sedative so that the bone could be removed. The bone was situated caudal to the lower canine teeth. Gigli wire was threaded through the hole in the center of the bone, and the bone was cut in two places to allow its removal. While the bone was being sawed, tissue trauma occurred to the skin of the mandible. The dog was sent home on antibiotics 1 T bid PO 7d.

1. yr _____

2. F/S _____

3. mandible _____

4. Hx _____

5. PE _____

6. tachycardic _____

7. tachypnic _____

8. rostral _____

9. IV _____

10. caudal _____

11. canine teeth _____

12. antibiotics _____

13. T _____

14. bid _____

15. PO _____

16. d _____

A 9-wk-old ♀ DSH kitten was presented to the clinic for inappetence. On PE, it was noted that the kitten had bilateral yellow-green mucopurulent ocular and nasal discharge. T = 103.8°F, HR = 170 BPM, RR = 40 breaths/min, MMs = pink, CRT = 2 sec. The kitten was alert. An audible wheeze was heard on thoracic auscultation; lungs had increased bronchial sounds and referred URT sounds. The conjunctiva was reddened and edematous. The abdomen palpated normally. Dx: URI; DDx: (1) rhinotracheitis virus, (2) calicivirus, (3) chlamydia.

17. wk _____

18. ♀ _____

19. DSH _____

20. inappetence _____

21. PE _____

22. bilateral _____

23. mucopurulent _____

24. ocular _____

25. nasal _____

26. T _____

27. °F _____

28. HR _____

29. BPM _____

30. RR _____

31. min _____

32. MMs _____

33. CRT _____

34. sec _____

35. wheeze _____

36. thoracic _____

37. auscultation _____

38. bronchial _____

39. URT _____

40. conjunctiva _____

41. edematous _____

42. abdomen _____

43. palpated _____

44. Dx _____

45. URI _____

46. DDx _____

47. rhinotracheitis _____

A 5-yr-old F/S DSH cat was presented to the clinic with stranguria and hematuria. T = 102.4°F, HR = 180 BPM, RR = 35 breaths/ min, MMs = pink and moist, CRT = 1 sec. Heart and lungs auscultated normally. Oral exam revealed mild tartar with grade II gingivitis. Abdominal palpation yielded normal kidneys, normal intestinal loops, a tense and painful caudal abdomen, and a turgid urinary bladder. Dx: cystitis; DDx: (1) FUS, (2) crystalluria.

48. yr _____

49. F/S _____

50. DSH _____

51. stranguria _____

52. hematuria _____

53. auscultated _____

54. oral _____

55. tartar _____

56. gingivitis _____

57. abdominal _____

58. palpation _____

59. caudal _____

60. turgid _____

61. Dx _____

62. cystitis _____

63. DDx _____

64. FUS _____

65. crystalluria _____

A 2-yr-old intact male golden retriever was presented with a 4" laceration with extensive hemorrhage on his right carpus. Pressure bandages were immediately applied for hemostasis. When the bleeding was under control, the dog was anesthetized so that the blood vessels could be ligated and the wound sutured.

66. intact _____

67. 4" _____

68. laceration _____

69. hemorrhage _____

70. carpus _____

71. hemostasis _____

72. anesthetized _____

73. ligated _____

74. sutured _____

A 6-mo-old F black Labrador retriever was presented to the clinic for OHE. A preanesthetic blood screen (PCV, ALT, BUN, GLU) and IV fluid line were done before surgery. The animal was anesthetized, clipped, and prepped for surgery. A ventral midline incision was made, and the reproductive tract was identified. The ovaries, uterine horns, and uterus were removed after proper ligation. When the abdominal incision was being closed, the veterinarian noted pooling of blood in the abdomen. The ligatures were rechecked and still in place. A large amount of blood was coming from the abdominal incision, and the veterinarian had the technician reassess the animal. The CRT was prolonged, the MMs were pale, and the animal was tachycardic and hypothermic. Blood was taken for another PCV, and the fluid rate was increased. The PCV was low normal. The owner was called to see whether the dog had been sick recently, and the owner stated that the dog was seen eating rat bait about 3 days earlier. Additional blood was collected in a heparin tube for assessment of clotting times, and the dog was given vitamin K1. The incision was closed, and the dog was closely monitored during recovery. The dog made a slow recovery and was hospitalized an additional night for observation. Clotting times from the lab demonstrated prolonged clotting times.

75. OHE _____

76. preanesthetic _____

77. blood screen   _____

78. PCV   _____

79. ALT   _____

80. BUN   _____

81. GLU   _____

82. IV   _____

83. anesthetized   _____

84. ventral midline incision   _____

85. ligation   _____

86. CRT   _____

87. MMs   _____

88. tachycardic   _____

89. hypothermic   _____

90. heparin   _____

A 10-yr-old M/N cockapoo was presented to the clinic for scooting (dog assumes a sitting position and drags the anal region along the ground) and licking the perianal region. On PE, it was noted that the dog was obese and had dermatitis of the tail head region and oily skin. The TPR were normal. The dog had an hx of tenesmus and reluctance to stand. Rectal palpation of the anal sacs revealed moderately enlarged sacs. Both anal sacs were expressed, and inspissated material was expressed. Both anal glands were flushed with an antiseptic. The dog was discharged with antibiotics, and an appointment was made to reassess the anal sacs in 7 days.

91. perianal   _____

92. dermatitis   _____

93. TPR   _____

94. hx   _____

95. tenesmus   _____

96. rectal palpation   _____

97. anal sacs   _____

98. inspissated   _____

99. antiseptic   _____

CHAPTER 19

HORSES, ETC.

Objectives

Upon completion of this chapter, the reader should be able to

- Recognize, define, spell, and pronounce terms related to horses, donkeys, mules, and ponies
- Analyze case studies and apply medical terminology in a practical setting

HORSES, DONKEYS, MULES, AND PONIES

Equine animals have been used for transportation, field work, pack work, and recreation. Horses, donkeys, mules, and ponies have been used as companion animals as well.

Many of the anatomy and physiology concepts and medical terms related to equine species have been covered in previous chapters. The lists in this chapter apply more specifically to the care and treatment of equine species.

MODE OF MOVEMENT

See Figure 19–1.

- **amble** (ahm-buhl) = lateral gait that is different from the pace by being slower and more broken in cadence.
- **back** = slowly trotting in reverse.
- **beat** (bēt) = time when the foot (or feet if simultaneous) touches the ground.
- **canter** (kahn-tər) = slow, restrained three-beat gait in which the two diagonal legs are paired.
- **dressage** (druh-sahzh) = method of riding in which a rider guides (rather than uses hands, feet, or legs) a trained horse through natural maneuvers.

- **equitation** (ehk-wih-tā-shuhn) = act and practice of riding a horse.
- **fox trot** (fohcks troht) = slow, short, broken type of gait in which the head usually nods.
- **gallop** (gahl-ohp) = fast four-beat gait in which the feet strike the ground separately (first = one hind foot, second = other hind foot, third = front foot on the same side as first step, and fourth = other front foot on the same side as second step); also called **run.**
- **jog** (johg) = slow trot.
- **pace** (pās) = fast two-beat gait in which the front and hind feet on the same side start and stop at the same time.
- **pointing** (poyn-tihng) = stride in which extension is more pronounced than flexion.

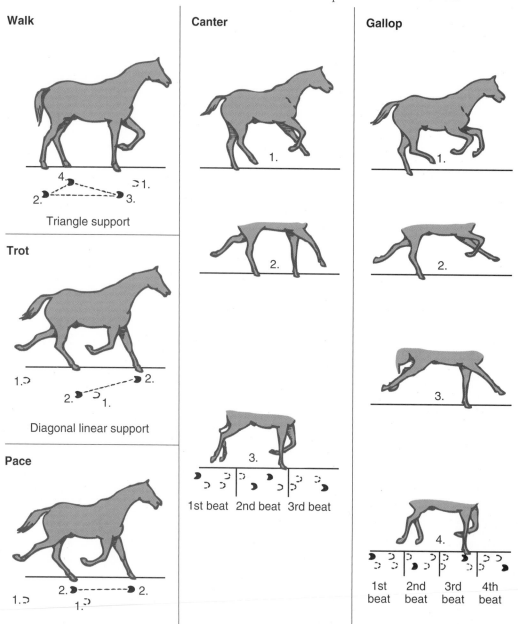

Figure 19–1 Basic gaits of equine.

- **rack** (rahck) = fast, flashy unnatural four-beat gait in which each foot meets the ground separately at equal intervals; also called **single-foot.**
- **rolling** (rō-lihng) = excessive side-to-side shoulder motion.
- **running** (ruhn-ihng) **walk** = slow four-beat gait intermediate in speed between a walk and rack.
- **stride** (strīd) = distance between successive imprints of the same foot.
- **suspension** (suh-spehn-shuhn) = time in which none of the feet are in contact with the ground.
- **swing** (swihng) = non–weight-bearing phase of a stride.
- **trappy** (trahp-pē) = short, quick, choppy stride.
- **trot** (troht) = natural, rapid two-beat diagonal gait in which the front foot and the opposite hind foot take off and hit the ground at the same time.
- **walk** = natural, slow flat-footed four-beat gait in which each foot takes off and strikes the ground at separate intervals.
- **Western** (wehs-tərn) = method of riding in which the stirrup length is long, the rider rides in an upright posture, and the rider has a one-handed hold on the reins.

ANATOMY, PHYSIOLOGY, AND DISEASE TERMS

- **bad mouth** = malocclusion in which the top and bottom teeth do not meet (Figure 19–2).
- **bag up** = development of mammary glands or udder near parturition; also called **bagging up.**
- **bars** (bahrz) = support structure that angles forward from the hoof wall to keep it from overexpanding; also the gap between a horse's incisors and molars; also the side points on the tree of a saddle.
- **bishoping** (bihsh-ohp-ihng) = artificial altering of teeth of an older horse to sell it as a younger horse.
- **check ligament** (chehck lihg-ah-mehnt) = one of two ligaments to the digital flexors of equine; maintains the limbs in extended position during standing.

Figure 19–3 Flehmen reaction in a stallion. (Courtesy of Dreamstime.)

- **chestnut** (chehst-nuht) = horny growths on the medial surface of the equine leg either above the knee in the front limb or toward the caudal area of the hock in the rear limb; also a coat color of horses.
- **cracks** (krahkz) = hoof wall defects that form because the hoof is too long and not trimmed frequently enough.
- **croup** (krūp) = top part of equine rump.
- **cups** (kuhpz) = deep indentations of the incisors in the center of the occlusal surface in young permanent teeth.
- **curb** (kərb) = enlargement on the caudal aspect of the hind leg below the hock.
- **dental stars** (dehn-tahl stahrz) = marks on the occlusal surface of the incisor teeth appearing first as narrow, yellow lines, then as dark circles near the center of the tooth.
- **flehmen** (fleh-mehn) **reaction** = response of a stallion to the scent of a female horse's urine in which he extends his head and curls his upper lip (Figure 19–3).
- **flexor tendon** (flehck-sər tehn-dohn) = tendon that causes the fetlock joint to flex.
- **foal heat** (fōl hēt) = first estrus that occurs shortly after parturition (usually not fertile).
- **full-mouthed** (fuhl mouthd) = having all of the permanent teeth and cups present.
- **Galvayne's groove** (gahl-vānz groov) = mark on labial surface of the equine tooth; used to determine age; usually appears around 10 years of age (Figure 19–4).
- **guttural pouch** (guht-ər-ahl powch) = large, air-filled ventral outpouching of the eustachian tube in equine (Figure 19–5).
- **hindgut** (hihnd-guht) = collective term for the cecum, small colon, and large colon.
- **in wear** = condition in which a tooth has risen to the masticatory level; when opposing teeth have reached sufficient height above the gum line to grind against one another.

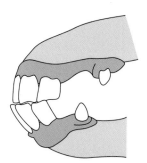

Monkey mouth
(lower incisors protrude
beyond upper incisors)

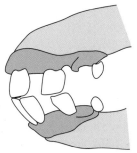

Parrot mouth
(upper incisors protrude
beyond lower incisors)

Figure 19–2 Types of bad mouth (malocclusions).

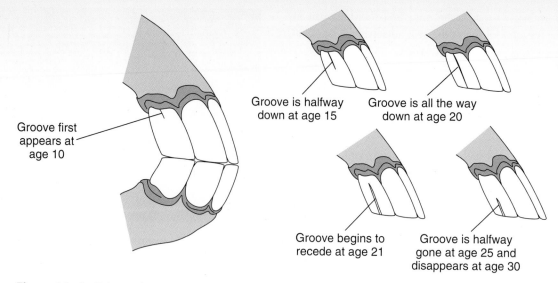

Figure 19–4 Galvayne's groove.

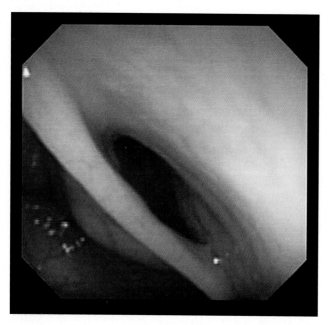

Figure 19–5 Endoscopic view of the guttural pouch. (Courtesy of Laura Lien, CVT, BS.)

- **lamina** (lah-mih-nah) = tissue that attaches hoof to the underlying foot structures.
- **laminitis** (lahm-ihn-ī-tihs) = inflammation of the sensitive laminae under the horny wall of the hoof; a sequela of laminitis is **founder** (fownd-ər) (Figure 19–6).
- **milk teeth** = first teeth that the animal develops.
- **monkey mouth** = condition in which the mandible is longer than the maxilla, causing the lower incisors to protrude beyond the upper incisors.
- **nippers** (nihp-pərz) = central incisors of equine; also a tool to remove excess hoof wall.
- **parrot mouth** = condition in which the maxilla is longer than the mandible, causing the upper incisors to protrude beyond the lower incisors.

- **periople** (pehr-ē-ō-puhl) = varnishlike coating that holds moisture in the hoof and protects the hoof wall.
- **quidding** (kwihd-ihng) = condition in which a horse drops food from the mouth while chewing.
- **quittor** (kwihd-ər) = festering of the foot anywhere along the border of the coronet.
- **scratches** (skrahch-ehz) = low-grade infection or scab in the skin follicles around the fetlock; also called **grease heel.**
- **smooth mouth** = condition in which no cups are present in the permanent teeth.
- **stay apparatus** (stā ahp-ahr-ah-tuhs) = anatomical mechanism of the equine limb that allows the animal to stand with little muscular effort; includes many muscles, ligaments, and tendons.
- **tush** (tuhsh) = canine tooth in a horse (usually found only in males).
- **waxed teats** (wahcksd tētz) = accumulation of sticky, clear to yellow-colored dried milk at the nipple openings that may occur before parturition.
- **winking** (wihnk-ihng) = opening of the labia to expose the clitoris while the female assumes a mating position.
- **wolf teeth** = rudimentary first upper premolars in equine that are usually shed in maturity.

MARKINGS

See Figure 19–7.

- **ankle** = white marking from the coronet to the fetlock on a horse's leg; also called **sock.**
- **bald face** = wide white marking that extends beyond both eyes and nostrils; also called **apron.**
- **banding** (bahn-dihng) = style of mane that is sectioned and fastened with rubber bands; seen in Western show horses.

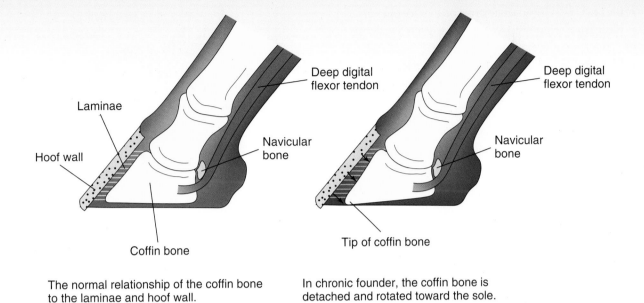

The normal relationship of the coffin bone to the laminae and hoof wall.

In chronic founder, the coffin bone is detached and rotated toward the sole.

Figure 19–6 Laminitis.

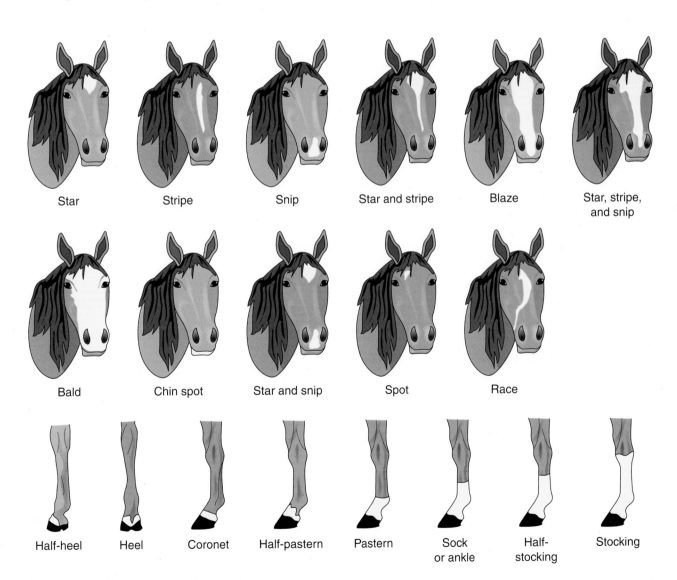

Figure 19–7 Natural face and leg markings.

- **blaze** (blāz) = broad white stripe on the face of a horse.
- **blemish** (blehm-ihsh) = unattractive defect that does not interfere with performance.
- **bloom** (bloom) = shiny coat for show horses.
- **chin spot** (chihn spoht) = white marking on the chin of a horse's face.
- **coronet** (kohr-ō-neht) = white marking covering the coronary band on a horse's leg.
- **distal** (dihs-tahl) **spots** = dark circles on a white coronet band.
- **half heel** (hahf hēl) = white marking on the medial or lateral aspect of the caudal region of the hoof.
- **half pastern** (hahf pahs-tərn) = white marking from the coronet to the middle of the pastern on a horse's leg.
- **half stocking** (hahlf stohk-ihng) = white marking from the coronet to the middle of the cannon.
- **heel** (hēl) = white marking across the entire heel.
- **pastern** (pahs-tərn) = white marking from the coronet to the pastern on the horse's leg.
- **points** (poyntz) = black coloration from the knees and hocks down in bays and browns (may include the ear tips).
- **race** (rās) = long wave or irregular stripe down a horse's face.
- **snip** (snihp) = small white marking on a horse's muzzle.
- **spot** (spoht) = small white marking on a horse's face.
- **star** (stahr) = white mark, often in a diamond shape, found between the eyes on the face of the horse.
- **stocking** (stohck-ihng) or **full stocking** = white marking from the coronet to the knee.
- **stripe** (strīp) = long, straight marking down a horse's nose; sometimes called **strip.**

EQUIPMENT

- **aids** (ādz) = means by which a rider communicates with a horse (e.g., voice, hands, legs, seat).
- **bit** (biht) = part of the bridle that is put in the horse's mouth to control the animal.
- **breeching** (brē-chihng) = part of a harness that passes around the rump of a harnessed horse.
- **bridle** (brī-duhl) = part of a harness that includes the bit, reins, and headstall (Figure 19–8).
- **calks** (kawkz) = grips on the heels and the outside of the front shoes of horses.
- **cinch** (sihnch) = part of a saddle used to hold it onto the horse; placed around the girth area.
- **clinch cutter** (klihnch kuht-tər) = tool used to remove horseshoes.
- **cradle** (krā-duhl) = device used to prevent an animal from licking or biting an injured area.
- **halter** (hahl-tər) = device used to lead and tie a horse; also called a **head collar.**

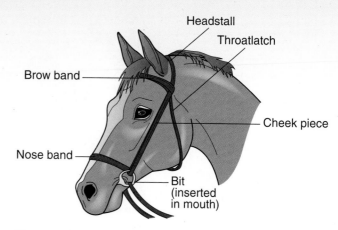

Figure 19–8   Parts of a bridle.

- **hobble** (hohb-uhl) type of restraint in which the front feet or hind feet are placed in straps to keep them from moving.
- **hoof pick** (huhf pihck) = instrument used to clean the sole, frog, and hoof wall.
- **hoof testers** (huhf tehs-tərs) = instrument used to test sensitivity in the equine foot.
- **pincher** (pihn-shər) = tool used to remove horseshoes; pincers are central incisors.
- **puller** (puhl-ər) = tool used to remove worn horseshoes.
- **rasp** (rahsp) = tool used for leveling a horse's foot (Figure 19–9).
- **Scotch hobble** (skohtch hohb-uhl) = type of restraint in which all four feet are tied in an *X* pattern to keep them from moving or kicking.
- **shoe** (shū) = plate or rim of metal nailed to the palmar or plantar surface of an equine hoof to protect the hoof from injury or to aid in hoof disease management.
- **tack** (tahck) = equipment used in riding and driving horses.
- **throatlatch** (thrōt-lahtch) = bridle part that connects the bridle to the head located under the horse's throat; also area under throat where the head and neck are joined and where the harness throatlatch fits (Figure 4–1).

MANAGEMENT TERMS

- **as-fed basis** (ahs-fehd bā-sihs) = amount of nutrients in a diet expressed in the form in which it is fed.
- **bedding** (behd-ihng) = material used to cushion the animal's shelter.
- **birth date** (bərth dāt) = for racing or showing, a foal's birthday is considered January 1 regardless of the actual month it was born.
- **blistering** (blihs-tər-ihng) = application of an irritating substance to treat a blemish.
- **board** (bōrd) = to house.
- **bolt** (bōlt) = to eat rapidly or to startle.

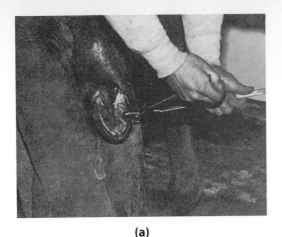
(a)

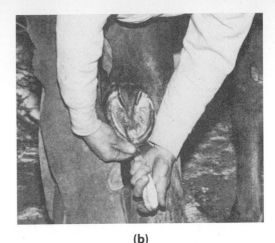
(b)

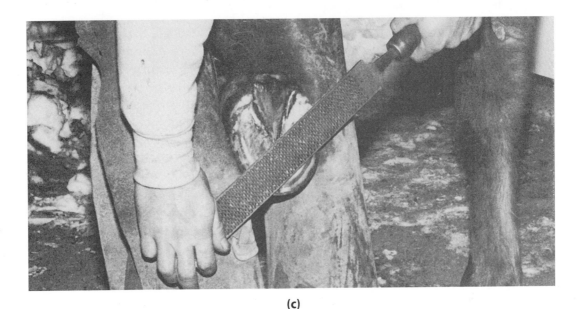
(c)

Figure 19–9 Trimming a horse's hoof. (a) Use of nippers to cut the horny wall to a proper length and angle. (b) Use of hoof knife to pare dead and flaky tissue from the sole. (c) Use of rasp to eliminate jagged edges and sharp corners.

- **bots** (bohtz) = larvae of the bot fly, *Gastrophilus;* occur in the stomach.
- **bowline knot** (bō-līn noht) = type of nonslippable knot.
- **box stall** (bohcks stahl) = enclosure where a horse can move freely.
- **cast** (kahst) = to be caught in a recumbent position and unable to rise.
- **casting** (kahs-tihng) = restraint method using ropes to place animals in lateral recumbency.
- **cribbing** (krihb-ihng) = vice in which an equine grasps an object between the teeth, applies pressure, and swallows air (Figure 19–10).
- **cross-tying** (krohs-tī-ihng) = method of using two ropes to secure a horse so that the head is level.
- **diluter** (dĭ-loo-tər) = type of fluid that is used to increase the volume of semen (thus diluting the sample).

- **driving** (drī-vihng) = harnessing and controlling horses from behind.
- **dry matter basis** (drī mah-tər bā-sihs) = method of expressing concentration of a nutrient based on absence of water in the feed.
- **extender** (ehcks-tehn-dər) = additive used to extend the lifespan of sperm cells.
- **farrier** (făr-ē-ər) = person who cares for equine feet, including trimming and shoeing.
- **feathering** (feh-thər-ihng) = fringe of hair around an equine foot just above the hoof; also used to describe the fringe of hair on caudal aspects of canine limbs.
- **firing** (fihr-ihng) = making a series of skin blisters with a hot needle over an area of lameness.
- **flighty** (flī-tē) = nervous.

Figure 19–10   Cribbing.

- **floating** (flō-tihng) = filing off the sharp edges of equine teeth (Figure 19–11).
- **get** (geht) = sire's offspring.
- **grade** (grād) = animal that is not registered with a specific breed registry.
- **hand** = unit used to measure height of an equine at the withers; equal to 4 inches.
- **heaving** (hē-vihng) = extra contraction of the flank muscles during respiration; caused by loss of lung elasticity.

- **hunters** (huhn-tərz) = horses that are judged while jumping fences or chasing fox.
- **jumpers** (juhm-pərz) = horses that compete at shows by jumping and are judged on height, time, and faults.
- **lather** (lah-thər) = accumulation of sweat on a horse's body.
- **leg cues** (lehg kūz) = signals given to the horse through movement of the rider's legs.
- **longe** (luhng) = act of exercising a horse on the end of a long rope, usually in a circle (Figure 19–12); also spelled *lunge*.

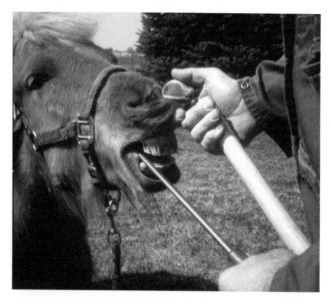

Figure 19–11   Twitch applied to a horse's upper lip to allow teeth to be floated. (Courtesy of Ron Fabrizius, DVM, Diplomate ACT.)

Figure 19–12   Longeing a horse. In longeing, a horse on a long strap or line travels in a large circle around the handler. Longeing helps train young horses, exercises horses, and improves balance and development of stride. Longeing also is spelled *lungeing*. (Courtesy of iStock Photo.)

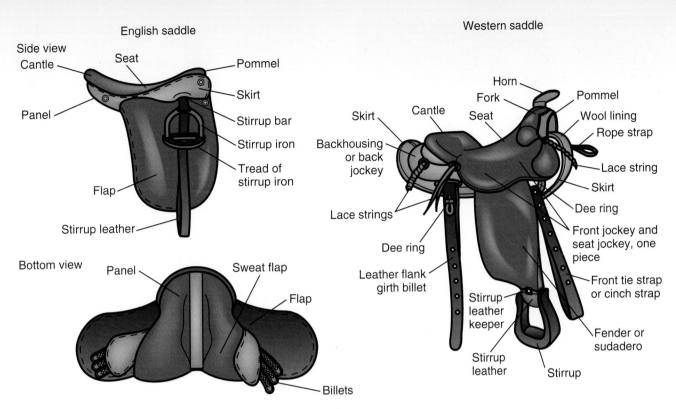

Figure 19–13 depicts the English saddle (Side view and Bottom view) and Western saddle.

English saddle — Side view: Cantle, Seat, Pommel, Skirt, Stirrup bar, Stirrup iron, Tread of stirrup iron, Panel, Flap, Stirrup leather.

Bottom view: Panel, Sweat flap, Flap, Billets.

Western saddle: Skirt, Cantle, Horn, Fork, Seat, Pommel, Wool lining, Rope strap, Backhousing or back jockey, Lace string, Skirt, Dee ring, Front jockey and seat jockey, one piece, Lace strings, Dee ring, Front tie strap or cinch strap, Leather flank girth billet, Stirrup leather keeper, Fender or sudadero, Stirrup leather, Stirrup.

Figure 19–13 Parts of the English saddle and Western saddle.

- **near side** = left side of horse.
- **off side** = right side of horse; also called **far side.**
- **paddock** (pah-dohck) = small fenced-in area; also called **corral.**
- **pasture** (pahs-chər) = area for grazing animals; also means grass or other forage that grazing animals eat.
- **pasture mating** (pahs-chər mā-tihng) = natural breeding; also called **natural cover.**
- **plumb** (pluhm) **line** = line formed when a weight is placed on the end of a string to measure the perpendicularity of something (used to detect straightness of a horse leg).
- **quick-release knot** = knot that breaks loose easily.
- **saddle** (sahd-uhl) = piece of tack placed over the back of an equine for riding, draft, or pack (Figure 19–13).
- **settle** (seht-uhl) = to breed successfully; said of a mare when she becomes pregnant.
- **shod** (shohd) = equine with horseshoes.
- **strike** (strīk) = defensive or aggressive movement of a horse in which the front leg is moved quickly and cranially.
- **tease** (tēz) = act of determining whether a mare is in heat (estrus) by presenting a stallion to her.
- **teaser** (tē-zər) = stallion used to determine which mares are in heat (estrus).
- **twitch** (twihtch) = mode of restraint in which a device is twisted on the upper lip or muzzle (see Figure 19–11).
- **waxing** (wahcks-ihng) = accumulation of colostrum on the nipples of mares usually before foaling; also called **waxed teats.**

TYPES OF HORSES

- **draft** (drahft) **horse** = large breed of working horse (usually over 17 hands) (Figure 19–14).

Figure 19–14 Belgian horses are examples of draft horses. (Courtesy of iStock Photo.)

Figure 19–15 Standardbred horses are examples of light horses. (Courtesy of iStock Photo.)

- **light horse** = breed of horse that is intermediate in size and stature (usually greater than 14.2 hands) (Figure 19–15).
- **miniature horse** = small breed of horse (usually less than 8.2 hands).
- **pony** (pō-nē) = small breed of equine (usually about 14 hands).
- **pony of Americas** = breed of pony (usually between 11.5 and 14 hands) that originated from a cross between a Shetland pony stallion and an Arab/Appaloosa mare.

TERMS FOR UNSOUNDNESS IN HORSES

See Figure 19–16.

- **bog spavin** (bohg spah-vihn) = enlargement of proximal hock caused by distention of the joint capsule; **spavin** means swelling.
- **bone spavin** (bōn spah-vihn) = bony enlargement at the base and medial surface of the hock.
- **bowed tendons** (bōd tehn-dohnz) = thickening of the caudal surface of the leg proximal to the fetlock.
- **capped hock** (kahpd hohck) = thickening of the skin or large callus at the point of the hock.
- **fistulous withers** (fihs-tyoo-luhs wih-thərz) = inflammation of the withers.
- **grease heel** (grēs hēl) = infection or scab in the skin around the fetlock; also called **scratches** (skrahch-ehz).
- **osselets** (ohs-eh-lehts) = soft swellings on the cranial (and sometimes sides) of the fetlock joint.
- **poll evil** (pōl ē-vihl) = fistula on the poll that does not heal easily.
- **quarter crack** (kwahr-tər krahck), **toe crack** (tō krahck), or **heel crack** (hēl krahck) = cracks in quarters, toe, or heel, respectively, of hoof wall caused by poor management.

- **quittor** (kwihd-ər) = festering of the foot along the border of the coronet.
- **ringbone** (rihng-bōn) = bony enlargement on the pastern bones; high ringbone occurs at the pastern joint; low ringbone occurs at the coffin joint.
- **splints** (splihntz) = inflammation of the interosseous ligament that holds the splint bones to the cannon bone.
- **stifled** (stī-flehd) = displaced patella.
- **sweeney** (swē-nē) = atrophy of the shoulder muscles.
- **thoroughpin** (thər-ə-pihn) = fluctuating enlargement located in the hollows proximal to the hock; thoroughpins can be pressed from side to side, hence the name (inflammation of deep digital flexor tendon sheath as it crosses the planter surface of the hock).

EQUINE VACCINATIONS

- **equine encephalomyelitis** (ehn-sehf-ah-lō-mī-ih-lī-tihs) = mosquito-transmitted infectious alphaviral disease of horses that is associated with motor irritation, paralysis, and altered consciousness; there are three types: Eastern, Western, and Venezuelan; also known as sleeping sickness.
- **equine influenza** (ihn-flū-ehn-zah) = myxovirus infection of horses that is associated with mild fever, watery eyes, and persistent cough; commonly called **flu.**
- **equine protozoal myelitis** (mī-eh-lī-tihs) = protozoal infection caused by *Sarcocystis neurona* (and perhaps other protozoa) that causes weakness, ataxia (especially in the hindquarters), weight loss, seizures, and other CNS signs; abbreviated EPM.
- **equine viral arteritis** (ahr-tər-ī-tihs) = togavirus infection of horses that is associated with upper respiratory disease signs, abortion, and lesions in small arteries; abbreviated EVA.
- **equine viral rhinopneumonitis** (rī-nō-nū-mohn-ī-tihs) = herpesvirus infection of horses that is associated with signs of upper respiratory disease and with abortion; abbreviated EVR; also called equine herpesvirus.
- **Potomac** (pō-tō-mihck) **horse fever** = rickettsial bacterial disease of horses that is associated with fever, anorexia, incoordination, diarrhea, and edema of the extremities; also called **equine ehrlichiosis** (ehr-lihck-ē-ō-sihs).
- **rabies** (rā-bēz) **virus** = fatal zoonotic rhabdovirus infection of all warm-blooded animals that causes neurologic signs; transmitted by a bite or infected body fluid; abbreviated RV.
- **strangles** (strān-guhlz) = contagious bacterial disease of horses caused by the bacteria *Streptococcus equi*; signs include high fever, nasal discharge, anorexia, and swollen and abscessed mandibular lymph nodes.

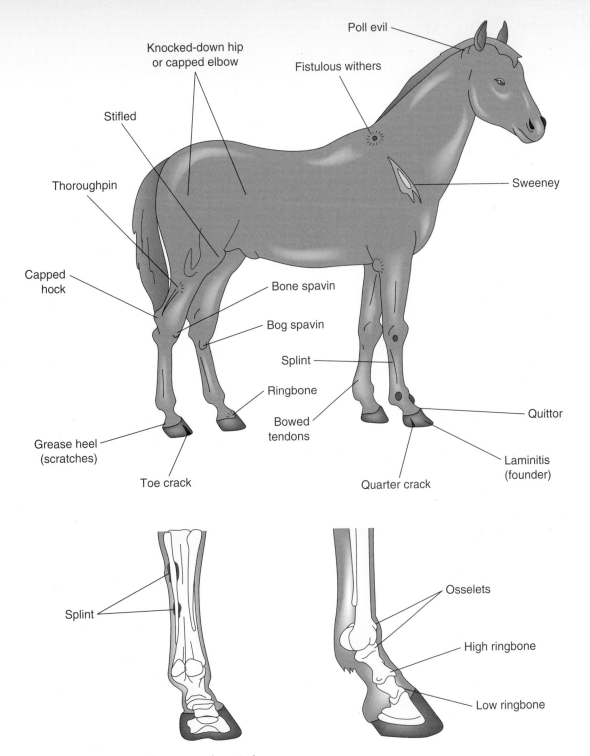

Figure 19–16   Terms for unsoundness in horses.

- **tetanus** (teht-ah-nuhs) = highly fatal bacterial disease caused by the toxin of *Clostridium tetani* that is associated with hyperesthesia, tetany, and convulsions; also called **lockjaw.**

- **West Nile virus** = mosquito-borne viral encephalitis that entered the United States (probably from the Middle East) in 1999. Affected horses may exhibit ataxia, paralysis, depression, head pressing, head tilt, seizures, and death; abbreviated WNV.

REVIEW EXERCISES

Multiple Choice

Choose the correct answer.

1. A person who cares for equine feet, including trimming and shoeing, is called a(n)

   a. equestrian
   b. farrier
   c. hobble
   d. quittor

2. A natural, rapid two-beat diagonal gait of horses is called the

   a. pace
   b. trot
   c. canter
   d. gallop

3. In horses, a rapid two-beat gait in which the front and hind feet on the same side start and stop at the same time is called the

   a. pace
   b. trot
   c. canter
   d. gallop

4. What is the term for the large, air-filled outpouching of the eustachian tube in equine?

   a. Galvayne's groove
   b. guttural pouch
   c. nippers
   d. periople

5. A common sequela of laminitis is

   a. scratches
   b. stay apparatus
   c. breeching
   d. founder

6. The term given to the broad white stripe on the face of a horse is its

   a. star
   b. stripe
   c. race
   d. blaze

7. For racing or showing, a foal's birthday is considered to be

   a. January 1
   b. April 1
   c. June 1
   d. December 1

8. To house a horse is called

   a. blistering
   b. bolting
   c. boarding
   d. bedding

9. The part of the bridle that is located under the horse's throat is called the

   a. tack
   b. throatlatch
   c. chinstrap
   d. halter

10. The tool used for leveling the foot of an equine is the

   a. pincher
   b. hoof pick
   c. rasp
   d. nippers

11. The act of exercising a horse on the end of a long rope is

   a. longeing
   b. driving
   c. handing
   d. plumbing

12. The term for the white marking from the coronet to the knee of a horse is

   a. pastern
   b. stocking
   c. sock
   d. coronet

13. The vice in horses in which they grasp an object between the teeth, apply pressure, and swallow air is

    a. flighting
    b. lathering
    c. cribbing
    d. cross-typing

14. The unit used to measure the height of an equine at the withers is

    a. grade
    b. get
    c. cast
    d. hand

15. The part of the bridle that is put in the horse's mouth is the

    a. bit
    b. cinch
    c. cradle
    d. rasp

16. The canine tooth in a horse that is usually found only in males also is called the

    a. wolf tooth
    b. tush
    c. Galvayne's groove
    d. nipper

17. The act and practice of riding a horse is

    a. equitation
    b. Western
    c. dressage
    d. ambling

18. Teeth that have risen to the masticatory (chewing) level are said to be

    a. cracked
    b. curbed
    c. in swing
    d. in wear

19. The development of the mammary glands or udder near parturition is called

    a. bishoping
    b. cupping
    c. bagging up
    d. curbing

20. The first teeth that an animal develops are its

    a. nippers
    b. milk teeth
    c. wolf teeth
    d. cups

CROSSWORD PUZZLE

Horse Terms Puzzle

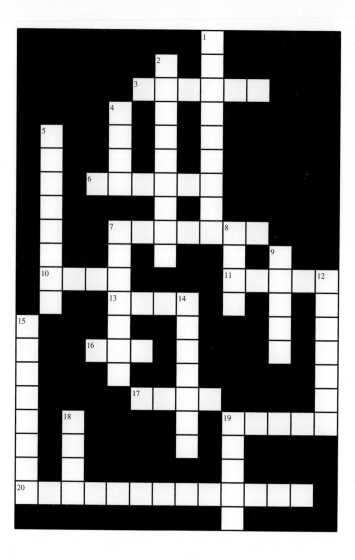

Across

3 fast, four-beat gait
6 distance between successive imprints of the same foot
7 nervous
10 left side of horse
11 top part of the equine rump
13 tool used for leveling the foot
16 offspring
17 equine with horseshoes
19 broad white stripe on the face of a horse
20 large, air-filled ventral outpouching of the eustachian tube in equine

Down

1 first teeth that an animal develops
2 founder is a sequela of this condition
4 time when the foot touches the ground
5 condition in which a horse drops food from the mouth while chewing
7 person who cares for equine feet
8 equipment used in riding and driving horses
9 to house
12 natural breeding
14 small fenced-in area
15 white marking from the coronet to the knee
18 to eat rapidly or startle
19 shiny coat

CASE STUDIES

Define the underlined terms in each case study.

A 6-yr-old quarter horse <u>mare</u> was presented for signs of pawing at the abdomen, <u>flank</u>-watching, <u>anorexia</u>, and lack of <u>stool</u> production. On PE, she was <u>tachycardic</u> and <u>hyperpnic</u>, <u>MMs</u> were tacky and discolored, and the <u>CRT</u> was 2 seconds. Her ears and limbs were cold to the touch. The gut was <u>auscultated</u>, and <u>borborygmus</u> was noted. The veterinarian did a <u>rectal palpation</u> of the horse and noted gas-filled segments of bowel that felt <u>proximal (oral)</u> to an <u>impaction</u>. <u>Ventral midline abdominocentesis</u> was performed, and blood was collected for a <u>CBC</u>. Laboratory results revealed few <u>peritoneal</u> fluid changes on the abdominocentesis sample, and the CBC was unremarkable. The veterinarian thought the impaction was mild and opted for medical treatment. A <u>nasogastric tube</u> was passed to relieve the gas distention, and mineral oil was given via the <u>NG tube</u> (Figure 19–17). An <u>analgesic</u> was given <u>IV</u> for pain relief. The horse recovered uneventfully.

1. mare _____

2. flank _____

3. anorexia _____

4. stool _____

5. tachycardic _____

6. hyperpnic _____

7. MMs _____

8. CRT _____

9. auscultated _____

10. borborygmus _____

11. rectal palpation _____

12. proximal (oral) _____

13. impaction _____

14. ventral midline abdominocentesis _____

15. CBC _____

16. peritoneal _____

17. nasogastric tube _____

18. NG tube _____

19. analgesic _____

20. IV _____

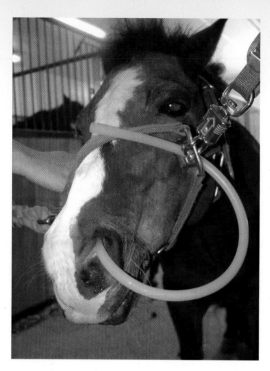

Figure 19–17   Nasogastric tube placement in a horse. (Courtesy of Laura Lien, CVT, BS.)

A 2-yr-old Arabian <u>colt</u> was anorectic and reluctant to stand. The owner called the veterinarian for an examination. The veterinarian walked the horse, and it showed a short-striding gait. The horse was <u>febrile</u> and <u>tachycardic</u>. Upon <u>palpation</u> of the hoof, the area near the <u>coronary band</u> was warm and a <u>pulse</u> could be palpated. The veterinarian suspected <u>laminitis</u> and recommended <u>radiographs</u> of the hoof. Radiographs revealed rotation of the <u>coffin bone</u> from the hoof wall. The <u>Dx</u> of laminitis (<u>founder</u>) was made. Treatment consisted of <u>NSAIDs</u> and hoof trimming and reshoeing by a <u>farrier</u>.

21. colt   _____

22. febrile   _____

23. tachycardic _____

24. palpation   _____

25. coronary band _____

26. pulse   _____

27. laminitis   _____

28. radiographs _____

29. coffin bone _____

30. Dx   _____

31. founder   _____

32. NSAIDs   _____

33. farrier   _____

CHAPTER 20

[Make Room for the Ruminants]

Objectives

Upon completion of this chapter, the reader should be able to

- Recognize, define, spell, and pronounce terms related to cattle, sheep, goats, and llamas
- Analyze case studies and apply medical terminology in a practical setting

RUMINANTS

A **ruminant** (roo-mihn-ehnt) is a cud-chewing animal that has a forestomach that allows fermentation of ingesta. Cattle, sheep, and goats have four stomach compartments. The first three—the rumen, reticulum, and omasum—are actually outpouchings of the esophagus. The abomasum is considered the true, or glandular, stomach. Llamas have three stomach compartments and are called pseudoruminants.

CATTLE

Cattle provide humans with meat, milk, hides, and other by-products. There are basically two types of cattle: dairy and beef. Dairy cattle are bred for their milk-producing qualities, whereas beef cattle are bred for meat. Some breeds are considered dual purpose, which means they have both dairy and beef traits.

Many of the anatomy and physiology concepts and medical terms related to cattle have been covered in previous chapters. The lists in this chapter apply more specifically to the care and treatment of cattle.

Industry Terms

- **artificial insemination** (ahr-tih-fih-shahl ihn-sehm-ihn-ā-shuhn) = breeding method in which semen is collected, stored, and deposited in the uterus or vagina without copulation taking place (Figure 20–1); abbreviated AI.
- **balling** (bahl-ihng) **gun** = tool used to administer pills, boluses, or magnets to livestock (Figure 20–2); also called **bolus gun.**
- **barren** (bār-ehn) = animal that was not bred or did not conceive.
- **body capacity** (boh-dē kah-pah-siht-ē) = heart girth and barrel.
- **brand** (brahnd) = method of permanently identifying animal by scarring the skin with heat, extreme cold, or chemicals (Figure 20–3).
- **bred** (brehd) = said of an animal that has mated and is pregnant.

Figure 20–3   Freeze branding kills the hair pigment, thus marking the animal. (Courtesy of Gary Farmer.)

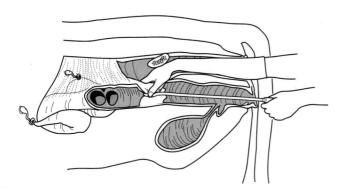

Figure 20–1   During artificial insemination, semen is placed in the female reproductive tract, allowing the process of fertilization to take place just as in natural mating.

Figure 20–2   Balling gun used to administer a magnet to a Holstein cow. (Courtesy of Ron Fabrizius, DVM, Diplomate ACT.)

- **breed** (brēd) = group of animals that are genetically similar in color and conformation so that when mated to each other, they produce young identical to themselves; also the act of breeding.
- **calving interval** (kahv-ihng ihn-tər-vahl) = amount of time between the birth of a calf and birth of the next calf from the same cow.
- **carcass** (kahr-kuhs) = body of animal after it has been slaughtered; usually has head, hide, blood, and offal removed.
- **casting** (kahs-tihng) = restraint method using ropes to place an animal in lateral recumbency.
- **cattle** (kah-tuhl) = more than one member of the genus *Bos.*
- **chute** (shoot) = mechanical device that is used to restrain cattle (Figure 20–4).
- **cleaning a cow** = common term for removal of a retained placenta; also called **cleansing a cow.**
- **cod** (kohd) = remnants of steer scrotum.
- **colostrum** (kō-lah-struhm) = first milklike substance produced by the female after parturition; it is thick, yellow, and high in protein and antibodies.
- **conformation** (kohn-fōr-mā-shuhn) = shape and body type of an animal.
- **corium** (kōr-ē-uhm) = specialized, highly vascular cells that nourish the hoof and horn.
- **crossbred** (krohs-brehd) = having resulted from the mating of two different breeds within the same species.
- **cull** (kuhl) = to remove an animal from the rest because it does not meet a specific standard or is unproductive.
- **dehorn** (dē-hōrn) = to remove horns or horn buds by mechanical, thermal, or chemical means.
- **dual purpose** (dool pər-puhs) = bred and used for both meat and milk production.
- **ear tagging** (ēr tahg-ihng) = placement of identification device in the ear.

Figure 20-4 Headgate chute.

- **embryo transfer** (ehm-brē-ō trahnz-fər) = removal of an embryo from a female of superior genetics and placing it in the reproductive tract of another female.
- **F₁ generation** (F-1 jehn-ər-ā-shuhn) = first offspring from purebred parents of different breeds or varieties; F1 stands for first filial.
- **feeders** (fē-dərz) = beef cattle that are placed in a feedlot based on age and weight.
- **feedlot** (fēd-loht) = confined area where an animal is fed until it is slaughtered (Figure 20–5a).
- **flushing** (fluhsh-ihng) = act of increasing feed before breeding or embryo transfer to increase the number of ova released.
- **fly strike** (flī strīk) = infestation with maggots.
- **free stall** (frē stahl) = stall for cattle in which each animal is free to lie down, feed, move, or seek out other animals.
- **gomer** (gō-mər) **bull** = bull used to detect female bovines in heat; bull may have penis surgically deviated to the side, may be treated with androgens, or may be vasectomized so as not to impregnate females; also called **teaser bull.**
- **halter** (hahl-tər) = head harness worn by animals for restraint that extends behind the head and over the nose.

(a)

(b)

(c)

Figure 20-5 Cattle housing. (a) Feedlot; (b) stanchion; (c) hutch.

- **heart girth** (hahrt gərth) = circumference around the thoracic cavity used to estimate an animal's weight and capacity of the heart and lung.
- **hutch** (huhch) = individual housing pen for calves (and other small animals such as rabbits) (Figure 20–5c).
- **hybrid** (hī-brihd) = offspring resulting from mating of two different species.
- **hybrid vigor** (hī-brihd vihg-ohr) = mating of dissimilar breeds to increase productivity and performance in the F1 generation of crossbred animals over that shown by either parent; also called heterosis (heh-tər-ō-sihs).
- **inbred** (ihn-brehd) = resulting from the mating of two closely related animals (e.g., son to dam, sire to daughter); also called **close breeding.**
- **lead rope** (lēd rōp) = piece of rope, leather, or nylon that is attached via a clasp to a halter.
- **magnet** (māg-neht) = charged metal device that is used to prevent hardware disease (traumatic reticuloperitonitis); it is given orally and placed in the reticulum.
- **malpresentation** (mahl-prē-sehn-tā-shuhn) = abnormal position of a fetus just before parturition.
- **marbling** (mahr-blihng) = streaks of fat interdispersed throughout meat that increase its tenderness.
- **offal** (aw-fuhl) = inedible visceral organs and unusable tissues removed from the carcass of a slaughtered animal.
- **parturient paresis** (pahr-too-rē-ahnt pahr-ē-sihs) = hypocalcemic metabolic disorder of ruminants seen in late pregnancy or early lactation; also called **milk fever.**
- **pinch** (pihnch) = common term for a bloodless castration using an emasculatome.
- **proven** (proov-ehn) = said of an animal whose ability to pass on specific traits is known and predictable.
- **rectal palpation** (rehck-tahl pahl-pā-shuhn) = method of determining pregnancy, phase of the estrous cycle, or disease process by inserting a gloved arm into the rectum of the animal and feeling for a specific structure (Figure 20–6).
- **render** (rehn-dər) = to melt down fat by heat.
- **replacement** (rē-plās-mehnt) = animal that is raised for addition to the herd (one that replaces a less desirable animal).
- **ruminating** (roo-mihn-ā-tihng) = cud-chewing process (see page 117).
- **scurs** (skərz) = underdeveloped horns that are not attached to the skull.
- **somatic** (sō-mah-tihck) **cell count** = determination of number of cells (e.g., leukocytes and epithelial cells) in milk to test for mastitis; abbreviated SCC.
- **spotter** (spoh-tər) **bull** = vasectomized male bovine used to find and mark female bovines in estrus.
- **springing** (sprihng-ihng) = anatomic changes in a ruminant that indicate parturition is near.

Figure 20–6 Rectal palpation in a cow. (Courtesy of Ron Fabrizius, DVM, Diplomate ACT.)

- **stall** (stahl) = small compartment to house an animal.
- **stanchion** (stahn-chuhn) = restraint device that secures cattle around the neck to allow accessibility for milking, feeding, and examining (Figure 20–5b).
- **standing heat** = phase of estrus in which a female bovine will stand to be mounted.
- **switch** (swihtch) = distal part of a bovine tail that consists of long, coarse hairs (Figure 20–7).
- **tailing** (tā-lihng) = restraint technique used in cattle in which the tailhead is grasped and raised vertically; also called a **tail jack** (Figure 20–8).
- **tankage** (tahnk-ahj) = animal residues left after fat is rendered in a slaughterhouse; used for feed or fertilizer.
- **tattoo** (taht-too) = permanent identification of an animal using indelible ink that is injected under the skin.
- **tie stall** (tī stahl) = stall large enough for only one animal, which is usually tied by a neck chain (cattle) or halter (horse).
- **traumatic reticuloperitonitis** (traw-mah-tihck reh-tihck-yoo-loh-pehr-ih-tō-nī-tihs) = relatively common disease in adult cattle caused by ingestion and migration of a foreign body into the reticulum; occasionally the foreign body penetrates the diaphragm and pericardium causing pericarditis; abbreviated TRP; commonly called hardware disease.
- **veal** (vēl) = confined young dairy calf that is fed only milk or milk replacer to produce pale, soft, and tender meat.
- **wean** (wēn) = to remove young from their mother so they can no longer nurse.
- **windbreak** (wihnd-brāk) = shelter in which an animal can stand and be protected from the wind.

Milk-Related Terms

- **alveoli** (ahl-vē-ō-lī) = milk-secreting sacs of mammary gland; also used to describe gas exchange sac of respiratory system.
- **dry** (drī) = not lactating.

Figure 20–7 The distal part of a bovine tail is the switch.

Figure 20–8 Tail jack or tailing a cow. (Courtesy of Ron Fabrizius, DVM, Diplomate ACT.)

- **drying off** = ending the production of milk when milk yield is low or before freshening.
- **gland cistern** (sihs-tərn) = area of udder where milk collects before entering the teat cistern (Figure 20–9).
- **milk solids** = portion of milk that is left after water is removed; includes protein and fat.
- **milk veins** = veins found near the ventral midline of a cow (Figure 20–10); also called **mammary veins.**

- **milk well** = depression in the cow's ventral underline where milk veins enter the body.
- **milk yield** = amount of milk produced in a given period.
- **milking** = process of drawing milk from the mammary glands (Figures 20–11a and b).
- **streak canal** (strēk kah-nahl) = passageway that takes milk from the teat cistern to the outside of the body; also called the **papillary duct** or **teat canal.**

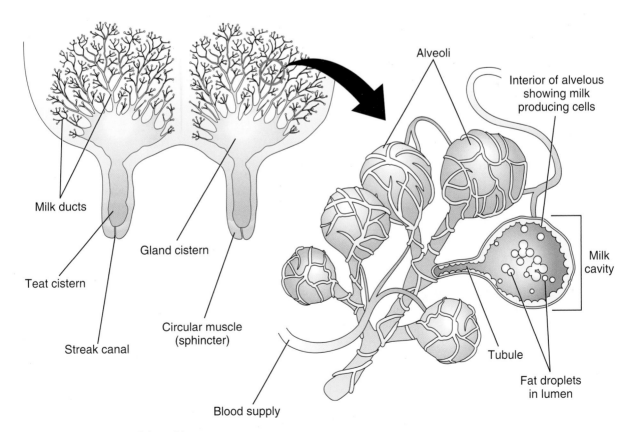

Figure 20–9 Parts of the udder.

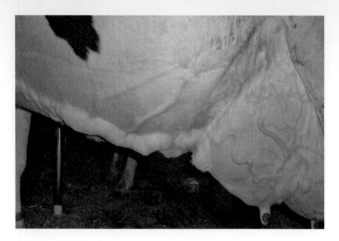

Figure 20–10 The veins found on a cow's ventral midline are the milk veins. (Courtesy of Jim Meronek, DVM, MPH.)

- **strip cup** = metal cup with a lid that is used for detecting mastitis.
- **supernumerary teats** (soo-pər-nū-mahr-ehr-ē tētz) = more than the normal number of teats.
- **teat** (tēt) = nipple, especially the large nipples of ruminants and equine.
- **teat cannula** (tēt kahn-yoo-lah) = short and narrow, round-pointed metal or plastic tube used to pass from the exterior through the teat canal and into the teat cistern to relieve teat obstructions (Figure 20–12).
- **teat cistern** (sihs-tərn) = cavity in the udder where milk is secreted before leaving the teat.
- **teat dipping** = submerging or spraying the nipple with antiseptic to prevent the development of mastitis.

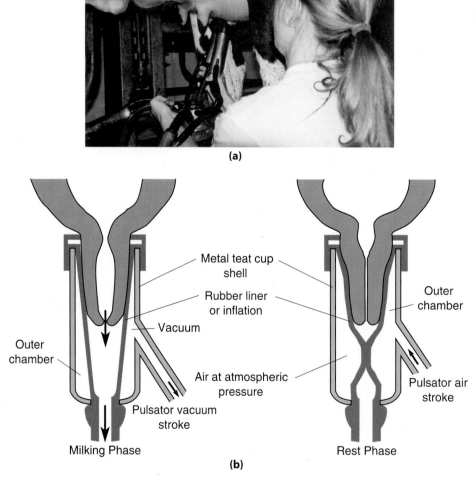

Figure 20–11 (a) Teat cups are placed on the cow's teats to collect milk. (b) Milking is accomplished using a vacuum system that pulsates on the teats. [(a) Courtesy of Gary Farmer; (b) Courtesy of the Cooperative Extension Service, The University of Georgia.]

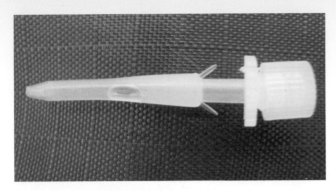

Figure 20–12 Teat cannula.

Figure 20–13 Sheep and cattle have a dental pad instead of upper teeth. (Courtesy of Dreamstime.)

- **teat stripping** = removing milk from the teat before or after milking by occluding the teat at the top between the thumb and forefinger and then pulling downward to express all of the milk; also called **stripping.**
- **three quartered** (thrē kwar-tərd) = condition in cows in which one quarter is no longer used due to previous damage or infection.
- **udder** (uh-dər) = milk production organ in ruminants and equine species.

Feeding-Related Terms

- **bypass protein** (bī-pahs prō-tēn) = protein that is heat or chemically treated so that it is not altered in the ruminant stomach.
- **concentrate** (kohn-sehn-trāt) = type of feed that is high in total digestible nutrients and low in fiber.
- **creep feed** (krēp fēd) = high-energy feed that is fed to young animals in special feed devices so that adult animals cannot gain access to the feed.
- **cudding** (kuhd-ihng) = act of chewing cud; **cud** (kuhd) is regurgitated food particles, fiber, rumen fluid, and rumen microorganisms.
- **dental pad** (dehn-tahl pahd) = hard surface of the upper mouth of cattle that serves in place of upper teeth (Figure 20–13).
- **ensiling** (ehn-sī-lihng) = process in which a forage is chopped, placed in a storage unit that excludes oxygen, and fermented to allow longer preservation of feed.
- **feedstuff** (fēd-stuhf) = any dietary component that provides a nutrient; also called **feed.**
- **finishing** (fihn-ihsh-ihng) = act of feeding beef cattle high-quality feed before slaughter to increase carcass quality and yield.
- **graze** (grāz) = to eat grasses and plants that grow close to the ground.

- **legumes** (lehg-yooms) = roughage plants that have nitrogen-fixing nodules on their roots; examples include alfalfa and clover.
- **premix** (prē-mihx) = ration mixed with various feedstuffs at the feedmill.
- **ration** (rah-shuhn) = amount of food consumed by animal in a 24-hour period.
- **roughage** (ruhf-ahj) = type of feed that is high in fiber and low in total digestible nutrients; examples include pasture and hay; also called **forage.**
- **silage** (sī-lahj) = type of roughage feed that is produced by fermenting chopped corn, grasses, or plant parts under specific moisture conditions to ensure preservation of feed without spoilage (Figures 20–14a and b).
- **supplement** (suhp-lah-mehnt) = additional feed product that improves and balances a poorer ration; also called **additive.**
- **sweetfeed** (swēt-fēd) = food that consists of grains and pellets mixed with molasses to increase palatability.

SHEEP

Sheep are raised for wool, meat, and research models. As in cattle, some breeds are better known for their wool production and others are better known for their meat quality. Sheep usually give birth to twins rather than single lambs (Figure 20–15).

Many of the anatomy and physiology concepts and medical terms related to sheep have been covered in previous chapters. The following list of words is used more specifically for sheep and sheep production.

- **band** (bahnd) = large group of range sheep.
- **carding** (kahr-dihng) = process of separating, straightening, and aligning wool fibers.

(a)

(b)

Figure 20–14 Silage is a type of roughage feed produced by fermenting chopped plant material to ensure preservation of feed without spoilage. (a) Corn silage is fermented chopped corn. (b) Haylage is fermented chopped hay. (Courtesy of Jim Meronek, DVM, MPH.)

Figure 20–15 Twins are common in sheep. (Courtesy of Ron Fabrizius, DVM, Diplomate ACT.)

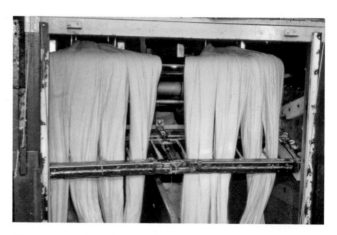

Figure 20–16 In the combing process, the fibers are untangled and smoothed. (Courtesy of American Sheep Industry Association.)

- **clip** (klihp) = one season's wool yield.
- **combing** (kō-mihng) = separating long fibers from short fibers and then arranging and laying fibers parallel by length prior to further treatment (Figure 20–16).
- **crimp** (krihmp) = amount of wave in wool.
- **crutching** (kruhtch-ihng) = process of clipping wool from dock, udder, and vulva of sheep before lambing; also called **tagging.**
- **docking** (dohck-ihng) = removal of the distal portion of the tail; also means reducing in value (Figure 20–17).

- **felting** (fehl-tihng) = property of wool fibers interlocking when rubbed together under heat, moisture, or pressure.
- **fleece** (flēs) = another term for wool.
- **grease wool** = wool that has been shorn from a sheep and has not been cleaned.
- **lamb** (lahm) = young sheep meat.
- **lanolin** (lahn-ō-lihn) = fatlike substance secreted by the sebaceous glands of sheep.
- **mutton** (muh-tihn) = adult sheep meat.
- **rumping** (ruhm-pihng) = method of restraining sheep by placing them in a sitting position with the front legs elevated; also called **tipping.**
- **scouring** (skow-ər-ihng) = cleaning wool (Figure 20–18); also diarrhea in livestock.
- **shear** (shēr) = to shave off wool, hair, or fur.
- **singleton** (sihn-guhl-tohn) = one offspring born.

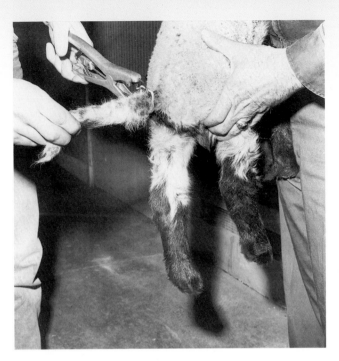

Figure 20–17 Docking. Tail docking of a lamb with an elastrator.

Figure 20–18 Scouring removes material from the wool. (Courtesy of American Sheep Industry Association.)

GOATS

Goats are raised for milk, meat, wool, and research models and as pets.

Many of the anatomy and physiology concepts and medical terms related to goats have been covered in previous chapters. The following list applies more specifically to the care and treatment of goats. Additional terms pertaining to goats are listed under the cattle section because they apply to both animals.

- **cabrito** (kah-brē-tō) = meat of young goats under 50 pounds.
- **cashmere** (kahzh-mər) = fine undercoat of goats (can be harvested from any goat).
- **chevon** (shehv-ehn) = adult goat meat.
- **clip** (klihp) = hair harvested from one animal in one shearing.
- **disbud** (dihs-buhd) = to remove horn growth in kids or calves by use of a hot iron or caustic substance; also called **debudding.**
- **wattle** (waht-tuhl) = appendage suspended from the head (usually the chin) in chickens, turkeys, and goats (Figure 20–19).

CAMELIDS

Camelids are large, herbivorous animals that have slender necks and long legs. Camelids do not have horns or antlers. Camelids are considered pseudoruminants because they

Figure 20–19 Wattle in a goat. (Courtesy of iStock Photo.)

have three stomach compartments (reticulum, omasum, and abomasum) instead of four. There are three genera of camelids; the true camels of Asia (genus *Camelus*); the wild guanaco and the domesticated alpaca and llama of South America (genus *Lama*); and the vicuña of South America (genus *Vicugna*). Llamas are the larger species with less desirable wool; alpacas are smaller than llamas and have high-quality wool; vicunas are undomesticated and are the smallest and rarest with fine, high-quality wool; and guanacos are slightly smaller than llamas and are undomesticated. Llamas are becoming popular pets and are used as pack animals and for fiber production (Figure 20–20).

Many of the anatomy and physiology concepts and medical terms related to llamas have been covered in previous chapters.

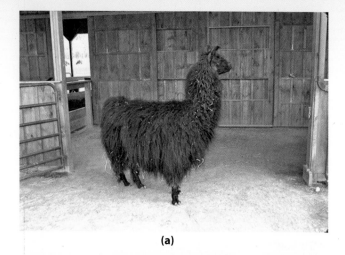

(a)

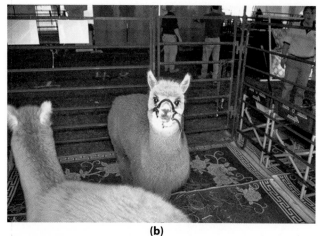

(b)

Figure 20–20 (a) Llama. (b) Alpaca. (Courtesy of USDA.)

The following list applies more specifically to the care and treatment of llamas.

- **Berserk Male Syndrome** (bər-sərk māl sihn-drōm) = undesirable behaviors sometimes seen in male llamas, especially those that are hand-raised; these animals may be overly aggressive and frequently need to be euthanized because they are unsafe; abbreviated BMS.
- **cushing** (kuhsh-ihng) = common term for copulation of llamas; also the position camelids take when lying down.
- **fighting teeth** = set of six teeth in llamas that include an upper vestigial incisor and an upper and lower canine on each side (Figure 20–21).
- **fleece** (flēs) = hair of llama or alpaca; also known as fiber.
- **guard llama** = animal that lives with livestock (mainly sheep) to protect them from predatory animals.
- **harem breeding** (hār-uhm brē-dihng) = style of breeding where a male is left with females most of the year.
- **orgle** (ohr-gehl) = call of a male llama or alpaca.

Figure 20–21 Fighting teeth of a llama.

- **spitting** (spiht-ihng) = behavior in which an animal spits saliva at an animal or a person who is perceived to be a threat or an annoyance; spitting also may be used to establish hierarchy.

VACCINATIONS OF RUMINANTS

- **bovine respiratory syncytial** (sihn-sihsh-ahl) **virus** = paramyxovirus infection of bovine that is associated with fatal pneumonia; abbreviated BRSV or RSV.
- **bovine viral diarrhea** = togavirus infection of bovine that is associated with acute stomatitis, gastroenteritis, and diarrhea; abbreviated BVD.
- **brucellosis** (broo-sehl-ō-sihs) = bacterial infection caused by *Brucella abortus* that causes abortion in cattle.
- **clostridial** (klohs-trihd-ē-ahl) **disease** = group of bacterial infectious conditions of ruminants caused by various species of *Clostridium,* which includes blackleg, malignant edema, pulpy kidney, enterotoxemia, and overeating disease.
- **coronavirus** (kō-rohn-ah-vī-ruhs) = corona virus infection that causes diarrhea in calves.
- **E. coli** (ē kō-lī) = bacterial infection that typically causes scours in calves; also known as colisepticemia and septicemic colibacillosis.
- **enzootic abortion** = bacterial infection caused by *Chlamydophila abortus* (formerly known as *Chlamydia psittaci)* that causes abortion in sheep.
- **Hemophilus somnus** (hē-moh-fihl-uhs sohm-nuhs) = bacterial infection that causes reproductive, urinary, respiratory, and septicemic disease in cattle.
- **infectious bovine rhinotracheitis** (rī-nō-trā-kē-ī-tihs) = herpesvirus infection of bovine that is associated with fever, anorexia, tachypnea, and cough; abbreviated IBR.

- **leptospirosis** (lehp-tō-spī-rō-sihs) = bacterial disease caused by various serotypes of *Leptospira*; signs include renal failure, jaundice, fever, and abortion.
- **parainfluenza** (pār-ah-ihn-flū-ehn-zah) = paramyxovirus infection of ruminants that is associated with fever, cough, and diarrhea; one part of the shipping fever complex; abbreviated PI-3.
- **pasteurellosis** (pahs-tər-ehl-ō-sihs) = bacterial infection caused by *Pasteurella multocida* and *Mannheimia hemolytica* (formerly known as *Pasteurella hemolytica*) that causes respiratory disease in cattle, sheep, and goats.
- **rotovirus** (rō-tō-vī-ruhs) = reovirus that causes scours in calves.

- **soremouth** (sōr-mowth) = poxvirus infection of sheep, goats, and camelids that causes mouth lesions; also called contagious ecthyma and orf.
- **tetanus** (teht-ah-nuhs) = highly fatal bacterial disease caused by the toxin of *Clostridium tetani* that is associated with hyperesthesia, tetany, and convulsions; also called lockjaw.
- **vibriosis** (vihb-rē-ō-sihs) = *Campylobacter fetus* bacterial infection that is associated with infertility and irregular estrous cycles; bulls are vaccinated; also called campylobacteriosis.

REVIEW EXERCISES

Multiple Choice
Choose the correct answer.

1. The common term for removing a retained placenta in a cow is
   a. coding
   b. culling
   c. cleaning a cow
   d. chuting a cow

2. The inedible visceral organs and unusable tissues removed from the carcass of a slaughtered animal are the
   a. marbling
   b. scurs
   c. render
   d. offal

3. What is the name of the passageway that takes milk from the teat cistern to the outside of the udder?
   a. milk duct
   b. gland cistern
   c. alveoli
   d. streak canal

4. What is the term for the restraint device that secures cattle around the neck to allow accessibility for milking, feeding, and examining?
   a. switch
   b. stanchion
   c. corium
   d. lead rope

5. An individual housing pen for calves is called a
   a. halter
   b. stanchion
   c. hutch
   d. barren

6. The term for shaving off wool, hair, or fur is
   a. cutting
   b. scouring
   c. carding
   d. shearing

7. Another term for wool is
   a. fur
   b. curl
   c. fleece
   d. crimp

8. The common term for llama copulation is
   a. tupping
   b. cushing
   c. felting
   d. finishing

9. The amount of food consumed by livestock in a 24-hour period is its

   a. graze
   b. ration
   c. roughage
   d. silage

10. The appendage suspended from the head of goats is the

   a. wattle
   b. comb
   c. crimp
   d. clip

11. The behavior in which a camelid propels saliva at an animal or a person who is perceived to be a threat is called

   a. regurgitation
   b. vomiting
   c. spitting
   d. drooling

12. The term for adult goat meat is

   a. mutton
   b. veal
   c. lamb
   d. chevon

13. The milk production organ in ruminants is called the

   a. nipple
   b. udder
   c. milk well
   d. gland cistern

14. The type of roughage feed produced by fermenting chopped plant parts is

   a. concentrate
   b. sweetfeed
   c. silage
   d. hay

15. The hypocalcemic metabolic disorder of ruminants that is seen in late pregnancy or early lactation is

   a. parturient paresis
   b. malpresentation
   c. rendering
   d. brucellosis

16. The mechanical device that is used to restrain cattle is a

   a. cast
   b. chute
   c. cod
   d. cull

17. Removing an animal from the rest because it does not meet a specific standard is known as

   a. coding
   b. culling
   c. cleaning
   d. cross-breeding

18. The mating of dissimilar breeds to increase productivity and performance in the next generation is known as

   a. heart girth
   b. inbred
   c. casting
   d. hybrid vigor

19. A large group of range sheep is a

   a. band
   b. card
   c. clip
   d. crimp

20. Roughage plants that have nitrogen-fixing nodules on their roots are

   a. feedstuff
   b. legumes
   c. roughages
   d. sweetfeeds

CROSSWORD PUZZLE

Ruminant Terms Puzzle

Across

3 remove an animal from the rest
5 another term for wool
6 metal cup with a lid that is used for detecting mastitis
7 large group of range sheep
8 milk production organ in ruminants and equine
10 chewing cud
12 distal part of the bovine tail with long, coarse hairs
13 young llama
16 adult sheep meat
17 animal that is not lactating
18 nipples of ruminants and equine
19 not able to reproduce

Down

1 copulation in llamas
2 one season's wool yield
3 common name for removal of a retained placenta
4 appendages suspended from the chin of goats
9 removal of horn growth in kids
11 mechanical device used to restrain cattle
14 amount of food consumed in 24-hour period
15 California mastitis test

CASE STUDIES

Define the underlined terms in each case study.

A 2-yr-old French alpine <u>doe</u> was presented with recurrent abdominal distention, decreased milk production, and <u>dyspnea</u>. On PE, the <u>MMs</u> were pink, <u>CRT</u> was normal, <u>TPR</u> was normal, breathing was labored but no abnormal respiratory sounds were <u>auscultated</u>, and the mammary glands were normal. The herd that this doe is in is closed, and the rest of the animals were normal. Urine was tested and revealed a trace amount of glucose and negative ketones. Rumen fluid had a <u>pH</u> of 6.9 and a healthy population of small and medium <u>protozoal</u> organisms. The majority of the large protozoal population was dead. A <u>rumenostomy</u> was performed, and administration of 0.5 <u>L</u> warm water was initiated.

1. doe _____

2. dyspnea _____

3. MMs _____

4. CRT _____

5. TPR _____

6. auscultated _____

7. pH _____

8. protozoal _____

9. rumenostomy _____

10. L _____

A 2-yr-old Suffolk <u>ram</u> was examined for <u>lethargy</u>, <u>anorexia</u>, and weight loss. The <u>PE</u> was <u>WNL</u> except for pale <u>mucous membranes</u> and a distended abdominal area. The owner was questioned about grazing, nutritional, and deworming practices on the farm. The veterinarian took a blood sample for a <u>hematocrit</u> and a stool sample for parasite testing. The hematocrit revealed that the ram was mildly <u>anemic</u>, and the fecal exam revealed *Haemonchus* eggs. The owner was advised to administer a broad-spectrum <u>anthelmintic</u>, to rotate pastures, and to monitor animals for signs of anemia.

11. ram _____

12. lethargy _____

13. anorexia _____

14. PE _____

15. WNL _____

16. mucous membranes _____

17. hematocrit _____

18. anemic _____

19. anthelmintic _____

CHAPTER 21

HOG HEAVEN

Objectives

Upon completion of this chapter, the reader should be able to

- Recognize, define, spell, and pronounce terms related to swine
- Analyze case studies and apply medical terminology in a practical setting

PIGS

Pigs have been domesticated since 6,500 BC and have been used as livestock both as a forager in developing countries and as confinement-raised animals in commercial operations in industrialized nations. Today pigs are used for their meat and hides, as research models, and for pharmaceutical production. Some pigs, such as potbellied pigs, have more recently been housed as pets.

Many of the anatomy and physiology concepts and medical terms related to pigs have been covered in previous chapters. The lists in this chapter and in Figure 21–1 apply more specifically to the care and treatment of pigs.

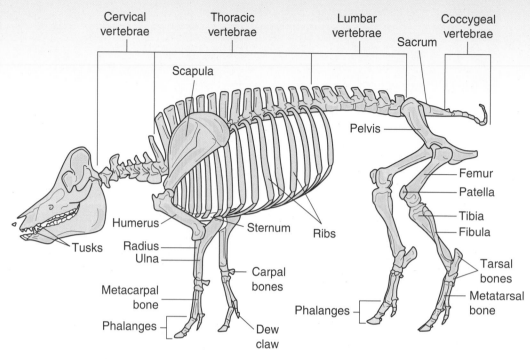

Cervical vertebrae

Thoracic vertebrae

Lumbar vertebrae

Coccygeal vertebrae

Sacrum

Scapula

Pelvis

Femur

Patella

Tibia

Fibula

Humerus

Sternum

Ribs

Tusks

Radius

Ulna

Carpal bones

Tarsal bones

Metatarsal bone

Metacarpal bone

Phalanges

Phalanges

Dew claw

Figure 21–1 Pig skeleton.

EQUIPMENT AND INDUSTRY TERMS

- **backfat** (bahk-faht) = thickness of fat along the dorsum of the pig.
- **bacon** (bā-kohn) = meat product from the sides of the pig.
- **boar taint** (bōr tānt) = odor of pork that is harvested from an adult boar; as boar ages, boar taint becomes more pronounced and results in an unpalatable product.
- **casting** (kahs-tihng) = restraint method using ropes to place animal in lateral recumbency.
- **checkoff** (chehck-ohf) = system where a portion of the sale price of every pig goes to the National Pork Board to promote and improve the pork industry.
- **cross-fostering** (krohs-fohs-tər-ihng) = moving piglets from one litter to another to balance litter size (done in the first day or two of life).
- **dunging pattern** (duhn-gihng pah-tərn) = tendency for animals to eliminate wastes in a particular location.
- **ear notching** (ēr nohtch-ihng) = identification method used in swine in which notches of various patterns are cut in the ear (Figures 21–2a and b).
- **farrowing crate** (făr-ō-ihng krāt) = holding pen that limits sow movement before and during parturition (Figure 21–3a).
- **farrowing house** (făr-ō-ihng hows) = building dedicated to the delivery and raising of piglets to weaning.
- **farrowing pen** (făr-ō-ihng pehn) = sow holding area that has guardrails and floor junctures that allow young pigs to escape; used before and during parturition; pen is larger than a crate (Figure 21–3b).

- **finish** (fihn-ihsh) = degree of fat on an animal that is ready for slaughter.
- **hide** (hīd) = pig skin.
- **hog hurdle** (hohg hər-duhl) = portable partition used to move swine by blocking the area in which the pig should not go (Figure 21–4a).
- **hog snare** (hohg snār) = restraint method in which the pig's snout is secured by a loop tie that is attached to a long handle; also called **snare** (Figure 21–4b).
- **hog-tight** (hohg-tīt) = fencing that prevents animal escape.
- **lard** (lahrd) = soft, white fat that is the product of rendering pig fat.
- **needle teeth** = eight temporary incisors and canine teeth of young swine (Figure 21–5).
- **piles** (pīlz) = common term for a prolapsed rectum in swine.
- **ringing** = act of implanting a wire ring through a pig's nose to discourage rooting.
- **sling** (slihng) = swine restraint device with four leg holes and an additional hole under the neck for blood collection; device looks like a hammock.
- **snout** (snowt) = upper lip and apex of nose of swine.
- **specific pathogen free** (speh-sihf-ihck pahth-ō-jehn frē) = national system of accrediting free of specific diseases; animals are obtained by cesarean section and raised in isolation to prevent certain infectious diseases; abbreviated SPF; not disease-free.
- **tusk** (tuhsk) = overgrown canine tooth of boar.
- **wallow** (wahl-ō) = natural or artificial wading area to cool swine.

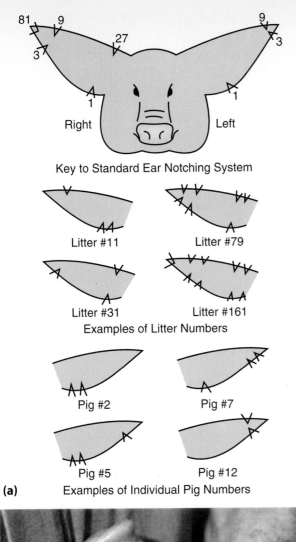

Key to Standard Ear Notching System

Litter #11   Litter #79

Litter #31   Litter #161

Examples of Litter Numbers

Pig #2   Pig #7

Pig #5   Pig #12

(a) Examples of Individual Pig Numbers

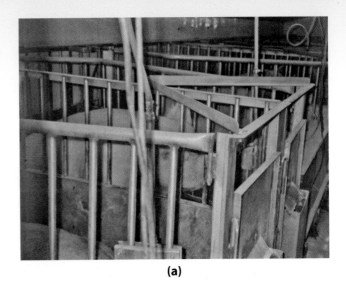

(a)

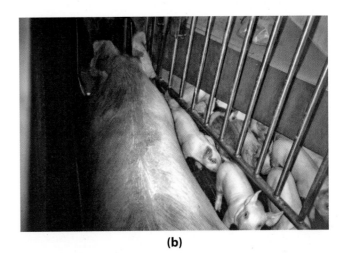

(b)

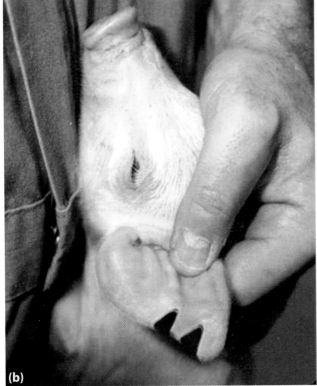

(b)

(c)

Figure 21–3 Housing of swine. (a) Farrowing or gestation crate restrains sows during farrowing. (b) Farrowing pen allows baby pigs to move away from the sow to prevent being crushed. (c) Larger pigs are housed away from the sow. (Courtesy of Ron Fabrizius, DVM, Diplomate ACT.)

Figure 21–2 (a) Ear notching identification in swine. (b) Pig with notched ears. (Courtesy of iStock Photo/Kathryn Bell.)

(a)   (b)

Figure 21–4   Restraint of pigs. (a) Hog hurdle. (b) Hog snare. (Courtesy of Ron Fabrizius, DVM, Diplomate ACT.)

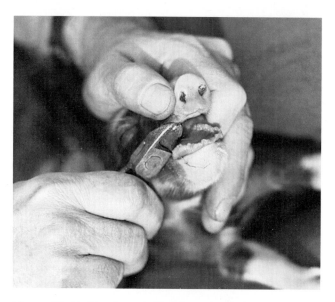

Figure 21–5   Needle teeth. Needle teeth of baby pigs are clipped to prevent injury to the sow during nursing and to minimize wounds from sibling pigs.

Swine Management Terms

- **closed herd** = group of animals that restricts entrance of new animals.
- **farrow-to-finish** (făr-ō too fihn-ihsh) = full-service swine operation that houses breeders, newborns, weanlings, and feeder stock.
- **farrow-to-wean** (făr-ō too wĕn) = swine operation that houses breeding sows and offspring until they reach weaning age or weight.
- **feeder-to-finish** (fē-dər too fihn-ihsh) = swine operation that raises weanling pigs to market weight.

- **finisher pig** (fihn-ihsh-ər pihg) = swine over 100 pounds to slaughter.
- **grower pig** (grō-ər pihg) = swine from about 40 to 100 pounds.
- **open herd** = group of animals in which animals from other groups are allowed to join the existing group.
- **starter pig** (stahr-tər pihg) = swine from about 10 to 40 pounds; also called feeder pig.

Swine Vaccinations

- ***Bordetella*** (bōr-dih-tehl-ah) = bacteria causing respiratory disease; atrophic rhinitis pathogen.
- ***Clostridium perfringens type C*** (klohs-trih-dē-uhm pər-frihn-jehns tīp C) = bacteria causing enterotoxemia that results in diarrhea and high mortality.
- ***E. coli*** (ē-kō-lī) = bacteria causing colibacillosis that causes edema, ataxia, and death.
- ***Erysipelas*** (ehr-ih-sihp-eh-lahs) = bacteria causing acute septicemia, skin lesions, chronic arthritis, and endocarditis.
- ***Haemophilus*** (hē-mohf-ih-luhs) = bacteria causing respiratory disease associated with acute onset, pyrexia, and reluctance to move.
- **leptospirosis** (lehp-tō-spī-rō-sihs) = bacterial (*Leptospira*) disease of swine associated with pyrexia, anorexia, neurologic signs, and abortion.
- ***Mycoplasma*** (mī-kō-plahz-mah) = bacteria causing respiratory disease seen largely in young pigs with a severe cough.
- **parvovirus** (pahr-vō-vī-ruhs) = parvovirus infection affecting mainly gilts and associated with abortion.

- *Pasteurella* (pahs-too-rehl-ah) = bacteria causing respiratory disease that sometimes leads to pericarditis and pleuritis.
- **porcine proliferative enteropathy** (por-sīn prō-lihf-ər-ah-tihv ehn-tər-oh-pah-thē) = bacterial disease (caused by *Lawsonia intracellularis*) that causes ileitis.
- **porcine** (por-sīn) **respiratory and reproductive syndrome** = arteriviral disease of swine characterized by reproductive failure in sows and respiratory disease in young and growing pigs; abbreviated PRRS.
- **pseudorabies** (soo-dō-rā-bēz) = herpesvirus infection associated with pyrexia and neurologic signs.

- **rotavirus** (rō-tə-vī-ruhs) = rotavirus associated with villous destruction in the intestine, malabsorption, and diarrhea.
- *Streptococcus suis* (strehp-tō-kohk-kuhs sū-his) = bacterial infection that causes meningitis.
- **swine influenza** (swīn ihn-flū-ehn-zah) = orthomyxoviral infection that causes respiratory disease.
- **transmissible gastroenteritis** (trahnz-mihs-ih-buhl gahs-trō-ehn-tehr-ī-tihs) = coronaviral disease of swine characterized by villous destruction of jejunum and ileum, malabsorption, diarrhea, and dehydration; abbreviated TGE.

REVIEW EXERCISES

Multiple Choice

Choose the correct answer.

1. The eight temporary incisors and canine teeth of young swine are known as
    a. piles
    b. needle teeth
    c. finishing teeth
    d. tusks

2. A swine over 100 pounds to slaughter is known as a(n)
    a. finisher pig
    b. grower pig
    c. starter pig
    d. open pig

3. Swine weighing 10 to 40 pounds are called
    a. finisher pigs
    b. grower pigs
    c. starter pigs
    d. open pigs

4. What identification method is used for swine?
    a. tattooing
    b. branding
    c. ear notching
    d. tail notching

5. What is the name of the restraint method that uses ropes to place swine in lateral recumbency?
    a. hog-tighting
    b. hog-tying
    c. slinging
    d. casting

6. The overgrown canine tooth of the boar is its
    a. needle tooth
    b. molar
    c. tusk
    d. snout

7. A swine operation that raises weanling pigs to market weight is what type of management system?
    a. farrow-to-finish
    b. farrow-to-wean
    c. feeder-to-finish
    d. feeder-to-farrow

8. The upper lip and apex of the nose of swine is its
    a. tusk
    b. wallow
    c. snout
    d. lard

9. What type of restraint method has the pig's snout secured via a loop tie that is attached to a long handle?
    a. hog hurdle
    b. snare
    c. sling
    d. crate

10. A holding pen that limits sow movement before and during parturition is known as a
    a. farrowing crate
    b. farrowing pen
    c. finishing pen
    d. starter pen

11. The common term for a prolapsed rectum in swine is

    a. ringing
    b. sling
    c. piles
    d. backfat

12. A group of animals that restricts entrance of new animals is a(n)

    a. open herd
    b. closed herd
    c. farrowing herd
    d. finisher herd

13. Moving piglets from one litter to another to balance litter size is

    a. checkoff
    b. boar taint
    c. piglet transfer
    d. cross-fostering

14. A portable partition to move swine by blocking the area is a

    a. hog snare
    b. hog hurdle
    c. hog ring
    d. hog sling

15. Pig skin also is called

    a. hide
    b. dung
    c. piles
    d. lard

16. The tendency for animals to eliminate wastes in a particular location is known as a

    a. notching pattern
    b. farrowing pattern
    c. slinging pattern
    d. dunging pattern

17. The odor of pork that is harvested from an adult boar is called

    a. pork taste
    b. boar taint
    c. boar odor
    d. pork taint

18. White fat that is the product of rendering pig fat is

    a. wallow
    b. piles
    c. lard
    d. bacon

19. Swine from about 40 to 100 pounds in weight are

    a. finisher pigs
    b. grower pigs
    c. starter pigs
    d. farrowing pigs

20. SPF pigs are

    a. completely disease-free
    b. obtained by cesarean section
    c. raised in large intermingling groups
    d. part of a nonaccredited system

CROSSWORD PUZZLE

Swine Terms Puzzle

Across

4 restraint method in which the pig's snout is secured
   via a loop tie
6 intact male pig
7 common term for prolapsed rectum in swine
9 giving birth to swine
10 female pig

Down

1 overgrown canine tooth of boar
2 eight temporary incisors and canine teeth of young swine
3 restraint method using ropes to place animal in lateral
   recumbancy
5 wading area to cool swine
6 castrated male pig
8 implanting a wire ring through a pig's nose

CASE STUDIES

Define the underlined terms in each case study.

A group of 4-<u>mo</u>-old <u>barrows</u> was presented with <u>clinical</u> <u>signs</u> of sneezing, <u>purulent</u> <u>nasal</u> discharge, and decreased weight gain. Most of the barrows had a mild to moderate <u>deviation</u> of the <u>snout</u>. The farmer sacrificed one pig for <u>necropsy</u> and found that the nasal <u>turbinates</u> were <u>atrophied</u> and <u>asymmetrical</u>. <u>Dx</u> was atrophic <u>rhinitis</u>, which is a common disease of pigs caused by two types of bacteria. Control measures such as better ventilation and improved hygiene were discussed with the farmer (Figure 21–6).

1. mo

2. barrows

3. clinical

4. signs

5. purulent

6. nasal

7. deviation

8. snout

9. necropsy

10. turbinates

11. atrophied

12. asymmetrical

13. Dx

14. rhinitis

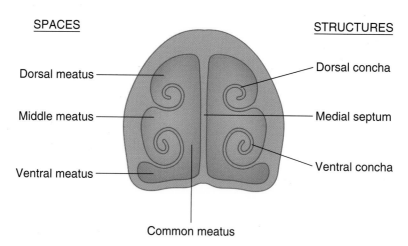

SPACES   STRUCTURES

Dorsal meatus ——   — Dorsal concha

Middle meatus ——   — Medial septum

Ventral meatus ——   — Ventral concha

Common meatus

Figure 21–6 Atrophic rhinitis. Atrophic rhinitis causes atrophy of the turbinates and distortion of the nasal septum. The turbinates soften and erode due to infection by bacteria.

A 3-yr-old sow was seen by the area veterinarian for dystocia. She had farrowed two litters normally before this but had problems with mastitis and agalactia. Because the pig had uterine inertia, the veterinarian had to perform an emergency C-section to retrieve the remaining pigs.

15. yr   _____

16. sow   _____

17. dystocia   _____

18. farrowed   _____

19. mastitis   _____

20. agalactia   _____

21. uterine inertia   _____

22. C-section   _____

Provide the medical term for the underlined definitions.

A litter of 2-wk-old pigs was presented with vomiting, abnormal frequency and liquidity of fecal material, incoordination, and elevated body temperature. A blood sample was taken, and the pigs tested positive for pseudorabies, also called Aujeszky's disease, which is a herpesvirus infection. In adult pigs, pseudorabies can cause respiratory disease and termination of pregnancy. All positive animals were isolated and kept under observation. Pseudorabies control measures, such as adding only serologically negative animals to the herd, avoiding visiting infected premises, keeping wild animals away from swine, and providing separate equipment for each group of animals, were discussed with the owner.

23. vomiting   _____

24. abnormal frequency and liquidity of fecal material   _____

25. incoordination   _____

26. elevated body temperature   _____

27. termination of pregnancy   _____

28. wild   _____

CHAPTER 22

[BIRDS OF A FEATHER]

Objectives

Upon completion of this chapter, the reader should be able to

- Recognize, define, spell, and pronounce terms related to birds
- Analyze case studies and apply medical terminology in a practical setting

BIRDS

Birds are two-legged, egg-laying, warm-blooded animals that have a unique anatomy because of their ability to fly. To allow flight, their respiratory and skeletal systems differ greatly from those of other vertebrates (Figure 22–1). Birds have thin skin that consists of two layers: the epidermis (superficial layer) and dermis (deep layer). From the dermis emerge feathers that may be a variety of different types. Birds have beaks that are varied in shape to accommodate their various dietary preferences. Birds that are seed eaters have short, strong bills that can crack open a hard-shelled seed easily, while birds that eat flying insects have flat bills with a wide base so that they can catch a small moving target. Raptors such as hawks and eagles have hooked bills that are sharp, making it easy to rip apart their prey. The skeletal system of birds is modified to enable the bird to fly. These modifications include fusion of bones (provides a rigid structure) and the development of wing bones that are hollow, lack bone marrow, and contain air (reduce weight to enhance flight).

Birds and the terms associated with them vary depending on the types of birds people keep. Some birds, such as psittacines, are popular as pets. Other birds, such as poultry and ratites, are used as livestock. The terms related to birds typically refer to one or the other of these uses for birds.

Some anatomy and physiology concepts and medical terms related to avian species have been covered in previous chapters. Additional anatomy and physiology concepts are covered in this chapter.

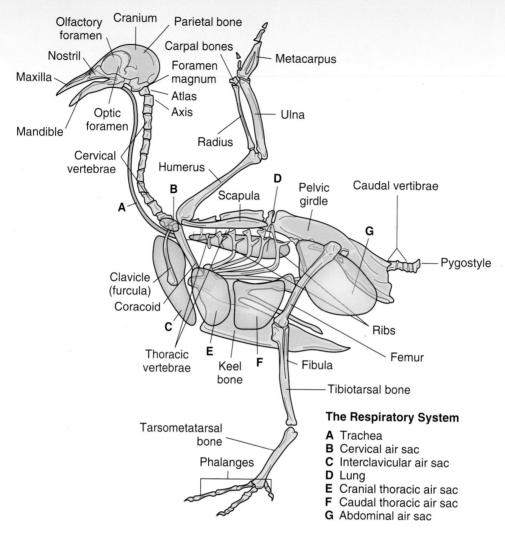

Figure 22–1   Skeletal and respiratory system of birds. Air flow through the lungs and air sacs. (1) First inspiration: Air flows into the trachea and through the primary bronchi and goes to the caudal and abdominal air sacs. Air already in the caudal air sacs moves to the cranial air sacs. (2) First expiration: Air travels back to the parabronchi and gas exchange occurs. (3) Second inspiration: Air moves from the parabronchi to the cranial air sacs. (4) Second expiration: Air moves out of the cranial air sacs, into the parabronchi, and out of the trachea.

The respiratory system labels:

The Respiratory System
A Trachea
B Cervical air sac
C Interclavicular air sac
D Lung
E Cranial thoracic air sac
F Caudal thoracic air sac
G Abdominal air sac

ANATOMY AND PHYSIOLOGY TERMS

Respiratory System

- **air sacs** = thin-walled sacs in the respiratory tract of birds that store air and provide buoyancy for flight.
- **choana** (kō-ā-nah) (or **choanal space**) = caudal naris; the cleft in the hard palate of birds (Figure 22–2).
- **nasal gland** = gland that allows sea birds to drink saltwater; found in the rostral portion of the beak.
- **parabronchi** (pahr-ah-brohnck-ī) = tiny passages in bird lungs that are the primary sites of gas exchange between air and blood; birds do not have alveoli.
- **syrinx** (sehr-ihncks) = voice organ of birds located at the tracheal bifurcation (where the trachea splits into bronchi).

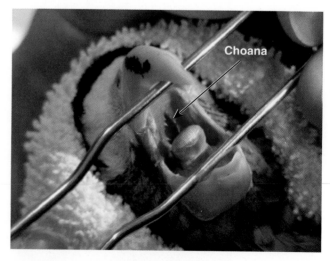

Figure 22–2   Choana of a parrot.

Integumentary System

- **apterium** (ahp-tehr-ē-uhm) = area or tract of skin without feathers or down; plural is **apteria** (ahp-tehr-ē-ah).
- **barb** (bahrb) = one of the parallel filaments projecting from the main feather shaft (rachis); forms the feather vane.
- **barbule** (bahr-byool) = one of the small projections fringing the edges of the barbs of feathers; attach to adjacent barbules to give the vane rigidity.
- **beak** (bēk) = hard mouth part of bird that is a modified epidermal structure that covers the rostral part of the maxilla and mandible.
- **calamus** (kahl-ah-muhs) = hollow shaft at the proximal end of the feather shaft; "quill."
- **central tail feathers** = primary feathers at the center of the tail.
- **cere** (sēr) = thickened skin at base of external nares of birds; may be different colors in some birds to denote sex (Figure 22–3).
- **cheek** (chēk) = area of the face below the eye of a bird.
- **contour feathers** (kohn-tər fehth-ərz) = body or flight feathers arranged in rows.
- **coverts** (kuh-vərhtz or kō-vehrtz) = small feathers that cover the bases of large feathers on the body; named based on their location.

- **feather** (fehth-ər) = epidermal structure analogous to hair; used for insulation and thermoregulation, in courtship displays, and for flight (Figure 22–4).
- **filoplume** (fihl-ō-ploom) = type of feather that resembles a bristle feather topped by a down feather.
- **lateral tail feathers** = large tail feathers to each side of the central tail feathers.
- **mantle** (mahn-tuhl) = feathers across the dorsum (back) (Figure 22–3).
- **molt** (mōlt) = process of casting off feathers before replacement feathers appear.
- **orbital ring** (ohr-bih-tahl rihng) = ring of unfeathered skin around the eye.
- **pin feathers** = developing feathers that have blood flowing through them and can grow as new feathers or to replace molted feathers; also called blood feathers.
- **plume** (ploom) = down feathers.
- **primary feathers** = longer, thinner remiges connected to the carpometacarpus and phalanges that project along the outer edge of a bird's wing; also called primaries and primary quills.
- **pteryla** (tehr-ih-lah) = feather tract of birds; plural is **pterylae** (tehr-ih-lā).
- **rachis** (rā-kuhs) = distal end of the feather shaft.
- **rectrices** (rehck-trih-sehz) = tail flight feathers.

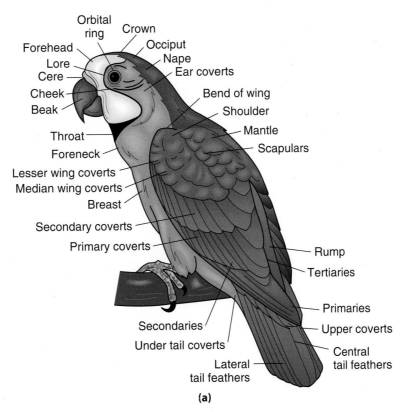

Figure 22–3 (a) External parts of a bird.

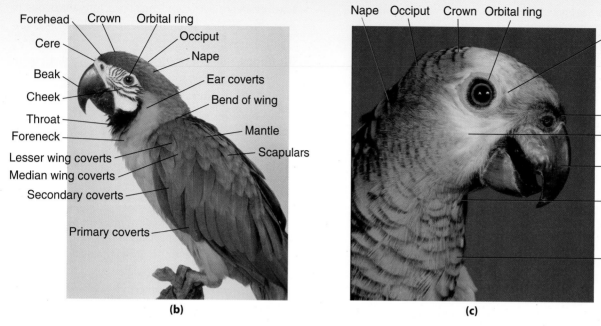

Figure 22–3   (b) External parts of a bird on a blue and gold macaw. (c) External parts of a bird's head on a blue-fronted Amazon parrot. [(b) and (c) Courtesy of Isabelle Francais.]

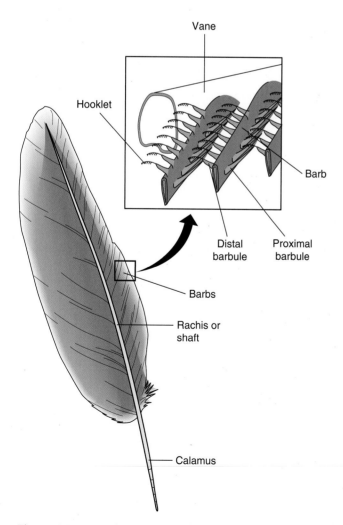

Figure 22–4   Feather parts.

- **remiges** (rehm-ih-jehz) = primary wing feathers; singular is **remix.**
- **secondary feathers** = smaller, lighter remiges than the primary feathers that are connected to the ulna; also called secondaries.
- **setae** (sē-tā) = sensitive bristles that grow on the heads of many birds; also called **bristles.**
- **shaft** (shahft) = quill or central part of a contour feather; also called **scapus** (skā-puhs).
- **snood** (snood) = long, fleshy extension at the base of a turkey's beak.
- **spurs** (spərz) = projecting body (as from a bone) or a sharp, horn-covered, bony projection from the shank of male birds of some species.
- **tertiary feathers** = shorter remiges that are connected to the humerus in some bird species; also called tertiaries and tertials.
- **uropygial** (yoor-ō-pihj-ē-ahl) **gland** = gland located laterally to the tail feather attachment that secretes oil used to waterproof or preen feathers; also called the **preen** (prēn) **gland.**
- **wattle** (waht-tuhl) = appendage suspended from the head (usually the chin) in chickens, turkeys, and goats.

External Anatomy Terms

- **breast** (brehst) = chest area of birds.
- **crown** (krown) = area caudal to the forehead to the cranial portion of the neck; the top of the head.
- **forehead** (fōr-hehd) = part of the head rostral to the eyes.

- **foreneck** (fōr-nehck) = area cranial to the breast where the clavicle (wishbone) is located.
- **lore** (lōr) = lateral area between the cranial portion of the eyes and bill of birds.
- **nape** (nāp) = caudodorsal portion of the neck of birds.
- **occiput** (ohck-sihp-uht) = sloping caudal part of the head of birds.
- **rump** (ruhmp) = area that overlies the pelvis cranial to the tail.
- **throat** (thrōt) = area caudal to the head and cranial to the chest.

Gastrointestinal System

- **cloaca** (klō-ā-kah) = common passage for fecal, urinary, and reproductive systems in birds and lower vertebrates (Figure 22–5).
- **coprodeum** (kōp-rō-dē-uhm) = rectal opening into the cloaca.
- **crop** (krohp) = esophageal enlargement that stores, moistens, and softens food in some birds (Figure 22–6).
- **droppings** = composite of feces and urine in birds.
- **Meckel's diverticulum** (mehck-ehlz dī-vər-tihck-yoo-luhm) = structure at the terminal end of the jejunum that functions as a lymphatic organ.
- **proventriculus** (prō-vehn-trihck-yoo-luhs) = elongated, spindle-shaped glandular stomach of birds.

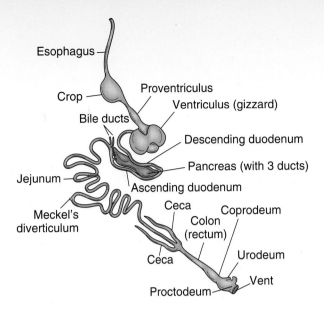

Figure 22–6   Gastrointestinal tract of the bird.

- **vent** (vehnt) = external opening of the cloaca of birds.
- **ventriculus** (vehn-trihck-yoo-luhs) = muscular stomach of birds; also called the gizzard.

Musculoskeletal System

- **columella** (kohl-uhm-eh-lah) = bony structure that replaces the malleus, incus, and stapes in the middle ear.
- **furcula** (fuhr-kuh-lah) = fused clavicle in birds; the wishbone.
- **keel** (kēl) = sternum, or breastbone, of birds (excluding ratites) (Figure 22–7).
- **pygostyle** (pihg-ō-stīl) = bony termination of the vertebral column in birds where tail feathers attach; also called **rump post.**
- **scleral** (skleh-rahl) **ring** = overlapping bony plate encircling the eye at the corneal–scleral junction.

Urogenital System

See Figure 22–8.

- **infundibulum** (ihn-fuhn-dihb-yoo-luhm) = portion of the oviduct located closest to the ovary that captures ovulated eggs.
- **isthmus** (ihs-muhs) = portion of the oviduct located farthest from the ovary that adds the shell membranes.
- **magnum** (mahg-nuhm) = middle portion of the oviduct that separates the albumin and chalaza over the egg yolk and sperm.
- **sperm nests** = clusters of spermatozoa held in readiness in the infundibulum to fertilize the egg as it comes from the ovary.
- **urodeum** (yoo-rō-dē-uhm) = area of the cloaca in which the ureters and vagina open.

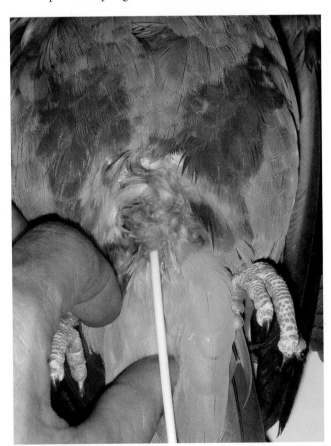

Figure 22–5   Cloacal swab.

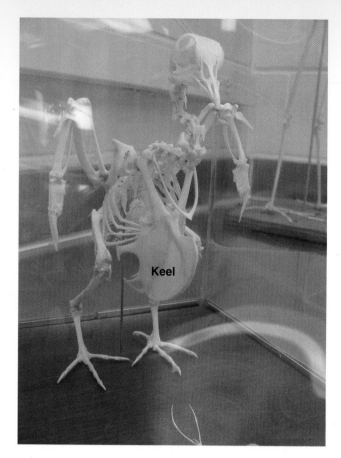

Figure 22–7   Keel bone of a bird.

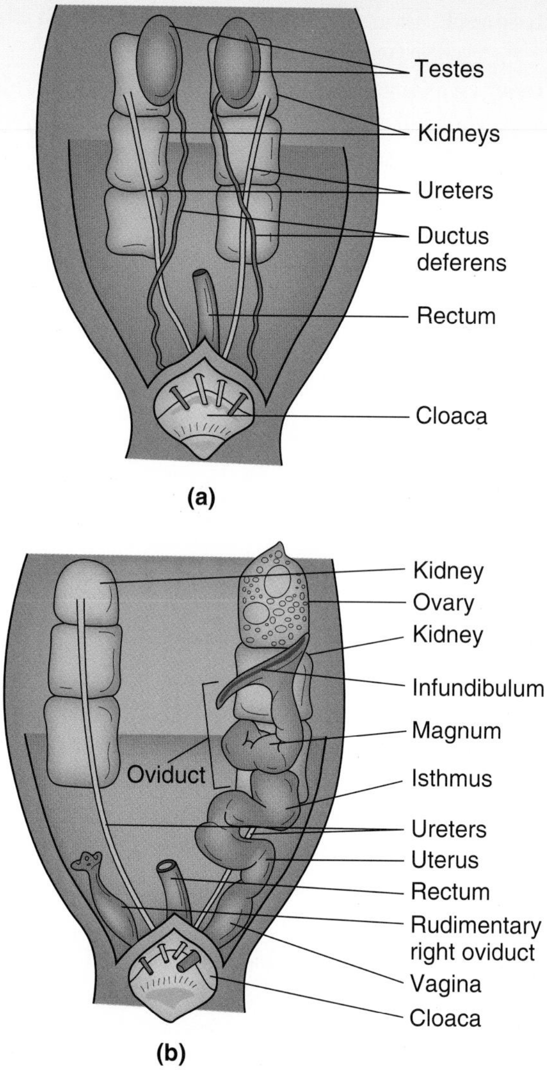

- **uterus** (yoo-tər-uhs) = portion of the female reproductive tract in birds that produces the shell and shell pigments; also the area in which the egg turns around so that it is laid blunt end first.
- **vagina** (vah-jī-nah) = portion of the female reproductive tract in birds that directs the egg to the cloaca.

POULTRY TERMS

- **broiler** (broy-lər) = young chicken approximately 8 weeks old weighing 1.5 kg or more; also called fryer or young chicken.
- **brood** (brūd) = young that are hatched or cared for at the same time.
- **brooder** (brū-dər) = housing unit for rearing birds after hatching.
- **cage operation** = method of raising chickens in which the hens are kept in confinement as they produce eggs.
- **candling** (kahn-dlihng) = process of shining a light through an egg to check embryo development (Figure 22–9).
- **chalaza** (kahl-ā-zah) = ropelike structure that holds the yolk to the center of the egg (Figure 22–10).

Figure (a) labels: Testes, Kidneys, Ureters, Ductus deferens, Rectum, Cloaca

(a)

Figure (b) labels: Kidney, Ovary, Kidney, Infundibulum, Magnum, Isthmus, Ureters, Uterus, Rectum, Rudimentary right oviduct, Vagina, Cloaca, Oviduct

(b)

Figure 22–8   Urogenital system of the bird. (a) Male bird; (b) female bird. Only the left ovary and oviduct are functional in female birds.

Figure 22–9   Candling is a process by which eggs pass over an intense light to reveal cracks in the shell or to check on embryo development. (Courtesy of USDA.)

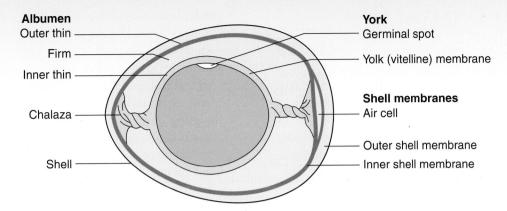

Figure 22–10 Parts of an egg.

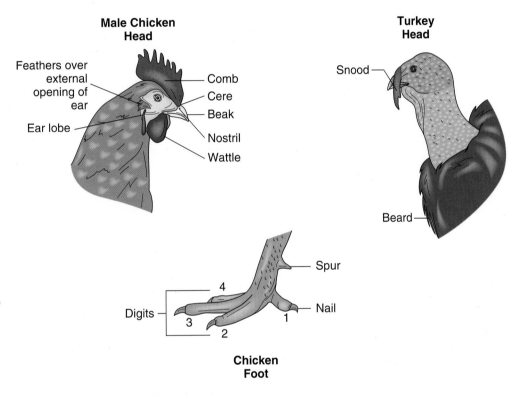

Figure 22–11 Head and feet of poultry.

- **comb** (kōm) = in domestic fowl, the vascular, red cutaneous structure attached in a sagittal plane to the dorsum of the skull (Figure 22–11).
- **debeaking** (dē-bēk-ihng) = removing about one-half of the upper beak and a small portion of the lower beak in poultry to prevent feather picking, cannibalism, and fighting; also called **beak trimming** in poultry (Figure 22–12).
- **hatch** (hahtch) = emergence of a baby bird from the shell.
- **incubation** (ihn-kyoo-bā-shuhn) = process of a fertilized poultry egg developing into a newly hatched bird.
- **layer** (lā-ər) = chicken raised for egg production.
- **poultry** (pōl-trē) = any domesticated fowl raised for meat, eggs, or feathers.
- **yolk** (yōk) = yellow part of the egg that contains the germinal cells.

Figure 22–12 Debeaking a chicken.

PET BIRD TERMS

- **beak trimming** = trimming the tip of the beak to keep the beak properly aligned.
- **clipping** (klihp-ihng) = trimming wings of birds to alter their flight; also called **pinioning** (pihn-yehn-ihng) (Figures 22–13 and 22–14).
- **columbiformes** (kō-luhm-bih-fōrmz) = group of dove-like birds that includes doves and partridges that have short beaks and two forward-facing toes and two rear-facing toes.

- **cuttlebone** (kuht-uhl-bōn) = shell of a cuttlefish that is typically provided in the cage for the bird to use in wearing down its beak.
- **feather picking** = undesired behavior in which birds remove their own feathers mainly due to stress or disease; also called feather plucking.
- **fledgling** (flehdj-lihng) = young bird that has recently acquired its flight feathers and typically is out of the nest but not eating on its own (Figure 22–15).
- **hand-raised** = refers to a bird that has been raised by humans.

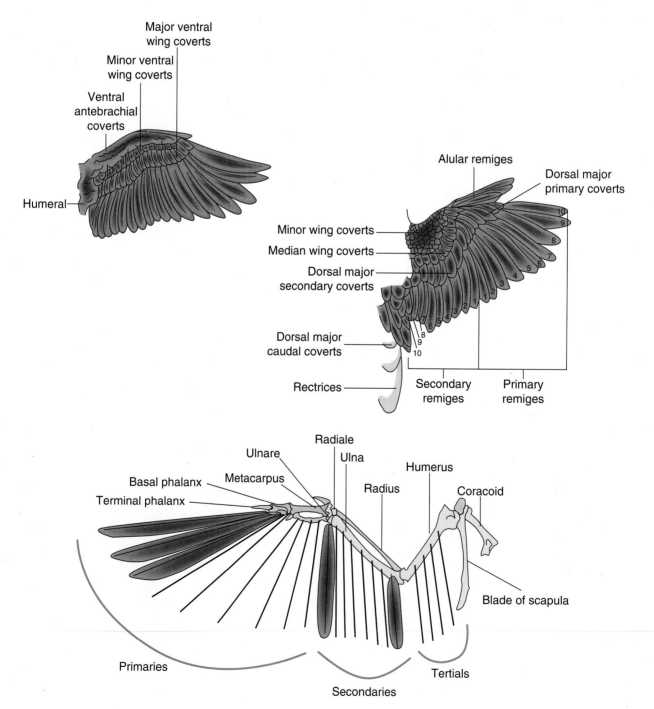

Figure 22–13 Parts and types of feathers of a wing.

- **passeriformes** (pahs-ər-ih-fŏrmz) = group of perching birds that includes most songbirds (such as finches, sparrows, mynahs, and canaries) that have three forward-facing toes and one rear-facing toe (see Figure 22–16b).

- **perch** (pərch) = stick or dowel provided for the bird to sit on; some perches have rough surfaces that assist the bird in wearing down its toenails (Figures 22–16a, b, and c).

- **psittacine** (siht-ah-sēn) = group of parrotlike birds that includes parrots, macaws, budgerigars, cockatiels, cockatoos, conures, lovebirds, and parakeets that have a strong curved beak and two forward-facing toes and two rear-facing toes (Figure 22–16a).

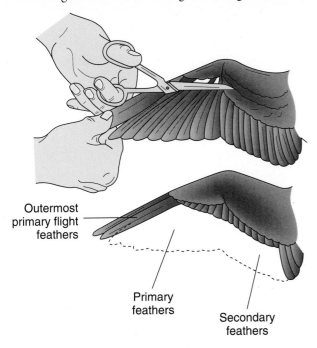

Outermost primary flight feathers

Primary feathers

Secondary feathers

Figure 22–14 Wing clipping in a pet bird.

Figure 22–15 Fledgling budgerigar. (Courtesy of Isabelle Francais.)

(a)

(b)

Parrot/2.5 cm (1 in) square

Finch/1.25 cm (0.5 in) round

Budgie/1.25 cm (0.5 in) oval

(c)

Figure 22–16 (a) African gray parrots on a wooden perch. (b) Gouldian finch (red-headed) on a wooden perch. (c) Ideal perch size and shape for a variety of birds. [(a) and (b) Courtesy of Isabelle Francais.]

RATITE TERMS

- **breeding animals** = animals purchased, often as pairs, to produce offspring.
- **emu** (ē-mū) = large, fast flightless bird native to Australia that has been imported to zoos and used in animal agriculture; males are slightly smaller than females and have gray, black, and brown feathers (Figure 22–17).
- **emu oil** = oil produced from emu fat that is used in pharmaceuticals and cosmetics.
- **meat** = ratite muscle that is a red meat yet is lower in fat and cholesterol than beef.
- **ostrich** (ow-strihch) = large, flightless bird native to Africa that is the largest ratite (can weigh over 400 pounds) and may be black, blue, or red (Figure 22–18).
- **ostrich feathers** = feathers of ostriches that are used commercially for feather dusters and decoration.
- **ostrich skin** = skin from ostriches that is tanned into leather and used to make boots and other leather products.
- **ratite** (rah-tīt) = class of large, flightless birds that are raised for their meat and hides.
- **rhea** (rē-ah) = large, flightless bird native to South America that may be pale gray to brown in color and lack tail feathers; females lay their eggs in the same nest, and the male incubates the eggs (Figure 22–19).
- **rhea skin** = skin of rhea that is tanned into leather and typically used to make boots.

Figure 22–18 Ostrich. (Courtesy of iStock Photo.)

Figure 22–19 Rhea. (Courtesy of iStock Photo.)

Figure 22–17 Emu. (Courtesy of USDA.)

REVIEW EXERCISES

Multiple Choice

Choose the correct answer.

1. What is the term for the spaces in the bird respiratory tract that store air and provide buoyancy for flight?

    a. nasal sacs

    b. air sacs

    c. apteria

    d. ceres

2. The caudal nares (also called the cleft in the hard palate of birds) are called the

    a. choana

    b. cloaca

    c. calamus

    d. crop

3. The sternum, or breastbone, of some birds is the

    a. pygostyle

    b. crop

    c. spur

    d. keel

4. The esophageal enlargement that stores, moistens, and softens food in some birds is the

    a. coprodeum

    b. cloaca

    c. crop

    d. choana

5. A young bird that has recently acquired its flight feathers is known as a

    a. clutch

    b. fledgling

    c. brooder

    d. snood

6. The glandular stomach of birds is the

    a. vent

    b. Meckel's diverticulum

    c. ventriculus

    d. proventriculus

7. A body or flight feather is known as a

    a. down feather

    b. contour feather

    c. lume

    d. filoplume feather

8. The voice organ of birds is the

    a. syrinx

    b. cere

    c. choana

    d. setae

9. The primary wing feathers are known as

    a. rectrices

    b. rachis

    c. pteryla

    d. remiges

10. The process of casting off feathers before replacement feathers appear is known as

    a. pluming

    b. molting

    c. shafting

    d. feathering

11. Large, flightless birds used for their meat and hides are

    a. columbiformes

    b. passeriformes

    c. ratites

    d. psittacines

12. Finches are examples of which type of bird?

    a. columbiformes

    b. passeriformes

    c. ratites

    d. psittacines

13. Cockatiels and cockatoos are examples of which type of bird?

    a. columbiformes

    b. passeriformes

    c. ratites

    d. psittacines

14. Doves are examples of which type of bird?

    a. columbiformes

    b. passeriformes

    c. ratites

    d. psittacines

15. A stick or dowel on which a caged bird sits is called a

    a. scleral ring

    b. barbule

    c. barb

    d. perch

16. Which type of feather has a blood supply running through it that if broken can lead to heavy bleeding?

    a. covert

    b. plume

    c. pin

    d. bristle

17. The common passage for fecal, urinary, and reproductive systems in birds is the

    a. crop

    b. cloaca

    c. choana

    d. coprodeum

18. The ropelike structure that holds the yolk to the center of the egg is the

    a. chalaza

    b. yolk tendon

    c. isthmus

    d. snood

19. The wishbone in birds is the

    a. keel bone

    b. pygostyle

    c. carpus

    d. furcula

20. The external opening of the cloaca in birds is the

    a. ventriculus

    b. vent

    c. diverticulum

    d. infundibulum

CROSSWORD PUZZLE

Avian Terms Puzzle

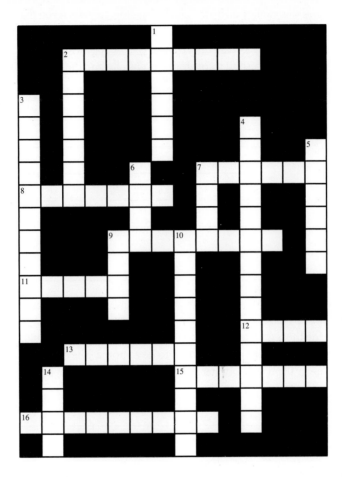

Across

2 preen gland
7 common passage for fecal, urinary, and reproductive systems in birds and lower vertebrates
8 primary wing feathers
9 trimming wings of birds to alter flight
11 chicken raised for egg production
12 red cutaneous structure attached in a sagittal plane to the dorsum of the skull in domestic fowl
13 posterior naris of birds
15 ropelike structure that holds the yolk to the center of the egg
16 young bird that has recently acquired its flight feathers

Down

1 voice organ of birds
2 area of the cloaca in which the ureters and vagina open
3 muscular stomach of birds (gizzard)
4 elongated, glandular stomach of birds
5 distal end of the feather shaft
6 sternum of birds
7 esophageal enlargement in some birds
9 thickened skin at base of external nares of birds
10 group of parrot-like birds
14 process of casting off feathers

CASE STUDIES

Define the underlined terms in each case study.

A <u>neonatal</u> greater sulfur-crested cockatoo was presented with a history of being tube-fed with a red rubber tube. The tube became dislodged from the feeding syringe and was swallowed by the chick. A <u>radiograph</u> was taken that confirmed that the tube was swallowed and was located extending from the <u>crop</u> to the <u>proventriculus</u>. An <u>ingluviotomy</u> was considered. Because it would be a less traumatic procedure, it was decided to retrieve the feeding tube with an <u>endoscope</u>. With the cockatoo under general <u>anesthesia</u> with isoflurane, the tube was successfully removed. The bird recovered uneventfully, and the owner was counseled on hand-feeding with a spoon instead of a feeding tube.

1. neonatal _____

2. radiograph _____

3. crop _____

4. proventriculus _____

5. ingluviotomy (also known as cropotomy) _____

6. endoscope _____

7. anesthesia _____

A 3-yr-old male budgerigar (parakeet) was presented for a growth on its <u>thorax</u> near the <u>thoracic inlet</u>. The bird had been kept singly and was fed a predominantly seed diet. On physical examination, the bird was noted to have a 2.5 <u>cm</u> diameter mass located <u>subcutaneously</u> on the <u>rostral</u> portion of the thorax. Differential <u>diagnoses</u> included <u>neoplasia</u> resulting in <u>lipoma</u> or a <u>xanthoma</u>. A <u>fine needle aspirate (FNA)</u> was performed, which confirmed the mass to be a lipoma. The owner declined surgical removal. The owner did elect to administer Lugol's iodine in the drinking water because this condition may be related to <u>hypothyroidism</u> (caused by iodine deficiency relating to an all-seed diet).

8. thorax _____

9. thoracic inlet _____

10. cm _____

11. subcutaneously _____

12. rostral _____

13. diagnoses _____

14. neoplasia _____

15. lipoma _____

16. xanthoma _____

17. fine needle aspirate (FNA) _____

18. hypothyroidism _____

A commercial <u>flock</u> of chickens developed disease in <u>cockerels</u> and <u>pullets</u> that included birds that were pale, were <u>anorexic</u>, were <u>emaciated</u>, had bloody <u>droppings</u>, and were dehydrated. Chicks were <u>asymptomatic</u>. The <u>layers</u> had a reduction in the rate of egg production. A diagnosis of coccidiosis was made from identification of coccidia protozoa by fecal examination. The flock was treated with a coccidiostat added to the feed, and facility management issues such as instituting <u>brooder</u> sanitation and switching to a <u>cage operation</u> were discussed with the owners.

19. flock   _____

20. cockerels   _____

21. pullets   _____

22. anorexic   _____

23. emaciated   _____

24. droppings   _____

25. asymptomatic   _____

26. layers   _____

27. brooder   _____

28. cage operation   _____

An <u>ostrich</u> <u>flock</u> had <u>chicks</u> developing neurologic signs including limb <u>paralysis</u>, <u>torticollis</u>, <u>ataxia</u>, and <u>opisthotonus</u>. In five days after initial signs appeared, many of the chicks had died. The outbreak occurred during the summer, producing a high <u>mortality</u> rate. <u>Histological</u> examination of tissues was inconclusive; however, equine encephalitis virus and Newcastle disease virus infection were suspected. <u>Serology</u> was performed on these birds, and the results are pending.

29. ostrich   _____

30. flock   _____

31. chicks   _____

32. paralysis   _____

33. torticollis   _____

34. ataxia   _____

35. opisthotonus   _____

36. mortality   _____

37. histological   _____

38. serology   _____

CHAPTER 23

ALL THE REST

Objectives

Upon completion of this chapter, the reader should be able to

- Recognize, define, spell, and pronounce terms related to laboratory animals
- Recognize, define, spell, and pronounce terms related to exotic animals or pocket pets
- Analyze case studies and apply medical terminology in a practical setting

LABORATORY ANIMALS, POCKET PETS, AND REPTILES

Laboratory animals include a wide range of species, even species some people consider to be pets and livestock. Pocket pets are small, nontraditional pets that were once thought of solely as laboratory animals or that are exotic to many areas of the country. Reptiles are cold-blooded vertebrates that are gaining popularity as nontraditional pets. Many terms used in the laboratory animal field have already been covered in other chapters. New terms pertaining to laboratory animals are related to scientific studies and the facilities in which they occur and are used in the care of pocket pets (Table 23–1).

Table 23–1 Common Laboratory and Pocket Pet Terms

Term	Pronunciation	Definition
acclimatization	ahck-lih-mah-tih-zā-shuhn	adjustment of an animal to a new environment
agouti	ah-goo-tē	naturally occurring coat color pattern that consists of dark-colored hair bands at the base of the hair and lighter increments of hair color toward the tip
albino	ahl-bī-nō	animal with a white coat and pink eyes; devoid of melanin (Figure 23–1)
ambient	ahm-bē-ahnt	surrounding
analogous	ahn-ahl-oh-guhs	refers to structures that differ anatomically but have similar functions
anogenital distance	ā-nō-jehn-ih-tahl	area between the anus and genitalia; females have a shorter anogenital distance than males, which is used to determine the sex of animals (Figures 23–2a through d)
antivivisectionist	ahnt-ih-vihv-ih-sehck-shuhn-ihst	person who opposes surgery on live animals for research or educational purposes
autosome	aw-tō-zōm	non–sex-determining chromosome
axenic	ā-zehn-ihck	germ-free
barbering	bahr-bər-ihng	behavioral disorder in which dominant animals bite or chew the fur of subordinate animals
barrier sustained	bār-ē-ər suh-stānd	gnotobiotic animals that are maintained under sterile conditions in a barrier unit
biohazard	bī-ō-hahz-ahrd	substance that is dangerous to life
calvarium	kahl-vahr-ē-uhm	top of the skull
cannibalism	kahn-ih-bahl-ihz-uhm	devouring one's own species
cesarean derived	sē-sā-rē-ahn	animal is delivered via cesarean section into a sterile environment to avoid possible contamination
cheek pouch		space in oral cavity of hamsters that carries food and bedding
chromodachryorrhea	krō-mō-dahck-rē-ō-rē-ah	shedding of colored (blood-colored) tears
contact bedding		substrate with which animal comes into direct contact; also called **direct bedding**
control		standard normal against which experimental results are compared; also called **experimental control**
crepuscular	krē-puhs-kuh-lahr	becoming active at twilight or before sunrise
data	dah-tah	mass of accumulated information or results of an experiment
dusting	duhs-tihng	cleaning method in which chinchillas roll in dust
emission	ē-mihsh-uhn	discharge

Table 23–1   Common Laboratory and Pocket Pet Terms (continued)

Term	Pronunciation	Definition
estivate	ehs-tih-vāt	to reduce body temperature, heart and respiration rates, and metabolism to dormancy in summer
exsanguination	ehcks-sahn-gwih-nā-shuhn	removal of blood or blood loss from the body
fomite	fō-mīt	inanimate carrier of disease
fur-slip	fər-slihp	shedding of hair patches from rough handling of chinchillas
genotype	jē-nō-tīp	genetic makeup of an individual for a particular trait
gnotobiotic	nōt-ō-bī-ah-tihck	said of germ-free animals that have been introduced to one or two known nonpathogenic microorganisms
heterozygous	heht-ər-ō-zī-guhs	having two different genes for a given genetic trait; usually one gene is dominant over the other
hibernate	hī-bər-nāt	to reduce body temperature, heart, and respiration rates and metabolism to dormancy in winter
homologous	hō-mohl-ō-guhs	having a common origin but different functions in different species
homozygous	hō-mō-zī-guhs	having two identical genes for a given genetic trait
hooded	huhd-ehd	refers to a rat having a white coat with a black "hood" over the head and shoulders and pigmented eyes (Figure 23–3)
horizontal transmission	hōr-ih-zohn-tahl trahnz-mihs-shuhn	disease transfer from one animal to the other
hybrid	hī-brihd	strain resulting from mating two inbred strains
hypothesis	hī-pohth-eh-sihs	statement of research supposition
hystricomorph	hihs-trihck-ō-mōrf	type of rodent that includes guinea pigs, chinchillas, and porcupines
inbred	ihn-brehd	resulting from at least 20 brother–sister or parent–offspring matings
in situ	ihn sih-too	at the normal site
in vitro	ihn vē-trō	outside living organisms; in test tubes or other laboratory glassware
in vivo	ihn vē-vō	inside living organisms
latent infection	lā-tehnt ihn-fehck-shuhn	condition that may not be clinically noticed but under stress or poor health will develop into a recognizable disease state
macroenvironment	mahck-rō-ehn-vī-rən-mehnt	surroundings above the cellular level
metanephric	meht-ah-nehf-rihck	embryonic-like kidney
microenvironment	mīk-rō-ehn-vī-rən-mehnt	surroundings at the cellular level
monogamous	moh-noh-goh-muhs	pairing with one mate
murine	moo-rēn	of mice and rats

(continued)

Table 23-1   Common Laboratory and Pocket Pet Terms (continued)

Term	Pronunciation	Definition
outbred	owt-brehd	from unrelated parents; also called **random bred**
phenotype	fē-nō-tīp	physical characteristics of an individual
phylogeny	fī-lohj-eh-nē	developmental history of a species
pithing	pihth-ihng	destroying the brain and spinal cord by thrusting a blunt needle into the cranium or vertebral column
polygamous	poh-lihg-ah-muhs	having multiple mates
polytocous	poh-liht-ō-kuhs	giving birth to multiple offspring
prehensile	prē-hehn-sihl	adapted for grasping and seizing
progenitor	prō-jehn-ih-tōr	parent or ancestor
progeny	proh-jehn-ē	offspring or descendants
propagate	proh-pah-gāt	to reproduce
protocol	prō-tō-kawl	written procedure for carrying out experiments
rack	rahck	metal device that supports caging units
reduction	rē-duhck-shuhn	theory of using the minimal number of animals for a project that will yield valid results; one of the three *R*'s principles of Russell and Burch
refinement	rē-fīn-mehnt	theory of inflicting minimal stress and pain to animals in research; one of the three *R*'s principle of Russell and Burch
replacement	rē-plās-mehnt	theory of using cell or tissue culture or mathematical models instead of animals in research if possible; one of the three *R*'s principle of Russell and Burch
ringtail	rihng-tāl	abnormal condition in which annular lesions form on the tails of rats housed in low-humidity environments
rodent	rō-dehnt	class of animal that has chisel-shaped incisor teeth
rosette	rō-seht	swirl hair growth pattern in Abyssinian guinea pig
rudimentary	roo-dih-mehn-tār-ē	incompletely developed
sable	sā-buhl	color pattern that has cream-colored undercoat and black guard hairs on feet, tail, and mask
scurvy	skər-vē	common term for vitamin C deficiency
sexual dimorphism	sehcks-yoo-ahl dī-mōrf-ihzm	physical or behavioral differences between females and males of a given species
shoebox	shoo-bohcks	caging that has solid-bottom flooring
suspended cage	suh-spehnd-ehd	caging that hangs from a metal rack and has wire flooring
teratology	tehr-aht-ohl-ō-jē	study of embryo development

Table 23–1   Common Laboratory and Pocket Pet Terms (continued)

Term	Pronunciation	Definition
test group		collection of animals used for experimental manipulation
transgenic	trahnz-jehn-ihck	refers to removing or synthesizing specific genes from one strain and injecting them into the cells of another strain
vector	vehck-tər	something that carries disease from one animal to another
vertical transmission	vər-tih-kahl trahnz-mihs-shuhn	disease transfer from mother to fetus
vestigial	vehs-tih-jē-ahl	said of a structure that has lost a function it previously had

Figure 23–1   Albino mouse. (Courtesy of iStock Photo.)

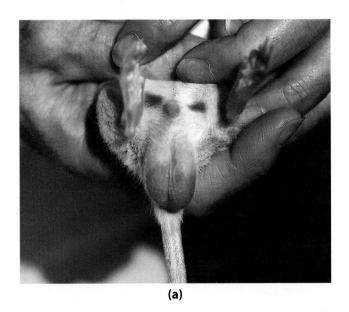

(a)

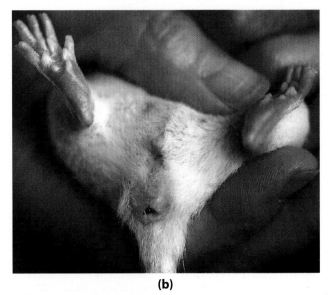

(b)

Figure 23–2   Anogenital distance. (a) Genital area of a male rat. (b) Genital area of a female rat.

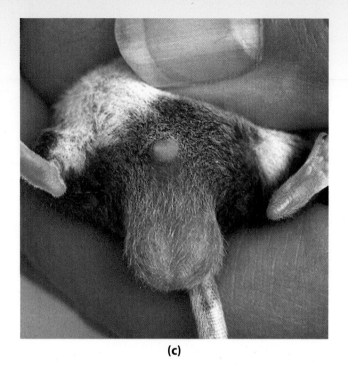

(c)   **(d)**

Figure 23–2   (c) Genital area of a male mouse. (d) Genital area of a female mouse. (Courtesy of Dean Warren.)

Figure 23–3   Rat with hooded markings. (Courtesy of iStock Photo.)

RODENTS

Rodents are gnawing animals that have continuously growing upper and lower incisors. The most common examples of rodents are hamsters, gerbils, mice, rats, guinea pigs, and chinchillas. The order Rodentia is divided into three suborders based on the musculature of the jaw and other skull structures (Figure 23–4). These suborders are the Sciuromorpha (squirrel-like rodents), the Myomorpha (ratlike rodents), and the Hystricomorpha (porcupine-like rodents). The two suborders

of primary importance to the veterinary field are the Hystricomorpha, which includes chinchillas and guinea pigs, and the Myomorpha, which includes rats, mice, hamsters, and gerbils (Figure 23–5).

Mice and Rats

Mice and rats are rodents that are classified in the subfamily Murinae. Rats and mice were once used mainly as research animals but are also popular pets. Both are clean and quiet animals to keep and like to be housed with others of their species. Rats are larger than mice and have more rows of scales on their tails.

There are many rat species; however, only some species, such as the black rat (*Rattus rattus*) and the brown rat (*Rattus norvegicus*), have been domesticated and used for research and as pets. The black rat (sometimes called the roof rat) is believed to have originated in Southern Asia and was the major reservoir of the Black Plague in Europe in the 1200s. The brown rat (sometimes called the Norway rat) is believed to have originated in Eastern Asia and is well established in the United States. Rats have the ability to adapt to many different habitats, environments, and food sources, which explains why rats are found in all parts of the world. Rats are agile climbers and excellent swimmers and are very curious.

They do best when kept with other rats and are primarily nocturnal. Rats lack a gallbladder or tonsils (Figure 23–6). Rats have 16 teeth, and the dental formula is 2(I 1/1, C 0/0, P 0/0, M 3/3).

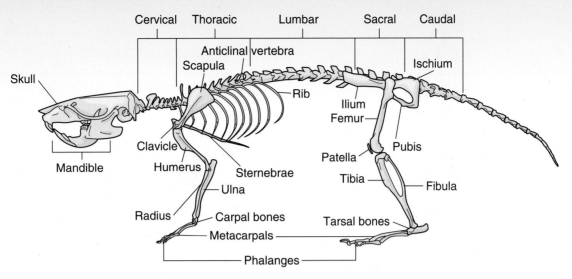

Figure 23–4 Rodent skeleton.

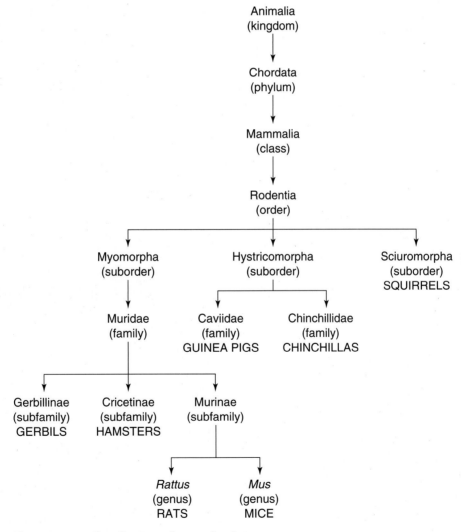

Figure 23–5 Classification scheme of rodents.

Mice are believed to have originated from Asia and have spread throughout the world. The best-known species of mouse is the common house mouse (*Mus musculus*), which dwells wherever humans live. Mice have a pointed nose and slit upper lip. Mice have 16 teeth, and the dental formula of the mouse is 2(I 1/1, C 0/0, P 0/0, M 3/3). Mice have a perfect visual field because of the placement and shape of their eyes; however, their detailed vision is poor. To compensate, mice

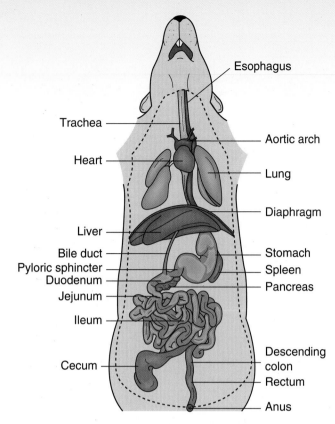

Figure 23–6 Internal structures of a rat.

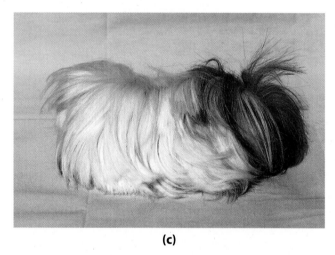

Figure 23–7 Examples of guinea pig varieties. (a) Abyssinian; (b) American short-haired; (c) Peruvian. [(a) and (b) Courtesy of Isabelle Francais; (c) Courtesy of Dean Warren.]

have large ears and a highly developed sense of hearing. Mice have a highly developed sense of smell as well.

Mice do best when kept with other mice and are primarily nocturnal. A mouse colony is led by one head male. The head male is the only one allowed to mate with the females. Fighting between males for the head male status is common in mouse colonies.

Guinea Pigs

Guinea pigs, also called cavies, are rodents that are believed to have originated in South America. Guinea pigs have short bodies; stocky legs; and short, sharp claws. Guinea pigs tend to be gentle, easy-to-handle rodents. There are several common varieties of guinea pigs (Figure 23–7a through c). The male guinea pig is larger than the female guinea pig (Figures 23–8a and b). Guinea pigs have very sensitive senses of hearing and smell.

Guinea pigs cannot synthesize vitamin C as most other mammals can; therefore, vitamin C must be supplied in their diet. Feeding fresh pellets with supplemental fresh produce can provide the guinea pig with enough vitamin C. Guinea pigs have 20 teeth, and their dental formula is 2(I 1/1, C 0/0, P 1/1, M 3/3). The guinea pig has a large cecum with numerous pouches; therefore, feeding adequate roughage is important.

Hamsters

Hamsters are rodents that are small (about 5 to 6 inches long), have short stump tails, and originated from the desert areas of Syria. The Syrian Golden, or Golden, hamster (Figures 23–9a and b) was bred from wild hamsters in the 1930s, and all domes-

ticated hamsters sold are descendants of this original breeding. Hamsters are used as pets and research animals. Hamsters are rodents that are nocturnal and solitary. A few hamster species exist; however, the Golden hamster is most common. Hamsters have almost no tails. Hamster skin is abundant, loose, and pliable (Figures 23–10a and b).

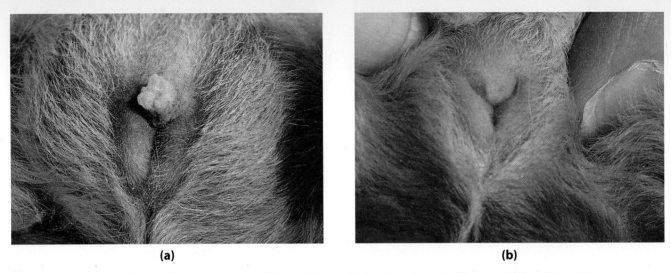

Figure 23–8   (a) Genital area of a male guinea pig. (b) Genital area of a female guinea pig. (Courtesy of Dean Warren.)

Figure 23–9   Hamsters. (a) Short-haired hamster. (b) Long-haired hamster. (Courtesy of iStock Photo.)

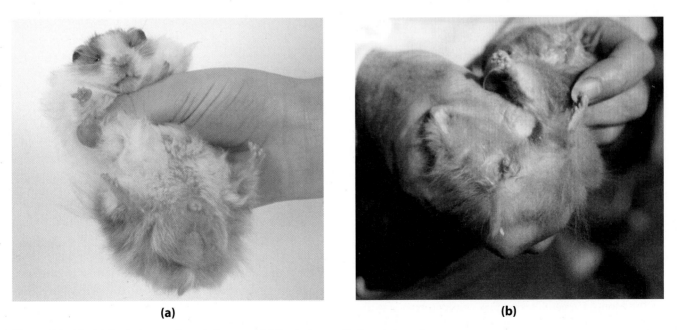

Figure 23–10   (a) Genital area of a male hamster. (b) Genital area of a female hamster.

Cheek pouch

Figure 23–11   Cheek pouch of a hamster.

Figure 23–12   Gerbil. (Courtesy of Isabelle Francais.)

Hamsters prefer a temperature around 70°F and will go into a deep estivation if the temperature rises above 80°F. Hamsters have prominent scent glands on their flanks. They use the secretions from these glands to mark territory. Hamsters also have cheek pouches to store and transport food (Figure 23–11). Hamsters have 16 teeth, and the dental formula is 2(I 1/1, C 0/0, P 0/0, M 3/3).

Gerbils

Gerbils, also known as jirds, are burrowing rodents with bodies that are short and have a hunched appearance (Figure 23–12). The tail is covered with fur, has a bushy tip, and is used for support while standing. The male gerbil has a scent gland on its abdomen. This scent gland allows the adult male to leave his scent by sliding his abdomen across an object (Figures 23–13a and b).

The Mongolian gerbil is the most common type of gerbil and is found in parts of China, in the former Soviet Union, and

(a)

(b)

Figure 23–13   (a) Genital area of a male gerbil. (b) Genital area of a female gerbil.

throughout most of Mongolia. Gerbils are kept as pets and are used as research animals. The Mongolian gerbil is the most common type of gerbil kept as pets.

Gerbils are quiet animals; however, they are quite active. Gerbils are hardy and resistant to disease. Gerbils have 16 teeth, and the dental formula is 2(I 1/1, C 0/0, P 0/0, M 3/3). The adrenal gland of the gerbil is large and contributes to the gerbil's ability to conserve water.

Figure 23–14 Standard chinchilla. (Courtesy of iStock Photo.)

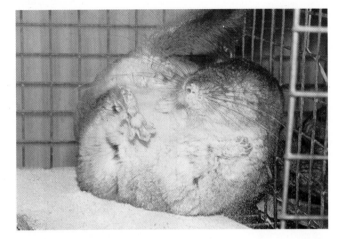

Figure 23–15 Dusting in a chinchilla. (Courtesy of Michael Gilroy.)

Chinchillas

Chinchillas are rodents that were originally used for their fur but are becoming increasingly popular as pets and research animals. Chinchillas are believed to have originated in South America and were used by the Incas as a source of fur. The native habitat of the chinchilla is the barren areas of the Andes Mountains at elevations up to 20,000 feet. Chinchillas resemble small rabbits, but have shorter ears and bushy tails. Their fur is thick and soft because they have fewer guard hairs than some of the other pocket pets (Figure 23–14).

Chinchillas are nocturnal, shelter in crevices and holes along rocks, and live in groups. Chinchillas clean their fur by rolling in dust (Figure 23–15). Chinchillas also can release their fur as a defense mechanism. If grabbed too roughly by a predator or handler, the chinchilla will leave a patch of fur behind. Chinchillas are hindgut fermenters with a large stomach, jejunum, and cecum. Chinchillas have 20 teeth, and the dental formula is 2(I 1/1, C 0/0, P 1/1, M 3/3). Male chinchillas have open inguinal rings. The testes are located in the inguinal canal, without a true scrotal sac (Figures 23–16a and b).

FERRETS

Ferrets (Figure 23–17) are mammals that belong to the family Mustelidae, which includes weasels, mink, and polecats. Ferrets have been in the United States for more than 300 years

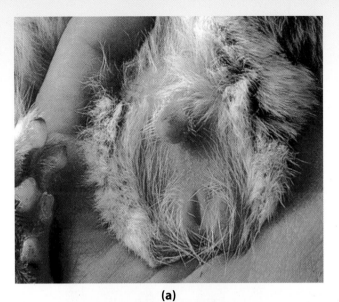

(a)

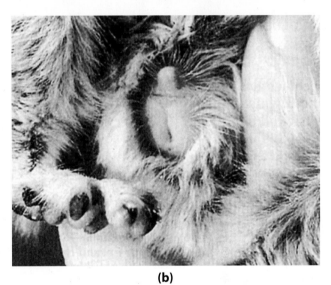

(b)

Figure 23–16 (a) Genital area of a male chinchilla. (b) Genital area of a female chinchilla. (Courtesy of Michael Gilroy.)

and were originally used for rodent control. Today ferrets are used primarily as pets; however, they are curious and like to get into small spaces and eat just about anything that is lying around.

Ferrets also are used in research and as working animals because of their ability to reach difficult places. Ferrets have long, slender bodies and long tails. Ferret tails are about one-half the length of the head and body. Ferrets have short legs and small, rounded ears. Ferrets are primarily nocturnal; they do not see well in bright light. Ferrets have highly developed senses of hearing, smell, and touch. Ferrets have 40 permanent teeth, and the dental formula is 2(I 3/3, C 2/2, P 4/3, M 1/2). Female ferrets are induced ovulators, which can cause health problems if the female does not come out of heat. **Hyperestrogenism** (hī-pər-ehs-trō-jehn-ihz-uhm) is elevated blood estrogen levels seen in

Figure 23–17 Fitch ferrets. Ferrets may be restrained by scruffing the skin of the dorsal neck. Male ferrets (right) are larger than female ferrets (left).

intact cycling female ferrets if not bred (Figures 23–18a, b, and c).

RABBITS

Rabbits are mammals classified as lagomorphs. Rabbits have four upper incisor teeth, whereas rodents have two incisor teeth (Figures 23–19 and 23–20). Rabbits may be dwarf, mini, standard, or giant in size and may be lop-eared (Figures 23–21a through e). Hares are in the same family as rabbits, usually are larger than rabbits, and have longer ears than rabbits. Hares do not build nests as rabbits do, and young hares are born fully furred with their eyes open. Hares live above ground and do not dig tunnels as rabbits do. Rabbits are kept indoors or outdoors as pets, are raised for pelts and food, and are used for research.

Proper ventilation is important in keeping rabbits disease-free. Rabbits are hindgut fermenters, so adequate roughage in the diet is important. Rabbits have a simple glandular stomach and a large cecum. Rabbits cannot vomit. Rabbits also need to gnaw to keep the length of their incisors

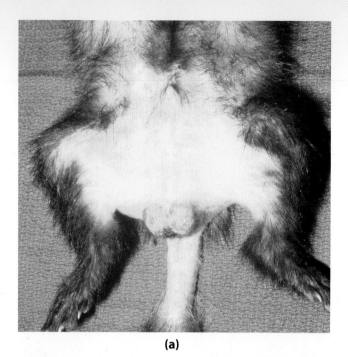

(a)

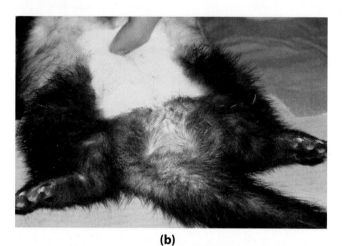

(b)

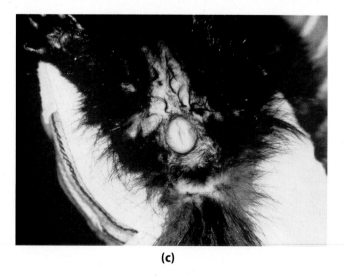

(c)

Figure 23–18 External genitalia of ferrets. (a) Genital area of a male ferret. (b) Genital area of a nonestrous female ferret. (c) Genital area of an estrous female ferret.

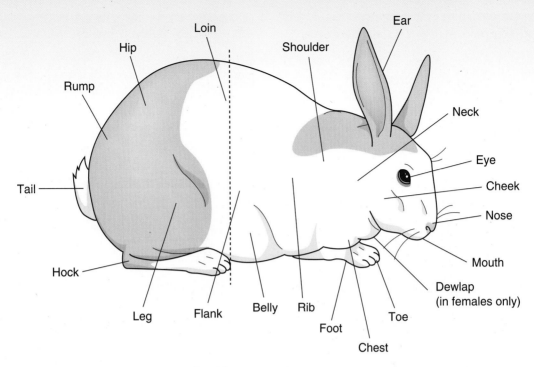

Figure 23–19 External anatomy of a rabbit.

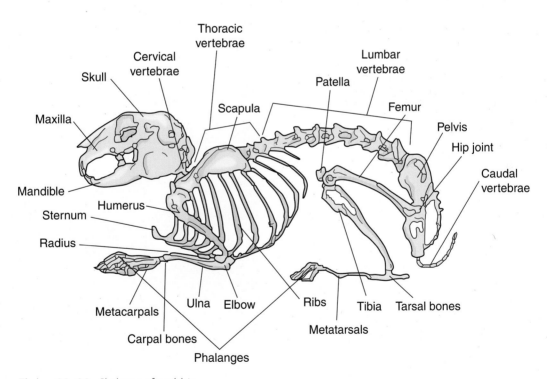

Figure 23–20 Skeleton of a rabbit.

in check. Rabbits have 28 permanent teeth, and the dental formula of the rabbit is 2(I 2/1, C 0/0, P 3/2, M 3/3).

Rabbits have powerful hindquarters for jumping. If not restrained properly, a rabbit can kick its rear legs and fracture its spine. In male rabbits, the inguinal canals remain open for life, the scrotum is hairless, and the testes descend at about 12 weeks of age.

- **night stool** (nīt stool) = rabbit nocturnal feces that is looser than normal and contains vitamins and nutrients that the rabbit consumes.
- **snuffles** (snuhf-uhlz) = common term for upper respiratory disease of rabbits caused by *Pasteurella multocida* (Figure 23–22).

Figure 23–21 Types of Rabbits. (a) Dwarf. (b) Mini. (c) Standard. (d) Giant. (e) Lop-eared. (Courtesy of Isabelle Francais.)

- **sore hocks** = ulceration of the foot pads and foot area caused by the animal's body weight pressing down on the foot; commonly seen in rabbits housed on wire cage floors.
- **torticollis** (tōr-tih-kō-luhs) = contracted state of the cervical muscles producing torsion of the neck; also called **wry neck** (Figure 23–23).

REPTILES

Reptiles are cold-blooded vertebrates that have lungs and breathe air. Reptiles have a body covering (bony skeleton, scales, or horny plates) and a heart that has two atria and in most species one ventricle.

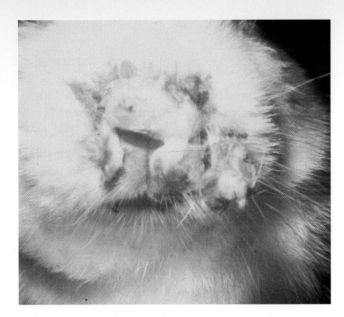

Figure 23–22 Snuffles in a rabbit. (Courtesy of USDA.)

Figure 23–23 Torticollis in a rabbit. (Courtesy of USDA.)

There are four different types of reptiles: Chelonia (turtles, tortoises, and terrapins), Serpentes (snakes, pythons, and boas), Squamata (iguanas and lizards), and Crocodilia (crocodiles and alligators) (Figures 23–24a through d).

Modern-day reptiles have three body types. One type, including lizards, has long bodies and clearly defined tails. The second type, including snakes, has long bodies that

Figure 23–24 Types of reptiles. (a) This Western painted turtle is an example of Chelonia. (b) This ball python is an example of Serpentes. (c) This Madagascar day gecko is an example of Squamata. (d) These American alligators are examples of Crocodilia. (Courtesy of Isabelle Francais.)

taper into tails. The third type, including turtles, has short, thick bodies encased in shells.

Movement in reptiles varies with the type of body structure in each group. Crocodiles and lizards have paired limbs that are attached to the body at right angles. This enables the animal to lift its body off the ground while moving. Crocodiles and alligators have strong limbs, whereas lizards have weaker limbs. Snakelike reptiles do not have limbs, and their movement is due to undulating movements of the body. The scales on the underside of the body project outward as the muscles are contracted and relaxed, allowing pressure to move the animal forward. Some turtles and tortoises have limbs that enable them to move on land using a creeping and crawling motion, yet also have modified limbs that enable them to swim.

Reptile skin has a horny surface layer. In lizards and snakes, this layer forms a hard, continuous covering of scales. These scales lie beneath the superficial layer of the epidermis so that the body can grow; this superficial layer is shed, allowing further growth. Crocodiles, alligators, and some lizards have bony dermal scales covered by a horny epidermal layer that also is shed to allow for body growth. Turtles and tortoises do not molt their thick epidermal skin; each year a new epidermal scale is formed beneath the old one. These epidermal scales (scutes) form rings that can be counted, allowing a person to estimate the animal's age.

Lizards and snakes have teeth that are fused into the jawbones; some snakes have teeth fused to the palate bones. Crocodile teeth are set in sockets. Turtles and tortoises do not have teeth; their jaws form sharp crushing plates.

The tongues of reptiles vary greatly. Some have short, fleshy tongues that have little movement, whereas other reptiles have tongues that are long, slender, and forked (some lizards and snakes).

Tortoises, crocodiles, and many lizards and snakes are oviparous (lay eggs that hatch after leaving the female body). Some species are ovoviviparous (eggs hatch inside the body of the female with the live young emerging from the female's body).

- **brille** (brī-uhl) = transparent layer that permanently covers the eyes of snakes (snakes cannot close their eyes and have brille instead of eyelids).
- **carapace** (kahr-ah-pās) = dorsal region of a turtle shell (Figure 23–25).
- **chin glands** = secretory organs located on the throats of turtles; also called **mental glands.**
- **dysecdysis** (dihs-ehck-dī-sihs) = difficult or abnormal shedding.
- **ecdysis** (ehck-dī-sihs) = shedding or molting.
- **femoral pores** (fehm-ōr-ahl poorz) = sexually dimorphic glands prominent in mature lizards that are located on the ventral surface of lizard thighs; also called **femoral glands.**
- **head gland** = small secretory organ located on the head of snakes.
- **musk glands** = four secretory organs that open lateroventrally near the carapace edge of turtles.
- **plastron** (plahs-trohn) = ventral region of a turtle shell.
- **scent gland** = saclike secretory organ located at the base of the tail in snakes.
- **scute** (skoot) = any scalelike structure such as those found on tortoise or turtle shells or on the heads of snakes.

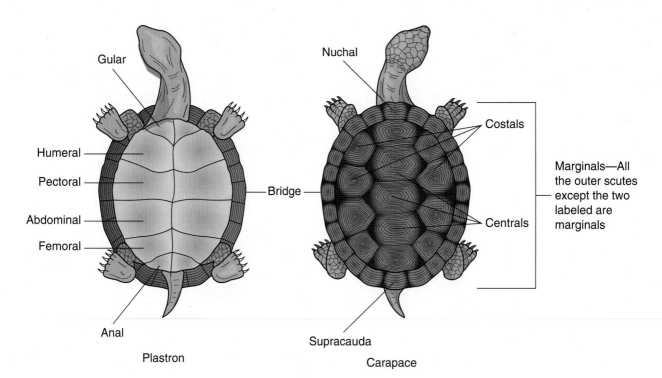

Figure 23–25 Scutes (scalelike structures) on a turtle shell.

- **spectacle** (spehck-tah-kuhl) = transparent, highly vascular, unshed, abnormal covering over the cornea of some reptiles; also called an **eyecap** or a **brille** (brī-uhl) (Figures 23–26a, b, and c).
- **tail autotomy** (tā-uhl aw-toh-tō-mē) = lizard's ability to lose its tail.

Figure 23–26 (a) Normal spectacle, or eyecap, in a snake. (b) Spectacle in process of shedding. (c) Shed spectacle. (Courtesy of Kimberly Kruse Sprecher, CVT.)

AMPHIBIANS

Amphibians are animals that live the larval part of their lives in water and their adult lives partially or completely on land. They are cold-blooded and do not have scales. There are approximately 4,000 species of amphibians in three orders: Gymnophiona (wormlike amphibians called caecilians), Caudata (newts and salamanders), and Salientia (frogs and toads) (Figures 23–27a and b).

Amphibians have thin, moist skin that is used for breathing (cutaneous respiration) and for absorption of water. Amphibian skin is covered with fluid-secreting glands that produce a slimy mucus. This mucus helps conserve moisture; prevents excessive amounts of water from being absorbed into the body; and makes the animals slippery, which aids in their defense. The tongues of amphibians vary greatly from those who do not have tongues to those who have long tongues with sticky tips to aid in the capture of small insects. Amphibians typically eat insects; however, larger species may feed on mice and small rodents.

- **urostyle** (yoor-ō-stīl) = long caudal vertebra of some amphibians.

Figure 23–27 Types of amphibians. (a) American toad; (b) Green tree frog. (Courtesy of Isabelle Francais.)

REVIEW EXERCISES

Multiple Choice

Choose the correct answer.

1. The ventral region of a turtle shell is the
    a. plastron
    b. carpace
    c. dysecdysis
    d. ecdysis

2. A term for germ-free is
    a. barrier
    b. ambient
    c. axenic
    d. fomite

3. Biting or chewing the fur of subordinate animals is known as
    a. cribbing
    b. barbering
    c. cannibalism
    d. polytocous

4. A term meaning outside living organisms is
    a. in situ
    b. in vitro
    c. in vivo
    d. inbred

5. Rabbit nocturnal feces that is looser than normal is known as
    a. diarrhea
    b. night stool
    c. dark feces
    d. melena

6. The medical term for shedding or molting is
    a. dysecdysis
    b. ecdysis
    c. estivation
    d. transgenic

7. The space in the oral cavity of hamsters that allows them to carry food and bedding is the
    a. oral cavity
    b. cheek cavity
    c. cheek pouch
    d. calvarium

8. The area between the anus and genitalia used to determine the sex of animals is the
    a. analogous space
    b. cesarean space
    c. anogenital distance
    d. crepuscular area

9. Chinchillas clean themselves by
    a. fur-slip
    b. barbering
    c. dusting
    d. shining

10. Incompletely developed parts are known as
    a. homologous
    b. heterozygous
    c. homozygous
    d. rudimentary

11. An animal with a white coat and pink eyes is called a(n)
    a. albino
    b. axenic
    c. cashmere
    d. agouti

12. Something that carries disease from one animal to another is a
    a. test group
    b. vestigial
    c. vector
    d. vertical transmission

13. The standard normal against which experimental results are compared is the

   a. variable
   b. control
   c. fomite
   d. ambient

14. Adjustment of an animal to a new environment is

   a. ambient
   b. acclimatization
   c. calvarium
   d. gnotobiotic

15. Another term for brille is

   a. spectacle
   b. urostyle
   c. scute
   d. plastron

16. Wry neck, or the contracted state of the cervical muscles producing torsion of the neck, also is called

   a. ecdysis
   b. dysecdysis
   c. carapace
   d. torticollis

17. Intact cycling female ferrets may have elevated blood estrogen levels. This condition is called

   a. scurvy
   b. homozygous
   c. homologous
   d. hyperestrogenism

18. To reduce body temperature, heart and respiration rates, and metabolism to dormancy in winter is called

   a. calvariumism
   b. hibernation
   c. estivation
   d. metanephrism

19. To reduce body temperature, heart and respiration rates and metabolism to dormancy in summer is called

   a. calvariumism
   b. hibernation
   c. estivation
   d. metanephrism

20. The written procedure for carrying out experiments is the

   a. phenotype
   b. genotype
   c. protocol
   d. polytocous

CROSSWORD PUZZLE

Laboratory and Pocket Pet Terms Puzzle

Across

1 inanimate carriers of disease
3 cleaning method of chinchillas
6 contracted state of the cervical muscles producing torsion of the neck
8 dorsal region of a turtle shell
11 standard "normal" against which experimental results are compared
12 to reproduce
14 inside living organisms
16 study of embryo development
17 transparent, unshed, abnormal covering over the cornea of some reptiles
20 incompletely developed

Down

1 shedding of hair patches from rough handling of chinchillas
2 behavioral disorder in animals where dominant animals bite or chew the fur from subordinate animals
4 something that carries disease from one animal to another
5 offspring or descendants
7 statement of research supposition
9 germ-free
10 dormant state during summer
11 top of the skull
13 ventral region of a turtle shell
15 rabbit noctural feces
18 shedding
19 surrounding
21 mice and rats

CASE STUDIES

Define the underlined terms in each case study.

A 6-mo-old female green iguana was presented for examination for <u>lethargy</u>, <u>ataxia</u>, and limb twitching. It was kept in an aquarium with food, water, a hot rock, and a white light source. The diet consisted mainly of head lettuce and grapes. Upon <u>PE</u>, the iguana was noted to be underweight and undersized for its age, and limb tremors could be induced during handling. Based on the history, it was likely that this animal had secondary nutritional <u>hyperparathyroidism</u>, otherwise known as <u>metabolic</u> bone disease. <u>Radiographs</u> were taken, and blood was drawn for a <u>serum</u> <u>chemistry profile</u>. Results indicated poor bone density, <u>hypocalcemia</u>, and <u>hypoproteinemia</u>. The iguana was treated with calcium and vitamin D3 supplementation. The owners were counseled on proper husbandry and nutrition of green iguanas (Figure 23–28).

1. lethargy   _____

2. ataxia   _____

3. PE   _____

4. hyperparathyroidism _____

5. metabolic   _____

6. radiographs   _____

7. serum   _____

8. chemistry profile   _____

9. hypocalcemia   _____

10. hypoproteinemia   _____

Figure 23–28 Green iguana.

A 5-yr-old three-toed box turtle was presented with a swelling on the right side of its head. The turtle was housed in an aquarium with dirt substrate, a water dish, and a hide box. On PE, the right tympanic membrane was noted to bulge out laterally from the head. Oral examination revealed stomatitis characterized by yellow plaques on the oral mucosa. A diagnosis of otitis media and potential hypovitaminosis was made. The abscess in the ear was lanced and debrided. The oral plaques were swabbed for cytology and were found to demonstrate squamous metaplasia consistent with hypovitaminosis A. The turtle was treated with antibiotics and vitamin A supplementation. Husbandry and nutrition were discussed with the owner.

11. tympanic membrane _____

12. laterally _____

13. oral _____

14. stomatitis _____

15. plaques _____

16. mucosa _____

17. diagnosis _____

18. otitis media _____

19. hypovitaminosis _____

20. abscess _____

21. lanced _____

22. debrided _____

23. cytology _____

24. squamous _____

25. metaplasia _____

A 5-yr-old M/N ferret was presented for signs of alopecia on its dorsal lumbar and tail regions and dysuria. Although differential diagnoses of ectoparasites, dermatophytosis, atopy, cystitis, and urolithiasis were considered, the most likely dx was hyperadrenocorticism. This endocrine disease is caused by adrenal gland hyperplasia or neoplasia. Production of excessive androgen hormones results in cessation of the normal hair growth cycle and may cause prostatic hyperplasia, which in turn produces the clinical signs described. Treatment options include surgical excision of the affected adrenal glands and medical therapy aimed at reducing the functional tissue of the adrenal glands. Many ferrets respond very well to the therapeutic options.

26. alopecia _____

27. dorsal lumbar _____

28. dysuria _____

29. ectoparasites _____

30. dermatophytosis _____

31. atopy   _____

32. cystitis   _____

33. urolithiasis   _____

34. dx   _____

35. hyperadrenocorticism _____

36. hyperplasia   _____

37. neoplasia   _____

38. cessation   _____

39. prostatic hyperplasia _____

40. excision   _____

A 3-yr-old M chinchilla was noted to be limping on its left rear leg. He had taken a fall from the top of a bedroom dresser. Upon presentation to the veterinarian, a <u>midshaft</u>, <u>spiral</u> <u>femoral fx</u> was <u>diagnosed</u> with <u>radiographs</u>. Options included <u>coaptation</u> by <u>splinting</u> the leg, surgical repair of the <u>fx</u>, or restricted cage rest. The owner chose to have the leg splinted. The chinchilla was anesthetized with isoflurane, an <u>inhalant anesthetic</u>, via a face mask. A splint was fashioned with fiberglass casting and bandage materials. One week later it was noted that the splint had shifted, so the animal was returned to the veterinarian. Upon removal of the splint, it was discovered that the leg was no longer <u>viable</u> because <u>vascular</u> and soft tissue damage had occurred. The chinchilla was again anesthetized for surgical <u>amputation</u> of the leg. He recovered well over the next few weeks and <u>ambulated</u> well as a three-legged chinchilla.

41. midshaft   _____

42. spiral   _____

43. femoral fx   _____

44. diagnosed   _____

45. radiographs   _____

46. coaptation   _____

47. splinting   _____

48. fx   _____

49. inhalant anesthetic _____

50. viable   _____

51. vascular   _____

52. amputation   _____

53. ambulated   _____

A 2-yr-old M guinea pig was noted to have <u>hematuria</u> and <u>pollakiuria</u> from the owner's <u>hx</u>. On PE, the guinea pig was <u>BAR</u>. All systems examined were <u>WNL</u> other than <u>palpation</u> of the urinary bladder, which elicited <u>vocalization</u> as a result of assumed pain. The guinea pig was restrained for <u>cystocentesis</u>, and a <u>UA</u> was performed. The urine contained numerous <u>erythrocytes</u> and <u>leukocytes</u>, as well as calcium oxalate <u>crystals</u>. A <u>urolith</u> was suspected. Whole body <u>radiographs</u> confirmed the presence of a <u>radiopaque</u> density in the urinary bladder. A <u>cystotomy</u> was performed to remove the calcium oxalate stone. The owners were instructed to feed timothy hay instead of alfalfa hay to decrease dietary calcium intake.

54. hematuria _____

55. pollakiuria _____

56. hx _____

57. BAR _____

58. WNL _____

59. palpation _____

60. vocalization _____

61. cystocentesis _____

62. UA _____

63. erythrocytes _____

64. leukocytes _____

65. crystals _____

66. urolith _____

67. radiographs _____

68. radiopaque _____

69. cystotomy _____

A 1-yr-old F domestic mouse was presented for <u>pruritus</u> and a powdery substance on the <u>dorsal</u> aspect of the neck. PE revealed a pruritic mouse with white granular material on the <u>dorsal cervical region</u>. A cellophane tape preparation was performed on the neck <u>debris</u>. <u>Ectoparasitic</u> ascariasis was diagnosed, the most likely agent being *Myobia musculi*. The mouse was treated <u>topically</u> with the <u>parasiticide</u> ivermectin, which was placed on the <u>interscapular</u> skin. The owner was instructed to return for a second treatment in 10 days.

70. pruritus _____

71. dorsal _____

72. dorsal cervical region _____

73. debris _____

74. ectoparasitic _____

75. topically _____

76. parasiticide _____

77. interscapular _____

Socrates, a 1 1/2-yr-old hooded rat, was examined for what the owner described as a cold. <u>Conjunctivitis</u> and <u>sinusitis</u> were noted on PE. The owner was concerned that there was bleeding from the <u>nares</u>. He was informed that when rats become ill, they groom themselves less, which allows <u>nasal</u> discharge to accumulate on the fur. The red tinge that the owner was seeing was not blood, but a porphyrin pigment that is a normal component of the nasal secretions. Various viral and bacterial <u>etiologic</u> agents are responsible for respiratory disease in rats, including *Mycoplasma*. The <u>antibiotic</u> oxytetracycline was prescribed <u>PO</u> for 2 weeks. The owner was informed that this may be a <u>chronic</u>, recurrent disease.

78. conjunctivitis _____

79. sinusitis _____

80. nares _____

81. nasal _____

82. etiologic _____

83. antibiotic _____

84. PO _____

85. chronic _____

ABBREVIATIONS

IMAGING ABBREVIATIONS

a or amp	ampere
A/P	anterior/posterior
APMLO	anterior to posterior and medial to lateral oblique
Ba	barium
CT or CAT	computed (axial) tomography
DL-PaMO	dorsolateral-palmaromedial oblique
DL-PlMO	dorsolateral-plantaromedial oblique
DM-PaLO	dorsomedial-palmarolateral oblique
DP	dorsopalmar
DPl	dorsoplantar
DPr DDiO	dorsoproximal-dorsodistal oblique (often called the skyline)*
D/V	dorsal/ventral
f	frequency
Hz	Hertz (a unit of frequency)
λ	wavelength
IVP	intravenous pyelogram
kV	kilovolt
kVp	kilovolt peak
kW	kilowatt
LM	lateral to medial
LMO	lateral to medial oblique
mA	milliamperage
mAs	milliamperage in seconds

mc	millicurie
ML	medial to lateral
MLO	medial to lateral oblique
MRI	magnetic resonance imaging
O	oblique
P/A	posterior/anterior
Pa45Pr-PaDiO	palmaroproximal-palmarodistal oblique (often called the skyline navicular)
PALMO	posterior to anterior and lateral to medial oblique
PAMLO	posterior to anterior and medial to lateral oblique rad unit of measurement of absorbed dose of ionizing radiation
RT	radiation therapy
T	period of time (used in ultrasound)
v	velocity
V	volt
V/D	ventral/dorsal
W	watt
X-ray	roentgen ray

SPECIALIST OR TITLE ABBREVIATIONS

The letter *D* may appear in front of these specialties, which means that the person is a diplomate of that particular college. For example, DACVIM after someone's name means that the person is a diplomate in the American College of Veterinary Internal Medicine.

ABVP	American Board of Veterinary Practitioners
ABVT	American Board of Veterinary Toxicology
ACLAM	American College of Laboratory Animal Medicine
ACPV	American College of Poultry Veterinarians
ACT	American College of Theriogenologists

*Abbreviation may be preceded with *flexed* and may include a number referring to the deviation from perpendicular for a beam; for example, flexed D30Pr-DdiO is a flexed dorsoproximal-dorsodistal oblique taken at a 30-degree angle to the cassette.

ACVA	American College of Veterinary Anesthesiologists
ACVB	American College of Veterinary Behaviorists
ACVCP	American College of Veterinary Clinical Pharmacology
ACVD	American College of Veterinary Dermatology
ACVECC	American College of Veterinary Emergency and Critical Care
ACVIM	American College of Veterinary Internal Medicine
ACVM	American College of Veterinary Microbiologists
ACVN	American College of Veterinary Nutrition
ACVO	American College of Veterinary Ophthalmologists
ACVP	American College of Veterinary Pathologists
ACVPM	American College of Veterinary Preventive Medicine
ACVR	American College of Veterinary Radiology
ACVS	American College of Veterinary Surgeons
ACZM	American College of Zoological Medicine
AHT	animal health technician
AIMVT	Academy of Internal Medicine for Veterinary Technicians
ALAT	assistant laboratory animal technician
ASVDT	American Society of Veterinary Dental Technicians
AVDC	American Veterinary Dental College
AVECCT	American Veterinary Emergency and Critical Care Technicians
AZVT	Association of Zoo Veterinary Technicians
CH	certified herbalist
CVA	certified veterinary acupuncturist
CVT	certified veterinary technician
DC	doctor of chiropractic
DVM	doctor of veterinary medicine
LAT	laboratory animal technician
LATG	laboratory animal technologist
LVT	licensed veterinary technician
ND	doctor of naturopathy
RVT	registered veterinary technician
RVTG	registered veterinary technologist
SVBT	Society of Veterinary Behavior Technicians
VA	veterinary assistant
VMD	veterinary medical doctor (veterinariae medicinae doctor)
VNCA	Veterinary Nurses Council of Australia
VTS (Internal Medicine—Small Animal)	Veterinary Technician Specialist in small animal internal medicine
VTS (Internal Medicine—Large Animal)	Veterinary Technician Specialist in large animal internal medicine
VTS (Internal Medicine—Cardiology)	Veterinary Technician Specialist in Cardiology
VTS (Internal Medicine—Neurology)	Veterinary Technician Specialist in Neurology
VTS (Internal Medicine—Oncology)	Veterinary Technician Specialist in Oncology

ASSOCIATION ABBREVIATIONS

AAHA	American Animal Hospital Association
AALAS	American Association of Laboratory Animal Science
AKC	American Kennel Club
AO	Arbeitsgemeinschaft für Osteosyntesesfragen (Association for the Study of Fracture Treatment in Man, founded by a group of Swiss surgeons); used to describe specialized bone plates and instruments used in orthopedic repair
APHIS-VS	Animal & Plant Health Inspector Services—Veterinary Services
ASIF	Association of the Study of Internal Fixation
ASPCA	American Society for the Prevention of Cruelty to Animals
AVMA	American Veterinary Medical Association
CAAHT	Canadian Association of Animal Health Technologists & Technicians
CDC	Centers for Disease Control and Prevention (human)
CVTEA	Committee on Veterinary Technician Education and Activities
DEA	Drug Enforcement Adminstration
DHIA	Dairy Herd Improvement Association
DOT	Department of Transportation (used for OSHA regulation of hazardous material transfer)
FDA	Food and Drug Administration
FSIS	Food Safety and Inspection Services
IACUC	Institutional Animal Care and Use Committee
NADC	National Animal Disease Center
NAPCC	National Animal Poison Control Center
NAVTA	National Association of Veterinary Technicians in America
NVSL	National Veterinary Services Laboratories
OFA	Orthopedic Foundation for Animals
OSHA	Occupational Safety & Health Administration

USDA	United States Department of Agriculture
USP	United States Pharmacopeia
VTAS	Veterinary Technician Anesthetist Society

PHARMACOLOGY ABBREVIATIONS

ac	before meals (*ante cibum*)
ad lib	as much as desired (*ad libitum*)
bid	twice daily (*bis in die*)
BSA	body surface area
cal	calorie
cap	capsule
cc	cubic centimeter (same as mL)
cm	centimeter
conc	concentration
dr	dram; equal to $\frac{1}{8}$ oz or 4 mL
D_5W	5% dextrose in water
ED	effective dose
ED_{50}	median effective dose
fl oz	fluid ounce
g	gram
gal	gallon
gr	grain; unit of weight approximately 65 mg
gt	drop (*gutta*)
gtt	drops (*guttae*)
hr	hour
IA	intra-arterial
IC	intracardiac
ID	intradermal
IM	intramuscular
IP	intraperitoneal
IT	intrathecal
IU	International Unit or intrauterine
IV	intravenous
kg	kilogram
km	kilometer
L or l	liter
lb or #	pound
LD	lethal dose
LRS	lactated Ringer's solution
m	meter
MBC	minimum bactericidal concentration
mcg or μg	microgram
MED	minimal effective dose
mEq	milliequivalent
mg	milligram
MIC	minimum inhibitory concentration
MID	minimum infective dose
mL	milliliter (same as cc)
MLD	minimum lethal dose
mm	millimeter (also used for muscles)
NPO or npo	nothing by mouth (*non per os*)
NS	normal saline
NSAID	nonsteroidal anti-inflammatory drug
OTC	over the counter
oz	ounce
pc	after meals (*post cibum*)
PDR	*Physician's Desk Reference*
%	percent
pH	hydrogen ion concentration (acidity and alkalinity measurement)
PO or po	orally (*per os*)
ppm	parts per million
PR	per rectum
prn	as needed
pt	pint
PZI	protamine zinc insulin
q	every
qd	every day
qh	every hour
q4h	every 4 hours
q6h	every 6 hours
q8h	every 8 hours
q12h	every 12 hours
q24h	every 24 hours
qid	four times daily (*quater in die*)
qn	every night
qod or eod	every other day
qP	as much as desired
qt	quart
®	registered trade name (superscript next to drug name)
sid	once daily (q24h is more common abbreviation for once daily)
sig	let it be written as (used when writing prescriptions)
sol'n or soln	solution
SQ, SC, subq, or subcu	subcutaneous
T	tablespoon or tablet (or temperature)
tab	tablet
tid	three times daily (*ter in die*)
tsp or t	teaspoon
vol	volume
VPB	*Veterinary Pharmaceuticals and Biologicals*

Laboratory Abbreviations

ab	antibody
ABO	human blood groups
ag	antigen
alb	albumin
alk phos	alkaline phosphatase
ALT	alanine aminotransferase (formerly SGPT)
amyl	amylase
AST	aspartate aminotransferase (formerly SGOT)
BP	blood pressure
BUN	blood urea nitrogen
CBC	complete blood count
CFT	complement fixation test
CHOL	cholesterol
CK	creatine kinase
CMT	California mastitis test
CREA	creatine
diff	differential white blood count
EDTA	ethylenediaminetetraacetic acid; type of anticoagulant
ESR	erythrocyte sedimentation rate
GGT	gamma glutamyl transpeptidase
GLU	glucose
GTT	glucose tolerance test
H&E	hematoxylin and eosin stain
Hb or Hgb	hemoglobin
Hct or crit	hematocrit
HDL	high-density lipoprotein
HPF	high-power field
HW	heartworm
LDH	lactate dehydrogenase
LDL	low-density lipoprotein
LPF	low-power field
MCH	mean corpuscular hemoglobin
MCHC	mean corpuscular hemoglobin concentration
MCV	mean corpuscular volume
ME	myeloid–erythroid ratio
NRBC	nucleated red blood cell
PCV	packed cell volume
PMN	polymorphonuclear neutrophil leukocyte
PT	prothrombin time
PTT	partial thromboplastin time
qns	quantity not sufficient
qs	quantity sufficient
RBC	red blood cell
rpm	revolutions per minute
SAP	serum alkaline phosphatase
SCC	somatic cell count
sed or SR	sedimentation rate
SGOT	serum glutamic oxaloacetic transaminase; now abbreviated AST
SGPT	serum glutamic pyruvic transaminase; now abbreviated ALT
SPF	specific pathogen free
sp. gr.	specific gravity
Staph	*Staphylococcus* bacteria
Strep	*Streptococcus* bacteria
T_3	triiodothyronine (one type of thyroid hormone)
T_4	thyroxine (one type of thyroid hormone)
TB	tuberculin
TBIL	total bilirubin
TNTC	too numerous to count
TP	total protein
UA	urinalysis
WBC	white blood cell
WMT	Wisconsin mastitis test

Vaccination Abbreviations

BVD	bovine viral diarrhea
DHLPP	distemper, hepatitis, leptospirosis, parainfluenza, and parvovirus
DHLPP-CV	distemper, hepatitis, leptospirosis, parainfluenza, parvovirus, and coronavirus
EEE	eastern equine encephalitis
EIA	equine infectious anemia
EPM	equine protozoal myelitis
FeLV, FeLeuk, or FeLuk	feline leukemia virus
FIP	feline infectious peritonitis
FIV	feline immunodeficiency virus
FVRCP	feline viral rhinotracheitis, calicivirus, and panleukopenia
FVRCP-C	feline viral rhinotracheitis, calicivirus, panleukopenia, and chlamydia
IBR	infectious bovine rhinotracheitis
MLV	modified live vaccine
PHF	Potomac horse fever
PI-3	parainfluenza 3 virus
PRRS	porcine reproductive and respiratory syndrome
RV	rabies vaccine
TE	tetanus
TGE	transmissible gastroenteritis
WEE	western equine encephalitis
WNV	West Nile virus
VEE	Venezuelan equine encephalitis

PHYSICAL EXAMINATION, PHYSIOLOGY, AND PATHOLOGY ABBREVIATIONS

ACh	acetylcholine
ACH	adrenocortical hormone
AChE	acetylcholinesterase
ACTH	adrenocorticotrophic hormone
ADH	antidiuretic hormone
AI	artificial insemination
ANS	autonomic nervous system
ASAP	as soon as possible
BAR	bright, alert, responsive
BD/LD	big dog/little dog (used for dogs wounded in a fight)
BM	bowel movement
BPM	beats or breaths per minute
c̄	with
C	castrated
cath	catheter
CC	chief complaint
ChE	cholinesterase
CHF	congestive heart failure
CNS	central nervous system
CO	carbon monoxide
CO_2	carbon dioxide
CP	conscious proprioception
CPR	cardiopulmonary resuscitation
CRT	capillary refill time
C-section	cesarean section
CSF	cerebrospinal fluid
CSM	carotid sinus massage
CVP	central venous pressure
DA	displaced abomasum
DD	differential diagnosis
DDN	dull, depressed, nonresponsive
DIC	disseminated intravascular coagulation
DLH	domestic longhair (feline)
DNA	deoxyribonucleic acid
DOA	dead on arrival
DSH	domestic shorthair (feline)
ECG or EKG	electrocardiogram or electrocardiograph
EEG	electroencephalogram or electroencephalograph
EMG	electromyogram
F	Fahrenheit or female
FA	fatty acid
FLUTD	feline lower urinary tract disease
F/S	female spayed
FSH	follicle-stimulating hormone

FUO	fever of unknown origin
FUS	feline urological syndrome
GFR	glomerular filtration rate
GH	growth hormone
GI	gastrointestinal
GSW	gunshot wound
HBC	hit by car
hCG	human chorionic gonadotropin
HR	heart rate
ICSH	interstitial cell-stimulating hormone
ICU	intensive care unit
IVDD	intervertebral disc disease
K-9	canine
Ⓛ	left
LA	large animal
LDA	left displaced abomasum
LE	lupus erythematosus
lg	large
LH	luteinizing hormone
LOC	level of consciousness
LV	left ventricle
M	male
M/C	male castrated
MDB	minimum database
mm	millimeter
MM	mucous membrane
mm Hg	millimeters of mercury
M/N	male neutered
MS	mitral stenosis
MSDS	material data safety sheet
N	neutered (or normal on physical examination)
NA or N/A	not applicable
NPN	nonprotein nitrogen
OB	obstetrics
OHE or OVH	ovariohysterectomy
OR	operating room
p̄	after
P	pulse
PD	polydipsia
PDA	patent ductus arteriosus
PE	physical examination
pg	pregnant
PM	postmortem; also abbreviation for evening
PMI	point of maximal intensity
PNS	peripheral nervous system
POVMR	problem-oriented veterinary medical records
PU	polyuria
PVC	premature ventricular complex
Ⓡ	respirations or right

RDA	right displaced abomasum
RNA	ribonucleic acid
R/O	rule out
RP	retained placenta
RR	respiration rate
s̄	without
S	spayed
SA	sinoatrial or small animal
SOAP	subjective, objective, assessment, plan (record-keeping acronym)
stat	immediately (*statim*)
T	temperature (tablespoon or tablet)
TLC	tender loving care
TPN	total parenteral nutrition
TPO	triple pelvic osteotomy
TPR	temperature, pulse, and respiration
TSH	thyroid-stimulating hormone
TTA	transtracheal aspiration
TTW	transtracheal wash
TVT	transmissible venereal tumor
URI	upper respiratory infection
UTI	urinary tract infection
VM	vagal maneuver
VSD	ventricular septal defect
WNL	within normal limits
wt	weight

SYMBOLS

≅	approximately equal to
↑	increased
↓	decreased
+	positive (used to describe test results); may have multiple +s to indicate degree
−	negative (used to describe test results); may have multiple −s to indicate degree
·	times or multiplication sign
√	check
=	equal to
#	number (in front of number; for example, #1); pound (following number; for example, 50#)
≠	not equal to
<	less than

>	greater than
@	at or each
%	percent
♂	male
♀	female
o	degree

THE XS

bx	biopsy
dx	diagnosis
ddx	differential diagnosis
fx	fracture
hx	history
Rx	prescription
sx	surgery
tx	treatment

EYE AND EAR ABBREVIATIONS

AD	right ear
AS	left ear
AU	both ears
IOP	intraocular pressure
OD	right eye (also abbreviation for overdose)
OS	left eye
OU	both eyes

CHEMICAL ABBREVIATIONS

Fe	iron
H	hydrogen
H_2O	water
H_2O_2	hydrogen peroxide
HCl	hydrochloric acid
I	iodine
K	potassium
KCl	potassium chloride
N	nitrogen (also abbreviation for normal)
Na	sodium
NH_3	ammonia
O_2	oxygen

PLURAL FORMS OF MEDICAL TERMS

Many plural word forms are formed by adding *s* to the singular form. This is true for medical terms as well. For example, the plural of *laceration* is *lacerations* and the plural of *bone* is *bones*. However, some rules should be followed when plural forms of medical terms are used. These rules are presented in the following table.

Singular Ending	Change or Deletion from Singular Form	Add Plural Ending	Examples (singular)	Plural Form
s, ch, or sh		es	abscess	abscesses
			stitch	stitches
			brush	brushes
y	delete *y*	ies	capillary	capillaries
is	delete *is*	es	diagnosis	diagnoses
um	delete *um*	a	bacterium	bacteria
us*	delete *us*	i	alveolus	alveoli
a	delete *a*	ae	vertebra	vertebrae
ix	delete *ix*	ices	cervix	cervices
yx	delete *yx*	ices	calyx	calices
ex	delete *ex*	ices	cortex	cortices
ax	delete *ax*	aces	thorax	thoraces
oma#	delete *oma*	omata	stoma	stomata
nx	delete *nx*	nges	phalanx	phalanges
on†	delete *on*	a	spermatozoon	spermatozoa
en	delete *en*	ina	foramen	foramina
u	delete *u*	ua	cornu	cornua

*except plural of *virus* is *viruses* and plural of *sinus* is *sinuses*

#except plural of *carcinoma* may be *carcinomas*

†except plural of *chorion* is *chorions*

PREFIXES, COMBINING FORMS, AND SUFFIXES FOR MEDICAL TERMS

Word Part	Meaning	Word Origin (L = Latin, G = Greek)
A		
a-	away from, no, not, without	L, G
ab-	away from	L
abdomin/o	abdomen	L
able	capable of	L
abort/o	premature expulsion of a nonviable fetus	L
abrad/o or abras/o	scrape off	L
abrupt/o	broken away from	L
abs-	away from	L
absorpt/o	to suck up or in	L
ac-	toward	L
-ac	pertaining to	L, G
acanth/o	spiny	G
acetabul/o	hip socket	L
-acious	characterized by	L
acne/o	point	G
acous/o or acoust/o	hearing	G
-acousia	hearing	G
acr/o	extremities or top	G
acromi/o	point of the shoulder blade	G
acu/o	sharp, severe, sudden, needle	L
acuit/o or acut/o	sharp	L
ad-	toward	L
-ad	toward, to, in the direction of	L
aden/o	gland	G
adhes/o	stickiness or clinginess	L
adip/o	fat	L
adnex/o	bound to	L
adren/o or adrenal/o	adrenal glands	L
aer/o	air or gas	L, G
aesthet/o	sensation or feeling	G
af-	toward	L

Word Part	Meaning	Word Origin (L = Latin, G = Greek)
affect/o	exert influence on	L
ag-	toward	L
agglutin/o	clumping	L
aggress/o	attack or step forward	L
-ago	attack, diseased condition	G
-agra	excessive pain or seizure	G
-aise	ease or comfort	F
al-	similar	L
-al	pertaining to	L, G
alb/i, alb/o, or albin/o	white	L
albumin/o	albumin (one type of protein)	L
alg/e or alg/o	pain	G
algesi/o	pain or suffering	G
-algesia	pain or suffering	G
-algesic	painful	G
algi-	pain or suffering	G
-algia	pain or suffering	G
align/o	to bring in line or the correct position	L
aliment/o	to nourish	L
all-	other or different	G
all/o	other or different	G
alopec/o	baldness	G
alveol/o	small sac	L
amb-	both sides or double	L
amb/i	both sides or double	L
ambly/o	dim or dull	G
ambul/o or ambulat/o	to walk	L
ametr/o	out of proportion	G
-amine	nitrogen compound	L
amni/o	fetal membrane	G
amph-	around, on both sides, or doubly	G

Word Part	Meaning	Word Origin (L = Latin, G = Greek)
amput/o or amputat/o	to cut away or cut off	L
amyl/o	starch	G
an-	no, not, without	G
an/o	anus or ring	L
an/o	upward	G
ana-	up, apart, backward, excessive	G
andr/o	male	G
angi-	vessel	G
angi/o	vessel	G
angin/o	strangling	L
anis/o	unequal	G
ankyl/o	bent or stiff	G
anomal/o	irregularity	G
ante-	before or forward	L
anter/o	front	L
anti-	against	G
anxi/o or anxiet/o	uneasy or distressed	L
aort/o	aorta, or major artery in the body	L, G
ap-	toward	G
aphth/o	small ulcer or eruption	L, G
apic/o	apex or point	L
aplast/o	lack of or defective development	G
apo-	opposed or detached	G
aponeur/o	aponeurosis	G
aqu/i, aqu/o, aque/o	water	L
-ar	pertaining to	L
arachn/o	spider	G
arc/o	bow or arch	L
-arche	beginning	G
arrect/o	upright	L
arteri/o	artery	L, G
arthr/o	joint	G
articul/o	joint	L
-ary	pertaining to	L
as-	toward	L
-ase	enzyme	G
asphyxi/o	absence of a pulse	G
aspir/o or aspirat/o	to breathe in	L
asthen-	weakness	G
-asthenia	weakness	G
asthmat/o	gasping	L, G
at-	toward	L
atel/o	incomplete	G
ather/o	plaque or fatty substance	G
athet/o	uncontrolled	G
-atonic	lack of tone or strength	L
atop/o	out of place	G
atres/i	without an opening	G
atri/o	atrium or chamber	L
attenuat/o	diluted or weakened	L
aud-	hear or hearing	L
audi/o, audit/o, aur/i, or aur/o	hear or hearing	L
auscult/o	to listen	L
aut/o	self	G

Word Part	Meaning	Word Origin (L = Latin, G = Greek)
ax/o	axis or main stem	G
axill/o	armpit	L
azot/o	urea or nitrogen	G

B

Word Part	Meaning	Word Origin (L = Latin, G = Greek)
bacill/o	rod	L
bacteri/o	bacteria	G
bar/o	pressure or weight	G
bas/o	base (not acid)	G
bi-	twice, double, or two	L
bi/o	life	G
bifid/o	split into two parts	L
bifurcat/o	divide into two branches	L
bil/i	bile or gall	L
bilirubin/o	bilirubin, or breakdown product of blood	L
-bin	twice, double, or two	L
bio-	life	G
bis-	twice, double, or two	L
-blast	embryonic	G
blast/i	embryonic	G
blephar/o	eyelid	G
borborygm/o	rumbling sound	L
brachi/o	arm	L
brachy-	short	G
brady-	abnormally slow	G
brev/i or brev/o	short	L
bronch/i, bronch/o, or bronchi/o	bronchial tube	L
brux/o	to grind	G
bucc/o	cheek	L
bucca-	cheek	L
bulla-	blister	L
burs/o	sac of fluid near a joint (bursa)	L, G

C

Word Part	Meaning	Word Origin (L = Latin, G = Greek)
cac-	bad, diseased, or weak	G
cac/o	bad, diseased, or weak	G
cadaver/o	dead body	L
calc/i or calc/o	calcium or the heel	L
calcane/o	heel bone (calcaneus)	L
calcul/o	stone	L
cali/o or calic/o	cup	L
callos/o or call/i	hard	L
calor/i	heat	L
canalicul/o	little duct	L
canth/o	corner of the eye	G
capill/o	hairlike	L
capit/o	head	L
capn/o	carbon dioxide	G
-capnia	carbon dioxide	G
capsul/o	little box	L
carb/o	carbon	L
carcin/o	cancerous	G
cardi/o	heart	G
cari/o	decay	L
carot/o	sleep or stupor	G

Word Part	Meaning	Word Origin (L = Latin, G = Greek)
carp/o	carpus or joint of front limbs between the radius or ulna and metacarpals	L, G
cartilag/o	cartilage or gristle	L
caruncul/o	bit of flesh	L
cat-, cata-, or cath-	down, lower, under, or downward	G
catabol/o	breaking down	G
cathart/o	cleansing or purging	G
cathet/o	insert or send down	G
caud/o	toward the tail	L
caus/o, caust/o, cauter/o, or caut/o	burning	L, G
cav/i or cav/o	hollow	L
cavern/o	containing hollow spaces	L
cec/o	blind gut, or cecum	L
-cele	hernia, cyst, cavity, or tumor	L
celi/o	abdomen	G
cement/o	cementum	L
cent-	hundred	L
-centesis	surgical puncture to remove fluid or gas	G
cephal-	head	G
cephal/o	head	G
-ceps	head	L
cera-	wax	L
cerebell/o	cerebellum	L
cerebr/o	cerebrum, or largest part of brain	L
cerumin/o	cerumen, or earwax	L
cervic/o	neck or necklike	L
chalas/o	relaxation or loosening	G
-chalasis	relaxation or loosening	G
chalaz/o	hailstone, small lump	G
cheil/o	lip	G
chem/i, chem/o, or chemic/o	drug or chemical	G
chol/e	bile or gall	G
cholangi/o	bile duct	G
cholecyst/o	gallbladder	G
choledoch/o	common bile duct	G
cholesterol/o	cholesterol	G
chondr/o or chondri/o	cartilage	G
chord/o	cord or spinal cord	G
chori/o	chorion, or membrane	G
choroid/o	choroid layer of the eye	G
chrom/o or chromat/o	color	G
chym/o	juice	G
cib/o	meal	L
-cidal	death	L
-cide	killing or destroying	L
cili/o	microscopic hairlike projections or eyelashes	L
cine/o	movement	G
circi/o	ring or circle	L
circulat/o	to go around in a circle	L
circum-	around	L
circumscrib/o	confined or limited in space	L
cirrh/o	tawny, orange-yellow	G
cis/o	cut	L
-clasis or -clast	break	G
clav/i	key	L
clavicul/o	collarbone, clavicle	L
clitor/o	female erectile tissue	G
clon/o	violent action	G
-clysis	irrigation	G
co-	together, with	L
coagul/o or coagulat/o	clump or congeal	L
cocc/i or cocc/o	round	L
-coccus	round bacterium	L
coccyg/o	tailbone	G
cochle/o	snail-like or spiral	L
-coel	hollow	G
coher/o or cohes/o	to stick together	L
coit/o	coming together	L
col/o or colon/o	colon (part of large intestine)	L, G
coll/a	glue	G
coll/i	neck	L
colp/o	vagina	G
column/o	pillar	L
com-	together	L
comat/o	deep sleep	L, G
comminut/o	to break into pieces	L
communic/o	to share	L
compatibil/o	to sympathize with	L
con-	together or with	L
concav/o	hollow	L
concentr/o	to remove excess water or condense	L
concept/o	to receive or become pregnant	L
conch/o	shell	L, G
concuss/o	shaken together violently	L
condyl/o	knuckle or knob	L, G
confus/o	disorder or confusion	L
conjunctiv/o	pink mucous membrane of the eye or connected	L
consci/o	aware	L
consolid/o	to become solid	L
constipat/o	pressed together	L
constrict/o	to draw tightly together to narrow	L
contact/o	touched	L
contagi/o	touching of something or infection	L
contaminat/o	to pollute	L
contine/o or continent/o	to keep in or restrain	L
contra-	against or opposite	L
contracept/o	prevention of fertilization of egg with sperm	L
contus/o	bruise	L
convalesc/o	to become strong	L
convex/o	arched	L
convolut/o	twisted or coiled	L

Word Part	Meaning	Word Origin (L = Latin, G = Greek)
convuls/o	to pull together	L
copi/o	plentiful	L
copulat/o	joining together	L
cord/o	spinal cord	L, G
cordi/o	heart	L
cor/o or core/o	pupil	G
core-	pupil	G
cori/o	skin or leather	L
corne/o	transparent anterior portion of the sclera; cornea	L
coron/o	crown	L, G
corp/u or corpor/o	body	L
corpuscul/o	little body	L
cort-	covering	L
cortic/o	outer region or cortex	L
cost/o	rib	L
cox/o	hip or hip joint	L
crani/o	skull	L, G
-crasia	mixture	L, G
crepit/o or crepitat/o	crackling	L
crin/o	to secrete or separate	G
-crine	to secrete or separate	G
cris/o or critic/o	turning point	L
-crit	separate	G
cruci-	cross	L
cry/o	cold	G
crypt/o	hidden	G
cubit/o	elbow	L
cuboid/o	cubelike	G
cusp/i	pointed or pointed flap	L
cut-	skin	L
cutane/o	skin	L
cyan/o	blue	G
cycl/o ciliary	body of the eye or cycle	G
cyst-, -cyst	bag or bladder	G
cyst/o	urinary bladder	G
cyt/o	cell	G
-cyte or -cytic	cell	G
-cytosis	condition of cells	L

D

Word Part	Meaning	Word Origin (L = Latin, G = Greek)
dacry-	tear or tear duct	G
dacry/o	tear or tear duct	G
dacryocyst/o	lacrimal sac	G
dactyl/o	digits	G
dart/o	skinned	G
de-	from, not, down, or lack of	L
dec/i	ten or tenth	L
deca-	ten or tenth	L
decem-	ten	L
decidu/o	falling off or shedding	L
decubit/o	lying down	L
defec/o or defecat/o	free from waste	L
defer/o	carrying down or out	L
degenerat/o	breakdown	L
deglutit/o	swallow	L
dehisc/o	to burst open	L

Word Part	Meaning	Word Origin (L = Latin, G = Greek)
dek- or deka-	ten	G
deliri/o	wandering in the mind	L
delta	triangle	L
dem/o	population or people	G
-dema	swelling (fluid)	G
demi-	half	L
dendr/o	resembling a tree or branching	G
dent-	teeth	L
dent/i or dent/o	teeth	L
depilat/o	hair removal	L
depress/o	pressed or sunken down	L
derm/o or dermat/o	skin	G
derma-	skin	G
-desis	surgical fixation of a bone or joint or to bind	G
deteriorat/o	worsening	L
deut-, deuto-, or deuteron	second	G
dextr/o	right side	L
di-	twice	G
dia-	through, between, apart, or complete	G
diaphor/o	sweat	G
diaphragmat/o	wall across or muscle separating the thoracic and abdominal cavities	G
diastol/o	expansion	G
didym/o	testes or double	G
diffus/o	spread out	L
digest/o	divided or to distribute	L
digit/o	digit	L
dilat/o or dilatat/o	spread out	L
dilut/o	to dissolve or separate	L
diphther/o	membrane	G
dipl/o	double	G
dipla-	double	G
dips/o	thirst	G
-dipsia	thirst	G
dis-	negative, apart, or absence of	L
dis-	duplication or twice	G
dislocat/o	displacement	L
dissect/o	cutting apart	L
disseminat/o	widely scattered	L
dist/o	far	L
distend/o or distent/o	to stretch apart	L
diur/o or diuret/o	increasing urine output	G
divert/i	turning aside	L
domin/o	controlling	L
don/o	to give	L
dors/i or dors/o	back of body	L
drom/o	running	G
-drome	to run together	G
du/o	two	L
duct/o	to lead	L
-duct	opening	L
duoden/i or duoden/o	duodenum	L
dur/o	dura mater or tough, hard	L

Word Part	Meaning	Word Origin (L = Latin, G = Greek)
dy- or dyo-	two	L
-dynia	pain	G
dys-	difficult, bad, or painful	G

E

Word Part	Meaning	Word Origin (L = Latin, G = Greek)
e-	out of or from	L
-eal	pertaining to	L
ec-	out or outside	G
ecchym/o	pouring out of juice	G
ech/o	sound	G
eclamps/o or eclampt/o	flashing or shining forth	G
-ectasia or -ectasis	dilation or enlargement	G
ecto-	out or outside	G
-ectomy	surgical removal or excision	G
eczemat/o	eruption	G
edemat/o	swelling, fluid, or tumor	G
edentul/o	toothless	L
ef-	out	L
effect/o	to bring about a response	L
effus/o	pouring out	L
ejaculat/o	to throw out	L
electr/o	electricity	G
eliminat/o	to expel from the body	L
em-	in	G
emaciat/o	wasted away by disease or lean	L
embol/o	something inserted or thrown in	L, G
embry/o	fertilized egg	G
-emesis	vomiting	G
emet/o	vomit	G
-emia	blood or blood condition	G
emmetr/o	in proper measure	G
emolli/o	to soften	L
en-	in, into, or within	L, G
encephal/o	brain	G
end/o	within, in, or inside	G
endo-	within, in, or inside	G
endocrin/o	to secrete within	G
enem/o	to inject	G
ennea-	nine	G
enter/o	small intestine	G
ento-	within	G
enzym/o	leaven	G
eosin/o	red or rosy	G
epi-	upon, above, on, or upper	G
epidemi/o	among the people	G
epididym/o	tube at the upper part of each testis (epididymus)	G
epiglott/o	lid covering the larynx (epiglottis)	G
episi/o	vulva	G
epithel/i	outer layer of skin or external surface covering	G
equin/o	horse or horselike	L
erect/o	upright	L
erg/o	work	G

Word Part	Meaning	Word Origin (L = Latin, G = Greek)
eruct/o or eructat/o	to belch	L
erupt/o	to burst forth	L
erythem/o or erythemat/o	flushed or redness	G
erythr/o	red	G
es-	out of, outside, or away from	L
-esis	abnormal condition or state	G
eso-	inward	G
esophag/o	tube that connects mouth to the stomach (esophagus)	G
esthesi/o	sensation or feeling	G
-esthesia	sensation or feeling	G
esthet/o	feeling or sense of perception	G
estr/o	female	L
ethm/o	sieve	G
eti/o	to cause	G
eu-	well, easy, or good	G
evacu/o or evacuat/o	to empty out	L
eviscer/o or eviscerat/o	disembowelment	L
ex-	out of, outside, or away from	L, G
exacerbat/o	to aggravate	L
excis/o	cutting out	L
excori/o or excoriat/o	to scratch	L
excret/o	separate or discharge	L
excruciat/o	intense pain	L
exhal/o or exhalat/o	to breathe out	L
-exia or -exis	condition	G
exo-	out of, outside, or away from	G
exocrin/o	to secrete out of	G
expector/o	to cough up	L
expir/o or expirat/o	to breathe out or die	L
exstroph/o	twisted out	G
extern/o	outside or outer	L
extra-	outside or beyond	L
extrem/o or extremit/o	outermost or extremity	L
extrins/o	contained outside	L
exud/o or exudat/o	to sweat out	L

F

Word Part	Meaning	Word Origin (L = Latin, G = Greek)
faci-	face or form	L
faci/o	face or form	L
-facient	producing	L
fasci/o	fibrous band (fascia)	L
fascicul/o	little bundle	L
fatal/o	death or fate	L
fauc/i	narrow pass	L
febr/i	fever	L
fec/i or fec/o	sediment	L
femor/o	thigh bone (femur)	L
fenestr/o	window	L
fer/o	to carry or bear	L
-ferent or -ferous	carrying	L
fertil/o	productive or fruitful	L
fet/i or fet/o	fetus	L
fibr/o	fiber	L
fibrill/o	muscular twitching	L
fibrin/o	threads of a clot (fibrin)	L

Word Part	Meaning	Word Origin (L = Latin, G = Greek)
fibros/o	fibrous connective tissue	L
fibul/o	small bone of distal hindlimb (fibula)	L
-fic	producing or forming	L
fic/o	producing or forming	L
filtr/o or filtrat/o	to strain through	L
fimbri/o	fringe	L
fiss/o or fissur/o	crack, split, or cleft	L
fistul/o	tube or pipe	L
flamme/o	flame-colored	L
flat/o	rectal gas	L
flex/o	bend	L
fluor/o	luminous	L
foc/o	point	L
foll/i	bag or sac	L
follicul/o	small sac (follicle)	L
foramin/o	opening	L
fore-	before or in front of	L
-form	form, figure, or shape	L
form/o	form, figure, or shape	L
fornic/o	arch or vault	L
foss/o	shallow depression	L
fove/o	pit	L
fract/o	break	L
fren/o	bridle or device that limits movement	L
frigid/o	cold	L
front/o	forehead or front	L
-fuge	to drive away	L
funct/o or function/o	to perform	L
fund/o	bottom, base, or ground	L
fung/i	fungus	L
furc/o	branching or forking	L
furuncul/o	boil or infection	L
-fusion	to pour	L

G

Word Part	Meaning	Word Origin (L = Latin, G = Greek)
galact/o	milk	G
gamet/o	sperm or egg	G
gangli/o or ganglion/o	ganglion	G
gangren/o	gangrene	L, G
gastr/o	stomach	G
gastrocnemi/o	calf muscle of hindlimb	G
gemin/o	twin or double	L
gen/o or genit/o	producing or birth	G
-gene	production, origin, or formation	G
-genesis or -genic	producing	G
genit/o	related to birth or the reproductive organs	L
genous	producing	G
ger/i or geront/o	old age	G
germin/o	bud or germ	L
gest/o or gestat/o	to bear or carry offspring	L
gigant/o	giant or very large	G
gingiv/o	gum	L
glauc/o	gray	G
glen/o	socket or pit	G
gli/o	glue	G

Word Part	Meaning	Word Origin (L = Latin, G = Greek)
globin/o	protein	L
globul/o	little ball	L
-globulin	protein	L
glomerul/o	grapelike cluster or clusters of capillaries in the nephron (glomerulus)	L
gloss/o	tongue	G
glosso-	tongue	G
glott/i or glott/o	back of tongue	G
gluc/o or glyc/o	sugar	G
glute/o	buttocks	G
glyc/o	sweet	G
glycer/o	sweet	L
glycogen/o	animal starch (glycogen)	G
gnath/o	jaw	G
goitr/o	enlargement of the thyroid gland	L
gon/e, gon/o, or goni/o	seed or angle	G
gonad/o	sex glands	L, G
gracil/o	slender	L
grad/i	to move, go, or step	L
-grade	to go	L
-gram	record or picture	G
granul/o	granule(s)	L
-graph	instrument for recording; occasionally used to describe the record or picture	G
-graphy	process of recording	G
grav/i	heavy or severe	L
gravid/o	pregnancy	L
guttur/o	throat	L
gynec/o	woman	L
gyr/o	turning or folding	G

H

Word Part	Meaning	Word Origin (L = Latin, G = Greek)
hal/o	breath	L, G
halit/o	breath	L
hallucin/o	to wander in the mind	L
hect-, hector-, or hecato	hundred	G
hem-	blood	G
hem/e	deep red iron-containing pigment	G
hem/o or hemat/o	blood	G
hemangi/o	blood vessel	G
hemi-	half	G
hepa- or hepar-	liver	G
hepat/o	liver	G
hept- or hepta-	seven	G
hered/o or heredit/o	inherited	L
herni/o	protrusion of a part through tissues normally containing it	L
herpet/o	creeping	L, G
heter/o	different	G
hetero-	different	G
hex- or hexa-	six	G
hiat/o	opening	L

Word Part	Meaning	Word Origin (L = Latin, G = Greek)
hidr/o	sweat	G
hil/o	notch or opening from a body part (hilus)	L
hirsut/o	hairy or rough	L
hist/o or histi/o	tissue	G
holo-	all	G
hom/o	same	G
home/o	unchanging	G
hormon/o	to urge on or excite	G
humer/o	long bone of the proximal forelimb (humerus)	L
hyal/o	glassy, clear	G
hydr/o	water	G
hydra-	water	G
hygien/o	healthful	G
hymen/o	membrane (hymen)	G
hyper-	over, above, increased, excessive, or beyond	G
hypn/o	sleep	G
hypo-	under, decreased, deficient, or below	G
hyster/o	uterus	G

I

Word Part	Meaning	Word Origin (L = Latin, G = Greek)
-ia	state or condition	L, G
-iac	pertaining to	G
-ial	pertaining to	L
-iasis	condition or abnormal condition	G
iatr/o	relationship to treatment or doctor	G
-ible	capable of being	L
-ic	pertaining to	L, G
ichthy/o	dry or scaly	G
icter/o	jaundice	L, G
ictero-	jaundice	L, G
idi/o	peculiar to an individual	G
idio-	peculiar to an individual	G
-iferous	bearing, carrying, or producing	L
-ific	producing	L
-iform	resembling or shaped like	L
-igo	diseased condition	L
-ile	capable of being or pertaining to	L
ile/o	ileum	L
ili/o	part of the pelvis (ilium)	L
illusi/o	deception	L
immun/o	protected, safe, or immune	L
impact/o	pushed against	L
impress/o	pressing into	L
impuls/o	pressure or urging on	L
in-	in, into, not, without	L
-in	substance, usually with specialized function	L
incis/o	cutting into	L
incubat/o	hatching or incubation	L
indurat/o	hardened	L

Word Part	Meaning	Word Origin (L = Latin, G = Greek)
infarct/o	filled in or stuffed	L
infect/o	tainted or infected	L
infer/o	below or beneath	L
infest/o	attack	L
inflammat/o	flame within or to set on fire	L
infra-	beneath, below, or inferior to	L
infundibul/o	funnel	L
ingest/o	to carry or pour in	L
inguin/o	groin	L
inhal/o or inhalat/o	to breathe in	L
inject/o	to force or throw in	L
innominat/o	nameless	L
inocul/o	to introduce or implant	L
insipid/o	tasteless	L
inspir/o or inspirat/o	to breathe in	L
insul/o	island	L
intact/o	whole	L
inter-	between or among	L
intermitt/o	not continuous	L
intern/o	within or inner	L
interstiti/o	space between things	L
intestin/o	intestine	L
intim/o	innermost	L
intoxic/o	to put poison in	L, G
intra-	within, into, inside	L
intrins/o	contained within	L
intro-	within, into, or inside	L
introit/o	entrance or passage	L
intussuscept/o	to receive within	L
involut/o	curled inward	L
ion/o	charged particle	G
ipsi-	same	L
ir/i or ir/o	iris of eye or rainbow	G
isch/o	to hold back	G
ischi/o	part of pelvis (ischium)	G
-ism	condition	G
iso-	equal	G
-itis	inflammation	G
-ium	structure	G

J

Word Part	Meaning	Word Origin (L = Latin, G = Greek)
jejun/o	jejunum	L
jugul/o	throat	L
juxta-	nearby	L

K

Word Part	Meaning	Word Origin (L = Latin, G = Greek)
kal/i	potassium	L
kary/o	nucleus	G
karyo-	nucleus	G
kata- or kath-	down	G
kel/o	growth or tumor	G
kera-	horn or hardness	G
kerat/o	hornlike, hard, or cornea	G
ket/o or keton/o	breakdown product of fat (ketones)	German
kinesi/o	movement	G
-kinesis	motion	G
-kinetics	pertaining to motion	G

Word Part	Meaning	Word Origin (L = Latin, G = Greek)
koil/o	hollow or concave	G
kraur/o	dry	G
kyph/o	bent or hump	G

L

Word Part	Meaning	Word Origin (L = Latin, G = Greek)
labi/o	lip	L
labyrinth/o	maze (used to describe the inner ear)	G
lacer/o or lacerat/o	torn	L
lacrim/o	tear or tear duct	L
lact/i or lact/o	milk	L
lactat/o	to secrete milk	L
lacun/o	pit	L
lamin/o	one part of the dorsal portion of the vertebra (lamina)	L
lapar/o	abdomen or flank	G
laps/o	to slip or fall downward	L
laryng/o	voice box	G
lat/i or lat/o	broad	L
later/o	side	L
lax/o or laxat/o	to loosen or relax	L
leiomy/o	smooth muscle	G
lemm/o	peel	G
lent/i	lens of the eye	L
lenticul/o	shaped like a lens	L
-lepsy	seizure	L, G
lept/o	thin	G
letharg/o	drowsiness or indifference	L
leuco-	white	G
leuk/o	white	G
lev/o or levat/o	to lift up	L
libid/o or libidin/o	sexual drive	L
ligament/o	fibrous tissue connecting bone to bone (ligament)	L
ligat/o	binding or tying off	L
lingu/o	tongue	L
lip/o	fat	G
-lite or -lith	stone or calculus	L, G
lith/o	stone or calculus	L, G
-lithiasis	presence of stones	L, G
lob/i or lob/o	lobe or well-defined part of an organ	L, G
loc/o	place	L
loch/i	confinement or birthing process	L, G
longev/o	long-lived	L
lord/o	bent backward	G
lumb/o	lower back, or loin	L
lumin/o	light	L
lun/o	moon	L
lunul/o	crescent	L
lup/i or lup/o	wolf	L
lute/o	yellow	L
lymph/o	lymphatic tissue	L
lymphaden/o	lymph gland	L
lymphangi/o	lymph vessel	L

Word Part	Meaning	Word Origin (L = Latin, G = Greek)
-lysis	breakdown, destruction, or separation	L, G
-lyst	agent that breaks down	L, G
lytic	to reduce or destroy	L, G

M

Word Part	Meaning	Word Origin (L = Latin, G = Greek)
macr/o	large or abnormal size	G
macro-	large or abnormal size	G
macul/o	spot	L
magn/o	great or large	L
major/o	larger	L
mal-	bad, poor, or evil	L
malac/o	abnormal softening	G
-malacia	abnormal softening	G
malign/o	bad or evil	L
malle/o	hammer	L
malleol/o	little hammer	L
mamm/o or mast/o	breast	L
man/i	rage or hand	L
man/o	hand	L
mandibul/o	lower jaw	L
-mania	obsessive preoccupation	L, G
manipul/o	handful or use of hands	L
manubri/o	handle	L
mastic/o or mastricat/o	to chew	L, G
mastoid/o	mastoid process of temporal bone	L, G
matern/o	of a mother	L
matur/o	ripe	L
maxill/o	upper jaw	L
maxim/o	largest	L
meat/o	opening or passageway	L
med-	middle	L
medi/o	middle	L
mediastin/o	in the middle	L
medic/o	medicine, healing, or doctor	L, G
medicat/o	medication or healing	L
medull/o	inner section, soft, or middle	L
mega-	large	G
megal/o	large	G
-megaly	large or enlargement	G
mei/o	less	L, G
melan/o	black	G
mellit/o	honey	L
membran/o	thin skin (membrane)	L
men/o	menstruation	G
mening/o or meningi/o	membrane lining the central nervous system	L, G
menisc/o	crescent	L, G
mens-	menstruation	L
mens/o	menstruation	L
ment/o	mind	L
mes-	middle	L, G
mes/o	middle	G
mesenter/o	mesentery or middle intestine	L, G
mesi/o	middle, or median, plane	G

Word Part	Meaning	Word Origin (L = Latin, G = Greek)
meta-	beyond, change, behind, after, or next	G
metabol/o	change	G
metacarp/o	beyond the carpus	L
metatars/o	beyond the tarsus	L
-meter	measuring device	G
metr/i, metr/o, or metri/o	uterus	G
mi/o	smaller or less	G
micr/o	small	G
micro-	small; one-millionth of the unit it precedes	G
mictur/o or micturit/o	to urinate	L
midsagitt/o	dividing into equal left and right halves	L
milli-	one-thousandth	L
-mimetic	mimic or copy	L
mineral/o	naturally occurring nonorganic solid substance (mineral)	L
minim/o	smallest	L
minor/o	smaller	L
mio-	smaller or less	G
mit/o	thread	G
mitr/o	miter having two points on top	L
mobil/o	capable of moving	L
mon/o	single or one	G
monil/i	string of beads	L
mono-	single or one	G
morbid/o	disease	L
moribund/o	dying	L
morph/o	shape or form	G
mort/i, mort/o, or mort/u	dead	L
mortal/i	death	L
mot/o or motil/o	movement	L
muc/o or mucos/o	mucus	L, G
multi-	many	L
muscul/o	muscle	L
mut/a	genetic change	L
my/o	muscle	G
myc/o or myc/e	fungus	G
mydri/o	wide	L, G
myel/o	spinal cord or bone marrow; "white substance"	G
myocardi/o	heart muscle	L
myom/o	muscle tumor	L
myos/o	muscle	L, G
myring/o	tympanic membrane (eardrum)	L
myx- or myxo-	mucus or slime	L, G
myx/o	mucus or slime	L, G
myxa-	mucus	L, G

N

Word Part	Meaning	Word Origin (L = Latin, G = Greek)
nar/i	nostril	L
narc/o	numbness or stupor	L, G
nas/i or nas/o	nose	L

Word Part	Meaning	Word Origin (L = Latin, G = Greek)
nat/i	birth	L
natr/o	sodium	L
nause/o	sick feeling to stomach	L, G
necr/o	death	G
-necrosis	death of tissue	G
ne/o	new	G
neo-	new	G
nephr/o	kidney	G
nephra-	kidney	G
nerv/o	nerve	L, G
neu-	nerves	G
neur/i or neur/o	nerve	G
neutr/o	neither or neutral	L
nigr/o	black	L
niter-	nitrogen	L, G
nitro-	nitrogen	G
noct/i	night	L
nod/o	knot or swelling	L
nodul/o	little knot	L
nom/o	control	G
non-	no	L
non-, nona-, or nonus-	nine	L
nor-	chemical term meaning normal or parent compound	Chemical term
norm/o	usual	L
not/o	back	G
novem-	nine	L
nuch/o	nape	L
nucle/o	nucleus	L
nulli-	none	L
numer/o	count or number	L
nunci/o	messenger	L
nutri/o or nutrit/o	nourishment or food	L
nyct/o or nyctal/o	night	G

O

Word Part	Meaning	Word Origin (L = Latin, G = Greek)
o-	egg	G
ob-	against	L
obes/o	extremely fat	L
obliqu/o	slanted or sideways	L
oblongat/o	oblong or elongated	L
obstetr/o or obstetr/i	one who receives	L
occipit/o	back of the skull	L
occlud/o or occlus/o	to close up	L
occult/o	hidden	L
octa-, octo-, or oct-	eight	G
ocul/o	eye	L
oculo-	eye	L
odont/o	tooth	G
-oid	like or resembling	G
olecran/o	proximal projection on the ulna	G
olfact/o	smell	L
oligo-	scanty	G

Word Part	Meaning	Word Origin (L = Latin, G = Greek)
-ologist	specialist	G
-ology	study of	G
-oma	tumor, mass, or neoplasm (usually benign)	G
oment/o	omentum, or fat	L
onc/o	tumor	G
onych/o	claw or nail	G
oo/o	egg	G
oophor/o	ovary	G
opac/o or opacit/o	shaded or dark	L
oper/o or operat/o	to work or perform	L
opercul/o	lid or cover	L
ophthalm/o	eye	G
-opia, -opsia, -opsis, or -opsy	vision	G
opisth/o	backward	G
opti/o, opt/i, or opt/o	eye	G
optic/o	vision	L, G
or/o	mouth	L
orbit/o	circle, bony cavity, or socket	L
orch/o, orchi/o, or orchid/o	testes	G
orex/i or orect/i	appetite	G
organ/o	organ	L, G
orth/o	straight, normal, or correct	G
ortho-	straight, normal, or correct	G
os-	mouth or bone	L
-osis	abnormal condition	G
osm/o	smell or odor; pushing	G
-osmia	smell or odor	G
oss/e or oss/i	bone	L
oste/o	bone	G
-ostomosis or -ostomy	surgically creating an opening	G
ot/o	ear or hearing	G
-otomy	cutting into	G
-ous	pertaining to	L
ov/o or ov/i	egg	L
ovari/o	ovary	L
ox/o, ox/y, or ox/i	oxygen	G
oxid/o	containing oxygen	L
oxy-	sharp, acid, quick, or oxygen	G

P

Word Part	Meaning	Word Origin (L = Latin, G = Greek)
pachy-	thick	G
palat/o	roof of mouth	L
pall/o or pallid/o	pale	L
palliat/o	hidden	L
palm/o	caudal surface of the manus (front paw) including the carpus	L
palpat/o	to touch	L
palpebr/o	eyelid	L
palpit/o	throbbing	L
pan-	all, entire, or every	G
pancreat/o	pancreas	L
papill/o or papill/i	nipplelike	L

Word Part	Meaning	Word Origin (L = Latin, G = Greek)
papul/o	pimple	L
par/o	apart from, beside, near, or abnormal; also reproductive labor	L, G
para-	apart from, beside, near, rear in quadrupeds, or abnormal	G
-para	to bear or bring forth	L
paralys/o or paralyt/o	to disable	G
parasit/o	near food	G
parathyroid/o	parathyroid glands	G
-paresis	weakness	L, G
paret/o or pares/i to	to disable	L, G
pariet/o	wall	L
parotid/o	parotid gland	G
paroxysm/o	sudden attack	G
part/o or parturit/o	birth or labor	L
-partum	birth or labor	L
patell/a or patell/o	patella	L
path/o	disease	G
-pathic	pertaining to disease	G
-pathy	disease	G
paus/o	stopping	L
pector/o	chest	L
ped/o	foot	L
ped/o or pedi/a	child	G
pedicul/o	louse	L
pelv/o or pelv/i	hip bone or pelvis	L, G
pen/i	penis	L
-penia	deficiency or reduction in number	G
pent- or penta-	five	G
pept/o or peps/i	digestion	L, G
per-	excessive or through	L
percept/o	to become aware	L
percussi/o	to tap or beat	L
peri-	around or surrounding	G
perine/o	membrane lining the abdominal cavity and its organs (peritoneum)	G
perme/o	to pass through	L
pernici/o	destructive or harmful	L
perone/o	fibular	L
pertuss/i	intense cough	L
petechi/o	skin spot	L
-pexy	surgical fixation	G
phac/o or phak/o	lens of the eye	G
phag/o or -phagia	eating or swallowing	G
-phage	one that eats	G
phalang/o	phalanges	G
phall/o	penis	G
pharmac/o	drug	G
pharyng/o	throat (pharynx)	G
phe/o	dusky	G
pher/o	to bear or carry	G
-pheresis	removal	G
phil/o	love or attraction to	L, G

Word Part	Meaning	Word Origin (L = Latin, G = Greek)
-philia	attraction for or increase in numbers	L, G
phim/o	muzzling or constriction of an orifice	G
phleb/o	vein	G
phlegm/o	thick mucus	L, G
phob/o	fear	G
phon/o	sound or voice	G
-phonia	sound or voice	G
phor/o	carry or movement	G
-phoresis	carrying or transmission	G
-phoria	to bear, feeling, or mental state	G
phot/o	light	G
phren/o	diaphragm	G
-phylactic	protective or preventive	G
-phylaxis	protection or prevention	G
physi/o or physic/o	nature	L
-physis	to grow	G
-phyte	plant	G
pigment/o	color	L
pil/o or pil/i	hair	L
pineal/o	pineal gland	L
pinn/i	external ear, or auricle	L
pituit/o	pituitary gland	L
plac/o	flat patch	G
placent/o	round, flat cake or placenta	L
-plakia	thin, flat layer or scale	G
plan/o	flat	L
plant/o or plant/i	caudal surface of the pes (rear paw) including the tarsus	L
plas/o or plas/i	growth, development, formation, or mold	L
-plasia	formation, development, or growth	G
-plasm	formative material of cells	G
plasm/o	something formed	G
-plast	primitive living cell	G
plast/o	growth, development, or mold	G
-plasty	surgical repair	G
-plegia or -plegic	paralysis	G
pleur/o	membrane lining the lungs (pleura)	G
plex/o	network	L
plic/o	fold or ridge	L
-pnea	breathing	G
-pneic	pertaining to breathing	G
pneu-	relating to air or lungs	G
pneum/o or pneumon/o	lung or air	G
pod/o	foot	G
-poiesis	formation	G
poikil/o	irregular	G
poli/o	gray matter of the brain and spinal cord	G
polio-	gray matter of the brain and spinal cord	G

Word Part	Meaning	Word Origin (L = Latin, G = Greek)
poly-	many	G
polyp/o	small growth	G
pont/o	part of the brain (the pons)	L
poplit/o	caudal surface of the patella	L
por/o	small opening (pore)	L, G
-porosis	passage or porous condition	G
port/i	gate or door	L
post-	after or behind	L
poster/o	toward the tail	L
potent/o	powerful	L
practic/o or pract/i	practice	G
prandi/o	meal	L
-prandial	meal	L
-praxia	action	G
pre-	before or in front of	L
pregn/o	pregnant or with offspring	L
prematur/o	too early	L
preputi/o	foreskin or prepuce	L
press/o	to press or draw	L
priap/o	penis	L, G
primi-	first	G
pro-	before or on behalf of	L, G
process/o	going forth	L
procreat/o	to reproduce	L
proct/o	anus and rectum	G
prodrom/o	precursor	L, G
product/o	to lead forward or yield	L
prolaps/o	to fall downward or slide forward	L
prolifer/o	to reproduce	L
pron/o or pront/o	bent forward	L
prostat/o	prostate gland	G
prosth/o or prosthet/o	addition or appendage	G
prot- or proto-	first	G
prot/o or prote/o	first, original, or protein	G
proxim/o	near	L
prurit/o	itch	L
pseud/o	false	G
psor/o or psor/i	itch	G
psych/o	mind	G
ptomat/o	a fall	G
-ptosis	drooping or sagging	G
-ptyalo	spit or saliva	G
-ptysis	spitting	G
pub/o	part of the hip bone (pubis)	L
pubert/o	adult	L
pudend/o	pudendum	L
puerper/i	labor	L
pulm/o or pulmon/o	lung	L
pulpos/o	fleshy or pulpy	L
puls/o	beating	L
punct/o	sting, puncture, or little hole	L
pupill/o	pupil of the eye	L
pur/o	pus	L
purpur/o	purple	L
pustul/o	blister	L
py/o	pus	G

Word Part	Meaning	Word Origin (L = Latin, G = Greek)
pyel/o	renal pelvis	G
pylor/o	gatekeeper or pylorus	G
pyr/o or pyret/o	fever or fire	G
pyramid/o	pyramid-shaped	G

Q

Word Part	Meaning	Word Origin
quadr/i or quadr/o	four	L
quart-	fourth	L
quinqu-	five	L
quint-	fifth	L

R

Word Part	Meaning	Word Origin
rabi/o	rage or madness	L
rachi/o	spinal column or vertebrae	G
radi/o	radiation or the radius (bone in distal forelimb)	L
radiat/o	giving off radiant energy	L
radicul/o	root or nerve root	L
ram/o	branch	L
raph/o	seam or suture	G
-raphy	suturing	G
re-	back or again	L
recept/o or recipi/o	receive	L
rect/o	rectum or straight	L
recuperat/o	to recover	L
reduct/o	to bring back together	L
refract/o	to bend back or turn aside	L
regurgit/o	to flood or gush back	L
remiss/o	to let go or give up	L
ren/o	kidney	L
restor/o	to rebuild	L
resuscit/o	to revive	L
retent/o	to hold back	L
reticul/o	network	L
retin/o	nervous tissue of the eye (retina) or net	L
retract/o	to draw back	L
retro-	behind or backward	L
rhabd/o	rod	G
rhabdomy/o	striated muscle	G
-rhage or -rhagia	bursting forth	G
-rhaphy	suture	G
-rhea	flow or discharge	G
rheum/o or rheumat/o	watery flow	L, G
-rhexis	rupture	G
rhin/o	nose	G
rhiz/o	root	G
rhonc/o	to snore	L, G
rhythm/o	rhythm	L, G
rhytid/o	wrinkle	G
rigid/o	stiff	L
roentgen/o	X-ray	Named for Wilhelm Conrad Röntgen
rostri	beak	L
rotat/o	to revolve	L

Word Part	Meaning	Word Origin (L = Latin, G = Greek)
-rrhage or -rrhagia	bursting forth	G
-rrhaphy	suture	G
-rrhea	flow or discharge	G
-rrhexis	rupture	G
rug/o	wrinkle or fold	L

S

Word Part	Meaning	Word Origin
sacc/o or sacc/i	sac	L
sacr/o	sacrum	L
sagitta	arrow	L
saliv/o	spit	L
salping/o	tube (uterine or auditory)	G
-salpinx	uterine tube, oviduct, or fallopian tube	G
sanguin/o or sangu/i	blood	L
sanit/o	soundness or health	L
saphen/o	clear or apparent	L
sapr/o	dead or decaying	G
sarc/o or sarcomat/o	flesh (connective tissue) or malignancy of connective tissue	G
sarco-	flesh	G
-sarcoma	malignant tumor	G
scalp/o	to carve or scrape	L
scapul/o	shoulder blade	L
-schisis	cleft or divided	G
schiz/o	cleft or divided	G
scirrh/o	hard	G
scler/o	white of the eye (sclera) or hard	G
-sclerosis	abnormal dryness or hardness	G
scoli/o	curved or crooked	G
-scope	instrument for visual examination	G
-scopic	pertaining to visual examination	G
-scopy	process of visually examining	G
scot/o	darkness	G
scrot/o	bag or pouch	L
seb/o	wax or sebum	L
sect/o	cutting	L
secti/o	to cut	L
secundi-	second	L
segment/o	in pieces	L
sell/o	saddle	L
semi-	half	L
semin/i	sperm, semen, or seed	L
sen/i	old	L
sens/i	feeling or sensation	L
sensitiv/o	affected by or sensitive to	L
seps/o	infection	G
sept- or septi-	seven	L
sept/o	infection or partition	L
ser/o	clear, fluid portion of blood (serum)	L
seros/o	serous	L
sesqui-	one and one-half	L
sex-	six	L
sial/o	saliva or salivary glands	G

Word Part	Meaning	Word Origin (L = Latin, G = Greek)
sialaden/o	salivary glands	G
sider/o	iron	G
sigmoid/o	sigmoid colon	L, G
sin/o or sin/u	hollow, sinus, or tubelike passage	L
sinistr/o	left	L
sit/u	place	L
skelet/o	bony framework of the body (skeleton)	G
soci/o	companion	L
solut/o	dissolved	L
solv/o	dissolved	L
soma-	body	G
somat/o	body	G
somn/o or somn/i	sleep	L
-somnia	sleep	L
son/o	sound	L
sopor/o	sleep	L
-spasm	sudden involuntary contraction or tightening	L, G
spasm/o or spasmod/o	sudden involuntary contraction or tightening	L, G
spec/i	to look at or a sort	L
specul/o	mirror	L
sperm/o or spermat/o	sperm cell or seed	G
sphen/o	wedge	G
spher/o	round, sphere, or ball	G
sphincter/o	tight band	L, G
sphygm/o	pulse	G
spin/o	backbone	L
spir/o	breathing	L
spirill/o	little coil	L
spirochet/o	coiled microorganism	L, G
splen/o	spleen	G
spondyl/o	vertebra	G
spontane/o	unexplained, voluntary	L
spor/o	seed	G
sput/o	spit	L
squam/o	scale	L
-stalsis	contraction	G
staped/o or stapedi/o	one of the middle ear bones (stapes or stirrup)	L
staphyl/o	cluster or bunch of grapes	G
stas/i or stat/i	controlling or stopping	G
-stasis	controlling or stopping	G
-static	controlling or stopping	L, G
steat/o	fat	G
sten/o	narrowing or contracted	G
-stenosis	narrowing, or stricture, of a duct or canal	G
ster/o	solid structure	G
stere/o	solid or three-dimensional	G
steril/i	barren	L
stern/o	breastbone	G
stert/o	to snore	L
steth/o	chest	G

Word Part	Meaning	Word Origin (L = Latin, G = Greek)
sthen/o	strength	G
-sthenia	strength	G
stigmat/o	point or spot	L, G
stimul/o	good or incite	L
stomat/o	mouth	G
-stomosis	making a new opening	G
-stomy	new opening	G
strab/i	to squint	G
strat/i	layer	G
strept/o	twisted chain	G
striat/o	stripe or groove	L
stric-	narrowing	L
strict/o	to draw tightly together or tie	L
strid/o	harsh sound	L
stup/e	stunned	L
styl/o	pointed instrument or pen	L
sub-	under, less, or below	L
subluxat/o	partial dislocation	L
sucr/o	sugar	L
sudor/i	sweat	L
suffoc/o or suffocat/o	to choke	L
sulc/o	furrow or groove	L
super-	above, excessive, or higher than	L
super/o	above, excessive, or higher than	L
superflu/o	overflowing or excessive	L
supin/o	lying on the back	L
supinat/o	bent backward or to place on the back	L
suppress/o	to press down	L
suppur/o or suppurat/o	to form pus	L
supra-	above or excessive	L
supraren/o	above or on the kidney	L
sutur/o	to stitch	L
sym-	with or together	G
symptomat/o	falling together or symptom	G
syn-	union or association	G
synaps/o or synapt/o	point of contact	G
syncop/o	to cut short or cut off	G
-syndesis	surgical fixation of the vertebrae	G
syndesm/o	ligament	G
syndrom/o	running together	G
synovi/o	synovial membrane or synovial fluid	L, G
syring/o	tube	G
system/o or systemat/o	entire body	G
systol/o	contraction	G

T

Word Part	Meaning	Word Origin
tachy-	abnormally fast	G
tact/i	touch	L
tars/o	tarsus, or edge of the eyelid	G
tax/o	order or coordination	G
techn/o or techni/o	skill	G
tectori/o	covering	L
tele/o	distant or far	G

Word Part	Meaning	Word Origin (L = Latin, G = Greek)
tempor/o	temple	L
ten/o or tend/o	tendon, strain, or to extend	G
tenac/i	sticky	L
tendin/o	tendon	L
tens/o	to stretch out, strain, or extend	L
terat/o	malformed fetus	G
termin/o	end	L
tert-	third	L
test/o or test/i	testis or testicle	L
tetan/o	rigid	G
tetra-	four	G
thalam/o	inner room or thalamus	G
thanas/o or thanat/o	death	G
the/o	to put or place	G
thec/o	sheath	L, G
thel/o	nipple	G
therap/o or therapeut/o	treatment	G
therm/o	heat	G
thio-	sulfur	G
thorac/o	chest	L, G
-thorax	pleural cavity or chest	G
thromb/o	clot	G
thym/o	thymus gland	G
thyr/o or thyroid/o	shield or thyroid gland	G
tibi/o	tibia	L
-tic	pertaining to	L, G
tine/o	gnawing worm or ringworm	L
tinnit/o	ringing	L
toc/o	birth	G
-tocia or -tocin	labor or birth	G
tom/o	to cut or section	G
-tome	instrument to cut	G
-tomy	cutting or incision	G
ton/o	tension or stretching	G
tone/o	to stretch	L, G
tonsill/o	tonsil or throat	L
top/o	place or location	G
tors/o	to twist or rotate	L
tort/i	twisted	L
tox/o or toxic/o	poison	L
trabecul/o	beams or little beam marked with crossbars	L
trache/o or trache/i	windpipe	G
trachel-	neck	G
tract/o	to draw, to pull, path, or bundle of nerve fibers	L
tranquil/o	quiet	L
trans-	across or through	L
transfus/o	to transfer or pour across	L
transit/o	changing	L
transvers/o	across or crosswise	L
traumat/o	injury	G
trem/o	shaking	L
tremul/o	fine tremor or shaking	L
treponem/o	coiled or turning microbe	G
-tresia	opening	G
tri-	three	G
trich/o	hair	G
trigon/o	triangle	L, G
-tripsy	crushing stone	G
trit- or trito-	third	G
-trite	instrument for cutting	L
trochle/o	pulley	G
trop/o	to turn or change	G
troph/o	nourishment, development, or growth	G
-trophy	nourishment, development, or growth	G
-tropia	to turn	G
-tropic or -tropin	having an affinity for or turning toward	G
tub/o or tub/i	tube or pipe	L
tubercul/o	little knot or swelling	L
tunic/o	covering or sheath	L
turbinat/o	coiled or spiral-shaped	L
tuss/i	cough	L
tympan/o	tympanic membrane, or eardrum	G

U

Word Part	Meaning	Word Origin (L = Latin, G = Greek)
ulcer/o	sore or ulcer	L
uln/o	one of the bones in the distal forelimb (ulna)	L
ultra-	beyond or excess	L
umbilic/o	naval	L
un-	not	L, G
ungu/o	nail or hoof	L
uni-	one	L
ur/o	urine or urinary tract	G
-uresis	urination	G
ureter/o	ureter	G
urethr/o	urethra	G
-uria	urination or urine	G
urin/o	urine or urinary organs	L, G
urtic/o	nettle, rash, or hives	L
-us	thing	L, G
uter/o or uter/i	uterus or womb	L
uve/o	vascular layer of the eye, iris, choroid, or ciliary body	L
uvul/o	little grape or uvula	L

V

Word Part	Meaning	Word Origin (L = Latin, G = Greek)
vaccin/o or vaccin/i	vaccine	L
vacu/o	empty	L
vag/o	wandering	L
vagin/o	sheath or vagina	L
valg/o	bent or twisted outward	L
valv/o or valvul/o	valve	L
var/o	bent or twisted inward	L
varic/o	swollen or dilated vein	L
vas/o	vessel, duct, or vas deferens	L
vascul/o	little vessel	L
vaso-	vessel	L
vast/o	great or extensive	L

Word Part	Meaning	Word Origin (L = Latin, G = Greek)
vect/o	to carry or convey	L
ven/o	vein	L
vener/o	sexual intercourse	L
venter-	abdomen	L
ventilat/o	to expose to air	L
ventr/o	belly side	L
ventricul/o	small chamber or ventricle of the brain or heart	L
venul/o	small vein	L
verg/o	to twist or incline	L
verm/i	worm	L
verruc/o	wart	L
vers/o or vert/o	turn	L
-version	to turn	L
vertebr/o	backbone	L
vertig/o or vertigin/o	turning around or revolution	L
vesic/o	urinary bladder	L
vesicul/o	seminal vesicle, blister, or little bladder	L
vestibul/o	entrance	L
vi/o	force	L
vill/i	tuft of hair or threadlike projections from a membrane	L
vir/o	poison or virus	L

Word Part	Meaning	Word Origin (L = Latin, G = Greek)
viril/o	masculine	L
vis/o	seeing or sight	L
visc/o or viscos/o	sticky	L
viscer/o	internal organ	L
vit/a or vit/o	life	L
viti/o	blemish or defect	L
vitre/o	glassy	L
viv/i	alive	L
voc/i	voice	L
vol/o	caudal surface of forelimbs (palm) or hindlimbs (sole)	L
volv/o	to turn or roll	L
vulgar/i	common	L
vulv/o	covering or vulva	L

X

Word Part	Meaning	Word Origin (L = Latin, G = Greek)
xanth/o	yellow	G
xen/o	strange or foreign	G
xer/o	dry	G
xiph/o or xiph/i	sword	G

Z

Word Part	Meaning	Word Origin (L = Latin, G = Greek)
zoo-	life, animal	G
zygomat/o	yoke or cheekbone	G
zygot/o	joined or yoked together	G

Note: Page numbers referencing figures are italicized and followed by an "*f*". Page numbers referencing tables are italicized and followed by a "*t*".